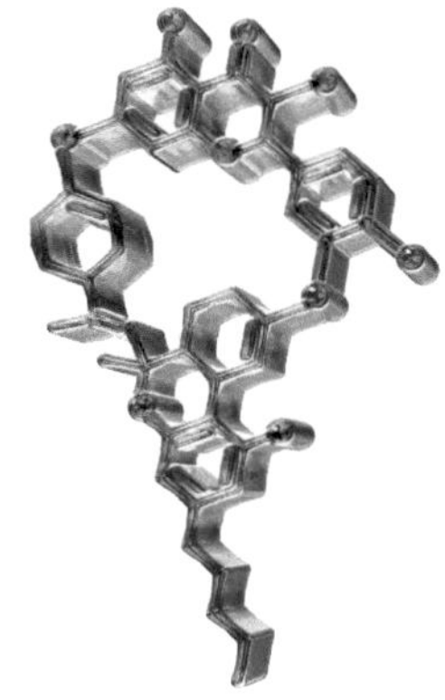

The Big Book of Terps, Edition 2
Understanding Terpenes and Synergy in Cannabis
Researched and written by Russ Hudson
Edited by Jacqueline Graddon, MBA

ISBN: 978-0-9863571-4-5

Never Stop Growing

Table of Contents

Foreword - By Mitch Earleywine, PhD

(*Professor of Psychology, University at Albany, SUNY; Fmr. Chair of the Executive Board, National Organization for the Reform of Marijuana Laws* [*NORML*])

A few weeks after Colorado first legalized the recreational cannabis market, I walked down a Denver street with friends from the NORML board. We turned a corner, and an unmistakable scent overwhelmed us all. Someone had obviously developed an elaborate home grow and we all would have guessed that the quality was outstanding. Of course, we could not have guessed the concentration of cannabinoids of any sort.

We smelled the delightful plant molecules known as terpenes.

Years later, a month before the COVID pandemic shut down our laboratory, my graduate students and I finally got approval to administer terpenes one at a time in a real experiment. These phytochemicals have had their psychoactive effects confirmed in aromatherapy research, and we had hoped to make the most of any relaxation, stimulation, or joy that they might generate. We literally suggested that we would run around to various vape shops in town with a single terpene in a little vape pen. Then we would beg folks to fill out a survey on their moods before and after taking multiple hits from an oil that contained only one of the classics. One pen had limonene, that citrus scent common to any strain named "Lemon." Another had linalool, the predominant component of lavender. And of course, we had a control pen with nothing but oil.

But once we saw that the ethics board was willing, we somehow lost motivation for the project. Nobody in the real world uses one terpene at a time. Nature bundles these molecules in delightful ways that seem to serve some grand evolutionary purpose. Looking at them one at a time seemed a bit naive. The entourage effect, that delightful impact that plant molecules have when they combine with each other and meet human physiology, is markedly more dramatic than the sum of its parts. Looking at them one at a time felt akin to taking a silvery metal like sodium and a toxic gas like chlorine and trying to explain table salt. The unique combination hardly compared to its components.

But that's often the scientific approach. Breaking things in pieces and understanding the pieces has helped some domains in some fields. Nevertheless, the approach is a far cry from appreciating the nature of novel aggregates that coalesce in unpredictable ways. As Russ Hudson details in this volume, the stories behind individual chemical compounds and their constituents can intrigue any student of plants, animals, history, ecology, psychology, or art. In a sense, these intriguing chunks of carbon and hydrogen reveal quite a bit about us and our world.

If all humans are siblings, then perhaps all living organisms are distant cousins. The idea that we evolved as a species just a bit behind the plant kingdom certainly has some intriguing implications. Although tens of thousands of organisms can survive with little more than sun and earth, ancient molecules, when given enough time, lined up to create mobile creatures. A few plants can move a bit, of course. The Venus Fly Trap and Bladderwort can capture small prey for their own ingestion thanks to some quick motion. Others toss pollen on flying insects in an effort to spread their offspring far and wide. But most movement is reserved for us animals. Those species who flourished around any edible organisms that could provide a nutrient or two had definite advantages. Those who couldn't function with the available foliage never had the chance to pass their genes along.

Although my mycologist friends would likely raise their hands to offer a caveat, I think it is fair to say that plants and humans literally grew up together. Humans benefitted from the available vegetation, and we eventually cultivated the species that we found most rewarding. But our explanations for these choices might be an example of what social psychologist Dick Nisbett called "telling more than we can know." The stories we recount about which plants we like and why we like them might be more myth than reality. Terpenes likely play a role outside our conscious awareness, an error that Russ Hudson is here to correct.

Few fans of cannabis ever question why our progenitors exerted effort to ensure that this genus never died, but the truth behind the narrative has more nuance than one might imagine. A great many unsaturated hydrocarbons sure seemed to play a role in this divine coincidence where plants and humans evolved together. These colorless, chiral molecules with strong and (frequently) pleasant scents just happened to have a superb impact on human physiology. The same class of substances that defended plants against insects and other herbivores also helped them thrive, in part, because humans found them so appealing. And as readers will note in each of these chapters, no single plant has the market cornered. No single molecule seems to be the lone contributor to one physiological effect.

This lesson about working together applies within our species as well. (We all benefit when we cooperate. No one of us can know what all of us know, and none of us can do what we could all do together.) But we see that the same is true across multiple species and kingdoms. If we take care of our Earth, we can relish the fact that it will take good care of us. Alas, the contrapositive is also true. If we neglect our environs, they will return the favor.

So let me caution readers before they continue: This book will change you. I have no doubt you'll find the prose remarkable and the content informative. But more than that, you risk finding yourself a different person when you finish any section. The details on each terpene underscore the recurring theme that our planet is phenomenal. That fact might seem banal in the abstract, and you've likely thought it before, but this work will make you BELIEVE it. Suddenly you'll note with any whiff of lavender or scent of citrus that the terpenes are more than a simple source of delight, analgesia, or sedation. These are the molecules we literally evolved with, stimulating the neurons in the most primitive part of our brain to fire, reminding us that our environs are richer and more luxurious than we might remember.

And these thoughts might inspire a new appreciation for the cannabis plant (of course), but also the entire plant kingdom, and the other kingdoms, and the planet, and the universe. So don't open this book. Don't turn a single page. Sure, you'll be entertained and instructed at first. (The first thrill is always free.) But down the line you'll find yourself eager to tell everyone the marvels of molecules. You'll note what a wild set of coincidences had to happen in order for our species and all these plants to develop into what we've become. And you'll have this nagging sense that you not only need to spread the word, but you also need to help take care of this planet, and the living things on it. I'm happy to join you, of course, but the task won't be effortless, or always enjoyable, even though we know it is the right thing to do.

~Mitch Earleywine, PhD
August, 2021. Albany, NY

A Note on Cannabis Education - By Rob Mejia, Ph.D.

(Adjunct Professor of Cannabis Studies at Stockton University, New Jersey. President of Our Community Harvest: A Minority-Owned Cannabis Education Company, and author of The Essential Cannabis Book.)

As a Stockton University cannabis professor, I recently gathered 50 cannabis professors from around the country to talk about what we teach and how we teach our cannabis courses at an event called a Cannabis Curriculum Convening. It is a fancy way of saying, "let's get together to share resources and discuss what we teach in our classes." Well, one of the panels which focused on teaching the Science of Cannabis course immediately got bogged down in a discussion about cannabis terms, definitions, and the functions of the plant.

This happened because there is a major disconnect between cannabis science and botany and the consumer market. That is, when you read most cannabis articles, listen to a podcast, or go into a dispensary, 99% of the time you will be asked what type of cannabis you want while being presented with a list of Sativa, Indica, and Hybrid strains. But what exactly are you being offered?

To begin with, most cannabis scientists will tell you that cannabis is a single plant (with some scientists saying that there are subspecies) but what makes each plant different and explains how it affects the body are the components or parts of each individual plant. In other words, if a plant has a high percentage of THC, it is likely to give you psychotropic ("high") effects. But this high can be influenced by the amount of CBD, which terpenes and flavonoids are present, the method of administration and even the setting of the consumption.

In the case of the dispensary employee who offers you Sativa, Indica, or a Hybrid strain, they are making assumptions about the effect of the cannabis plant based on a few main factors: how much THC the plant contains, how much CBD the plant contains, and possibly which terpenes are present. If they offer you a Sativa strain, it is assumed that this will be for daytime use, it will provide energy, and spur creativity. If you are offered an Indica strain, it is assumed that you are looking for relaxation, to tamp down anxiety, or even help sleeping. A hybrid strain is assumed to have a combination of the above effects. And many strains may be effective for pain relief which continues to be the number one reason consumers use cannabis.

This all means we have quite a bit of educating to do to get a cannabis user to understand the parts of the plant and how they affect the body. My hope is that as consumers become more familiar with the effects of THC, CBD, other minor cannabinoids, terpenes, and flavonoids -and track their consumption in a journal- they will become more educated. An educated cannabis consumer will come to understand what methods of responsible consumption work best for them, what proportions of THC and CBD are most effective, and how they respond to terpenes and flavonoids.

One classification system that is gaining some traction in dispensaries is using the terms chemovar and chemotype to help define user experience and to reflect the makeup of the plant more accurately. The chemotype system of classification focuses on the most abundant cannabinoid in the plant while the chemovar system of classification includes the top cannabinoids and terpenes. In other words, chemotypes or chemovars use the breakdown of a plant species according to its chemical composition. This classification system is essential for breeders, cultivators, and consumers. Chemical characteristics can determine, for example, what a cannabis Sativa plant's ratio of THC to CBD is and what terpenes it contains and in what quantities. And all of these elements will influence a user's experience.

In an effort to go beyond the "strain" definition, some states such as New Jersey are using the following classification system on cannabis flower packaging in dispensaries along with traditional strain names. We will see if other states follow suit and if this information helps adult cannabis users.

New Jersey's regulations specify that chemotypes shall be displayed as:

a. “High THC, Low CBD” where the THC to CBD ratio is greater than 5:1 and the total THC percentage is 15% or greater;
b. “Moderate THC, Moderate CBD” where the THC to CBD ratio is between 5:1 and 1:5 and the total THC percentage is between 5% and 15%;
c. “Low THC, High CBD” where the THC to CBD ratio is less than 1:5 and the total THC percentage is less than or equal to 5%;
or d. “Where usable cannabis does not conform to one of the three chemotypes, it shall be listed as the closest chemotype determined by mathematical analysis of the ratio of THC to CBD”.

This system of classification was implemented to try to help consumers determine which type of product(s) can be most effective for their health and wellness. It is thought that the traditional method of classifying cannabis by strain or cultivar neglected the chemical composition of the plant (ratios of THC to CBD) and especially the role of terpenes. With this new knowledge, we can expect a more educated populace in the future.

~Rob Mejia, PhD
July, 2023

Why Terpenes Matter - By Ed Rosenthal

(*Legacy guerilla cultivator, educator, and spokesperson for the cannabis industry for more than a half century, author of several cannabis books, most notably his 2021 work, Cannabis Growers Handbook.*)

Why terpenes matter can be summed up in one phrase; "If THC is the engine, then terpenes are the steering wheel."

Herbalists have been prescribing terpenes and flavonoids as supplements and medications for thousands of years. Mostly, they didn't know the particular oils but they concocted medications using all the parts of plants; roots, stems, leaves, flowers and seeds to make teas, salves and ingestibles. Plant based medicines continued as the primary medications throughout the 19th and early part of the 20th century. Early pharmaceuticals were standardized and quantified recipes made from natural ingredients. It was only in the 20th century that human-invented medications became available.

Herbalists prescribed medications made from whole plant concentrates and extracts to alter mental states: Rosemary (camphene and pinene) for concentration, lavender (linalool) for relaxation, hops tea (humulene), chamomile (α-bisabol) for stress relief and to become drowsy. Of course, herbal ingredients are used to treat physical problems, too, perhaps primarily. Caryophyllene, found in black pepper, cinnamon and oregano, can be used to treat anxiety, depression and stress.

The terpenes just described; a-bisabol, camphene, caryophyllene, humulene, linalool, and pinene, are all common constituents of the entourage of terpenes found in cannabis. As described in this book and Hudson's "Big Book of Terps," there are literally hundreds of terpenes found in cannabis flowers. (Don't get me started on the roots. Since the plant has interesting chemistry in the canopy, why wouldn't it also have some in the roots?)

When most people examine a bud, first they look at it and then smell it, looking for a pleasant or interesting sensation. A knowledgeable user may recognize some odors or the variety and have an idea of their general effects. But with so many varieties available from so many breeders and growers, it's often hard to know what kind of high you are about to inhale.

This gets more complicated, because buds contain more than one terpene. Most assays identify five or more. Imagine all of these terpenes and the cannabinoids attempting to pierce the blood-brain barrier, which determines what the head will be.

Grow Magazine and Humboldt Seeds both conduct pheno hunts in which they distribute identical clones to several growers. The resulting buds of each variety are not identical, nurture as well as nature plays a part in how the flower develops and the percentages of the different terpenes in each candidate. The heterogeneity of seeds of some varieties, especially when the product of complex hybrids, also factors into the equation.

There are solutions to the conundrum. Usually just a few terpenes comprise a high percentage of those present. Many shops list both cannabinoid and terpene percentages of the buds. This tendency will help you determine whether to purchase a particular variety or to look for the terpenes listed in the marketing or packaging. The more you sample the faster you'll become more familiar with the effects.

I'm mostly a pipe and joint smoker, so I devised a technique to sample the terpenes with cannabis. Small vials of the terpenes most commonly found in cannabis can be purchased inexpensively on the internet. First, I prepare the cannabis by baking it at 200 ° F (93° C) until the terpenes have boiled away leaving just the plant material and the cannabinoids. Then using a toothpick, applying directly from the small opening of the vial or another tool, the terpene or terpene combo is added to the cannabis sample. This is not totally scientific and contains considerably more terpenes than a bud sample would have naturally, but it gives me an approximation of the effects of both the individual terpenes or their entourage effects. The different combinations of terpenes and THC have profoundly different effects on consciousness.

Herbalists have already documented terpene effects on consciousness and this information is relevant to breeders. Rather than depending on just the "nose" to select breeding stock, using chemical analysis and the qualities of individual terpenes as well as the entourage effect can be used to create varieties with specific purposes, such as to induce relief from stress, help with sleep disorders, listening to music, partying, creativity, intimacy, routine work, or other activities.

Rather than breeding, another possibility is mixing buds or concentrates of different varieties to concoct terpene and cannabinoid recipes in much the same way as pigments are matched to make paint colors. Both medical conditions and the psyche could be addressed by a terpene / Cannabinoid entourage specifically designed for the individual. As a person becomes familiar with the combination it can be tweaked to perfection. An individual might use several formulas depending on time of day and activity. Brain scans might be used to determine the best recipes.

Imagine this conversation: "Try this. It was designed for me for hiking. I think you'll like it along the path."

~Ed Rosenthal
August 2023

PROLOGUE

This book was written for researchers, scientists, educators, cannabis industry professionals, medicinal cannabis patients, and for everyone interested in terpenes, flavonoids, and cannabinoids. Significant efforts were expended to make the work presentable and sustainably interesting to all demographics, including laypeople. For this reason, all citations in the text appear in-line for immediate reference. This formatting allows readers to instantly reference and verify information without breaking continuity by referring to an independent index or bibliography. Readers are encouraged to review the studies and papers referenced in this book in their entirety where necessity or special interest compels.

Please note that most studies cited here are not specific to cannabis. This is primarily because we simply do not have sufficient research available after decades of prohibitionist and research-restriction policies around the world, particularly in the United States. However, it should be understood that the molecules discussed herein are the same from plant to plant; for instance, the alpha-pinene that occurs in cannabis is the same alpha-pinene molecule that occurs in pine trees. The geraniol that occurs in cannabis strains is the same as the geraniol that occurs in roses. The flavonoid quercetin that occurs in hundreds of plants is the same molecule that occurs in cannabis.

We can presume that most plants use these compounds in a comparable way, and for the same or similar purposes, with exceptions for specialist plants and anomalies across species. Nevertheless, this presumption should be short-lived; we must continue to fight for legalization and scientific/research freedoms related to cannabis. Future editions of this book will be updated with emerging research, and readers that have unique or new research to share should contact the author directly at TheBigBookOfTerps.com.

Finally, it is important to consider that laboratory tests conducted as part of research for this book may be only partially accurate. This is because as of 2021, consistency in varieties or strains is sorely lacking. The Blue Dream acquired at a dispensary in California may be chemically entirely different than the Blue Dream acquired in Barcelona, Spain, or even in Oregon, Cali's northerly neighbor. Unfortunately, the power of strain name recognition often skews labeling and marketing efforts, making it extremely difficult to ascertain whether a particular batch of cannabis branded as any strain is in fact that stabilized strain. In most parts of the world, assurance of cannabis variety is challenging to impossible.

Because of these inconsistencies, breeders, geneticists, cultivators, dispensaries, legislators, investors, healthcare professionals, and all others involved in cannabis policy should advocate for more and better laboratory testing. It is not enough to only test for contaminants and cannabinoid content. Terpene and other phytochemical content is important and can directly impact the therapeutic effects of cannabinoid and other constituents such as alcohols, ketones, thiols, aldehydes, sulfurs, and esters. Cannabis users need to know the phytochemical profile of a product to make sound decisions for their health and well-being. Testing and subsequent dissemination and education is the only method of obtaining this confidence

INTRODUCTION

Terps, terps, everywhere; in the stain on wooded stair, in the shampoo in your hair,
In the magic of your bong, in the trichomes, thick and strong
Smell them here and taste them there; terps, terps, everywhere...

Terpenes are everywhere. In fact, chances are high that you have experienced, consumed, or otherwise ingested multiple terpenes at some point in the last 24 hours. Despite the worldwide pervasiveness of terpenes, most people have never heard of a 'terp.' Those who have likely learned about these compounds because of their experience with cannabis. But whether you're a 'toker or an abstainer, cannabis supporter or prohibitionist, we cannot escape terpenes and the effects that they have on us.

Your Nose Knows Terps

Did you smell roses, begonias, geraniums, or other flowers in the last day? Did you walk by a lawn filled with dandelions, clovers, goldenrod, or buttercups? Perhaps you encountered the sweet smell of lilacs, lupine, or lilies? Flowering thistle or burdocks? Orange, cherry, or apple blossoms? Those fragrances are often influenced by terpenes - thousands of different carbon-chain compounds that are found widely in the plant kingdom.

But terpenes are not just something that you smell or taste, without meaning. These volatile molecules interact with the human body in ways that affect our mood, our health, and even our decisions. Think about that if there are pine, spruce, aspen, or fir trees near where you live or work. That fresh, minty evergreen scent that pervades these trees is often caused at least in part by terpenes, and your olfactory senses are registering these aromatic molecules whether you are conscious of it or not.

Terps on the Tongue

Terpenes are also found in many of the foods we consume; so many foods, in fact, that it would be difficult to go more than a day without consuming some of these compounds.

For instance, in the last 24 hours:

Did you eat an orange, or take lemon in your water or tea? The smell and taste of both oranges and lemons are partly due to terpenes, while the vibrant colors of these fruits can often be attributed to flavonoids. (Flavonoids are covered in a companion text titled “The Formidable Book of Flavonoids, available in fall 2023)

Did you choose between one variety or another of pears, apples, berries, or other fruits? You made that decision based partly on which terpenes your senses detected.

Did you drink beer? The hops in your brew contained terpenes that impart potent sedative and relaxant effects to the beverage. These include myrcene, beta-caryophyllene, and, to a lesser degree, humulene.

Did you use tobacco? Terpenes are present in these products, as well - especially if you smoke menthols - that minty flavor is derived from terpenes, usually eucalyptol and menthol.

Are you the type of person that prefers to snack on candy, or do you prefer salad? Either way, you are consuming terpenes; in candy these compounds are often used as flavoring agents, and most salad items are rich in natural terpenes.

Did it rain near you? The fresh, earthy aroma that is present after a rain is caused by terpenes.

Although terpenes are naturally occurring compounds, this doesn't mean that you will only find them in natural or minimally processed foods. In fact, terpenes are used in a wide variety of food products, especially sweets like ice cream, puddings and gelatins, baked goods, candies, and even chewing gum. Terpenes are also used in enriched food products and beverages, often to improve the taste of less palatable ingredients.

Terpy Toiletries

Humans have been harnessing the power of terpenes through the production of essential oils and other extractions for thousands of years - so long, in fact, that most people don't even realize there are terpenes in use everywhere around them. Many types of personal care products feature one or more terpenes in the ingredient list. This includes an astounding variety of products found in the bathrooms of most modernized countries; perfumes and after-shave, deodorants, hair gels and creams, toothpaste and mouthwash, moisturizers and lotions, lipstick and lip balms, shampoos, conditioners, body soaps, and many more toiletries and hygiene products often contain terpenes.

Check the ingredient lists of the contents of your bathroom and kitchen for terpenes like limonene, linalool, cineole, geraniol, pinene, and others discussed later in this book - you might be surprised at how many of the products you use every day contain terpenes.

Terpenes in Industry & Medicine

Our use of terpenes extends far beyond food and personal care products. Terpenes are widely used in heavy industries; turpentine, a product used in thousands of industrial, agricultural, and medical applications, is comprised primarily of the terpenes alpha- and beta-pinene, as sourced from pine trees. Industries that work with turpentine, rubber, citronella, camphor, and many other extremely common base ingredients are all working directly with terpenes.

Other industrial and commercial uses for terpenes include as novel and experimental fuel sources, in paints, stains, solvents, and cleaning agents, in pesticides, herbicides, fungicides, and antibacterial applications, in the storage of grains and seeds, and in hundreds of household products.

One sector that is particularly dependent on terpenes is the fragrance and perfume industry, which has systematically categorized and characterized hundreds of aromatic compounds for use in thousands of different applications.

Beyond the fact that terpenes can improve our lives with the many products we use that are based on these compounds, terpenes can also be used to improve and even save human life. This is because many well-studied terpenes have been shown to exhibit anticancer, antiparasitic, antibacterial, antifungal, and antiviral properties. These volatile plant compounds also reduce swelling - a common cause of secondary disease - and may even be effective at fighting certain types of cancer cells. Some terpenes have significant analgesic or pain-relieving properties that can be harnessed at home and in the clinic by various plant essential oils and other therapeutic preparations.

Terpenes in Cannabis

Nearly all the things that people love about cannabis and marijuana are actually a love of the plant's phytochemicals including terpenes, flavonoids, esters, cannabinoids, and other compounds. That nose-hair-curling fragrance when you open a bag of Zombie Kush? Those are often terps talking. That crisp, fruity flavor of Super Lemon Haze that causes you to shut your eyes and work your tongue and lips like a suckling fish? Terps talking again.

If you are talking about any of the properties of cannabis that make your eyes light up like a kid at Christmas, then you are probably talking about terpenes and the other phytochemical constituents of cannabis. Whether you are a full-time cannabis connoisseur or a part-time 'toker, your ability to distinguish a particular cannabis variety (strain) depends partly on the terpene and other phytochemical content of the plant or product you are evaluating. Although you may not know exactly what the compound is called, when you identify Chocolope because of the variety's distinctive chocolate and melon aroma, what you've really done is recognize a particular terpene and phytochemical profile. Likewise, when you detect New York Diesel based on its unmistakable lemony-gasoline flavor, what you've really identified is a blend of terpenes and other compounds.

Terpenes and Us

Human physiology responds differently to different terpenes; among us, some are influenced more than others by these compounds. The reason one person might prefer the fragrance of the terpenes present in the Amnesia strain while another person disdains that strain and instead prefers the aroma of Afghan Kush terpenes is based on individual physiology. While much more research is needed to comprehend why terpenes affect people differently, the ultimate analysis and elucidation of this may be at least partially out of the realm of current scientific understanding, which makes terpenes nearly magical.

It is clear the human senses interact with terpenes on a daily basis, completely unrelated to cannabis. So, for those of us who do consume cannabis or hemp as part of our lifestyle, we need to understand that our more intimate relationship with these compounds requires careful consideration. To advance our personal use of this plant and to continue the international trends toward legalization of cannabis, it is critical that we understand why cannabis affects people the way it does, and, even more particularly, why the same strain or variety affects people differently. The answers to these questions are partly found in terpenes, and perhaps even more importantly, in the putative synergistic interactions between these compounds and cannabinoids.

Until recently, the cannabis industry has largely been a supporter and consumer of 'silver bullet' theories, with THC and CBD considered the holy grail of the plant. However, there is substantial evidence to suggest that this single molecule approach - reminiscent of pharmaceutical industry practice - may not be the best option for all situations, products, and people. Instead, a more holistic method is required; one where we critically examine all significant cannabinoids, terpenes, flavonoids, and other compounds - both individually, and as combinations based on stabilized chemovars - to assess for additive, agonistic, antagonistic, modulatory, and other synergistic effects.

TERP TSARS

This chapter recognizes and summarizes the work of some of the most important terpene and phytochemical researchers and scientists in the world.

The "Golden Era" of terpene research was in the mid to late 19th-century when German scientists dominated the scene. Today, there are far fewer terpene researchers, and only a handful of these are dedicated to terpenes in cannabis. However, there will likely be a resurgence of work in this field in the coming years, as the therapeutic and medical value of terpenes becomes clearer to the public, and as the roles of these constituents in cannabis are elucidated.

TIMELINE OF TERP TSARS

- 1830s: *Jean-Baptiste Dumas*. One of the first chemists to seriously investigate essential oil constituents
- 1830s: *Friedrich Wöhler*. The first to refute 'vital force' theory, discovered and proved isomerism, first chemist to discover and advance functional groups
- 1850s: *August Kekulé*. Conceptualized and advanced the theory of chemical structure and the hexagonal benzene ring nature of aromatic compounds
- 1860s: *Pierre-Eugène-Marcellin Berthelot*. Extracted the odor of fresh soil, refuted vital force, advanced synthesis of organic compounds, characterized pinene
- 1880s: *Otto Wallach*. Isoprene Rule, Wallach's Rule, numerous structural determinations of terpenes
- 1890s: *David Prain*. Early hemp researcher, proposed synergy in the form of more than one active constituent in cannabis
- 1900s: *Friedrich Heusler*. Wrote *The Chemistry of the Terpenes*, appears to be the first to propose that terpene oxidation products should be classified as terpenes

- 1920s-1930s: *Leopold Ruzicka*. Carried out the total synthesis of many terpenes, expanded upon the isoprene rule, defined terpenes according to number of isoprene units

- 1930s-1950s: *Walter Siegfried Loewe*. Identified the activity of cannabinol, invented the isobologram analysis method of detecting and measuring synergy in compounds

- 1940s-1960s: *Robert Burns Woodward*. Development of Woodward Rules, total synthesis of reserpine and chlorophyll

- 1950s: *Wilhelm Treibs*. The first to isolate and elucidate the structure of beta- caryophyllene - the top terpene in cannabis

- 1950s-1960s: *František Šorm*. Synthesized and deduced the structures of sesquiterpenes, investigated terpenes as analogues of insect hormones

- 1950s-1970s: *Sir Derek Harold Richard Barton*. Developed the Barton- McCombie deoxygenation concept, confirmed the structure of beta- caryophyllene

- 1960s-Present: *Raphael Mechoulam*. Discoverer of THC, total synthesis of this molecule as well as CBD and CBG, first to suggest entourage effect

- 1980s-Present: *Mahmoud A. ElSohly*. Synthesis of cannabispiran, identification of genes in δ9-tetrahydrocannabinol, various essential oil research, natural product structure determinations

- 1980s-Present: *John M. McPartland*. Researcher of synergy in cannabinoids and terpenes, cannabis botanical history researcher

- 1990s-Present: *Shimon Ben-Shabat*. Discovered 2-Arachidonoylglycerol, together with Mechoulam showed first evidence of the entourage effect in cannabis

- 1980s-Present: *Ethan Russo*. World's leading researcher on terpenes and synergy in cannabis

- 2000s-Present: *Sameksha Koul*. Author of *Comprehensive Treatise on the Terpenes*

- 2010-Present: *Colleen Quinn*. Clinical aromatherapist, terpene formulator

- 2000s-Present: *Susan Trapp*. Researcher; molecular evolution of terpenes, horizontal gene transfer of fungal terpenoids, founder of Terpedia.com

[Did we forget someone in this list? Let us know at www.TheBigBookOfTerps.com]

Jean-Baptiste Dumas 1800-1884
Important work: 1830's

One of the first chemists to systematically research essential oil constituents was Jean- Baptiste Dumas, a Frenchman whose name is inscribed among seventy-two other prominent scientists on the corners of the Eiffel tower.

In his early career, Dumas served as professor of chemistry at two prominent schools in Paris, then later served political posts under Charles-Louis Napoleon Bonaparte (Napoleon III) before returning to the field of chemistry.

Dumas investigated various hydrocarbons but is best known for his work with nitrogen- containing constituents - including the development of a method for analyzing the content of nitrogen in many compounds. Dumas also used innovative techniques to establish molecular and atomic weights by measuring vapor densities.

Friedrich Wöhler 1800-1882
Important work: late 1820s to 1860

Friedrich Wohler contributed more than 60 years of consistent, original scientific work to the field of organic chemistry. From Eschersheim, Germany, Wohler studied under the famous Swedish chemist Jons Jacob Berzelius, who later validated Wohler's discovery of and work on isomerism.

Prior to Wohler's discoveries, most chemists believed that organic compounds could not be created without a vital force or life force, which could only be formed or carried by a living thing. Wohler disproved this longstanding theory by synthesizing urea in his laboratory, radically changing the future of organic chemistry.

Wohler's synthesis of urea led to his discovery of isomers: chemical compounds that are identical in molecular content, but with slightly different arrangements of the atoms within. This work also led Wohler to pioneer and advance the concept of functional groups, in particular the benzoyl group, and spurred other researchers to discover many more functional groups that have significantly enhanced our understanding of organic chemistry.

Wohler's contributions extended beyond science and into academics. During Wohler's time, the predominant practice for instruction in organic chemistry consisted primarily of lectures, and, to a much lesser degree, limited scientific demonstrations performed in laboratories thronged with students. Wohler required all his students to conduct prescribed experiments in the lab as part of his curricula. This practice has since been adopted in organic chemistry programs throughout the world, and today is the standard of education for most stages of instruction in this field.

Friedrich August Kekulé 1829-1896
Important work: 1850s

Terpenes were named such by August Kekule, a German organic chemist who originally studied hydrocarbons that occurred in turpentine oil, among other essential oils. Just five years elapsed from the time Kekule earned his doctorate in 1852, to the time when he founded the theory of chemical structure - a foundational concept in modern chemistry. This theory focused on the concept of a molecular skeleton to which carbon and other atoms could be attached or bonded together.

Kekule's major work with terpenes also included the advancement of the theory that all aromatic compounds feature a hexagonal ring made of benzene; a fundamental concept in terpene science that remains undisputed.

Kekule went on to work with Otto Wallach - arguably the "father" of terpene science as mentioned below - and three of Kekule's students earned Nobel Prizes in chemistry in the early part of the 1900s.

Pierre Eugène Marcellin Berthelot 1827-1907

Important work: 1860s

Marcellin Berthelot first came to scientific prominence when he studied the aroma of fresh soil, successfully extracting via steam distillation the organic compounds responsible for this characteristic odor[1].

Berthelot earned his doctorate in 1854 based on his work synthesizing natural fats, and it was his continued work in this area that dealt the final killing blow to the concept of vital force; work begun decades earlier by Friedrich Wohler was finally completed by Berthelot.

Berthelot insisted that the vast majority of organic compounds could be synthesized, and successfully synthesized camphol and camphor from turpentine - a favorite essential oil of August Kekule, who later invited Berthelot to work with him in the second half of the 19th century. However, Berthelot chose to continue his work alone, carrying out thousands of experiments and ultimately publishing twenty-five books and more than sixteen hundred studies[2] before his death in 1907.

Berthelot's work modernized the classification of terpenes and illuminated the characterization of alpha- and beta-pinene. Additionally, some credit Berthelot as the first to advance the isoprene rule attributed to Otto Wallach.

Berthelot's influence in chemistry reverberated years after his death, with two of his students earning the Nobel Prize in chemistry in 1912.

Otto Wallach 1847-1931

Important work: 1880s to 1910s

Otto Wallach studied at the University of Gottingen under Friedrich Wohler, earning his doctorate in 1869. After various university appointments, Wallach went to work with August Kekule in Bonn, Germany, where the men remained in close cooperation for two decades. Wallach became a professor at the University of Bonn in 1876, and it was during this time that he launched a series of investigations into the constituents of wormseed and other essential oils.

Wallach's work showed that terpenes were based on a unit of five carbon atoms, and could be derived from isoprene. Much of this work was focused on obtaining crystalline compounds and studying the associated rearrangement reactions to make structural determinations. This led Wallach to the conclusion that racemic crystals are denser than those of the base enantiomer, a discovery known as 'Wallach's Rule'.

Wallach successfully characterized the terpenes pinene, camphene, limonene, dipentene, phellandrene, terpinolene, fenchene, a-terpineol, and others, and authored a comprehensive book about this work titled *Terpene und Campher* that was published in 1909. Additionally, of 179 original papers published by Wallach, more than 150 of them were on the subject of terpenes.

Wallach won the 1910 Nobel Prize for Chemistry, and the Davy Medal (Britain's highest award for chemists) in 1912. More than two hundred students earned their doctorate under his professorship[3] at the University of Bonn.

---

[1] Berthelot M, et al., C R Acad Sci 1891, 112, 598.

[2] Bernadette Bensaude-Vincent Pierre-Eugene-Marcellin Berthelot - French chemist. From: https://www.britannica.com/biography/Pierre-Eugene-Marcellin-Berthelot, Accessed November 20, 2020.

[3] Christmann, Mathias. Otto Wallach: Founder of Terpene Chemistry and Nobel Laureate 1910. Angewandte Chemie International Edition 49.50 (2010): 9580-9586.

David Prain 1857-1944

Important work: 1880s to 1890s

David Prain was a famous botanist who served as the director of the Royal Botanic Gardens of Calcutta and Kew, India, and as the professor of Botany at Calcutta Medical College. Prain carried out significant research in Indian hemp and produced a report for the government that - among other things - sought to determine the psychoactive compounds of the plant, in particular the active constituents of 'ganja', concluding that "the active principle of ganja is not a simple substance, but consists of at least two distinct things" (Prain, 1893.)

Prain's words from nearly 130 years ago serve as the first scientific claim that the active ingredients of cannabis cannot be relegated to a single compound, thus setting the stage for our case here, where we present the evidence and arguments for synergy in cannabis.

Friedrich Heusler 1866-1947

Important work: early 1900s

Friedrich Heusler, PhD, was the privatdocent of Chemistry at the University at Bonn. A German chemist who studied under Otto Wallach, Heusler was engaged in a major collaborative work where his section included covering the terpenes, when he realized the scope of terpene information was so vast that it required its own text. This led him to write *The Chemistry of the Terpenes*, a book that scientists of the time hailed as "well-nigh indispensable for the scientific investigations of terpenes and their derivatives".

In *The Chemistry of the Terpenes*, Heusler appears to be the first to propose that terpene oxidation products should be classified as terpenes;

"In fact, it is impossible to develop the chemistry of the terpenes unless these oxygen- containing compounds are considered as members of the terpene series" -Friedrich Heusler, PhD.

Heusler was influenced significantly by Wallach, with Heusler dedicating his book to Wallach;

"To Professor O. Wallach belongs the distinction of having elevated the methods of the terpene chemistry, by a series of superior experimental investigations, to such a plane, that the recognition and separation of the several terpene hydrocarbons have become relatively simple matters for the chemist" -Friedrich Heusler, PhD.

Leopold Ružicka 1887-1976

Important work: 1920s to 1930s

A Swiss scientist who received eight honorary doctorates in various fields, Leopold Ruzicka stumbled upon terpineol while investigating constituents of the herb Tanacetum cinerariifolium. This work led him to publish a hypothesis that the molecular skeleton of terpenes and other large molecules were comprised of linked isoprene units, resurrecting and expanding upon the work of Otto Wallach.

In Ruzicka's work for top perfumeries of the time, he carried out the total synthesis of fenchone, linalool, nerolidol, farnesol, and a partial synthesis of pinene[4]. Ruzicka also developed the biogenetic isoprene rule and was the first to define terpenes according to how many linked units of isoprene are in each molecule.

---

[4] Leopold Ruzicka - Biographical. NobelPrize.org. Nobel Media AB 2020. From: https://www.nobelprize.org/prizes/chemistry/1939/ruzicka/biographical/ Accessed 11/20/2020.

Ruzicka published nearly six hundred original papers in his career and was awarded the 1939 Nobel prize in chemistry for his work with higher terpenes, among other efforts.

Walter Loewe 1884-1963

Important work: 1930s to 1950s

Walter Loewe continued the work of David Prain by delving deeper into studies on the synergy of cannabis and other plants. He was the first to identify cannabinol as an active compound, although it would be years before researchers discovered the compound is a degradation product of tetrahydrocannabinol.

Loewe, a German pharmacologist, developed a method of measuring synergy among chemical constituents called isobologram analysis. This method is also capable of measuring additivity and antagonism, and "has been mathematically proven and widely used to evaluate drug interactions"[5], according to leading terpene and drug interaction researchers.

Robert Burns Woodward 1917-1979

Important work: 1940s to 1960s

Robert Burns Woodward intensively studied terpenes and made exemplary use of ultraviolet spectroscopy to determine their molecular structure, while also discovering the synthesis of complex natural compounds. After studying as a young man at Massachusetts Institute of Technology, Woodward went on to Harvard, where he remained for most of his career.

Woodward is most known for his 1956 total synthesis of reserpine; a substance derived from the Indian shrub Rauwolfia, as well as his 1960 synthesis of chlorophyll. His work led to the development of the Woodward Rules, which predict facets of stereochemistry and reactions in organic chemistry. He was awarded the Nobel Prize in Chemistry in 1965 for this and other work with natural products.

Wilhelm Treibs 1890-1978

Important work: 1950s

Wilhelm Triebs was a German chemist who received his doctorate under Otto Wallach. From 1951-1961, he served as chair of Leipzig University, earning high honors during his time at this esteemed school. Triebs was a member of the German Academy of Sciences as well as the Saxon Academy of Sciences.

Although Triebs primarily worked with azulenes (a molecular skeleton found in only two terpenoids), his work with the sesquiterpene and dietary cannabinoid beta- caryophyllene has earned his place as a Terp Tsar. Triebs was the first to successfully isolate the crystalline form of caryophyllene and elucidated its much-debated structure in 1952.

František Šorm 1913-1980

Important work: 1950s

Frantisek Sorm worked extensively with terpenes and terpenoids, particularly sesquiterpenes and sesquiterpenoids, synthesizing a large number of these compounds while also elucidating the structures of many. Sorm also worked with farnesyl and farnesol, using these compounds as insecticidal control agents to mimic the hormones of juvenile Galleria Mellonella (great wax moth), against both this and other species of insects[6].

---

[5] Huang, Ruo-Yue et al. "Isobologram Analysis: A Comprehensive Review of Methodology and Current Research." Frontiers in pharmacology vol. 10 1222. 29 Oct. 2019.

[6] Vaclav Jarolim, Karel Hejno, Frantisek Sehnal, Frantisek Sorm, Natural and synthetic materials with insect hormones activity 8. Juvenile activity of the farnesane-type compounds on Galleria Mellonella, Life Sciences, Volume 8, Issue 16, Part 2, 1969, Pages 831-841.

A Czech, Sorm founded the Institute for Organic Chemistry and Biochemistry arm of the Czechoslovak Academy of Sciences and was professor of organic chemistry at the University of Prague in the early 1950s.

DHR Barton 1918-1998

Important work: 1950s – 1970s

Sir Derek Harold Richard Barton validated work of Frantisek Sorm, particularly that regarding the structure of beta-caryophyllene. Although recognized worldwide for a variety of important discoveries, Barton is known among terpene researchers for developing what is referred to as the Barton-McCombie deoxygenation, where a hydroxyl functional group is replaced by an alkyl group, changing the function of the molecule. Barton published a comprehensive text detailing this and other work titled *Reason and Imagination: Reflections on Research in Organic Chemistry*, a nine- hundred-page collection of his papers meant to inspire originality in chemistry.

Barton served appointments at universities in England and the United States, was awarded the Nobel Prize in Chemistry in 1969, and was knighted in 1972.

Raphael Mechoulam

Important work: 1960s to 2023

Raphael Mechoulam is an Israeli scientist who discovered L9-tetrahydrocannabinol (THC) - the primary psychoactive constituent in cannabis - in 1964, paving the way for the later discovery of the endocannabinoid system. Mechoulam was successful in the first total synthesis of THC, cannabidiol (CBD), and cannabigerol (CBG), and discovered the cannabinoid anandamide, while his PhD student, Shimon Ben-Shabat discovered 2-arachidonoyl glycerol (2-AG).

The term 'entourage effect' was coined in 1988 by Mechoulam and his colleagues, although it was Ethan Russo who later correctly suggested that this synergistic action is due to the presence of terpenes and terpenoids, and, to a lesser but still important degree, flavonoids. Mechoulam's work primarily focuses on cannabinoids, which are based on diterpenes.

Russ Hudson contacted Dr. Mechoulam and asked whether he thought cannabis research should continue to focus on individual cannabinoid compounds, or on the synergy between compounds. His response:

"We reported the entourage effect many years ago, but unfortunately there is limited research on this effect. I believe that some but not all cannabinoid effects are enhanced (or somewhat modified) by some terpenes, but we lack solid data. I wish we had. Anyway, cannabis users generally prefer cannabis extracts rather than pure THC." -Dr. Raphael Mechoulam.

Mahmoud A. ElSohly

Important work: 1980 – Present

Mahmoud A. ElSohly is a cannabis researcher based in the United States, where he is the director of the Marijuana Research Project at the University of Mississippi. This has been the only legal source of cannabis for medical research purposes in the United States.

With a PhD from the University of Pittsburg, ElSohly's work has focused on determining the structure of various natural products and molecules in cannabis, the synthesis of cannabispiran, identification of genes in δ9-tetrahydrocannabinol, and the investigation of various essential oils.

ElSohly is the professor of pharmaceutics at the Research Institute of Pharmaceutical Sciences, and the president and laboratory director of ElSohly Laboratories, Inc.

John M. McPartland

Important Work: 1980s - Present

John M. McPartland has specialized in the botanical history of cannabis by studying fossil pollen, as well as studies involving synergy among cannabis oil constituents. McPartland has worked closely with Ethan Russo to determine the synergistic relationships between cannabinoids and terpenes.

McPartland founded GW Pharmaceuticals, a company that is "committed to the concept of synergy in whole plant extracts", producing one of the world's first cannabis-derived medicines approved for the public markets.

Shimon Ben-Shabat

Important Work: 1990s - Present

A student of Raphael Mechoulam, Shimon Ben-Shabat discovered an endogenous cannabinoid, 2-Arachidonoylglycerol (2-AG), and in the late 1990s was among the first scientists (together with Mechoulam) to demonstrate evidence of the entourage effect in cannabis[7].

Ben-Shabat has also studied the effect of flavone glycosides in the treatment of herpes[8].

Ethan Russo

Important Work: 1980s – Present

Dr. Ethan Russo is the world's leading researcher of synergy in cannabis, particularly the synergy between cannabinoids and terpenes. While David Prain was the first to suggest more than one active constituent in cannabis, and Mechoulam et al coined the term entourage effect to describe the same concept, it was Russo who proposed that the entourage effect or synergistic action is primarily focused between terpenes and cannabinoids, particularly CBD and THC.

Russo has collaborated closely with John M. McPartland at GW Pharmaceuticals, researching synergy in cannabis and the therapeutic effects of various terpenes. Russo is a neurologist and psychopharmacology researcher, and the director of research and development for the International Cannabis and Cannabinoids Institute.

Russ Hudson contacted Russo and asked about his thoughts for the future of cannabis and terpene research. Russo provided the following quote:

"It is due in large part to cannabis that there has been a recent focus on the therapeutic potential of terpenoids and essential oils. Hopefully, this will translate into better research funding, as this group of substances has tremendous potential for contributing to improved human and animal health."
-Dr. Ethan Russo.

[7] Ben-Shabat S, Fride E, Sheskin T, Tamiri T, Rhee MH, Vogel Z, Bisogno T, De Petrocellis L, Di Marzo V, Mechoulam R. An entourage effect: inactive endogenous fatty acid glycerol esters enhance 2-arachidonoyl-glycerol cannabinoid activity. Eur J Pharmacol. 1998 Jul 17;353(1):23-31.

[8] Yarmolinsky L, Huleihel M, Zaccai M, Ben-Shabat S. Potent antiviral flavone glycosides from Ficus benjamina leaves. Fitoterapia. 2012 Mar;83(2):362-7.

Sameksha Koul
Important Work: 2000s – Present

Sameksha Koul is the assistant professor of pharmacognosy and phytochemistry at Bahrat Institute of Technology in Meerut, India. Koul's book, *A Comprehensive Treatise on Terpenes,* is a thorough overview of the terpenes, suitable for both researchers and laypersons.

Colleen Quinn
Important Work: 2010 – Present

Colleen Quinn is a clinical aromatherapist specializing in the formulation of therapeutic terpene blends. Quinn trains the next generation of terpene blend formulators as an educator certifying professionals in the aromatherapy and terpene product formulations industry. Quinn has also worked extensively in oncology, providing therapy to cancer patients, and educating patients and their caregivers on the use of various terpene and essential oil formulations for therapies related to cancer treatment and general well-being. Russ Hudson contacted Quinn and asked her thoughts on the therapeutic nature of terpenes. Quinn replied with the following:

"Ultimately, the essence of plants that makes them therapeutic is terpenes, primarily because of the entourage effect or the theory that these phytochemicals work better together than as individual compounds" -Colleen Quinn.

Susan Trapp
Important Work: 2000s - Present

Susan Trapp has spent decades investigating terpenes, particularly the molecular evolution of terpenes and terpenoids, the horizontal gene transfer of trichothecenes, and in general the genomic architecture of plant terpenoids. Trapp is the founder of Terpedia.com, and, as of the summer of 2021, conducts bioassays and characterizations of cannabinoids, terpenes, and other phytochemicals for the Southwest College of Naturopathic Medicine.

Russ Hudson contacted Trapp and asked her thoughts on the viability of sourcing individual terpene constituents from cannabis. Trapp provided the following quote:

"Cannabis is certainly a viable industrial/medical source for individual terpene/enoids constituents, however, it is most likely not cost effective. There are other plant sources that may have higher quantities of individual terpene that will be much more cost effective and efficient to extract terpenes/enoid, certainly for the major abundant terpenes. For example, celery is high in myrcene and a much cheaper plant to grow than cannabis. Cannabis may be the only source for some of the minor terpenes/enoids being far less abundant and some potentially unique to cannabis.

Furthermore, given the field of bioengineering and biotechnology, it is a relatively straightforward problem to biosynthesize these compounds from microbes. This is already being pursued by a number of companies. Whether it is cost effective - is to be determined and individual terpene constituent dependent" -Susan Trapp, PhD.

Terp Tsars Review

Answer the following questions to test your knowledge of this section:

Question #1: Who was the first terpene scientist to refute vital force theory?

____

Question #2: Which German organic chemist named isoprene molecules 'terpenes?'

____

Question #3: Who advanced the theory that all aromatic compounds feature a hexagonal ring?

____

Question #4: Who studied the aroma of fresh soil?

____

Question #5: Which famous botanist first implied synergy in cannabis, stating: "the active principle of ganja is not a simple substance, but consists of at least two distinct things"?

____

Question #6: After studying under Otto Wallach, which German chemist went on to write *The Chemistry of the Terpenes*, where he proposed that terpene oxidation products should be classified as terpenes?

____

Question #7: Which German pharmacologist developed a method of measuring synergy among chemical constituents called isobologram analysis?

____

Question #8: Which Israeli scientist discovered delta-9-THC, the endocannabinoid system, and coined the term 'entourage effect'?

____

Question #9: Which researcher is known for their work on the synergy between THC and CBD?

____

For the answer key to Terp Tsars, please go here: http://thebigbookofterps.com/bbot-101-chapters-quiz-answer-key/

TERPENES 101

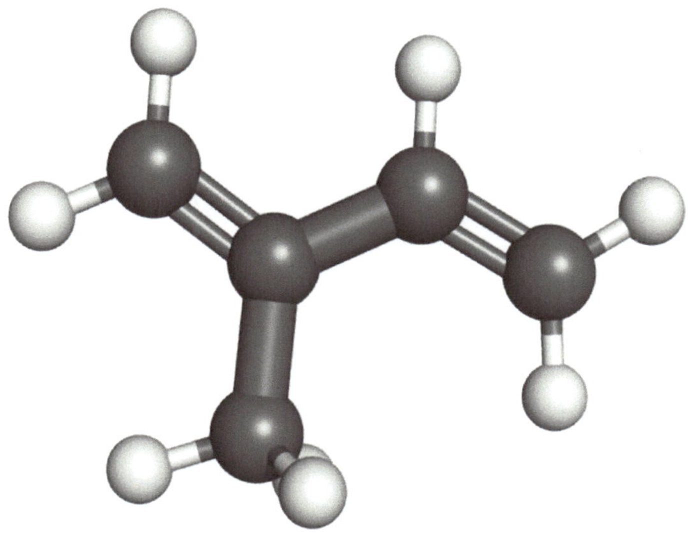

Figure 1: Isoprene is the basic building block of terpenes

TERPENES IN THE PLANT KINGDOM

Terpenes are small molecules produced primarily by plants (including marine plants), but also by fungi, insects, nematodes, and microorganisms like bacteria. For the duration of this book, we will deal mostly with plant-based terpenes, and particularly those that are found in detectable concentrations in cannabis. Terpenes are found in the volatile aromatic or essential oils of plants and serve an astounding array of functions. To begin, we will examine why and how plants use terpenes.

[Note: Although all the relevant concepts and terms are accounted for here and it is not a pre-requisite, readers are nevertheless encouraged to study or possess an understanding of basic organic chemistry to better grasp more complex information like chirality, isomerism, and isomeric determinations and designations.]

HOW DO PLANTS USE TERPENES?

Terpenes are tools of the plant world. Plants use these compounds as weapons of attack, implements of defense, to call for reinforcements, to heal and protect wounds, to promote propagation, as chemical weapons against insect, fungi, and bacteria attacks, to communicate with other plants and organisms in their environment, and as an effective sunscreen, among other uses discussed in this text. Cannabis likely uses terpenes for many of these purposes. However, as of the printing of this book, no studies have been conducted on how cannabis in particular uses specific terpenes. But because terpenes such as pinene, limonene, and linalool are the same from plant to plant, we can surmise that similarities in function and purpose exist between plant species.

Studying how plants use terpenes invariably leads to a new perspective of the natural environment. For instance, consider the terpenes pinene and myrcene, made by most varieties of conifer trees. These and several other terpenes that are also prominent in cannabis are used by evergreen trees to heal wounds. When a tree has been wounded - by bear claws, insects, human activity, or storm damage - terpenes are produced rapidly in such quantity that a dense and viscous sap forms. This sap is exuded to cover the wound and protect it, especially from bacterial, fungal, and viral infection, but also as a barrier to prevent the incursion of insects or other pests.

Consider what we as humans might do when we are injured; we cover our wounds securely and apply antiseptics or antibacterial medications. We allow our plasma and blood to fill in the wound and protect it. Comparatively, a tree's sap is its plasma or blood, and that blood is comprised mostly of terpenes.

The scenario above is just one of the ways that trees and other plants use terpenes. In other cases, as described below, plants use terpenes for a fascinating and diverse variety of purposes, showing that the natural world is highly complex and advanced in ways that cannot be completely understood without an understanding of terpenes.

Insect Traps

Plants like cannabis manufacture terpenes that are designed to passively defend against insect attack. One of the most effective methods for managing pests is to physically immobilize them. Cannabis plants do this by producing copious amounts of extremely sticky glandular trichomes containing terpenes like pinene and myrcene; the reality is that the 'sticky icky' produced by cannabis was never actually meant for humans. Instead, those cannabinoid- and terpene-packed trichomes act as a dangerous type of glue trap, meant to immobilize and eventually kill insects.

In fact, this method of insect control has been so effective in the natural world that pest control companies have mimicked it perfectly, often using terpenes to do so. For instance, if we were to examine the composition of a glue trap meant to deal with mice or rats, we might see terpenes like eugenol, which is made by plants such as cinnamon, nutmeg, and basil. In 2018 the Tomcat variety of mouse glue traps featured eugenol for "enhanced stickiness". When we examine the composition of natural glue traps made by plants like cannabis, we again see terpenes like pinene, myrcene, and eugenol, otherwise used by conifer trees to protect and heal wounds with a thick sap as discussed previously. Clearly, the same terpene can serve multiple functions in the same plant.

Insect Repellants

Some terpenes can function as insect repellants. Much like the case of the glue traps presented above, humans learned to use terpenes as insect repellants by studying the natural world of plants. For example, consider the wide proliferation of citronella products, which are designed to repel mosquitos and other biting pests. Citronella candles, torches, and oils can be purchased readily in most modern countries for use in managing mosquitos and other pests in outdoor areas. The central ingredient in these products is citronellal, a terpene.

Other popular mosquito, tick, flea, and biting fly repellants include the OFF! brand of insect sprays, which makes use of the terpenes found in the essential oil of lemon eucalyptus, including citronellal and geraniol; the latter of which is a terpene found in high concentrations among some varieties of cannabis, most notably Northern Lights.

In the natural world, plants can use terpenes to repel insect attacks, but some plants go beyond merely seeking to discourage insects with unpleasant aromas, textures, or flavors. For instance, plants that manufacture a variation of farnesene - a terpene found commonly but in low concentrations in cannabis - were shown to trick aphids into dispersing. Known specifically as (E)-b-farnesene, this compound was released by plants including maize during herbivore attacks, and served to mimic an alarm pheromone that is normally released by aphids to warn others nearby to flee[9].

Humans have long sought to better understand how plants use terpenes as insect repellants. One interesting study genetically engineered a variety of Arabidopsis thaliana (thale cress or mouse-ear cress) to overexpress the

---

[9] Bernasconi ML, Turlings TCJ, Ambrosetti L, Bassetti P, Dorn S: Herbivore-induced emission of maize volatiles repel the corn leaf aphid Rhopalosiphum maidis. Entomologia Experimentalis et Applicata 1998, 87:133-142.

monoterpene linalool, which is one of the most common terpenes in cannabis. The higher concentrations of linalool in the modified thale cress successfully repelled aphids (a bug that sucks the essential juices out of young leaves), whereas the unmodified plants did not enjoy such relief[10].

Other terpenes have been theorized to function merely by individual preference - some insects just don't like the smell of some terpenes, just as some people don't. Not everyone enjoys the earthy smell of the myrcene found in Kush Valley cannabis varieties, or the sharp, citrusy scent of Super Silver Haze, which has a high concentration of terpinolene.

*Insecticides

Plants can also produce terpenes that biologically attack insects and act directly as a pesticide. Limonene, pinene, cineole, linalool, myrcene, and pulegone (common terpenes in cannabis), have all been shown to possess fumigant properties[11].

In addition to toxic airborne terpenes, topical toxicity of monoterpenes such as limonene, pulegone, and variations of terpinene - all terpenoid molecules often found in cannabis - is potent against several insects including flies, corn borer larvae, and some mites[12]. This means that, for some pests, merely touching a plant that manufactures and releases these or other topically toxic terpene compounds can prove incapacitating or fatal.

Other natural terpene pesticides common to cannabis and other plants include beta-caryophyllene, myrcene, and pinene, which have been shown to be toxic to mosquitos, houseflies, some aphids, and tobacco cutworms[13].

*Insect Attractants – Pollinators

Certain terpenes are emitted by some plants to directly attract pollinators like bees, wasps, butterflies, hummingbirds, etc. Found in the volatile essential oils of cannabis and other plants, most terpenes are strongly aromatic molecules with low molecular weight, which explains why they evaporate so readily - a more volatile substance is easier to transfer from a plant to the air of the surrounding environment and into the scent receptors of beneficial pollinators.

For instance, insects that are attracted to terpenes like limonene, which has a strong citrus smell, are likely to land on cannabis, where these terpenes are often found in abundance, depending on the variety. Pollinators that harvest or otherwise transport pollen from lemon flowers would also likely be interested in exploring cannabis flowers, as both have high concentrations of limonene. While cannabis propagates quite well by transferring genetic material via airborne pollen alone, insect pollinators can help pollen reach flowering female plants over larger distances, and in places not easily reachable by wind currents carrying pollen. But most plants - especially those that flower and produce fruits - are much more dependent on attracting pollinators, which helps explain the wide variety of aromatic terpenes in the plant world. Aromatic terpenes help to lure beneficial pollinators via airborne scents. Terpenes like farnesene, ocimene, and myrcene are just a few examples of aromatic molecules that attract pollinators[14].

---

[10] Aharoni, A. et al. Terpenoid metabolism in wild-type and transgenic Arabidopsis plants. Plant Cell 15, 2866-2884 (2003).

[11] Dambolena, Jose S., Maria P. Zunino, Jimena M. Herrera, Romina P. Pizzolitto, Vanessa A. Areco, and Julio A. Zygadlo. Terpenes: Natural Products for Controlling Insects of Importance to Human Health-A Structure-Activity Relationship Study. Psyche: A Journal of Entomology 2016.2016 (2016).

[12] Lee, Sang Kyun, Insecticidal properties of monoterpenoids and their derivatives as a new natural means of crop protection (1997). Retrospective Theses and Dissertations. 11478.

[13] Giovanni Benelli, Roman Pavela, Riccardo Petrelli, Loredana Cappellacci, Giuseppe Santini, Dennis Fiorini, Stefania Sut, Stefano Dall'Acqua, Angelo Canale, Filippo Maggi. The essential oil from industrial hemp (Cannabis sativa L.) by-products as an effective tool for insect pest management in organic crops, Industrial Crops and Products, Volume 122, 2018, Pages 308-315.

[14] Hartmann, Marie-Andree. The way the dioecious plant Actinidia deliciosa attracts bees: critical role of volatile terpenes released from kiwifruit flowers of both genotypes. Journal of Experimental Botany 60.11 (2009): 2953- 2954.

Insect Attractants – Predators & Indirect Defense

When a plant uses a terpene or other chemical that works by attracting the predator of an herbivore, this is called indirect defense; a plant can call out to the enemies of its enemies and recruit assistance. The language spoken between plant and defender is often terpenes.

One example of this is when the roots of maize (corn) plants release a large amount of the sesquiterpene beta-caryophyllene in response to an attack by insect larvae. The rush of beta-caryophyllene attracts nematodes that subsequently prey on the insect larvae[15]. Bark beetles have also been shown to be attacked by carnivorous insects responding to the release of volatile terpenes from trees[16].

A variation of the sesquiterpene farnesene mentioned earlier, (E)-b-farnesene, is the exact terpene that some plants use to trick aphids into fleeing in alarm; this has also been shown to be used by pine trees in a form of indirect defense. In this case, the sesquiterpene is released into the air when a pine sawfly lays eggs on the extremities of a pine tree. The scent of this volatile terpenoid attracts the attention of a species of wasp that parasitizes the eggs[17].

Interesting research is now showing that some plants might call birds to their aid when attacked. When white birch trees were attacked by nests of autumnal moth caterpillars, the trees released a blend of terpenes including a variation of ocimene and linalool. This terpenoid blend attracted several species of birds that attacked and fed on the caterpillars[18].

Defense Against Competitors – Allelopathic Agents

Plants can use terpenes to influence the growth of other plants competing for the same resources. This can be accomplished by releasing terpenes or terpenoid mixtures into the ground via the root system, or by the evaporation of volatile terpene molecules into the atmosphere. When the chemicals released by one plant affect the growth or survival of competing plants, this is called allelopathy, and the terpenes and other chemicals involved are referred to as allelochemicals.

Examples of common plants that use terpenes to influence the growth of cohabitant plant species include at least one variety of marigold from South America, and the sage brush of Southern California. In the case of the marigold, the plant uses a variation of the terpene ocimene to prevent the germination of other plants[19]. Sage brush also prevents the germination of competing or cohabitating plants, but it uses a variation of the terpene cineole for this purpose[20]. Examples of other plants that produce and use terpenes to influence cohabitant species include the date

---

[15] Rasmann S, Kollner TG, Degenhardt J, Hiltpold I, Toepfer S, Kuhlmann U, Gershenzon J, Turlings TCJ: Recruitment of entomopathogenic nematodes by insect-damaged maize roots. Nature 2005, 434:732-737.

[16] Marco Michelozzi (1999) Defensive roles of terpenoid mixtures in conifers, Acta Botanica Gallica, 146:1, 73-84.

[17] Hilker, M., Kobs, C., Varama, M. & Schrank, K. Insect egg deposition induces Pinus sylvestris to attract egg parasitoids. J. Exp. Biol. 205, 455-461 (2002).

[18] Mantyla E, Alessio GA, Blande JD, Heijari J, Holopainen JK, Laaksonen T, Piirtola P, Klemola T: From plants to birds: higher avian predation rates in trees responding to insect herbivory. PLoS ONE 2008, 3:e2832 2810.1371/Journal.pone.0002832.

[19] M. Liza Lopez, Norma E. Bonzani, Julio A. Zygadlo. Allelopathic potential of Tagetes minuta terpenes by a chemical, anatomical and phytotoxic approach, Biochemical Systematics and Ecology, Volume 36, Issue 12, 2008, Pages 882-890.

[20] Jane Buckle PhD, RN, in Clinical Aromatherapy. Basic Plant Taxonomy, Basic Essential Oil Chemistry, Extraction, Biosynthesis, and Analysis (Third Edition), 2015.

palm,[21] and salvia. In the case of salvia (specifically s. leucophylla and s. mellifera), the plant releases a terpene blend of cineole and camphor into the air, which inhibits the germination of the seeds of other plants[22]. Allelopathic properties are also attributable to cannabis, and it is likely that the chemicals responsible for this are terpenes and terpenoids, although more research is needed to confirm this with specificity. One recent study showed that a water-based extract of cannabis sativa L. influenced the ability of rye, lupine, and rape seeds to germinate, and subsequently affected the vigor of the cohabitant species[23]. An earlier study showed that cannabis sativa significantly inhibited the growth of Parthenium hysterophorus[24] (Santa Maria plant) .

**Egg, Ova, & Larvae Control*

Some terpenes can inhibit, delay, or retard the growth of insect eggs or larvae, similarly to the way terpenes inhibit the development of cohabitant plant species. In fact, certain terpenes serve both purposes.

Terpenes that have been shown to possess significant ovicidal (ability to kill eggs) properties include limonene, linalool, geraniol, terpineol, pinene, cineole, nerolidol[25], and variations of caryophyllene[26] - all of which are terpenes that commonly occur in cannabis.

Other terpenes work by killing insect larvae directly. One study found that many monoterpenes possess larvicidal properties, but that the toxicity of these terpenes increased in modified compounds[27], such as sesquiterpenes and diterpenes.

**Herbivore Repellants – Bitter Terpenes*

Terpenes that are bitter like humulene (found in hops) and limonene (found in citrus fruits) help to ward off attack by hungry herbivores, which can include both insects and animals. Bitterness is a natural indicator that a plant may not be suitable for consumption, or perhaps that it could be poisonous, and plants use this to their advantage by producing bitter-tasting compounds like terpenes - even if the plant isn't actually toxic. The bitterness is cause enough for doubt.

In cannabis, monoterpenes have been shown to repel insect herbivores, while sesquiterpenes - the most bitter of the terpenes - are meant in part to deter grazing animals[28]. In humans, research has shown that bitterness is the most

---

[21] Hadrami A.E., Daayf F., Hadrami I.E. (2011) Secondary Metabolites of Date Palm. In: Jain S., Al-Khayri J., Johnson D. (eds) Date Palm Biotechnology. Springer, Dordrecht.

[22] Muller, C. H. Inhibitory terpenes volatilized from Salvia shrubs. Cornelius H. Muller Bull. Torrey Bot Club Bulletin of the Torrey Botanical Club 92(1):38 • January 1965.

[23] Krzysztof Pudelko, Leszek Majchrzak, Dorota Narozna. Allelopathic effect of fibre hemp (Cannabis sativa L.) on monocot and dicot plant species. Industrial Crops and Products, Volume 56, 2014, Pages 191-199.

[24] Singh, N.B. & Thapar, R. (2003). Allelopathic influence of Cannabis sativa on growth and metabolism of Parthenium hysterophorus. Allelopathy Journal. 12. 61-70.

[25] Toloza, Ariel. (2010). Bioactividad y toxicidad de componentes de aceites esenciales vegetales, en Pediculus humanus capitis (Phthiraptera: Pediculidae) resistentes a insecticidas piretroides. 10.13140/RG.2.1.4342.2480.

[26] Y.-C. Yang, S.-H. Lee, W.-J. Lee, D.-H. Choi, and Y.-J. Ahn, Ovicidal and adulticidal e ects of Eugenia caryophyllata bud and leaf oil compounds on Pediculus capitis, Journal of Agriculturaland Food Chemistry, vol. 51, no. 17, pp. 4884-4888, 2003.

[27] Lee, Sang Kyun, Insecticidal properties of monoterpenoids and their derivatives as a new natural means of crop protection, (1997). Retrospective Theses and Dissertations. 11478.

[28] S. Casano, G. Grassi, V. Martini and M. Michelozzi (2011) Variations in terpene profiles of different strains of Cannabis sativa L. Acta Horticulturae 925:115-121.

common cause of food rejection[29], suggesting that some of the terpenes we seek in cannabis are the same that seek to repel us.

Raw cannabis leaves are bitter because of terpenes like humulene, pinene, and limonene, but the bitterness isn't always enough to completely deter herbivores from munching on the plant. Given their preferred feed, most animals would likely leave cannabis alone. But if more desirable plants are not available, many animals will hungrily consume cannabis, including deer, elk, moose, cattle, horses, goats, and other ruminants, all of which can tolerate the bitterness of the plant's leaves. And because there is already a trend among humans to eat raw cannabis leaves in salads and juices, there is ample anecdotal evidence to suggest that the bitterness of cannabis leaves will not be enough to save the plant from salad bars of the future.

However, some varieties of cannabis may be better than others at repelling herbivores, depending on the phytochemical profile of the strain. For example, if we analyze the terpene profile of conifer trees such as the Sitka spruce or the twisted pine, we see that the trees use monoterpenes to repel the massive red deer of Europe and Asia[30]. Cannabis with similar terpene profiles may prove just as resistant to red deer or other animals, including humans. Of course, humans could endeavor to breed a variety of cannabis that is not so bitter.

*Terpenes as Communication

Chemical signaling is how much of the life on Earth communicates, and the medium of majority is terpenes[31]. This type of communication occurs between plants, insects, birds, fungi, bacteria, and invertebrates. Terpene communication appears to be easily understood between groups regardless of which entity is speaking and which is receiving.

For instance, when bark beetles attack conifer trees, a war occurs that is mostly based upon terpenes, as the tree attempts to amass enough volatile terpenes to repel or kill the invaders. Meanwhile, the beetles use these chemical signals to adjust their behavior, including cooperative behavior between beetles that can sometimes outpace the tree's ability to produce enough insecticidal terpenes[32] to defeat the insects. Both entities, regardless of the animosity and deadly intent between them, speak the same language.

Because of their sheer numbers, most of the world's communication in terpenes occurs between microorganisms. Recent research has demonstrated that bacteria and fungi communicate with each other in the language of terpenes[33], and it is already well understood that trees and other plants communicate with fungi and bacteria via terpenes and other compounds through the soil. Comparatively speaking, it is the languages of man that are odd and rare.

It is logical that this chemical communication extends to humans and plants, especially when it comes to cannabis. Consider that when you smell a particular variety of cannabis that you enjoy, you are receiving communication from the plant in the medium of terpenes. However, this is more than just an arbitrary smell - it tells us something about the plant, and it makes us feel a certain way, although it is often not easy to articulate precisely how. When

---

[29] Drewnowski, Adam, and Carmen Gomez-Carneros. "Bitter taste, phytonutrients, and the consumer: a review." The American Journal of Clinical Nutrition 72.6 (2000).

[30] Marco Michelozzi (1999) Defensive roles of terpenoid mixtures in conifers, Acta Botanica Gallica, 146:1, 73-84.

[31] Netherlands Institute of Ecology, Terpenes are the World's Most Widespread Communication Medium, from; https://phys.org/news/2017-04-terpenes-world-widespread-medium.html Accessed December 21, 2018.

[32] Blomquist GJ, Figueroa-Teran R, Aw M, Song M, Gorzalski A, Abbott NL, Chang E, Tittiger C. Pheromone production in bark beetles. Insect Biochem Mol Biol. 2010 Oct;40(10):699-712.

[33] Schmidt, Ruth, Victor de Jager, Daniela Zilhlke, Christian Wolff, Jorg Bernhardt, Katarina Cankar, Jules Beekwilder, Wilfred van Ijcken, Frank Sleutels, Wietse de Boer, Katharina Riedel, and Paolina Garbeva. Fungal volatile compounds induce production of the secondary metabolite Sodorifen in Serratia plymuthica PRI-2C. Scientific Reports 7 (2017).

looked at in this light, and with more research and study, it will become clear that humans can indeed receive the communication of plants through the language of terpenes.

The question that remains is, can we do more than just receive this communication?

*Terpenes as Sunscreen and Air Conditioning

Terpenes can help to keep plants cool. When it comes to plant air conditioning, the science behind this is simple: as terpenes volatilize or evaporate, the resulting airflow cools the plant and helps it retain more of its vital liquids, similarly to how "swamp coolers" work in certain climates.

Some terpenes can also act as sunscreen. As some plants are exposed to UVB radiation, they produce significantly more terpenes via resin accumulation in leaves[34], leading researchers to theorize that this increased production of resin and terpenes acts as a sunscreen and prevents cell damage[35].

One clue to the effectiveness of terpenes as sunscreen comes from a study conducted using mice, in which the terpene d-limonene protected the animals from skin damage due to UVB radiation[36], indicating that plants might use this isoprene compound similarly.

*Defense Against Fungi & Bacteria

If the number of terpene-based antibacterial and antifungal products on the market today are any indicator, humans are well-aware of the use of terpenes to fight bacteria and fungi, although they might not always understand the mode of action.

Variations of cineole, cymene, limonene, linalool, pinene, terpinene, and terpinolene, among others, have all been found to possess antifungal activity[37], and the monoterpene geraniol has been shown to be active against dermatophytes and yeasts[38]. These are among the top terpenes commonly found in cannabis.

Terpenes can work effectively against bacteria as well. In a study involving Pilgerodendron uviferum (cypress trees), variations of caryophyllene and humulene were shown to protect the trees against microbial attack[39]. The authors of the study concluded that the tree's resistance to bacterial attack was entirely due to sesquiterpenes and diterpenes. Other studies have concluded that the cell membrane of bacteria can be infiltrated, and the cell's contents spilled (effectively killing the bacteria cell) by terpenes, including nerolidol and menthol[40,41].

---

[34] Zavala, Jorge, and Damian Ravetta. The effect of solar UV-B radiation on terpenes and biomass production in Grindelia chiloensis (Asteraceae), a woody perennial of Patagonia, Argentina. Plant Ecology 161.2 (2004): 185-191.

[35] Wen Jing Zhang, Lars Olof Bjorn, The effect of ultraviolet radiation on the accumulation of medicinal compounds in plants, Fitoterapia, Volume 80, Issue 4, 2009, Pages 207-218.

[36] Uddin, Ahmed. (2014). d -limonene prevents ultraviolet irradiation: Induced cyclobutane pyrimidine dimers in Skh1 mouse skin. World Journal of Dermatology. 3. 64. 10.5314/wjd.v3.i3.64.

[37] Nazzaro, Filomena, Florinda Fratianni, Raffaele Coppola, and Vincenzo De Feo. "Essential Oils and Antifungal Activity." Pharmaceuticals 10.4 (2017).

[38] Diogo Miron, Fernanda Battisti, Fernanda K. Silva, Aline D. Lana, Bruna Pippi, Bruna Casanova, Simone Gnoatto, Alexandre Fuentefria, Paulo Mayorga, Elfrides E.S. Schapoval, Antifungal activity and mechanism of action of monoterpenes against dermatophytes and yeasts, Revista Brasileira de Farmacognosia, Volume 24, Issue 6, 2014, Pages 660-667.

[39] Solis, C. & Becerra, J. & Flores, C. & Robledo, J. & Silva, Mario. (2004). Antibacterial and antifungal terpenes from Pilgerodendron uviferum (D. Don) Florin. Journal of The Chilean Chemical Society - J CHIL CHEM SOC. 49.

[40] Inoue, Yoshihiro, Akiko Shiraishi, Toshiko Hada, Kazuma Hirose, Hajime Hamashima, and Jingoro Shimada. The antibacterial effects of terpene alcohols on Staphylococcus aureus and their mode of action. FEMS Microbiology Letters 237.2 (2004): 325-331.

[41] Trombetta, Domenico et al. Mechanisms of antibacterial action of three monoterpenes. Antimicrobial agents and chemotherapy vol. 49,6 (2005): 2474-8.

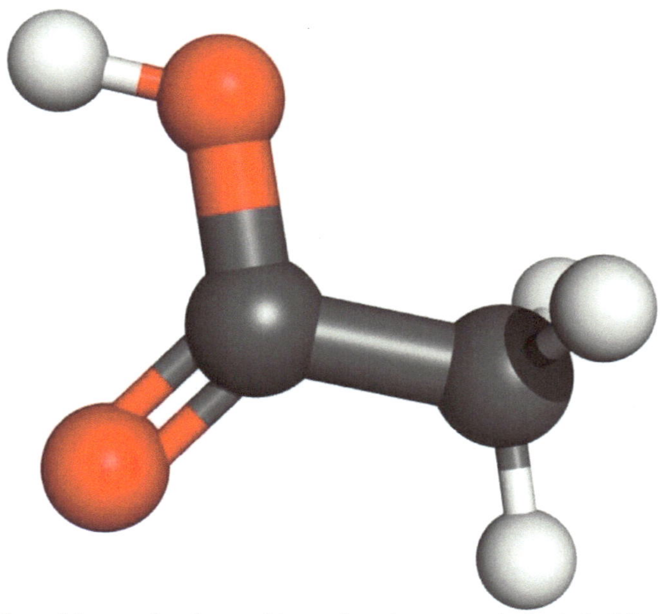

Figure 2: Acetic acid. Two of these molecules combine to form isoprene - the basic building block of most terpenes

WHEN DO PLANTS PRODUCE AND USE TERPENES?

Just as terpenes can be toxic to insects, fungi, and bacteria, they are also cytotoxic to plants - even the very ones that produce the terpenes in the first place. For this reason, terpenes are made and stored in protective enclosures within the plant, such as secretory idioblasts (isolated plant cells), and in trichomes[42]. If we take the pennyroyal plant as an example, we see that monoterpenes are formed in the young plant tissue, with more complex compounds like sesquiterpenes and diterpenes occurring in the older tissue[43]. Plant terpene production occurs at higher temperatures - usually when the sun is shining - and terpenes tend to evaporate as the daytime temperature rises. For anyone that has cultivated cannabis outside, you probably noticed that your plants smelled more strongly in the morning, before the temperature rose. That is likely because the terpenes in your plants hadn't started to evaporate off yet, which meant that the overall concentration of terpenes was higher and thus the smell was more potent. For this reason, experienced cannabis cultivators recommend harvesting outdoor plants in the morning.

This phenomenon of daytime terpene emissions has also been observed by researchers studying pollinators of kiwifruit; the evaporation of volatile terpene compounds primarily occurred during the day, when pollinators were most active[44]. As discussed previously, pollinators rely at least partly on the release of terpenes to find their favorite flowers, thus the close association and timing makes sense.

CHEMICAL NATURE OF TERPENES

Terpenes are formed in plants by combining two acetic acid molecules to form isoprene (also referred to as 2-methylbutane) via isopentenyl pyrophosphate as a precursor. This generally occurs within the cytosol or aqueous area of plant cells, where organelles and other cell components are suspended in the cell's fluid.

Each isoprene molecule and many of its derivatives are generally arranged in a head-to-tail fashion for most terpenes discussed herein. The head of each molecule is typically comprised of isopropyl, while the tail part is ethyl. Classified as hydrocarbons, terpenes are made from the mother hydrocarbon, isoprene, which has a formula of

[42] Marco Michelozzi (1999) Defensive roles of terpenoid mixtures in conifers, Acta Botanica Gallica, 146:1, 73-84.

[43] J. Battaile and W.D. Loomis, Biosynthesis of Terpenes, Biochim. Biophys. Acta, 51, 545 (1961).

[44] Hartmann, Marie-Andree. The way the dioecious plant Actinidia deliciosa attracts bees: critical role of volatile terpenes released from kiwifruit flowers of both genotypes. Journal of Experimental Botany 60.11 (2009): 2953- 2954.

C_5H_8. This denotes the fact that isoprene has five carbon atoms bound to eight hydrogen atoms. To form a terpene, two or more of these units or molecules of isoprene must join together. This process of interlinking isoprene units to form terpenes is known as Wallach's Law, and was first developed by Otto Wallach and Leopold Ruzicka.

Different terpene classes are produced by different biosynthetic pathways in plants, including the mevalonate (MVA) pathway, the methylerythritol phosphate (MEP) pathway, and the isopentenol pathway. Terpene classes are categorized according to how many C_5H_8 isoprene units are in the terpene molecule:

Hemiterpenes: One unit of isoprene has five carbon atoms and eight hydrogen atoms.

Monoterpenes: When two units of isoprene combine, the resulting $C_{10}H_{16}$ molecule (ten carbon atoms and sixteen hydrogen atoms create two units of isoprene) is called a monoterpene.

Sesquiterpenes: When three units of isoprene combine, the resulting $C_{15}H_{24}$ molecule (fifteen carbon atoms and twenty-four hydrogen atoms create three units of isoprene) is called a sesquiterpene.

Diterpenes: When four units of isoprene combine, the resulting $C_{20}H_{32}$ molecule (twenty carbon atoms and thirty-two hydrogen atoms create four units of isoprene) is called a diterpene.

Sesterterpene: When five units of isoprene combine, the resulting $C_{25}H_{40}$ molecule (twenty-five carbon atoms and forty hydrogen atoms create five units of isoprene) is called a sesterterpene.

TERPENES CLASSES CHART

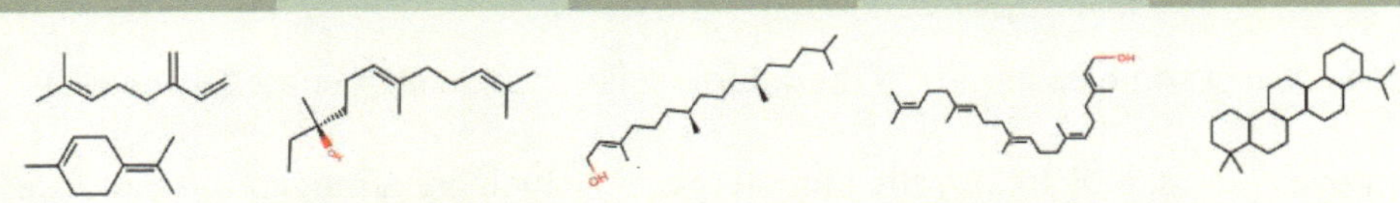

ISOPRENE C_5H_8	MONOTERPENES $C_{10}H_{16}$	SESQUITERPENES $C_{15}H_{24}$	DITERPENES $C_{20}H_{32}$	SESTERTERPENES $C_{25}H_{40}$	TRITERPENES $C_{30}H_{48}$
	pinene,	beta-caryophyllene,	phytol,	geranylfarnesol,	squalene,
	myrcene	humulene,	cannabinol,	dysidiolide,	β-amyrin,
	limonene,	bisabolol,	δ(9)-tetrahydroc-	astallatene,	maniladiol,
	linalool,	nerolidol,	annabinol,	boleracene,	lupane,
	ocimene,	guaiol,	cafestol,	caprudiene A,	oleanane,
	terpinolene	caryophyllene oxide,	kahweol,	manoalide	ursane,
	fenchol,	valence,	grayanotoxin,		glycrrhizin,
	camphene,	cedrene,	forskolin,		betulinic acid,
	borneol,	alpha-bergamotene,	marrubenol,		boswellic acid,
	carene,	elemene,	kaurane,		dammarane,
	phellandrene,	aromadendrene	casbene,		ginsenoide
	terpinene,		abietic acid,		
	1,8-cineole,		oridonin,		
	geraniol,		cembrene		
	sabinene,				
	camphor,				
	menthol,				
	geranyl acetate,				
	cymene,				
	pulegone				

Continuing the above classification system, triterpenes have six isoprene units, tetraterpenes have eight, and so on. There can be dozens, hundreds, or even thousands of isoprene units in a terpene chain. The larger isoprene chains are referred to generally as polyterpenes.

In the cannabis plant, the majority of terpenes consist of monoterpenes, sesquiterpenes, and diterpenes, examples of which include: myrcene, pinene, limonene, geraniol, eucalyptol, and linalool as monoterpenes; beta-caryophyllene, farnesol, and humulene as sesquiterpenes; CBN and THCA as diterpenes, and D8- and D9-THC as tricyclic terpenoids.

**Cannabinoids are Often Terpenes, too*

As mentioned above, cannabinoids can be terpenes, and terpenes can be cannabinoids. For instance, the most common sesquiterpene by weight in cannabis, beta-caryophyllene, is also a dietary cannabinoid, while cannabidiol (CBD) is based on a diterpene molecular skeleton. THCA is a diterpenoid, while Delta-8- and Delta-9-tetrahydrocannabinol are tricyclic terpenoids. Terpenes and cannabinoids are both produced in the trichomes or secretory glands of cannabis, and in fact cannabinoids are known as terpenophenolic compounds, derived in the same biosynthetic pathways as terpenes[45].

This text will cover cannabinoids and their relationship with the human body and interactions with the other compounds found in cannabis separately in the coming chapters, even though many important cannabinoids are, technically speaking, a more stable type of terpene.

**Terpene Structures*

Among the many monoterpenes, sesquiterpenes, and diterpenes, classifications of higher organizations of the basic isoprene structure are grouped into their own sub-classifications, based on arrangement and positions of the molecules in chains and hexagonal rings. At least one of these rings is often comprised of benzene, while others can be formed from different substances.

The following are just some examples of these molecular arrangement sub-classifications:

Acyclic: ocimene and linalool are acyclic monoterpenes, which are arranged in a linear fashion.

Non-head-to-tail: the isoprene units are joined together not head to tail, but instead tail or head of one, to the central carbon atoms of the other.

Monocyclic: limonene, a-phellandrene, and a-terpinene are monocyclic monoterpenes, which have a molecular structure containing one ring.

Bicyclic: pinene, borneol, and camphor are bicyclic monoterpenes, or those that have a molecular structure containing two rings.

Cyclopentanoid: terpenoid compounds based on a cyclopentanoid carbon skeleton where five carbon atoms have been arranged in a ring.

Rearranged skeletons: compounds where the original molecule has been rearranged as a structural isomer. More on this in the section below.

Tricyclic: terpenes that have a molecular structure containing three rings.

---

[45] Turner, C.E., ElSohly, M.A., Boeren, E.G., 1980. Constituents of Cannabis sativa L. XVII. A review of the natural constituents. J. Nat. Prod. 43, 169-234.

*Modified Terpenes

Generally speaking, terpenes contain only hydrogen and carbon atoms and nothing more. Terpenes that contain any modifications such as the addition of an oxygen atom or a functional group are referred to as terpenoids.

The broadest categorization of a terpene depends on whether the molecule contains oxygen atoms. Terpenes without oxygen molecules are simply called 'terpenes,' while terpenes that do contain one or more oxygen atoms are called **terpenoids**, where the 'oid' specifically denotes that oxygen atoms are present. For instance, when we refer to cannabinoids, terpenoids, or flavonoids, we know that all these compounds feature one or more oxygen atoms. Oxygen atoms can change the function of a molecule or the way it interacts with other molecules, or both.

*Functional Groups

Terpenoids can also contain functional groups. Functional groups are specific groups of atoms within a molecule that confer unique chemical reactivity and properties. Common functional groups in terpenoids include:

Hydroxyl group (-OH): A hydroxyl group is a single oxygen atom bonded to a hydrogen atom, forming an alcohol group. Terpenes containing hydroxyl groups are called **terpenols** or terpenoids. These functional groups can influence the solubility and biological activities of terpenes.

Carbonyl group (C=O): A carbonyl group consists of a carbon atom double-bonded to an oxygen atom. When a terpene contains a carbonyl group, it is referred to as a **terpenone** or ketone.

Ester group (RCOOR'): An ester group is formed by the reaction between a carboxylic acid and an alcohol. Terpenes with ester groups are called **terpenyl** esters. These groups can impart fruity or floral notes to the aroma of terpenes.

Aldehyde group (RCHO): An aldehyde group consists of a carbonyl group (C=O) bonded to a hydrogen atom and a carbon atom. Terpenes containing aldehyde groups are referred to as **terpenaldehydes**.

Epoxide group (R-O-R'): An epoxide group is a three-membered ring containing an oxygen atom. Terpenes with epoxide groups may exhibit unique biological activities due to the presence of this strained cyclic structure.

Double bonds (C=C): Terpenes may contain one or more double bonds between carbon atoms. These double bonds can occur within a ring (endocyclic) or outside the rings (exocyclic), and can influence the overall shape and reactivity of the molecule, affecting its biological properties.

Different terpenes can have combinations of these functional groups, leading to a wide range of diverse structures and functions. The presence and arrangement of functional groups are crucial in determining the specific properties and activities of individual terpenes.

*Isomerism

Isomerism is a phenomenon in chemistry where two or more chemical compounds have the same molecular formula, meaning they contain the same number and type of atoms, but have different structural arrangements or spatial configurations. The presence of isomers can significantly impact the properties and functions of substances in various chemical and biological processes. There are two main types of isomers:

Structural isomers: Also known as constitutional isomers, these isomers have the same molecular formula but differ in the way their atoms are connected. In terpenes, structural isomers can arise when the carbon skeleton is arranged differently.

Stereoisomers: Stereoisomers have the same molecular formula and the same connectivity of atoms, but their spatial arrangement of atoms in three-dimensional space differs. Terpenes can exhibit stereoisomerism due to the presence of double bonds or asymmetric centers (chiral centers) in their structures. There are two primary types of stereoisomers:

Geometric isomers (Cis-trans isomers): Geometric isomers arise when there is restricted rotation around a double bond, resulting in different spatial arrangements of substituents. In terpenes, if there is a double bond in the carbon chain, it can lead to geometric isomerism.

Optical isomers (Enantiomers): Optical isomers, also known as enantiomers, are a type of stereoisomers that are non-superimposable mirror images of each other. They have identical physical properties (e.g., boiling point, melting point) but interact differently with polarized light and exhibit different biological activities. Enantiomers are chiral molecules, meaning they have one or more chiral centers (asymmetric carbon atoms). Terpenes can form enantiomers if they contain one or more chiral centers. For instance, (+)-limonene and (-)-limonene are enantiomers. Both have the same molecular formula of $C_{10}H_{16}$, the same carbon skeleton, and the same functional groups, but they are mirror images of each other and rotate plane-polarized light in opposite directions.

Terpenes can exhibit all these types of isomerism, depending on their specific structures and functional groups present. Some terpenes have only two known isomers, such as alpha- and beta-pinene, while others can have many isomers, like selinene, which occurs as alpha, beta, delta, gamma, 6-epi, and 7-epi versions. Precise details explaining chirality, para, ortho, meta, trans, cis, R, S, D, and L designations, and the use of plane-polarized light to determine certain molecule characterizations are included in the following sections.

Chirality and Chiral Centers

Chirality describes the property of molecules and objects that are non-superimposable mirror images of each other. In the context of organic chemistry, chirality often arises due to the presence of chiral centers or asymmetric carbon atoms.

A chiral molecule is one that lacks a plane of symmetry, meaning it cannot be superimposed onto its mirror image. Chiral molecules exist as two distinct mirror-image forms called enantiomers. Enantiomers have identical chemical compositions but differ in their spatial arrangement of atoms. Chirality typically originates from the presence of chiral centers, which are carbon atoms bonded to four different substituents. These carbon atoms are often denoted as stereocenters or asymmetric carbon atoms.

Many terpenes exhibit chirality due to the presence of multiple chiral centers in their structures. For example, limonene is a common terpene found in cannabis. It is chiral and exists as two enantiomers: (d)-limonene and (l)-limonene.

The chirality in limonene arises from the presence of a single chiral center at one of the carbon atoms in the molecule. The two enantiomers of limonene are non-superimposable mirror images of each other. They have the same molecular formula and connectivity but differ in their optical activity and scent. (D)-Limonene has a citrusy, orange-like scent, while (l)-limonene has a lemon-like scent.

Chirality in terpenes is of great significance because it often influences their biological activities and sensory properties. Different enantiomers of terpenes can have different scents or flavors; for example, one enantiomer might have a pleasant aroma, while the other might have an unpleasant aroma. Chirality can also impact the biological activities of terpenes. Some enantiomers may exhibit distinct effects in biological systems. For instance, one enantiomer might have medicinal properties, while the other may not.

Chirality plays a pivotal role in the fragrance and flavor industry, where the use of enantiomerically pure terpenes is essential for achieving specific scents and tastes. This applies equally to the cannabis industry, where products are often not formulated based on chiral data. But by controlling the chirality of terpenes, chemists and formulators can tailor the sensory experience of various consumer products.

Para, Ortho, and Meta Isomers

Para-Isomers: a type of positional isomerism found in aromatic compounds, particularly benzene derivatives. They occur when two substituents are located on carbon atoms that are directly opposite each other across the benzene ring. In the case of a benzene ring, if two substituents are attached to the carbon atoms in positions 1 and 4, or 2 and 5, or 3 and 6, they are considered para-isomers. Para-isomers often have similar chemical properties due to their symmetrical placement on the benzene ring. They do not exhibit steric hindrance between the substituents.

Ortho-Isomers: a type of positional isomerism found in aromatic compounds, particularly benzene derivatives. They occur when two substituents are located on adjacent carbon atoms of the benzene ring. In the case of a benzene ring, if two substituents are attached to the carbon atoms in positions 1 and 2, or 2 and 3, or 4 and 5, etc., they are considered ortho-isomers. Ortho-isomers often have similar chemical properties due to their proximity on the benzene ring. They may exhibit steric hindrance when the substituents are bulky, impacting their reactivity.

Meta-Isomers: another type of positional isomerism found in aromatic compounds. They occur when two substituents are located on carbon atoms that are separated by one carbon atom (i.e., in the meta position) on the benzene ring. In the case of a benzene ring, if two substituents are attached to the carbon atoms in positions 1 and 3, or 2 and 4, they are considered meta-isomers. Meta-isomers generally have different chemical properties compared to ortho-isomers and para-isomers due to their distinct spatial arrangement on the benzene ring. This can lead to differences in reactivity and interactions with other molecules.

R and S Isomers

R and S isomers are descriptors used to specify the absolute configuration of chiral molecules. These descriptors are part of the Cahn-Ingold-Prelog (CIP) priority rules, which provide a systematic way to assign a unique label (R or S) to chiral centers in molecules.

R Isomer: The R configuration is assigned to a chiral center when the priority groups attached to the chiral atom rotate in a clockwise direction. This means that the group with the highest priority is located on the right side of the chiral center when viewed in a specific orientation.

S Isomer: The S configuration is assigned to a chiral center when the priority groups attached to the chiral atom rotate in a counterclockwise direction. In this case, the group with the highest priority is located on the left side of the chiral center when viewed in a specific orientation.

To determine whether a chiral center is R or S, priorities must be assigned to the substituent groups attached to that center. The priority assignments are made based on the atomic number of the atoms directly bonded to the chiral center. The higher the atomic number, the higher the priority. If two substituents have the same atoms directly bonded to the chiral center, you move outward in the molecule to compare the next set of atoms until a difference is found.

Once priorities are assigned, the molecule is oriented in such a way that the chiral center is in the foreground. After observing how the substituents are arranged around the chiral center, if the highest priority group is on the right side (clockwise), it's an R isomer. If it's on the left side (counterclockwise), it's an S isomer.

D and L Isomers

D and L isomers are also descriptors used to specify the absolute configuration of chiral molecules. These descriptors are important because they convey information about the spatial arrangement of chiral centers in biologically relevant molecules. D and L isomers are enantiomers of each other and are based on the optical activity of certain chiral molecules observed in polarized light.

D Isomer: The D configuration is assigned to a chiral molecule when its highest priority chiral center (usually the one farthest from the carbonyl group, in the case of carbohydrates or the alpha carbon in amino acids) has the substituent with the -OH or -NH2 group on the right side in a Fischer projection formula (the Fischer projection formula is a way of representing the 3D structure of chiral molecules in a 2D format. In this representation, horizontal lines represent bonds coming out of the plane (toward you), and vertical lines represent bonds going into the plane (away from you).

L Isomer: The L configuration is assigned to a chiral molecule when its highest priority chiral center has the substituent with the -OH or -NH2 group on the left side in a Fischer projection formula.

Determining Isomers via Plane Polarized Light

In organic chemistry, plane-polarized light plays a crucial role in determining the presence of enantiomers and characterizing their properties. Plane-polarized light is a type of light in which the electric field vectors oscillate in a specific plane, as opposed to regular unpolarized light in which the vectors oscillate in all directions randomly. Chiral molecules interact differently with plane-polarized light due to their asymmetry; this interaction is known as optical activity. When plane-polarized light passes through a solution of chiral molecules, the plane of polarization rotates. The extent and direction of this rotation depend on the properties of the chiral substance. If the plane of polarization rotates clockwise (to the right) when viewed along the direction of light propagation, the substance is said to be dextrorotatory, often denoted as (+) or (D). If the plane of polarization rotates counterclockwise (to the left), the substance is levorotatory, often denoted as (-) or (L).

When a mixture of enantiomers is exposed to plane-polarized light, the plane of polarization will rotate, but the degree of rotation will be less than if only one enantiomer is present. This is because the two enantiomers rotate the plane of polarization in opposite directions by equal amounts. By measuring the observed angle of rotation, chemists can determine the specific rotation of the compound. Specific rotation is a property of a pure substance and is directly related to the concentration of the enantiomer in a solution, the path length of the sample tube, and other experimental factors. Specific rotation values are often tabulated for known enantiomers under standardized conditions. By comparing the observed specific rotation to known values, chemists can identify which enantiomer is present in a mixture. However, a racemic mixture, which is an equal mixture of both enantiomers, will not rotate plane-polarized light because the rotations caused by one enantiomer are canceled out by the opposite rotation of the other enantiomer.

**General Stability of Terpenes – Volatility*

Terpenes are constructed with differing levels of stability. As the number of isoprene units in each terpene structure increases, the resulting terpene class becomes heavier and less volatile, with diterpenes including cannabinoids being considerably stable compared to monoterpenes and sesquiterpenes. This is important information for people who work with or consume cannabis, because it explains why the more volatile mono- and sesqui- terpenes like myrcene, pinene, linalool, and limonene vaporize at much lower temperatures than the diterpenes like cannabinoids. It is also worth noting that diterpenes are much less numerous in oil or extracts than the mono- and sesqui- terpenes.

**Concentration of Terpenes in Plants*

The terpene family is extremely large and complex, with many tens of thousands catalogued to date. The concentration of essential oils and the terpenes within them differs from plant to plant and is based on numerous environmental factors. In most cannabis samples, each laboratory-tested variety contains 35 to 50 identified terpenes (based mostly on gas chromatography testing), although most labs only actually test for about 10 common terpenes. Researchers have reported yields of cannabis essential oils ranging from .05 to .29% in buds[46], with researchers Ross and ElSohly reporting concentration of essential oils after curing to 0.80% within the bud.

This concentration occurs because the weight of the buds reduces dramatically through drying, thereby increasing the total net weight of essential oil within the dried bud, as most of the bud weight is water.

Modern cannabis chemovars can produce 3.5% or more terpene content in inflorescences (flowers)[47], or to 1.5% of inflorescence dry weight[48]. While the general content of these terpene profiles is inherited genetically, the total weight of essential oils and terpenes is much more dependent on environmental influences[49], including cultivation methods.

HOW DO HUMANS PERCEIVE TERPENES?

If – as many consumers and experts maintain – there are biochemical, pharmacological, and phenomenological distinctions between available cannabis 'strains', such phenomena are most likely related to relative terpenoid contents and ratios[50].

-Dr. Ethan Russo.

There are several theories regarding how humans perceive scent, but none of them are confirmed and in fact olfactory perception is widely regarded as an area where science is significantly lacking.

According to the locks and keys theory, how a terpene molecule is structured is important for human perception. Based on this previous prevailing theory, human olfactory receptors contain small depressions shaped to receive certain molecules. When those specifically shaped holes are filled with the correspondingly shaped volatile molecule, the human brain perceives this as a smell.

However, opposing theories and recent research indicate this prevailing theory may be wrong. Research published in 2021 based in part on computational modeling showed that differently shaped and sized odorant molecules were able to bind in many different orientations in the same receptors[51]. Researchers theorized that there were far too many odorant molecules and far too few receptors for the locks and keys theory to be viable. Their work surmised

---

[46] John M. McPartland DO, MS & Ethan B. Russo MD (2001) Cannabis and Cannabis Extracts, Journal of Cannabis Therapeutics, 1:3-4, 103-132.

[47] Fischedick JT, Hazekamp A, Erkelens T, Choi YH, Verpoorte R. Metabolic fingerprinting of Cannabis sativa L., cannabinoids and terpenoids for chemotaxonomic and drug standardization purposes. Phytochemistry. 2010 Dec;71(17-18):2058-73.

[48] Giese MW, Lewis MA, Giese L, Smith KM. Development and Validation of a Reliable and Robust Method for the Analysis of Cannabinoids and Terpenes in Cannabis. J AOAC Int. 2015 Nov-Dec; 98 (6): 1503-22.

[49] S. Casano, G. Grassi, V. Martini and M. Michelozzi (2011). Variations in terpene profiles of different strains of Cannabis sativa L. Acta Horticulturae 925:115-121.

[50] Russo, Ethan B. Taming THC: potential cannabis synergy and phytocannabinoid-terpenoid entourage effects. British journal of pharmacology vol. 163,7 (2011): 1344-64.

[51] del Mármol, J., Yedlin, M.A. & Ruta, V. The structural basis of odorant recognition in insect olfactory receptors. Nature 597, 126–131 (2021).

instead that there is one lock (a receptor) with many keys (odorant molecules), and those keys will fit inside the lock in different orientations.

Other theories for how humans perceive scent include the bond vibration-assisted olfactory theory, which posits that the perception of odors in humans is not shaped by the molecular structure of the odorant itself, but rather by the oscillations in which its electrons engage in quantum tunneling across energy gaps within the olfactory receptors. In simpler terms, the vibrational frequencies of groups of atoms within molecules is an electrical signal that can be translated by the receptor and brain as a smell. Appearing to support this theory at least partially is 2014 work where scientists easily trained honeybees to discriminate the smell of organic compounds and attributed intra-molecular vibrations to their odor discrimination abilities[52].

While more work is needed to elucidate how we perceive smell, what we do know is that the human nose is sensitive enough to detect even subtle differences in the same terpene molecule. One of the most recognizable examples of this is limonene, which is a top five terpene in cannabis. Limonene occurs as two enantiomers or mirror versions; one is commonly known as d-limonene, while the other is l-limonene, where the d isomer is perceived by human olfactory receptors as citrusy, and the l version as fuel or pine. Other differences in the same molecule, such as functional group substitutions, are also detectable by some, but not all, humans.

BIOAVAILABILITY OF TERPENES

Bioavailability refers to the proportion of a substance that enters the bloodstream when introduced into the body and becomes available for use or storage. Terpenes, in their pure form, can have low bioavailability when ingested orally. The bioavailability of terpenes can vary based on several factors including chemical structure, food matrix, and metabolism. Terpenes can be absorbed through various routes in the body, primarily through ingestion and inhalation:

Ingestion: When terpenes are consumed through food or dietary supplements, they must pass through the digestive system. In the stomach and intestines, they may undergo first-pass metabolism, which can reduce their bioavailability. Some terpenes are absorbed through the intestinal lining and enter the bloodstream, where they can circulate throughout the body.

Inhalation: Inhaling terpenes, such as through the aroma of essential oils or vaporized cannabis, allows for rapid absorption through the respiratory system. This method often bypasses the first-pass metabolism in the liver, leading to higher bioavailability.

Once terpenes are absorbed into the bloodstream, they can be distributed to various tissues and organs in the body. Terpenes are lipophilic, meaning they have an affinity for fat, which can influence their distribution and storage in adipose tissue (body fat). Terpenes can also interact with various biological receptors and enzymes in the body. For example, some terpenes may bind to receptors in the endocannabinoid system, influencing the activity of this system and potentially contributing to therapeutic effects. Terpenes may also enhance the bioavailability and therapeutic effects of other compounds, such as cannabinoids in cannabis (the Entourage Effect).

The metabolism of terpenes in the human body plays a crucial role in determining their bioavailability, distribution, and potential therapeutic effects. Terpenes can undergo various metabolic processes once they are absorbed into the bloodstream:

Phase I Metabolism: In the liver, terpenes may undergo Phase I metabolism, which involves enzymes known as cytochrome P450 (CYP) enzymes. These enzymes can oxidize terpenes, making them more water-soluble and

---

[52] Gronenberg W, Raikhelkar A, Abshire E, Stevens J, Epstein E, Loyola K, Rauscher M, Buchmann S. Honeybees (Apis mellifera) learn to discriminate the smell of organic compounds from their respective deuterated isotopomers. Proc Biol Sci. 2014 Jan 22;281(1778):20133089.

facilitating their excretion from the body. Phase I metabolism can alter the chemical structure of terpenes, potentially affecting their pharmacological activity.

Phase II Metabolism: Terpenes may also undergo Phase II metabolism, where they are conjugated with various molecules, such as glucuronic acid, sulfate, or glutathione. This conjugation process further increases water solubility and facilitates excretion through urine or bile.

Excretion: Once metabolized, terpenes are typically eliminated from the body through the kidneys in urine or through the liver in bile. The rate of excretion depends on factors such as the terpene's chemical structure, its degree of metabolism, and the individual's renal and hepatic function.

The metabolism of terpenes can also be influenced by the presence of other substances in the body, including drugs and dietary compounds. These interactions can affect the rate at which terpenes are metabolized and may impact their therapeutic effects. Terpene metabolism can also vary from person to person due to genetic factors, age, liver function, and the presence of underlying medical conditions. Some individuals may metabolize terpenes more efficiently than others, which can impact the duration and intensity of their effects.

It's important to note that while there is promising research on the potential therapeutic benefits of terpenes, more studies are needed to establish their efficacy, optimal dosages, and safety profiles. Additionally, individual responses to terpenes may vary, and interactions with other medications should always be considered. Finally, consumers who look to cannabis as a method of acquiring terpene compounds might be disappointed to know that eating foods and inhaling the volatile emissions of various plants is likely much more efficient than consuming whole-plant cannabis products or value-added products, which often feature added terpenes in a ratio incomparable to the original plant source.

Terpenes 101 Review

Answer the following questions to test your knowledge of this section:

Question #1: Name three ways that plants use terpenes:

Question #2: Terpenes are based on interlinked chains of what compound?

Question #3: Terpene molecules herein are mostly arranged:

a. Tail-to-tail
b. Head-to-sidechain
c. Head-to-tail
d. Sidechain-to-tail

Question #4: The chemical formula for monoterpenes is:

a. C_5H_8
b. $C_{10}H_{16}$
c. $C_{15}H_{24}$
d. $C_{10}H_{18}$

Question #5: The chemical formula for sesquiterpenes is:

a. $C_{15}H_{24}$
b. $C_{15}H_{16}$
c. $C_{15}H_{22}$
d. $H_{10}C_{18}$

Question #6: Bicyclic monoterpene molecules are defined by what key feature:

Question #7: What changes a terpene to a terpenoid?

Question #8: Describe at least one theory of how humans perceive terpenes: the answer on the website needs to change according to the new text written about opposing theories to locks and keys, etc.

For the answer key to all '101' chapters, please go here: http://thebigbookofterps.com/bbot-101-chapters-quiz-answer-key/

CANNABINOIDS 101

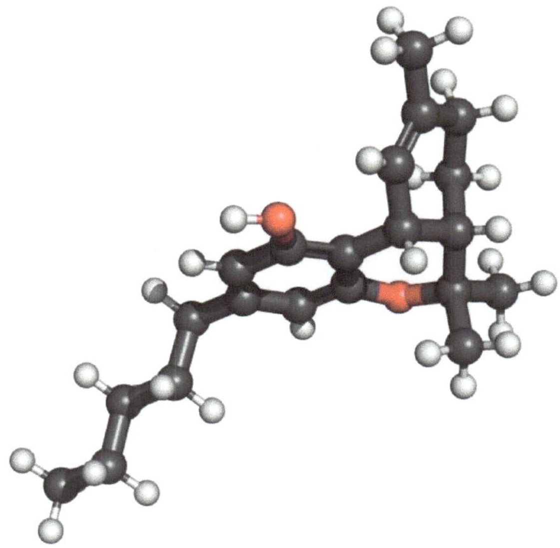

Figure 4: No cannabis molecule is more researched than this one, D9-THC

The scientific body of work covering cannabinoids has grown significantly since Mechoulam discovered the D9-THC molecule in 1964. Despite longstanding restrictions on cannabis research, there exists enough literature on these compounds that this text need not cover the subject exhaustively, nor does it need to cover cannabinoids individually; there are other books, texts, and papers for this purpose already popularized in the cannabis industry. Rather, this book is meant to substantially supplement the existing body of cannabis research and general knowledge (focused on cannabinoids) with an effort to develop and disseminate a more robust phytochemical profile of therapeutic constituents in cannabis.

Where most researchers and scientists have focused on cannabinoids, this text focuses on terpenes and the synergistic interactions between these compounds and other compounds. This section is meant to provide an overview of cannabinoids and the endocannabinoid system, correlated to the synergy of the aforementioned compounds.

CANNABINOIDS IN THE PLANT KINGDOM

Of the cannabinoids that are often featured in news stories and cannabis industry publications, nearly all are derived from the cannabis plant. However, most cannabinoids are also produced by other plants. The most obvious example of these is beta-caryophyllene, which is produced by hundreds of plant species; details of which can be found in the beta-caryophyllene individual compound chapter. Beta-caryophyllene is a dietary cannabinoid, but it is also a bicyclic sesquiterpene that selectively binds to the CB2 cannabinoid receptor.

Examples of non-cannabis plants that produce cannabinoids include black truffles and cacao, which both produce anandamide, an endocannabinoid; Acmella Oleracea (colloquially the toothache plant) is another example, which produces N- isobutylamides[53]; helichrysum italicum (a member of the daisy family) produces cannabigerol[54] (CBG); wairuakohu (Radula marginate) produces the cannabinoid perrotentinenic acid; and rhododendron

[53] Ramsewak, R S, A J Erickson, and M G Nair. Bioactive N-isobutylamides from the flower buds of Spilanthes acmella. Phytochemistry 51.6 (1999): 729-732.
[54] Bohlmann F, Hoffmann E. Cannabigerol-aehnliche Verbindunged aus Helichrysum umbraculigerum. Phytochemistry. 1979; 18:1371-1374.

anthopogonoides production includes three compounds: cannabichromene (CBC), cannabicyclol (CBL), and cannabicitran (CBT)[55].

HOW DO PLANTS USE CANNABINOIDS?

The biological uses of cannabinoids in plants are not well researched as of the publication of this book. However, if we again use the dietary cannabinoid beta-caryophyllene as an example, research has shown that the compound is used by plants for indirect defense by attracting predatory wasps, as an antibacterial agent against both gram-positive and gram-negative bacteria, as repellant against spider mites and ants, as insecticide against weevils, beetles, and moths, and as an oviposition deterrent (discourages or stops insects from depositing eggs on the plant).

Research indicates that additional cannabinoids might serve plants as insecticidal agents. For instance, cannabidiol (CBD) has been shown to significantly reduce growth and increase mortality in the tobacco hornworm[56]. Other work has indicated that cannabigerol (CBG) could protect plants from bacterial infection[57]. Meanwhile, cannabis consumers the world over might be interested to know that THC has been theorized to act as a sunscreen against UV-B radiation in cannabis[58].

These scant few examples indicate that more research is needed to elucidate the biological uses of cannabinoids in plants, particularly in cannabis.

CHEMICAL NATURE OF CANNABINOIDS

Cannabinoids are formed when geranyl pyrophosphate combines with olivetolic acid to produce cannabigerolic acid (CBGA). From this point, nearly all acid versions of classic cannabinoids (including THCA, CBDA, and CBCA) are derived with a small amount of the non-acid version of the parent compound remaining as cannabigerol.

Structurally related to tetrahydrocannabinol, classical cannabinoids may be diterpenes or triterpenes, and are much more stable than monoterpenes and sesquiterpenes. These compounds are typically based on a molecular skeleton containing twenty-one carbon atoms:

CBG (cannabigerol) $C_{21}H_{32}O_2$. Boiling Point: 52°C / 126°F

Although CBG binds to both CB1 and CB2 cannabinoid receptors, the compound is non-psychoactive. CBG derives directly from the acid version of the molecule.

THCA (tetrahydrocannabinolic acid) $C_{22}H_{30}O_4$. Boiling Point: 120°C / 248°F

Derived from CBGA and the precursor of THC, THCA is a diterpenoid with hydroxy, methyl, and pentyl groups. This molecule occurs as two isomers that differ in the position of the carboxylic acid group.

THCV (tetrahydrocannabivarin) $C_{19}H_{26}O_2$. Boiling Point: 220°C / 428°F

THCV is a propyl analogue of THC that interacts with both CB1 and CB2 receptors.

---

[55] Iwata N, Kitanaka S. New cannabinoid-like chromane and chromene derivatives from Rhododendron anthopogonoides. Chem Pharm Bull (Tokyo). 2011;59(11):1409-12.

[56] Park, SH., Staples, S.K., Gostin, E.L. et al. Contrasting Roles of Cannabidiol as an Insecticide and Rescuing Agent for Ethanol-induced Death in the Tobacco Hornworm Manduca sexta. Sci Rep 9, 10481 (2019).

[57] Aqawi M, Sionov RV, Gallily R, Friedman M, Steinberg D. Anti-Bacterial Properties of Cannabigerol Toward Streptococcus mutans. Front Microbiol. 2021 Apr 22. 12:656471.

[58] Pate, David. Possible role of ultraviolet radiation in evolution of Cannabis chemotypes. Economic Botany 37.4 (2008): 396-405.

D8/D9 THC (delta-8 / delta-9 tetrahydrocannabinol) $C_{21}H_{30}O_2$. Boiling Point: 157°C / 315°F

Delta-8 and delta-9 are isomers of the tricyclic terpenoid THC, differing in the position of their carbon double bond.

CBN (cannabinol) $C_{21}H_{26}O_2$. Boiling Point: 185°C / 365°F

Cannabinol is a metabolite (in most cases a degradation product) of THC, which preferentially binds to the non-psychoactive CB2 receptor, but also activates CB1 receptors.

CBDA (cannabidiolic acid) $C_{22}H_{30}O_4$. Boiling Point: 130°C / 266°F

Cannabidiolic acid is the precursor to CBD.

CBDV (cannabidivarin) $C_{19}H_{26}O_2$. Boiling Point: 165°C / 329°F

Cannabidivarin has seven isomers and is non-psychoactive.

CBD (cannabidiol) $C_{21}H_{30}O_2$. Boiling Point: 180°C / 356°F

Widely considered non-psychoactive, CBD shows activity at both CB1 and CB2 receptors and other key pathways.

CBC (cannabichromene) $C_{21}H_{30}O_2$. Boiling point: 220°C / 428°F

Cannabichromene does not show activity at CB1 or CB2 receptor sites, instead acting on transient receptor potential (TRP) channels and preventing the scheduled breakdown of endocannabinoids.

HOW DO HUMANS PERCEIVE CANNABINOIDS?

Humans (and other mammals) primarily perceive cannabinoids through the endocannabinoid system (ECS) and the CB1 and CB2 cannabinoid receptors, but also through other receptors and pathways such as TRP channels (transient receptor potential ion channels modulate temperature, taste, pressure, pain, and other senses).

**Endocannabinoids and the Endocannabinoid System (ECS)*

The endocannabinoid system (ECS) is a cell-signaling neuromodulatory system discovered by Mechoulam and colleagues while researching THC. This system is theorized to regulate many physiological processes, including cognitive, locomotive, immune system, pain modulation, appetite, homeostasis, and other functions. There are three parts of the ECS: endocannabinoids, endocannabinoid receptors, and enzymes.

Endocannabinoids:

Endogenous cannabinoids refer to cannabinoids produced by the human body; these are retrograde neurotransmitters that bind to cannabinoid receptors. There are at least six types of endocannabinoid ligands, the first two of which are well-studied, while the others are comparatively unknown: arachidonoylethanolamine (Anandamide or AEA), 2-arachidonoylglycerol (2-AG), 2-arachidonyl glyceryl ether (noladin ether), N-arachidonoyl dopamine (NADA), Virodhamine (OAE), and Lysophosphatidylinositol (LPI).

Exogenous cannabinoids, on the other hand, consist of phytocannabinoids and synthetic cannabinoids. Endocannabinoid receptors have cavities or docking sites that match the structural shape of the endocannabinoid ligands listed above. Phytocannabinoids like CBG, CBD, THC, CBC, and synthetic cannabinoids closely match these shapes and corresponding cavities and therefore activate the receptors, interrupt synaptic signaling, and interact with receptor sites and presynaptic and postsynaptic neurons. Cannabinoids bind reversibly to receptor sites, and different cannabinoids elicit different effects on these and other ECS pathways and channels. For instance, the effect of THC on these receptor sites is a feeling of euphoria or high.

Endocannabinoid Receptors:

Cannabinoid Receptor Type 1 (CB1)

CB1 receptors are G protein-coupled cannabinoid receptors; the proteins involved activate cellular responses involved in many physiological functions. Activated by endocannabinoids, phytocannabinoids, and synthetic cannabinoids, CB1 is encoded by the CNR1 gene, and appears to receive and interact primarily with the endocannabinoid anandamide (AEA) - the same chemical produced by cacao and black truffles mentioned previously. CB1 receptors most commonly occur in the central and peripheral nervous system.

Cannabinoid Receptor Type 2 (CB2)

CB2 receptors are expressed in immune tissue, the gastrointestinal system, and peripheral tissue. CB2 receptors receive and interact primarily with the endocannabinoid 2-arachidonoylglycerol (2-AG), and are encoded by the CNR2 gene. Scientists and researchers theorize that other cannabinoid receptors exist, separately from CB1 and CB2.

Enzymes:

The third part of the ECS features enzymes that break down endocannabinoids. In the classical ECS, the decomposition of endogenous cannabinoids is carried out by these enzymes automatically after the endocannabinoid has served its function. For instance, the decomposition of anandamide is carried out by fatty acid amino hydrolase, while several other enzymes - most notably monoacylglycerol acid lipase - carry out the destruction of 2-arachidonoylglycerol.

To summarize, the endocannabinoid system - which regulates many primal human processes - interfaces with cannabinoids produced by plants, particularly cannabis. One could speculate whether humans evolved for this design, or whether phytocannabinoids interface with the ECS by accident, merely because the molecules are shaped similarly to - and fit cavities seemingly meant for - endogenous endocannabinoids.

Whatever the case may be, the body of evidence supporting the therapeutic value of cannabinoids produced by cannabis is mounting, and the opportunity to use these compounds synergistically with terpenes and other phytochemicals is one that should not be overlooked.

MEDICAL AND THERAPEUTIC VALUE OF CANNABINOIDS

As mentioned previously, this text is not meant to comprehensively cover cannabinoids, nor is it intended to provide a review of the literature covering each compound and its medical or therapeutic uses. Rather, the information in this book will ideally lead to a realization of and commitment to the improvement of the existing and emerging therapeutic uses and related efficacy of cannabinoid-based medicines by using these compounds synergistically with terpenes and other phytochemicals.

The therapeutic potential of cannabinoids is remarkable, but research in this field is in its infancy. Cannabinoids are actively researched and used in the treatment of:

Addiction

Alzheimer's disease
Anorexia
Anxiolytic disorders
Arthritis
Cancer, numerous types
Crohn's disease
Epilepsy
Glaucoma
HIV / AIDS
Huntington's disease
Inflammation
Irritable bowel syndrome
Migraine
Multiple sclerosis
Nausea and vomiting
Neurological conditions
Ocular diseases
Parkinson's disease
Peripheral neuropathy
PTSD
Tourette syndrome

For each of these listed conditions where cannabinoids have been or could be used in treatment, it is likely that terpenes could be used synergistically to improve treatment efficacy.

Cannabinoids 101 Review

Answer the following questions to test your knowledge of this section:

Question #1: Name two plants other than cannabis that produce cannabinoids:

1 ______________________ 2 ______________________

Question #2: Cannabinoids are formed from which two compounds:

a. Geranyl pyrophosphate and olivetolic acid
b. Geranyl pyrane acid and oliveoil fraction
c. Geranyl acetate and ollitol
d. Geranyl acid and perrotentinenic acid

Question #3: What is the base compound in the production of nearly all cannabinoids?

a. THCV
b. CBNA
c. CBCA
d. CBNV
e. CBGA
f. THCA

Question #4: Delta-8-tetrahydrocannabinol is a/an ______________ of delta- 9-tetrahydrocannabinol:

a. Isotoner
b. Isopropyl
c. Analogue
d. Isomer
e. Homologue
f. Isologue

Question #5: What are the three parts of the Endocannabinoid System (ECS)?

1. ______________ 2. ______________ 3. ______________

Question #6: What is the difference between endogenous cannabinoids and exogenous cannabinoids?

____

____

____

Question #7: What is the function of enzymes in the ECS?

____

____

Question #8: Name three medical conditions that could be treated using cannabinoid- based medicine:

1. ________________ 2. ________________ 3. ________________

For the answer key to all '101' chapters please go here: http://thebigbookofterps.com/bbot-101-chapters-quiz-answer-key/

SYNERGY

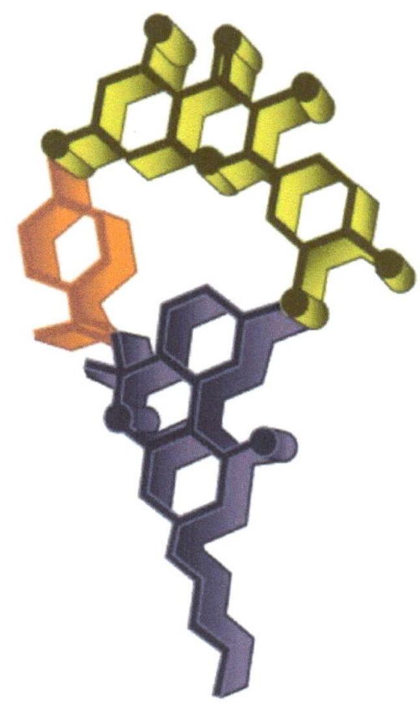

Compounds that may affect the pharmacology of cannabinoids are abundant in nature, and so we may dangerously and mistakenly consider their presence to be trivial. If so, this could cause us to lose sight of the subtlety and efficiency of their design when applied in combination[59].

- Dr. Ethan Russo.

Oxford Language Definition of Synergy:

"The interaction or cooperation of two or more organizations, substances, or other agents to produce a combined effect greater than the sum of their separate effects."

INTRODUCTION

Synergy as we define it today was first proposed as a concept by Shimon Ben-Shabat in 1998[60], who soon thereafter refined the theory together with Raphael Mechoulam[61] and mutual colleagues. At the time, Ben-Shabat referred to synergistic phenomena observed between compounds in the endocannabinoid system as the entourage effect, although it is only quite recently that this concept has been trending in the cannabis industry. However, evidence for synergy in plants has been known for at least a half century, and this is particularly true of the evidence for synergy among phytochemicals like terpenes and other compounds.

---

[59] Russo, Ethan & Marcu, Jahan. (2017). Cannabis Pharmacology: The Usual Suspects and a Few Promising Leads.

[60] Ben-Shabat S., Fride E., Sheskin T., Tamiri T., Rhee M. H., Vogel Z., et al. (1998). An entourage effect: inactive endogenous fatty acid glycerol esters enhance 2-arachidonoyl-glycerol cannabinoid activity. Eur. J. Pharmacol. 353 23-31.

[61] Mechoulam R., Ben-Shabat S. (1999). From gan-zi-gun-nu to anandamide and 2-arachidonoylglycerol: the ongoing story of cannabis. Nat. Prod. Rep. 16 131-143.

Initially, Ben-Shabat and Mechoulam referred primarily to synergy between endogenous and exogenous cannabinoids (see "Cannabinoids 101" chapter for more information on the endocannabinoid system), and later Dr. Ethan Russo extended the concept to include terpenes, terpenoids, and potentially other compounds including flavonoids. The principal idea behind the entourage effect is that pure compounds in cannabis - particularly the primary psychoactive compounds like THC and CBD - are modulated or supplemented by other compounds, mostly in a beneficial and additive way, resulting in therapeutic effects that are more potent than those elicited by single molecules like D9-THC.

In recent years, the cannabis industry has seized on the concept of the entourage effect, but with strikingly little understanding of the mechanisms behind it. Cannabis products and marketing materials in the United States, Canada, Europe, Israel, South America, and other places around the world tout the entourage effect as a proven, well-established, and unflinchingly positive aspect of their much-venerated plant.
Unfortunately, many of the benefits claimed by these products, groups, and marketing materials are untested and disseminated without sound science - or even basic knowledge of chemistry - to support them. Perhaps worse is that potential negative or detrimental synergies are ignored.

This chapter will discuss the evidence for and against synergy in cannabis, and in non-cannabis plants, the latter of which there is ample and well-established albeit neglected evidence for. Understanding synergy and how it applies to cannabis is critical for the nascent medical and recreational markets, and especially for thousands of ill people and the healthcare workers who treat them, the latter of which often are not aware that cannabinoid monotherapies may not be as efficacious as blended preparations or formulations[62].

SYNERGY 101

To begin, we must define three similar yet distinct concepts:

1. *Polypharmacology*

In polypharmacology, individual molecules are applied therapeutically to affect multiple targets or pathways, or in some cases, multiple drugs to affect a specific target[63], but without the concept of implied synergy.

2. *The Entourage Effect*

The entourage effect describes multiple components in cannabis all acting together, with one or two primary active compounds being supported by multiple inactive compounds, all taking place within the endocannabinoid system.

3. *Synergy*

Synergy describes the enhanced effects caused by interactions between two or more compounds, with most compounds involved being considered active. This synergy can exist outside of the endocannabinoid system.

While the general concept of synergy in phytochemicals can be explained simply by the wisdom that it is nutritionally better to eat whole fruit than it is to take vitamins or supplements of singular compounds[64], the difference between this and the entourage effect needs clearer delineation. Apart from existing outside of the endocannabinoid system, the most significant difference is the active nature of the compounds involved in synergy, versus the inactive view of the compounds in the entourage effect. For instance, the terpenes geraniol, beta-pinene,

---

[62] Sanchez-Ramos, Juan. The entourage effect of the phytocannabinoids. Annals of Neurology 77.6 (2015): 1083- 1083.

[63] Reddy, A Srinivas, and Shuxing Zhang. Polypharmacology: drug discovery for the future. Expert review of clinical pharmacology vol. 6,1 (2013): 41-7.

[64] Liu, R. H. (2013). Health-promoting components of fruits and vegetables in the diet. Adv. Nutr. Int. Rev. J. 4, 384S-392S.

alpha-humulene, and linalool have been shown to be cannabimimetic (cannabinoid-like), causing 75% of the classic CB1- response behaviors in mice, including antinociception, catalepsy, hypolocomotion, and hypothermia, while also selectively enhancing cannabinoid activity[65]. These actions indicate active versus inactive compounds.

Furthermore, consider that certain terpenes already have well-established actions outside of cannabis. Myrcene, for example, is a well-known and potent sedative, while beta-caryophyllene has affinity for the CB2 receptor, in both cases indicating active rather than inactive compounds.

Perhaps the most obvious example of synergy in cannabis - and the one that the majority of cannabis consumers can understand and relate to - is the aroma and flavor of each chemovar. This distinctive feature is caused mostly by the primary terpenes and terpenoids in the plant's inflorescences (flowers) volatilizing together, as opposed to being the product of a singular constituent.

*Synergy Implications

The viewpoint that synergy is always positive or beneficial is relative, and often leaves equally important negative synergies ignored. Terpenes, flavonoids, cannabinoids, and other compounds may interact synergistically, but whether this action is beneficial or detrimental depends on the individual situation and desired outcome.

*Synergy Characters

Terpenes and cannabinoids can potentially act synergistically with a virtually unlimited number of other compounds. These interactions can include activity and engagement between terpenes and chemotherapeutics, terpenes and cannabinoids, terpenes and antimalarial drugs, flavonoids and terpenes, terpenes and neuroprotective agents, and many other combinations.

*Mechanisms and Types of Synergy

As documented often in the individual compound chapters of this text, plants can produce specific blends of terpenes in response to a variety of stimuli. These blends appear to act synergistically, and can serve an astonishing number of complex functions.

Terpene formulations created in the pharmacological setting can also serve many functions, with synergy occurring via one or more of several different mechanisms including additive and competitive agonism or antagonism, regulation of cells, allosteric effects, multi-target effects, pharmacokinetic effects, modulation of adverse events[66], and other effects.

DIRECT EVIDENCE FOR NON-CANNABIS SYNERGY

Many studies have found evidence of synergy between terpenes and other compounds, including man-made drugs. The following 48 quotes taken directly from peer-reviewed studies or research papers offer a distinct and fascinating glimpse into the synergistic nature of these compounds:

---

[65] LaVigne, Justin E., Ryan Hecksel, Attila Keresztes, and John M. Streicher. Cannabis sativa terpenes are cannabimimetic and selectively enhance cannabinoid activity. Scientific Reports 11.1 (2021).

[66] Wagner H, Ulrich-Merzenich G. Synergy research: approaching a new generation of phytopharmaceuticals. Phytomedicine. 2009 Mar;16(2-3):97-110.

"Based on the dose-dependent antennal and behavioral responses of E. obliqua to these bioactive compounds, myrcene, terpinene, linalool, cis-verbenol, camphor and verbenone were found to play a key role in repelling the moths, and the mixture that included all eight compounds was significantly more effective[67]."

"However, significant cytotoxicity was observed when cells were cotreated with limonene and linalyl acetate whereas no other associations were effective. Only cotreatment, but not the single exposure to limonene and linalyl acetate, replicated distinctive morphological and biochemical changes induced by BEO (bergamot essential oil)[68]."

"Overall, the present investigation revealed that the major components of Prasophyllum roseum and specially the whole essential oil could be helpful in developing novel and safe mosquito control tools and also offer an environmentally safe and cheap tool for reducing Cx. pippins mosquito populations[69]."

"The three compounds alone inhibited cell proliferation in a dose-dependent fashion, while their pairwise combination produced synergistic antiproliferative effects in both cell lines[70]."

"In this study, diverse essential oils were initially investigated for anti-biofilm activity against C. albicans strains, and cascarilla bark oil and helichrysum oil and their components alpha-longipinene (a major constituent of both) and linalool were found to markedly inhibit biofilm formation without affecting planktonic cell growth. Moreover, alpha-longipinene and linalool were found to synergistically reduce biofilm formation[71]."

"On the other hand, the nematicidal activity of T. mastichina EO did not correlate with that of its major components alone (1,8-cineole, terpineol, linalool and beta-pinene) or in binary mixtures of 1,8-cineole with other compounds. Therefore, the contribution of each component to the activity of an EO follows a complicated pattern of interactions including synergistic and/or antagonistic interactions that contribute to the overall toxicity of the oil[72]."

"In S. aureus, the linalool with Imipenem association showed a synergistic effect (FIC = 0.0625); while with ciprofloxacin, the linalool showed additivity (FIC = 0.75). In Pseudomonas aeruginosa, the Imipenem/linalool association was synergistic for both the ATCC and clinical strains (FIC = 0.0625). We conclude that Linalool associated with existing standard antibiotics may increase antibacterial effectiveness, resulting in synergistic activity against bacterial strains of clinical importance[73]."

---

[67] Zhang, Zhengqun, Lei Bian, Xiaoling Sun, Zongxiu Luo, Zhaojun Xin, Fengjian Luo, and Zongmao Chen. Electrophysiological and behavioural responses of the tea geometrid Ectropis obliqua (Lepidoptera: Geometridae) to volatiles from a non-host plant, rosemary, Rosmarinus officinalis (Lamiaceae). Pest Management Science 71.1 (2015): 96-104.

[68] Russo R, Ciociaro A, Berliocchi L, Cassiano MG, Rombola L, Ragusa S, Bagetta G, Blandini F, Corasaniti MT. Implication of limonene and linalyl acetate in cytotoxicity induced by bergamot essential oil in human neuroblastoma cells. Fitoterapia. 2013 Sep; 89:48-57.

[69] Tabari MA, Youssefi MR, Esfandiari A, Benelli G. Toxicity of 8-citronellol, geraniol and linalool from Pelargonium roseum essential oil against the West Nile and filariasis vector Culex pipiens (Diptera: Culicidae). Res Vet Sci. 2017 Oct; 114:36-40.

[70] Rodenak Kladniew, Boris, Monica Polo, Sandra Montero Villegas, Marianela Galle, Rosana Crespo, and Margarita Garcia de Bravo. Synergistic antiproliferative and anticholesterogenic effects of linalool, 1,8-cineole, and simvastatin on human cell lines. Chemico-Biological Interactions 214 (2014): 57-68.

[71] Manoharan RK, Lee JH, Kim YG, Kim SI, Lee J. Inhibitory effects of the essential oils a-longipinéne and linalool on biofilm formation and hyphal growth of Candida albicans. Biofouling. 2017 Feb;33(2):143-155.

[72] Andres, Maria, Azucena Gonzalez-Coloma, Jesus Sanz, Jesus Burillo, and Paula Sainz. Nematicidal activity of essential oils: a review. Phytochemistry Reviews 11.4 (2012): 371-390.

[73] Silva V.A., Sousa, J.P., Guerra F.Q.S., Pessoa H.L.F., Freitas A.F.R., Coutinho, H.D.M., Alves L.B.N, Lima E.O Antibacterial Activity of the Monoterpene Linalool: Alone and in Association with Antibiotics Against Bacteria of Clinical Importance, International Journal of Pharmacognosy and Phytochemical Research 2015; 7(5); 1022-1026.

"The isolated sesquiterpenes are the major active principles responsible for the potent mosquitocidal activity, but the synergistic action of other constituents of the oil cannot be disregarded, as the whole oil is more effective than the individual compounds[74]."

"When the oils (from Laurus nobilis leaves and seeds) were added to a suboptimal concentration of the commercial drug, cytosine arabinoside, a clear synergic effect was observed[75]."

"These results suggest that the TPL/DCF (terpinolene/diclofenac) association had a synergistic anti-inflammatory and analgesic effect without causing apparent gastric injury, and that the serotonergic system may be involved in the antinociceptive effect of this association[76]."

"Compounds were also used as binary mixtures and tested for synergy, using toxicity and feeding inhibition parameters. The data showed that thymol and a-terpineol synergized the effects of both linalool and 1,8-cineole. The combination of compounds is more desirable because the insecticidal spectrum of some binary mixtures is increased[77]."

"Limonene exhibited synergistic interaction with citral, caryophyllene oxide and Benznidazole (decreasing by 17 times its IC50) and was the most effective and selective treatment[78]."

"None of the other compounds were toxic to adults, but caryophyllene oxide and sesamin exhibited moderate larvicidal effects (LD50>150 μg/ml). A mixture of the four compounds in the same ratio as in the hexane extract showed higher toxicity (LD50 34 ng/mg insect) towards adult insects than the pure compounds[79]."

"The in vitro trypanocidal activity of a 1:4 mixture of lupenone and caryophyllene oxide confirmed a synergistic effect of the terpenoids against epimastigotes forms of T. cruzi[80]."

"Oils with high sesquiterpene concentrations received a low rating, meaning that they smelled badly. In contrast, oils with high monoterpene percentages (but a low alpha- humulene or caryophyllene oxide concentration) got a high rating. Surprisingly, a mixed oil from different strains received the best rating[81]."

"Comparison of the deterrent activity of 'full mixtures' with respective artificial blends missing individual constituents demonstrates that, for most oils, minor constituents in a mixture can be as important as major

---

[74] Ravi Kiran, S., and P. Sita Devi. Evaluation of mosquitocidal activity of essential oil and sesquiterpenes from leaves of Chloroxylon swietenia DC. Parasitology Research 101.2 (2007): 413-418.

[75] Saab AM, Tundis R, Loizzo MR, Lampronti I, Borgatti M, Gambari R, Menichini F, Esseily F, Menichini F. Antioxidant and antiproliferative activity of Laurus nobilis L. (Lauraceae) leaves and seeds essential oils against K562 human chronic myelogenous leukaemia cells. Nat Prod Res. 2012;26(18):1741-5.

[76] Macedo, E M A et al. Association of terpinolene and diclofenac presents antinociceptive and anti-inflammatory synergistic effects in a model of chronic inflammation. Brazilian journal of medical and biological research = Revista brasileira de pesquisas medicas e biologicas vol. 49,7 (2016): e5103.

[77] Singh, Rajwinder, Opender Koul, Pushpinder Rup, and Jawala Jindal. Toxicity of some essential oil constituents and their binary mixtures against Chilo pattellas (Lepidoptera: Pyralidae). International Journal of Tropical Insect Science 29.2 (2009): 93-101.

[78] Moreno, Erika, Sandra Leal, Elena Stashenko, and Liliana Garcia. Induction of programmed cell death in Trypanosoma cruzi by Lippia alba essential oils and their major and synergistic terpenes (citral, limonene and caryophyllene oxide). BMC Complementary and Alternative Medicine 18.1 (2018): 1-16.

[79] Moussavi, Nastaran, Karl Egil Malterud, Bertin Mikolo, Dag Dawes, Fabrice Chandre, Vincent Corbel, Daniel Massamba, Hans J. Overgaard, and Helle Wangensteen. Identification of chemical constituents of Zanthoxylum heitzii stem bark and their insecticidal activity against the malaria mosquito Anopheles gambiae. Parasites & Vectors 8 (2015).

[80] Polanco-Hernandez, Glendy, Fabiola Escalante-Erosa, Karlina Garcia-Sosa, Maria E. Rosado, Eugenia Guzman- Marin, Karla Acosta-Viana, Alberto Gimenez-Turba, Efrain Salamanca, and Luis M. Pefia-Rodriguez. Synergistic Effect of Lupenone and Caryophyllene Oxide against Trypanosoma cruzi. Evidence-Based Complementary and Alternative Medicine 2013.

[81] Vito Mediavilla and Simon Steinemann Essential oil of Cannabis sativa L. strains International Hemp Association from: http://www.internationalhempassociation.org/jiha/jiha4208.html Accessed April 10, 2019.

constituents with respect to the overall feeding deterrent effect. The role of any individual compound is mixture-specific and depends on the interaction between the constituents of the mixture[82]."

"Natural thyme oil showed greater toxicity than any single constituent or blend of constituents. The toxicity of blends of selected constituents indicated a synergistic effect among the putatively active and inactive constituents, with the presence of all constituents necessary to reach the highest toxicity[83]."

"However, the observed activity might be due to synergism between compounds present in the plant extract. The synergism among these compounds which contribute to the cytotoxic activity, is not only dependent on the concentration of the compounds, but also on the structure and interaction(s) between the 27 compounds. This can explain the differences in the cytotoxic effect between crude extracts and isolated compounds against the same cell lines, as shown in our earlier report[84]."

"The effectiveness of a binary 1:2 mixture of thymol and vanillin (0.05:0.1 microl per cm2) was found to be significantly more effective than thymol alone for a period of 120 min[85]."

"Although none of these was active when presented individually at the levels present in the entrainment sample...the six-component blend and a blend of acetophenone, (Z)-3- hexenyl acetate, and 3-carene, in the same ratio and concentration as in a natural sample, was as attractive to female S. mosellana as the whole air entrainment sample[86]."

"The slow onset of inhibition of butyrylcholinesterase was also shown by individual constituents, such as 3-carene and beta-pinene. Analyses of the chemical composition of the oils and anti-butyrylcholinesterase activity of their constituents revealed that none of the compounds tested would account for the total activity of the oils and that synergy is likely[87]."

"Furthermore, the combination of borneol and menthol significantly enhanced the permeation rate in the same concentration, compared with their effects when used alone. Consequently, menthol combined with borneol can highly enhance fluconazole permeation through the ex vivo cornea[88]."

"These results demonstrated the synergic effects between TMPP (tetramethylpyrazine phosphate) and BO (borneol) for treating ischemia-reperfusion injury in the cortex and hippocampus regions[89]."

---

[82] Akhtar, Yasmin, Emmanuelle Pages, Alex Stevens, Rod Bradbury, Claudio A. G. da Camara, and Murrat B. Isman, Effect of chemical complexity of essential oils on feeding deterrence in larvae of the cabbage looper. Physiological Entomology 37.1 (2012): 81-91.

[83] Wu, Lipeng et al. Acaricidal Activity and Synergistic Effect of Thyme Oil Constituents against Carmine Spider Mite (Tetranychus Cinnabarinus (Boisduval)). Molecules (Basel, Switzerland) vol. 22,11 1873. 1 Nov. 2017.

[84] Malek, Sri Nurestri Abdul et al. Cytotoxic components of Pereskia bleo (Kunth) DC. (Cactaceae) leaves. Molecules (Basel, Switzerland) vol. 14,5 1713-24. 6 May. 2009.

[85] Park YU, Koo HN, Kim GH. Chemical composition, larvicidal action, and adult repellency of Thymus magnus against Aedes albopictus. J Am Mosq Control Assoc. 2012 Sep;28(3):192-8.

[86] Birkett, Michael, Toby Bruce, Janet Martin, Lesley Smart, Jon Oakley, and Lester Wadhams. Responses of Female Orange Wheat Blossom Midge, Sitodiplosis mosellana, to Wheat Panicle Volatiles. Journal of Chemical Ecology 30.7 (2004): 1319-1328.

[87] Savelev, Sergey U., Edward J. Okello, and Elaine K. Perry. Butyryl- and acetyl-cholinesterase inhibitory activities in essential oils of Salvia species and their constituents. Phytotherapy Research 18.4 (2004): 315-324.

[88] Liu, Jingjing, Shaoying Fu, Nan Wei, Yongsheng Hou, Xiaoning Zhang, and Hao Cui. The effects of combined menthol and borneol on fluconazole permeation through the cornea ex vivo. European Journal of Pharmacology 688.1 (2012): 1-5.

[89] Yu, Bin, Ming Ruan, Tao Liang, Shi-Wen Huang, Yun Yu, Hai-Bo Cheng, and Xiang-Chun Shen. The Synergic Effect of Tetramethylpyrazine Phosphate and Borneol for Protecting Against Ischemia Injury in Cortex and Hippocampus Regions by Modulating Apoptosis and Autophagy. Journal of Molecular Neuroscience 63.1 (2017): 70-83.

"These results suggested that the blend provided more useful information for female wasps than the individual compounds...the blend attracted the wasps at all of the tested concentrations, while the response windows for each pure compound were narrow. Additionally, at the 0.01 ng level, none of the individual compounds attracted A. ervi, while their mixtures attracted at a ten times lower level[90]."

"Radical scavenging activity of Carum oil was than that of ferula oil. These differential activities between essential oils could be reflected to the main components and strong synergism between them[91]."

"Our third case of study, the mixture of (R)-pulegone and citronellal, showed quite different results from those predicted by MM, with observed LC50 values being markedly lower than expected. Moreover, CI [combination index] values were quite far below one, suggesting a real synergism between these two terpenes, which was supported also by the isobolograms[92]."

"Terpinen-4-ol and gamma-terpinene exhibited a significant insecticidal activity against both insects, but c-terpinene was more toxic than terpinen-4-ol. When tested in a binary mixture with the synthetic insecticides profenofos and methomyl, it was found that both compounds enhanced the insecticidal activity of these insecticides by two- to threefold. These results show that terpinen-4-ol and c-terpinene have a synergistic effect on the insecticidal activities of synthetic insecticides profenofos and methomyl[93]."

"Otherwise, sabinene and terpinen-4-ol were strongly repellent against T. castaneum as well as the essential oil, while c-terpinene exhibited weaker repellency against T. castaneum compared with the positive control, DEET (N, N-diethyl-3- methylbenzamide). Moreover, only the essential oil exhibited strong repellency against L. serricorne, the three compounds exhibited weaker repellency against L. serricorne relative to DEET[94]."

"Our results showed imidacloprid treatment alone can significantly decrease aphid populations, and that combining insecticide with Ef3f (farnesene) further reduced numbers of apterous aphids...Our results demonstrate that imidacloprid can be effective in reducing the abundance of aphids in Chinese cabbage fields, but the degree of control can be even stronger in the presence of Ef3f[95]."

"The essential oil fraction of Pelargonium graveolens and its main components, geraniol and citronellol, exhibited additive effects with amphotericin B and with ketoconazole against both Aspergillus species. The FIC index calculated from these results was 0.562, which indicates the strong additive effects of geraniol and amphotericin B[96]."

---

[90] Takemoto, Hiroyuki, and Junji Takabayashi. Parasitic Wasps Aphidius ervi are More Attracted to a Blend of Host-Induced Plant Volatiles than to the Independent Compounds. Journal of Chemical Ecology 41.9 (2015): 801- 807.

[91] Kavoosi, Gholamreza, Asad Tafsiry, Ali Asghar Ebdam, and Vahid Rowshan. Evaluation of Antioxidant and Antimicrobial Activities of Essential Oils from Carum copticum Seed and Ferula assafoetida Latex. Journal of Foo Science 78.2 (2013): T356-T361.

[92] Esteban Scalerandi, Guillermo A. Flores, Marcela Palacio, Maria Teresa Defag6, Maria Cecilia Carpinella, Graciela Valladares, Alberto Bertoni, and Sara Maria Palacios. Understanding Synergistic Toxicity of Terpenes as Insecticides: Contribution of Metabolic Detoxification in Musca domestica. Front Plant Sci. 2018; 9: 1579.

[93] Mostafa A. Abbassy, Samir A. M. Abdelgaleil, & Rasha Y. A. Rabie. Insecticidal and synergistic effects of Majoranahortensis essential oil and some of its majorconstituents. Entomologia Experimentalis et Applicata. 131(3):225 - 232. May 2009.

[94] Y. Wang, C.X. You, K. Yang, Y. Wu, R. Chen, W.J. Zhang, Z.L. Liu, S.S. Du, Z. W. Deng, Z.F. Geng, and J. HanBioactivity of Essential Oil of Zingiber purpureum Rhizomes and Its Main Compounds against Two Stored Product Insects Journal of Economic Entomology; Lanham Vol. 108, Iss. 3, (Jun 2015): 925-932.

[95] Liang-LiangCui, Jie Dong, Frederic Francis, Ying-Jie Liu, Stephanie Heuskin, Georges Lognay, Ju-Lian Chen, Claude Bragard, John F. Tooker, Yong Liua. E-f3-farnesene synergizes the influence of an insecticide to improve control of cabbage aphids in China. Crop Protection Volume 35, May 2012, Pages 91-96.

[96] Shin, Seungwon. Anti- Aspergillus activities of plant essential oils and their combination effects with ketoconazole or amphotericin b. Archives of Pharmacal Research 26.5 (2003): 389-393.

"With sorbitol, geraniol MIC was increased by 64-fold and citronellol by 32-fold. The drugs caused leakage of intracellular material and inhibited ergosterol biosynthesis[97]."
"Toxicity of these four major constituents in the same proportion as the natural oil, was greater than whole oil and each individual component. Removal of any four constituents produced a decreased in effectiveness[98]."

"A strong synergism of DOX (doxorubicin) with NER (nerolidol) was found in the A2780 cells, while DOX acted synergistically with HUM (humulene) and CAO (b- caryophyllene) in the SKOV3 (ovarian cancer) cells. In the CCRF/CEM cells, a synergism of DOX with CAO and NER was observed[99]."

"CAO, NER and VAL synergistically potentiated the efficacy of DOX in cancer cell killing[100]."

"The presence of all constituents together in the artificial mixture caused a significant decrease in the number of eggs laid by females, at 0.26 mg liter (-1) (11 eggs), compared with the control (50 eggs). The toxicity of blends of selected constituents indicated that the presence of all constituents was necessary to reproduce the toxicity level of the natural oil[101]."

"Indian cassia was non-toxic alone, but the mixture of India cassia + green cardamom showed toxicity, clearly demonstrating a synergistic effect[102]."

"This study indicated that the combination of LM (l-menthol) and reboxetine, as well as LM and FLU (fluoxetine) showed synergistic effects[103]."

"Our results indicate that two of the three geranium oil constituents tested exhibit excellent anti-Candida activity and significant synergistic activity with fluconazole[104]."

"Our results indicated that EO's induced a selective antimicrobial activity. A synergistic effect of EO's and tetracycline (TET) was noticed with a reduction rate ranged from 2 to 8-fold (of bacteria)[105]."

---

[97] Pereira Fde O, Mendes JM, Lima IO, Mota KS, Oliveira WA, Lima Ede O. Antifungal activity of geraniol and citronellol, two monoterpene alcohols, against Trichophyton rubrum involves inhibition of ergosterol biosynthesis Pharm Biol. 2015 Feb;53(2):228-34.

[98] Gallardo A, Picollo MI, Gonzalez-Audino P, Mougabure-Cueto G. Insecticidal activity of individual and mixed monoterpenoids of geranium essential oil against Pediculus humanus capitis (Phtiraptera: Pediculidae). J Med Entomol. 2012 Mar;49(2):332-5.

[99] Ambroz, Martin et al. The Effects of Selected Sesquiterpenes from Myrica rubra Essential Oil on the Efficacy of Doxorubicin in Sensitive and Resistant Cancer Cell Lines. Molecules (Basel, Switzerland) vol. 22,6 1021. 20 Jun. 2017.

[100] Ambroz, Martin et al. The Influence of Sesquiterpenes from Myrica rubra on the Antiproliferative and Pro- Oxidative Effects of Doxorubicin and Its Accumulation in Cancer Cells. Molecules (Basel, Switzerland) vol. 20,8 15343-58. 21 Aug. 2015.

[101] Attia, S., K. L. Grissa, G. Lognay, S. Heuskin, A. C. Mailleux, and T. Hance. Chemical Composition and Acaricidal Properties of Deverra scoparia Essential Oil (Araliales: Apiaceae) and Blends of Its Major Constituents Against Tetranychus urticae (Acari: Tetranychidae). Journal of Economic Entomology 104.4 (2011): 1220-1228.

[102] Tripathi, Aran, Anil Singh, and Shikha Upadhyay. Contact and fumigant toxicity of some common spices against the storage insects Callosobruchus maculatus (Coleoptera: Bruchidae) and Tribolium castaneum (Coleoptera: Tenebrionidae). International Journal of Tropical Insect Science 29.3 (2009): 151-157.

[103] Wang, Weidong, Yuanyuan Jiang, Enbo Cai, Bingchen Li, Yan Zhao, Hongyan Zhu, Lianxue Zhang, and Yugang Gao. L-menthol exhibits antidepressive-like effects mediated by the modification of 5-HTergic, GABAergic and DAergic systems. Cognitive Neurodynamics 13.2 (2018): 191-200.

[104] Zore GB, Thakre AD, Rathod V, Karuppayil SM. Evaluation of anti-Candida potential of geranium oil constituents against clinical isolates of Candida albicans differentially sensitive to fluconazole: inhibition of growth, dimorphism and sensitization. Mycoses. 2011 Jul;54(4):e99-109.

[105] Miladi, Hanene, Tarek Zmantar, Bochra Kouidhi, Yasir Mohammed A Al Qurashi, Amina Bakhrouf, Yassine Chaabouni, Kacem Mahdouani, and Kamel Chaieb. Synergistic effect of eugenol, carvacrol, thymol, p-cymene and - terpinene on inhibition of drug resistance and biofilm formation of oral bacteria. Microbial pathogenesis 112 (2018) 156-163.

"When thymol and cymene were applied simultaneously, it resulted in a greater antibacterial effect than when the compounds were applied separately[106]."

"This study indicates a synergistic interaction between essential oils and PBO (piperonyl butoxide) in inhibiting the cytochrome P450 and GST detoxification enzymes in Ae. aegypti[107]."

"Data also suggest that β-Elemene shows the synergistic effect when treated with hyperbaric oxygen in TBI (traumatic brain injury)[108]."

"Interestingly, the synergistic interactions of these 10 substances of the TA (essential oil of cajeput leaves) exhibit excellent inhibition into the ACE2 and PDB6LU7 proteins[109]."

"The major constituents of each oil were also tested separately, and showed lower activity compared to the oils. Moreover, mixtures of the main constituents, in the same proportions found in G. friesiana and G. hispida oils, did not show the same activity as the original oils[110]."

"When co-administered at fixed ratios based on individual A50s, CBD and BCP (beta-caryophyllene) produced enhanced dose-dependent reduction in allodynic responses with synergistic effects observed for cold hypersensitivity in both sexes[111] [of rats]."

"We analyzed whether the long-term intake of honey and/or D-limonene...can counteract the neurodegeneration occurring in high fat diet (HFD)-induced obesity. The honey and D-limonene intake counteracted these alterations; however, they did so in a stronger manner when in combination[112]."

"The number of attracted beetles in the Methyl benzoate-nerolidol blend was greater than the combined sum of beetles attracted to these compounds alone, suggesting a synergistic interaction[113]."

"It appears that even 25 extracts are not sufficient to clearly delineate the active ingredient(s) in cannabis or [the] modulating effects of the other minor components. This suggests the complexity of all the ingredients, perhaps indicating that many of them are active and have modulation capacities[114]."

---

[106] Begofia Delgado, Alfredo Palop, Pablo S. Fernandez & Paula M. Periago. Combined effect of thymol and cymene to control the growth of Bacillus cereus vegetative cells. European Food Research and Technology volume 218, pages188-193(2004).

[107] Waliwitiya R, Nicholson RA, Kennedy CJ, Lowenberger CA. The synergistic effects of insecticidal essential oils and piperonyl butoxide on biotransformational enzyme activities in Aedes aegypti (Diptera: Culicidae). J Med Entomol. 2012 May;49(3):614-23.

[108] Xiangen Meng, Na Li, Yu Zhang, Danfeng Fan, Chen Yang, Hang Li, Dazhi Guo, and Shuyi Pan. Beneficial Effect of B-Elemene Alone and in Combination with Hyperbaric Oxygen in Traumatic Brain Injury by Inflammatory Pathway Translational Neuroscience 2018; 9: 33-37.

[109] My TTA, Loan HTP, Hai NTT, Hieu LT, Hoa TT, Thuy BTP, Quang DT, Triet NT, Anh TTV, Dieu NTX, Trung NT, Hue NV, Tat PV, Tung VT, Nhung NTA. Evaluation of the Inhibitory Activities of COVID-19 of Melaleuca cajuputi Oil Using Docking Simulation. ChemistrySelect. 2020 Jun 8;5(21):6312-6320.

[110] Costa EV, Teixeira SD, Marques FA, Duarte MC, Delarmelina C, Pinheiro ML, Trigo JR, Sales Maia BH. Chemical composition and antimicrobial activity of the essential oils of the Amazon Guatteriopsis species. Phytochemistry. 2008 Jun;69(9):1895-9.

[111] Eeswara A, Pacheco-Spiewak A, Jergova S, Sagen J. Combined non-psychoactive Cannabis components cannabidiol and β-caryophyllene reduce chronic pain via CB1 interaction in a rat spinal cord injury model. PLoS One. 2023 Mar 13;18(3):e0282920.

[112] Terzo S, Calvi P, Nuzzo D, Picone P, Allegra M, Mulè F, Amato A. Long-Term Ingestion of Sicilian Black Bee Chestnut Honey and/or D-Limonene Counteracts Brain Damage Induced by High Fat-Diet in Obese Mice. Int J Mol Sci. 2023 Feb 9;24(4):3467.

[113] Favaris AP, Túler AC, Silva WD, Pec M, Rodrigues SR, Maia ACD, Bento JMS. Methyl benzoate and nerolidol attract the cyclocephaline beetle Cyclocephala paraguayensis to trumpet flowers. Naturwissenschaften. 2023 Jan 26;110(1):3.

[114] Li D, Ilnytskyy Y, Ghasemi Gojani E, Kovalchuk O, Kovalchuk I. Analysis of Anti-Cancer and Anti-Inflammatory Properties of 25 High-THC Cannabis Extracts. Molecules. 2022 Sep 16;27(18):6057.

"These results indicate that the [Ibuprofen] + cineol, not only improve cognitive function and reduce anxiety, also reduce oxidative stress, therefore, the simultaneous use of these two compounds may be useful in improving HA-induced cognitive disorders and anxiety[115]."

CANNABIS SYNERGY

Despite the prolific use of the term entourage effect incorrectly used interchangeably with synergy, the latter concept is distinct and generally poorly understood; in fact, it is a literal work in progress absconded by enthusiastic but underinformed consumers and businesspeople. The reality is that there is a significant and often contentious divide between scientists who support and those who doubt the existence of synergy.

Nevertheless, if the phytochemical synergies listed previously can apply to many different plant-generated blends and many different terpene blends created by humans, then it is logical to assume that the same molecules as produced by cannabis can have similar effects.

**THC and CBD*

Most of the research concerning the existence of synergy in cannabis centers around THC and CBD, although the first mention of the entourage effect did not concern these cannabinoids. Interestingly, CBD and THC have been the subject of significant scientific debate, with some researchers attempting to classify cannabis species - proposed as indica, sativa, and ruderalis - using the ratios of these two cannabinoids as chemotaxonomic markers[116] (a naming system categorized by chemical content).

In 1998 Shimon Ben-Shabat and colleagues including Raphael Mechoulam discovered that the endogenous ligand 2-arachidonoyl-glycerol (2-Ara-Gl) was accompanied by several related esters at the site of CB1 and CB2, significantly potentiating the binding of 2-Ara-Gl at the receptor sites, and thereby increasing the biological activity of the ligand[117]. This led the researchers to theorize that the compounds must interact synergistically, and thus the concept of the entourage effect was born.

With THC and CBD at the forefront of cannabis science in the late 90s and early 2000s, Russo expanded upon the work of Ben-Shabat and Mechoulam, proposing that the two cannabinoids acted synergistically. According to Russo, "CBD possesses the unique ability to counteract the intoxicating and adverse effects of cannabis, such as anxiety, tachycardia, hunger, and sedation in rats and humans." Subsequently, Russo has authored three important papers that have been seized by the cannabis industry as indisputable proof of the entourage effect and synergy in cannabis:

- **2006**: *A Tale of two Cannabinoids: The Therapeutic Rationale for Combining Tetrahydrocannabinol and Cannabidiol*

In this paper, Russo asserts that CBD can mitigate undesirable side effects of THC, recommending that the two cannabinoids be combined in the therapeutic treatment of Multiple Sclerosis, cancer pain, rheumatoid arthritis, and several other conditions[118].

---

115 Bahrami T, Yaghmaei P, Yousofvand N. The effects of Ibuprofen and 1, 8- cineol on anxiety and spatial memory in hyperammonemic rats. Metab Brain Dis. 2023 Feb;38(2):613-620.

116 Small, E., Beckstead, H.D., 1973b. Cannabinoid phenotypes in Cannabis sativa. Nature 245, 147-148.

117 Ben-Shabat, Shimon & Fride, Ester & Sheskin, Tzviel & Tamiri, Tsippy & Rhee, Man & Vogel, Zvi & Bisogno Tiziana & De petrocellis, Luciano & Di Marzo, Vincenzo & Mechoulam, Raphael. (1998). An entourage effect: inactive endogenous fatty acid glycerol esters enhance 2-arachidonoyl-glycerol cannabinoid activity. European Journal of Pharmacology. 353. 23-31.

118 Russo E, Guy GW. A tale of two cannabinoids: the therapeutic rationale for combining tetrahydrocannabinol and cannabidiol. Med Hypotheses. 2006;66(2):234-46.

- **2011**: *Taming THC: Potential Cannabis Synergy and Phytocannabinoid- terpenoid Entourage Effects*

This is Russo's quintessential paper, proposing that the entourage effect or synergy observed in cannabis could extend beyond THC and CBD to other compounds, including terpenes[119].

- **2019**: *The Case for the Entourage Effect and Conventional Breeding of Clinical Cannabis: No Strain, no Gain*

In this paper, Russo again makes the case for the entourage effect, and discusses the chemotaxonomic classification of cannabis, recommending the cannabis industry refer to strains as the more accurate term 'chemovars'[120].

Note: Readers are encouraged to review in their entirety the three articles mentioned above after studying Terpenes 101 and Cannabinoids 101 in this book.

Prior to the work of Ben-Shabat, Mechoulam, and Russo, evidence of synergy in cannabis first came to light in 1974, when researchers demonstrated that two types of whole Brazilian cannabis induced psychological and other effects in human subjects that were 2-4 times as potent as THC alone[121]. Other early evidence of synergy in cannabis came in 1981, when experiments showed that whole cannabis elicited much higher anticataleptic activity in mice than those exposed to THC alone[122].

In 2001, McPartland and Russo published a review of the scientific literature on medical cannabis, finding again that non-cannabinoid compounds in cannabis enhanced the beneficial effects of THC, with the authors suggesting that terpenes and flavonoids play a role in the combined activity[123]. Two years later, Wilkinson et al. found that a cannabis extract performed better than pure D9-THC in a mouse model of multiple sclerosis, and in a rat model of epilepsy[124]. The extract resulted in faster onset of therapeutic effects and a reduction in the time to achieve maximum effect, while also exhibiting unexpected anticonvulsant activity.

One particularly interesting account of in-plant use of synergy by cannabis compounds was published in 2005 by Sirikantaramas, who noted an obvious synergy in trichomes between sticky terpenes which trap insects, and acidic cannabinoids which then act as insecticides[125]. The following year, an anonymously authored report found that terpenes including limonene and myrcene enhanced user experiences when combined with THC[126].

---

[119] Russo, Ethan B. Taming THC: potential cannabis synergy and phytocannabinoid-terpenoid entourage effects. British journal of pharmacology vol. 163,7 (2011): 1344-64.

[120] Russo, Ethan B. The Case for the Entourage Effect and Conventional Breeding of Clinical Cannabis: No "Strain," No Gain. Frontiers in plant science vol. 9 1969. 9 Jan. 2019.

[121] Carlini EA, Karniol IG, Renault PF, Schuster CR. Effects of marihuana in laboratory animals and in man. Br J Pharmacol. 1974 Feb;50(2):299-309.

[122] Fairbairn, J W, and J T Pickens. Activity of cannabis in relation to its delta'-trans-tetrahydro-cannabinol content. British journal of pharmacology vol. 72,3 (1981): 401-9.

[123] McPartland, J.M., Russo, E.B., 2001. Cannabis and Cannabis extracts: greater than the sum of their parts? J. Cannabis Therap. 1, 103-132.

[124] Wilkinson JD, Whalley BJ, Baker D, Pryce G, Constanti A, Gibbons S, Williamson EM. Medicinal cannabis: is delta9-tetrahydrocannabinol necessary for all its effects? J Pharm Pharmacol. 2003 Dec;55(12):1687-94.

[125] Sirikantaramas, S., Taura, F., Tanaka, Y., Ishikawa, Y., Morimoto, S., and Shoyama, Y. (2005). Tetrahydrocannabinolic acid synthase, the enzyme controlling marijuana psychoactivity, is secreted into the storage cavity of the glandular trichomes. Plant Cell Physiol. 46, 1578-1582.

[126] Name Withheld. (2006) Studying the Effects of Terpenes. O'Shaughnessy's Spring 2006.

In 2010, Johnson et al. published a multicenter, double-blind, randomized, placebo-controlled, parallel-group study of the efficacy, safety, and tolerability of a blended THC and CBD extract, and a pure THC extract in patients with intractable cancer-related pain. The researchers found that the THC-only extract was comparable to placebo, while the blend containing THC and CBD relieved advanced cancer pain that no longer responded to opioids[127]. The same year, Fischedick et al. proposed that synergistic effects observed in cannabis compounds could be due to processes unrelated to the CB1 receptor[128], while a year later Efferth and Koch theorized multiple synergistic mechanisms among various phytochemicals[129].

In 2015, Gallily, Yekhtin, and Hanus set out to find a way to overcome the biphasic dose response pure CBD produces when used in the treatment of pain (CBD is analgesic to a point, then stops working despite increased doses). They found that a full spectrum cannabis extract with CBD produced a linear analgesic effect with no observed ceiling[130] on increased doses. The authors also found that both pure cannabinoid compounds and cannabis extracts can induce anxiolytic (anxiety-relieving) and anxiogenic (anxiety-causing) effects, suggesting that the mechanism of action responsible for these opposing actions may include various entourage effects. The same year, Iseger and Bossong found that CBD could alleviate the negative effects of THC, including cognitive impairment, and reduce the risk of developing psychoses[131] after long-term use. In 2018, researchers found that a botanical drug preparation containing, among other compounds, small amounts of cannabigerol (CBG) and tetrahydrocannabinolic acid (THCA), was more potent than pure D9-THC in producing antitumor activity in several types of human breast cancer. The authors stated that "combination of cannabinoids with estrogen receptor- or HER2-targeted therapies (tamoxifen and lapatinib, respectively) or with cisplatin, produced additive antiproliferative responses in cell cultures[132]."

In a meta-analysis of clinical studies on epilepsy published the same year, Pamplona et al. concluded that two thirds of patients treated with a CBD-rich extract reported reduced seizure frequencies, and lower average dose than patients treated with pure CBD, while also suffering fewer adverse events[133]. The authors theorized that these differences are caused by synergistic effects of the primary cannabinoids with other phytocompounds.

Also in 2018, Kamal made several observations: that pure THC causes more anxiety than whole cannabis, that THC and nerolidol are correlated with increased anxiolytic activity, and that guaiol, eucalyptol, terpinene, phellandrene, and other compounds were associated with decreased anxiolytic activity[134]. The authors suggested that these actions may be caused by modulatory effects of terpenes and other compounds.

---

[127] Johnson JR, Burnell-Nugent M, Lossignol D, Ganae-Motan ED, Potts R, Fallon MT. Multicenter, double-blind, randomized, placebo-controlled, parallel-group study of the efficacy, safety, and tolerability of THC:CBD extract and THC extract in patients with intractable cancer-related pain. J Pain Symptom Manage. 2010 Feb;39(2):167-79.

[128] Fischedick J, Van Der Kooy F, Verpoorte R. Cannabinoid receptor 1 binding activity and quantitative analysis of Cannabis sativa L. smoke and vapor. Chem Pharm Bull (Tokyo). 2010 Feb;58(2):201-7.

[129] Efferth, T., and Koch, E. (2011). Complex interactions between phytochemicals. The multi-target therapeutic concept of phytotherapy. Curr. Drug Target. 12, 122-132.

[130] Gallily, R., Yekhtin, Z. and Hanus, L.O. (2015) Overcoming the Bell-Shaped Dose-Response of Cannabidiol by Using Cannabis Extract Enriched in Cannabidiol. Pharmacology & Pharmacy, 6, 75-85.

[131] Iseger, T. A., and Bossong, M. G. (2015). A systematic review of the antipsychotic properties of cannabidiol in humans. Schizophr. Res. 162, 153-161.

[132] Blasco-Benito S, Seijo-Vila M, Caro-Villalobos M, Tundidor I, Andradas C, Garcia-Taboada E, Wade J, Smith S, Guzman M, Perez-Gomez E, Gordon M, Sanchez C. Appraising the "entourage effect": Antitumor action of a pure cannabinoid versus a botanical drug preparation in preclinical models of breast cancer. Biochem Pharmacol. 2018 Nov;157:285-293.

[133] Pamplona FA, da Silva LR, Coan AC. Potential Clinical Benefits of CBD-Rich Cannabis Extracts Over Purified CBD in Treatment-Resistant Epilepsy: Observational Data Meta-analysis. Front Neurol. 2018 Sep 12;9:759.

[134] Kamal Brishna S., Kamal Fatima, Lantela Daniel E. Cannabis and the Anxiety of Fragmentation-A Systems Approach for Finding an Anxiolytic Cannabis Chemotype Frontiers in Neuroscience. VOLUME 12 2018 Pg 730.

Most recently, Anderson showed in 2021 that oral administration of a cannabidiolic acid (CBDA) extract in mice produced plasma CBDA concentrations that were fourteen times higher than in mice administered pure CBDA[135].

ANTI-SYNERGY ARGUMENTS AND EVIDENCE

Anti-synergy arguments and evidence against the entourage effect have been presented in the last several years. In 2019, Santiago et al. found that none of the six most common terpenes in cannabis activated CB1 or CB2, nor did these compounds modulate D9-THC[136]. A year later the same authors went on to find that the seven most common terpenes in cannabis do not modulate the actions of cannabinoids on human transient receptor potential ankyrin 1 (hTRPA1) or human transient receptor potential vanilloid 1 (hTRPV1) channels[137], both of which had been proposed as alternative options for synergy targets.

Other work addressed similar findings regarding CB1 and CB2 binding, with Finlay cautioning that "general lack of terpenoids effects on binding is not sufficient to completely rule out allosteric effects on function, as binding and functional modulation are separate in theory; receptor functional modulation may not necessarily be predicted by altered binding and vice versa[138]."

The same year, Ferber studied the entourage effect in cannabinoids used for the treatment of mood and anxiety disorders, concluding only that "Further research is warranted to investigate the potential therapeutic value of adding terpenes to treatment with CBD, with or without additional THC. The effects of combining this original treatment with the conventional pharmacological approach also require further investigation[139]."
Other researchers have stated that, because of low levels of volatile monoterpenes in most cannabis products to begin with, and because of processes such as first-pass metabolism, high rates of clearance and short half-lives for elimination, phytonutrients like terpenes most likely could not accumulate to therapeutic levels[140].

Perhaps the strongest arguments and evidence casting doubt on synergy in cannabis originate from Peter S. Cogan, PhD (pharmaceutical sciences). In his March 2020 review article and meta-analysis titled *'The 'Entourage Effect' or 'Hodge-podge Hashish': the Questionable Rebranding, Marketing, and Expectations of Cannabis Polypharmacy*, Cogan methodically and ruthlessly attacks many of the articles and studies presented in the preceding "Cannabis Synergy" section of this book, concluding with his opinion on the entourage effect[141] (EE) in cannabis:

"In the opinion of this author, solicitations invoking the EE should be regulated as any other explicit therapeutic claim, as they are enticing the buyer to make a purchase based on the promise of benefit and.are leveraging assumptions of pharmacologic necessity (e.g. small molecule synergy) to do so.

---

[135] Anderson, Lyndsey L., Maia G. Etchart, Dilara Bahceci, Taliesin A. Golembiewski, and Jonathon C. Arnold. Cannabis constituents interact at the drug efflux pump BCRP to markedly increase plasma cannabidiolic acid concentrations. Scientific Reports 11.1 (2021).

[136] Santiago, Marina, Shivani Sachdev, Jonathon C. Arnold, Iain S. McGregor, and Mark Connor. Absence of Entourage: Terpenoids Commonly Found in Cannabis sativa Do Not Modulate the Functional Activity of L9-THC at Human CB1 and CB2 Receptors. Cannabis and Cannabinoid Research 4.3 (2019): 165-176.

[137] Heblinski, Marika, Marina Santiago, Charlotte Fletcher, Jordyn Stuart, Mark Connor, Iain S. McGregor, and Jonathon C. Arnold. Terpenoids Commonly Found in Cannabis sativa Do Not Modulate the Actions of Phytocannabinoids or Endocannabinoids on TRPA1 and TRPV1 Channels. Cannabis and Cannabinoid Research 5.4 (2020): 305-317.

[138] Finlay, David B., Kathleen J. Sircombe, Mhairi Nimick, Callum Jones, and Michelle Glass. Terpenoids From Cannabis Do Not Mediate an Entourage Effect by Acting at Cannabinoid Receptors. Frontiers in Pharmacology 11 (2020).

[139] Ferber, Sari Goldstein, Dvora Namdar, Danielle Hen-Shoval, Gilad Eger, Hinanit Koltai, Gal Shoval, Liat Shbiro, and Aron Weller. The "Entourage Effect": Terpenes Coupled with Cannabinoids for the Treatment of Mood Disorders and Anxiety Disorders. Current Neuropharmacology 18.2 (2020): 87-96.

[140] Duggan, Peter J. The Chemistry of Cannabis and Cannabinoids. Australian Journal of Chemistry 74.6 (2021): 369-387.

[141] Cogan PS. The 'entourage effect' or 'hodge-podge hashish': the questionable rebranding, marketing, and expectations of cannabis polypharmacy. Expert Rev Clin Pharmacol. 2020 Aug;13(8):835-845.

Moreover, instead of treating entourage complexity with overarching assumptions regarding its clearly unpredictable effects, it seems more appropriate to test every proposed entourage product to see if it actually lives up to expectations."

ALTERNATIVES TO SYNERGY

Interestingly, there are emerging alternatives to synergy theory. One fascinating alternative has been put forth after researchers working with houseflies observed that the insects processed and neutralized primary terpenes in an insecticidal essential oil blend, but not minor or secondary terpenes. This led researchers to theorize that, because the insects were unable to neutralize minor terpenes, these constituents then exhibited much more severe and potent action against the flies[142]. Essentially, it appeared that the flies were more affected by the minor terpenes because they had no defenses left against them, and not because of the synergistic action of the constituents.

As Raphael Mechoulam mentioned in his commentary in the "Terp Tsars" chapter of this text, not all terpenes or other compounds are synergistic with all other compounds, in all other circumstances. However, although there is significant therapeutic value in individual compounds, when it comes to terpenes, the secondary general action of these compounds is often synergistic. Synergistic action of terpenes and other phytochemicals is logical and well-supported in the literature outside of cannabis, as documented herein.

The ever-contentious subject of cannabis lives up to this expectation even in the scientific journals. Nevertheless, the literature supporting synergy in plants other than cannabis is strong, and it seems logical to conclude that we can make at least some inferences about this body of evidence that subsequently cross over into the figurative and literal 'field' of cannabis.

WHAT'S NEXT FOR SYNERGY IN CANNABIS?

More Research / Better Dissemination / Better Oversight

While reasonable inferences made from non-cannabis phytochemicals will have to suffice for now, we must swiftly conduct more research into the combined or synergistic effects of cannabinoids, terpenes, and other compounds. This research must be targeted and collaborative, and dissemination of results should be integrated in such a way that cannabis media and publications have ready access to it. Finally, as Cogan pointed out, we must also achieve better oversight of therapeutic claims and marketing materials, which must be based on sound science alone, and not the mythology that often swirls and clouds the cannabis industry.

[142] Scalerandi, Esteban et al. Understanding Synergistic Toxicity of Terpenes as Insecticides: Contribution of Metabolic Detoxification in Musca domestica. Frontiers in plant science vol. 9 1579. 30 Oct. 2018.

Synergy Review

Answer the following questions to test your knowledge of this section:

Question #1: Match the term with the correct general definition:

a. Synergy	1. One compound affecting multiple targets
b. Polypharmacology	2. A nonsensical term
c. Synergistic bonding	3. Synergy occurring within the ECS
d. Entourage Effect	4. Two or more compounds acting together

Question #2: Geraniol, beta-pinene, alpha-humulene, and linalool are:

a. Bicyclic
b. Tricyclic
c. Cyclized
d. Cannabimimetic
e. Antispasmoidal

Question #3: Synergy characters can include:

a. Terpenes and flavonoids
b. Terpenes and cannabinoids
c. Terpenoids, flavones, and anticancer agents
d. Antibacterial agents, terpenoids, and cannabinoids
e. Flavonols, analgesics, terpenes, and benzodiazepines
f. All of the above

Question #4: Name the endogenous ligan that led Ben-Shabat and Mechoulam to coin the term 'entourage effect':

a. 2-Arachidonoyl-glycerol
b. 2-Delta-arachidonoyl glycol
c. Arachadonoyl-2-glycophosphate
d. 2-Arachidonoyl-geranoyl

Question #5: Overall, cannabis consumers prefer whole plant products versus pure THC:

a. True
b. False

Question #6: While Russo, McPartland, and Mechoulam are in favor of the existence of synergy, these three scientists are against the concept:

a. Fairborn, Childs, McDonald
b. Wendy, Rose, McLintock
c. Burger, Sorensen, Dumname
d. Cogan, Duggan, Santiago

Question #7: Name one obvious example of synergy in cannabis:

____

For the answer key to Synergy, please go here: http://thebigbookofterps.com/bbot-101-chapters-quiz-answer-key/

PINENE

Type: Bicyclic monoterpene
Chemical Formula: $C_{10}H_{16}$
Molecular Weight: 136.23 g/mol
Boiling Point: *Alpha:* 155 to 156 °C. @ 760 mm Hg[143]
Beta: 163 to 166 °C. @ 760 mm Hg[144]
Flash Point: *Alpha:* 90 °F. TCC (32.22 °C.)
Beta: 95 °F. TCC (35 °C.)
Melting Point: *Alpha:* -64 to -62 °C. @ 760 mm Hg
Beta: -61.54 °C (-78.77 °F)
Solubility: *Alpha:* Soluble in alcohol, fixed oils, mostly insoluble in water
Beta: Soluble in alcohol, mostly insoluble in water
Oral LD50: 3,700 mg/kg (rat)[145]
Biological Role: Insecticide, allelopathic agent, fungicide
Therapeutic Role: Anticancer agent, antioxidant, anti-inflammatory agent
Commercial Use: Perfumes and fragrances, solvents, flavoring agent
Occurrence in Cannabis: Top three
Aroma/Flavor: *Alpha:* camphor, pine, earthy, woody, terpenic, herbal
Beta: pine, green, dry, woody, eucalyptus, fresh, minty, balsamic, spicy

Occurs in Cannabis Varieties: The majority of cannabis varieties will contain an isomer or racemic mixtures of pinene, but specifically these varieties have been shown to contain pinene in high amounts: ACDC, Banana OG, Blue Dream, Bubba Kush, Cannatonic, Chem D, DJ Short Blueberry, Dutch Treat, Girl Scout Cookies, Gorilla Glue #4, Harlequin, Haze Berry, Island Sweet Skunk, Jack Herer, Obama Kush, OG Kush, Purps, Romulan, Royal Jack, Skywalker OG, Strawberry Cough, Trainwreck.

INTRODUCTION

Likely found in high concentrations in the majority of cannabis strains, pinene is also the most widely distributed natural terpene on earth[146]. In fact, pinene is the major constituent of the aroma of coniferous forests, especially those with pine trees. In Japan, there is a rapidly growing practice of forest bathing, which began in the 1980s and has spread worldwide today, where the concept amounts to deliberately inhaling pinene and other terpene emissions in a natural forest environment. The idea behind forest bathing is that natural compounds like pinene, when inhaled, promote good health and healing in humans. There may be significant truth to this concept, as evidenced in the "Potential Uses in Medicine", and "Implications for Human Health & Nutrition" sections below.

Because pinene vapors are heavier than air, these emissions linger below the forest canopy, saturating the lower environment with scented air rich in monoterpenes and other phytochemicals. The amount of these emissions is staggering; in the United States, total emissions of alpha-pinene from forests were around 6.6 mega tons annually in the 1970s. Today, it is likely this number is smaller considering the rate of global deforestation.

---

[143] Alpha-pinene, The Good Scents Company, from: http://www.thegoodscentscompany.com/data/rw1006351.html. Accessed July 21, 2023.

[144] Beta-pinene, The Good Scents Company, from: http://www.thegoodscentscompany.com/data/rw1344431.html. Accessed July 21, 2023.

[145] Agilent Technologies, Inc. Safety Data Sheet - Alpha-pinene standard. Published 03/30/2019 From: https://www.agilent.com/cs/library/msds/TRP-160-1_NAEnglish.pdf. Accessed July 21, 2023.

[146] Noma Y, Asakawa Y (2010). Biotransformation of monoterpenoids by microorganisms, insects, and mammals. In: Baser KHC, Buchbauer G (eds). Handbook of Essential Oils: Science, Technology, and Applications. CRC Press: Boca Raton, Florida, pp. 585-736.

Pinene is also the major constituent of pine sap and other coniferous resins, appearing as a clear to yellow or amber sap in the natural state, and a colorless, pungent liquid as a commercial product. This isoprene-based compound lends many cannabis varieties like Blue Dream, Gorilla Glue #4, Jack Herer, and OG Kush their terpenic, piney, woody, and sharp characteristics.

Interestingly, larvae of the sawfly species Susana cupressi have been found to sequester Cypress tree terpenes in their foregut, midgut, and oesophageal diverticula, and use these compounds – including alpha-pinene and beta-pinene – as defensive compounds against predators[147].

CHEMICAL STRUCTURE

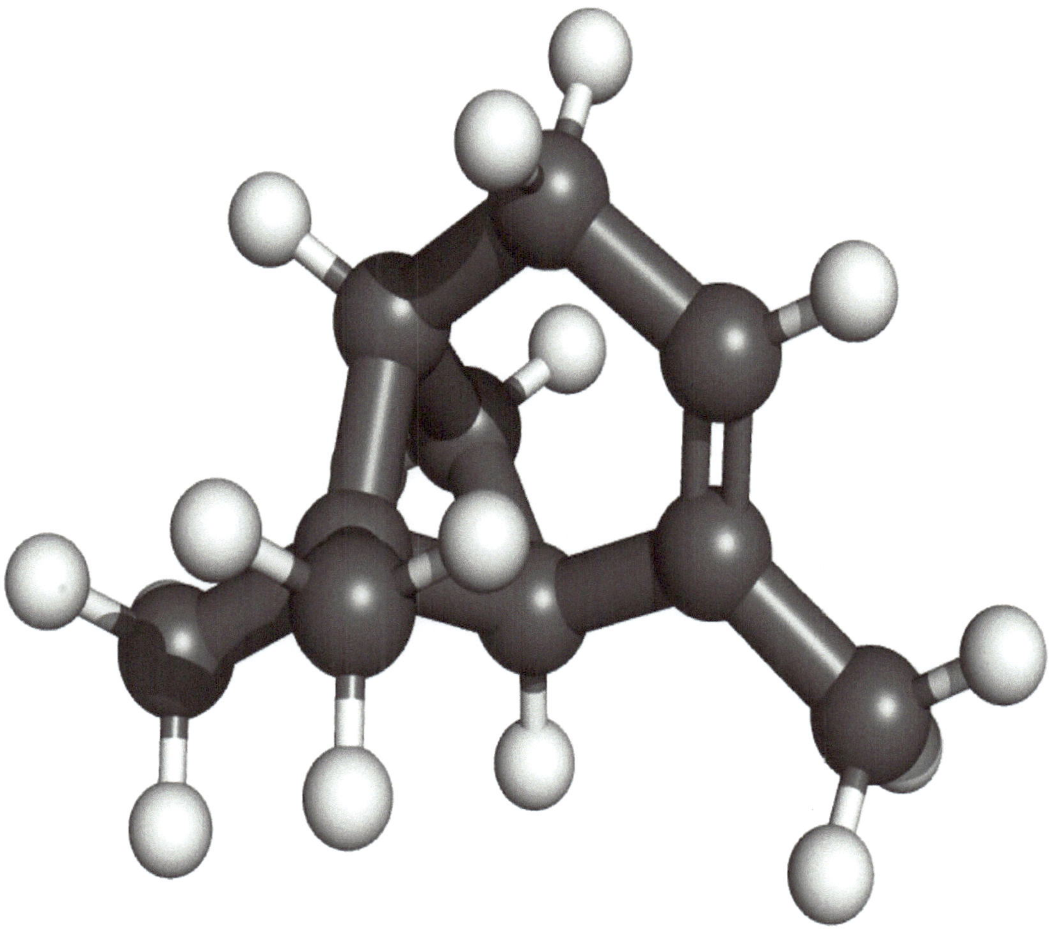

Pinene is a hydrocarbon alkene containing twenty-six atoms; ten carbon atoms, and sixteen hydrogen atoms, notated as $C_{10}H_{16}$. This terpene is produced in the mevalonate pathway of plants, and is derived from isopentenyl pyrophosphate (isoprene). The ten carbon atoms in this compound's skeleton denote the fact that pinene is comprised of two units of isoprene, each with five carbon atoms, which classifies this compound as a monoterpene.

Pinene is further classified as a bicyclic monoterpene because it has two rings fused together in the molecular structure. Of the two rings, the 4-membered ring is reactive, at least in part because of its contorted nature, which means that it is prone to skeletal rearrangements resulting in different functionality.

---

[147] Boevé JL, Sonet G, Jacobson HR, Angeli S. Cypress terpenes in sawfly larva of Susana cupressi (Hymenoptera: Symphyta: Tenthredinoidea). Naturwissenschaften. 2023 Mar 27;110(2):13.

There are two structural isomers (the same compound, but with different arrangements of the atoms within the molecule) of this hydrocarbon, both of which occur commonly in nature; alpha-pinene and beta-pinene. Alpha-pinene is substituted by methyl groups at positions 2, 6 and 6-1 respectively, and features an endocyclic double bond in the 5-membered ring. Conversely, beta-pinene contains an exocyclic double bond. In some plants, both alpha-pinene and beta-pinene occur in equal amounts in the same species; this is referred to as a racemic mixture.

OCCURRENCE IN PLANTS

Pinene can be found at some level in hundreds of different plant species. In fact, this terpene derived its name from pine trees, as it is the primary constituent of the sap or resin of many different varieties of pine. However, while the compound takes its name from the pine tree, it occurs in a wide variety of plants, including:

Alpha-pinene:

Abies holophylla
Ajowan seed
Anise
Artemisia
Basil
Black pepper
Blackberry
Blueberry
Cajuput
Cannabis
Carqueja
Cassia
Champaca
Citronella
Coriander
Cubeb
Cypress
Eriocephalus punctulatus
Fennel
Fleabane
Geranium
Guava
Hinoki
Horsemint
Juniper
Labdanum
Lavender
Lime
Mandarin
Marjoram
Microstrobos
Myrtle
Neroli
Oregano
Patchouli
Petitgrain
Pine
Pomelo
Achillea tenuifolia lam.
Allspice
Apple
Aspen
Basswood
Bell pepper
Black currant
Boldus
Calamus
Caraway
Carrot
Celery
Cinnamon
Clary Sage
Cotton
Cumin
Dill
Eucalyptus
Fig
Frankincense
Ginger
Gum
Ho leaf
Hyssop
Kachur
Laurel
Lemon
Lovage
Mango
Makrut lime
Mikan
Narcissus absolute
Nutmeg
Origanum osmanthus
Pennyroyal
Pimenta
Pinus
Pangos
Agathosma crenulata
Angelica
Armoise
Ayou
Bay
Bergamot
Black walnut
Buchu
Canaga
Cardamom
Cascarilla
Chamomile
Cilantro
Clove
Croton
Curry
Echinophora cinerea
Featherfew
Fir
Galangal
Grapefruit
Heracleum
Horse chestnut
Jambu
Kewda
Lavandin
Lemongrass
Mace
Manuka
Melissa
Mugwort
Nepeta
Orange
Parsley
Peppermint
Pimento
Plum
Protium

Rosemary
Snake root
Strawberry
Tangerine
Taxodium distichum
Tomato
Verbena
Wormwood
Yuzu
Zinnia

Saffron
Spearmint
Tagete
Tansy
Tea tree
Turmeric
Water mint
Yarrow
Zanthoxylum

Sage
Spruce
Tamarind
Tarragon
Thyme
Valerian
Wild orange
Ylang ylang
Zingiber

Beta-pinene:

Achillea tenuifolia
Allspice
Anise
Ayou
Bay laurel
Blood orange
Cajuput
Cananga
Carqueja
Cassia
Cayenne
Chamomile
Cinnamon
Close
Cornmint
Cumin
Cypress
Elemi
Fennel
Fig
Frankincense
Ginger
Guava
Ho leaf
Hyptis species
Kachur
Laurel
Lavender
Lemongrass
Litsea cubeba
Mandarin
Marjoram
Mentha longifolia
Mikan
Myrtle
Nutmeg
Oregano

Agathosma crenulata
Alpinia galanga
Apple
Baccharis coidifolia
Bergamot
Bois de rose
Calamintha nepeta
Caraway
Carrot
Cassinia
Cedarleaf
Champaca
Citronella
Coriander
Costmary
Curcuma amada
Dill
Eucalyptus species
Ferula
Fir
Galangal
Glycosmis pentaphylla
Heracleum paphlagonicum
Horsemint
Hyssop
Kewda
Lavandin
Layana
Levisticum officinale
Lovage
Mango
Mastic
Michelia alba
Myristica fragrans
Nepeta species
Ocimum gratissimum
Parsley

Ajowan
Angelica
Artemisia species
Basil
Black pepper
Bombycilaena erecta
Calamus
Cardamom
Cascarilla
Cassis
Celery
Chenopodium ambrosioides
Clary sage
Corn silk
Croton species
Curry
Echinophora cinerea
Featherfew
Feverfew
Fleabane
Geranium
Grapefruit
Hinoki
Hypericum dogonbadanicum
Juniper
Labdanum
Lavandula
Lemon
Lime
Mace
Manuka species
Melissa
Microstrobos species
Myrrhinium atropurpureum
Neroli
Orange
Parsnip

Patchouli	Pepper tree	Peppermint
Persea americana	Peucedanum petiolare	Phlomis fruticosa
Phoebe	Pimento	Pine species
Piper cernum	Plectranthus glandulosus	Plum
Pomelo	Prangos uechtritzii	Protium heptaphyllum
Pteronia	Ravintsara	Rosemary
Saffron	Sage	Salvia
Satureja species	Savin	Savory
Sideritis syriaca	Snake root	Spearmint
Spikenard	Stachys recta	Sugandha kokila
Sunflower	Tagete	Tanarind
Tangerine	Tansy	Tarragon
Taxodium distichum	Tejpat	Thyme
Tomato	Turmeric	Valerian
Vassoura	Water mint	Wormseed
Wormwood	Yarrow	Ylang ylang
Yuzu	Zanthoxylum species	Zingiber

BIOLOGICAL ACTIVITY IN PLANTS

Insecticidal: Jose S. Dambolena, et al showed that pinene can function as a powerful toxic fumigant, while Giovanni Benelli, et al concluded that pinene also works as a contact toxin for some common insect pests. In Salvia tomentosa (balsamic sage), the essential oil of the plant featured beta-pinene and alpha-pinene as the primary constituents, which caused complete mortality against acanthoscelides (bean weevils), and nearly complete mortality against tribolium castaneum (red flour beetles)[148].

***Ovicidal agent**:* Pinene has been shown to function as an ovicidal agent (attacking or inhibiting the growth of insect ova) against Pediculus humanus capitis (head lice)[149], and acts similarly against some other insect larvae.

Nematicidal: As the primary constituent of the essential oil of Dryobalanops aromatica (borneo camphor) and mentha haplocalyx var. piperascens (Japanese mint), alpha-pinene exhibited strong nematicidal properties and hatching inhibition activities against meloidogyne icognita[150] (the southern root-knot nematode, a parasitic roundworm).

***Antifungal agent**:* Filomena Nazzaro et al (2017) showed that pinene can function as an antifungal agent against several types of pathogenic or parasitic fungi. As primary constituents in the essential oil of liquidambar formosana hance (Formosan sweet gum), both alpha-pinene and beta-pinene exhibited inhibitory activity against L. betulina (a type of mushroom that typically grows on birch trees) and L. sulphureus[151] (a bracket fungus known as chicken of the woods). As the primary constituent of the essential oil of zanthoxylum bungeanum (Chinese-pepper), alpha-pinene controlled dry rot in potatoes inoculated with fusarium sulphureum (a common fungal pathogen of

---

148 Ulukanli, Zeynep, Salih Karaborklli, Menderes Cenet, Osman Sagdic, Ismet Ozturk, and Mehmet Balcilar. "Essential oil composition, insecticidal and antibacterial activities of Salvia tomentosa Miller." Medicinal Chemistry Research 22.2 (2012): 832-840.

149 Toloza, Ariel. (2010). Bioactividad y toxicidad de componentes de aceites esenciales vegetales, en Pediculus humanus capitis (Phthiraptera: Pediculidae) resistentes a insecticidas piretroides.

150 Kim, Soon-Il, Jae-Kook Lee, Young-Eun Na, Seong Tak Yoon, and Young Ju Oh. "Nematicidal and ovicidal activities of Dryobalanops aromatica and Mentha haplocalyx var. piperascens -derived materials and their formulations against Meloidogyne incognita second-stage juveniles and eggs." Nematology 16.2 (2014): 193-200.

151 Chien, Shih-Chang, Jun-Hong Xiao, Yen-Hsueh Tseng, Yueh-Hsiung Kuo, and Sheng-Yang Wang. "Composition and antifungal activity of balsam from Liquidambar formosana Hance." Holzforschung 67.3 (2013).

cannabis), strongly inhibiting the growth of this fungus while significantly inhibiting spore germination, dropping germination rates by as much as 84% and decreasing cell membrane integrity[152]. In all tests conducted in the Chinese-pepper study, the EO was more effective than a-pinene alone, suggesting synergy.

**Insect Repellant*: In the essential oil state, pinene has repelled insects[153], including those that are not necessarily herbivores of the species in question. As a primary constituent in the essential oil extracted from the fruits of evodia lenticellata, beta-pinene showed fumigant and contact toxicity and was strongly repellant against the red flour beetle and liposcelis bostrychophila (the common booklouse), exhibiting repellent properties comparable to that of DEET[154], one of the most common commercial insect repellants in the world. As the primary constituent of several coniferous tree saps, pinene also acts as a direct insect repellant and trapping system.

**Allelopathic Agent:* alpha-pinene has been shown to exhibit allelopathic properties against ipomoea triloba L. (little bell morning glory, considered a weed) by inducing oxidative stress during germination and the early stages of seedling growth[155]. Alpha-pinene also inhibited the radicle growth (the first part of a seedling to emerge from the seed) of five common weeds: cassia occidentalis, amaranthus viridis, triticum aestivum, pisum sativum, and cicer arietinum, while simultaneously inducing oxidative stress, enhancing production of ROS, and decreasing cell membrane integrity[156].

**Larvicidal:* As part of the essential oils of several different plants, beta-pinene and alpha-pinene exhibited strong toxicity against the larvae of Culex pipiens[157], a type of mosquito that is a known disease vector.

**Antifeedant:* African elephants have been shown to deliberately avoid beta-pinene, and in fact are deterred in general by monoterpenes[158], which likely act as antifeedants in a variety of plants. Interestingly, the sole monoterpene that the elephants readily tolerated was alpha-pinene.

USES IN INDUSTRY

Both alpha-pinene and beta-pinene are used frequently in the food, perfume, and chemical solvent industries. Most samples of industrial pinene are produced by fractional distillation of turpentine, presenting as a clear liquid, with up to 95% purity.

Oxidation products of pinene produced in the laboratory setting are used extensively in the perfume and fragrance industry. In this industry, pinene is used in the same way as myrcene in that these terpenes are generally processed as intermediate or precursor chemicals, used to produce more aromatic hydrocarbons like linalool, limonene, and geraniol. Pinene in also used as an intermediate in synthetic camphor manufacture, a product that is used in dozens of different applications.

---

[152] Xing-dong, Li, and Xue Hua-li. Antifungal activity of the essential oil of Zanthoxylum bungeanum and its major constituent on Fusarium sulphureum and dry rot of potato tubers. Phytoparasitica 42.4 (2014): 509-517.

[153] Nerio LS, Olivero-Verbel J, Stashenko E (2010). Repellent activity of essential oils: a review. Bioresour Technol. 101 (1): 372-378.

[154] Cao, Ju-Qin, Shan-Shan Guo, Yang Wang, Xue Pang, Zhu-Feng Geng, and Shu-Shan Du. Toxicity and repellency of essential oil from Evodia lenticellata Huang fruits and its major monoterpenes against three stored- product insects. Ecotoxicology and Environmental Safety 160 (2018): 342-348.

[155] Pergo, Erica, and Emy Ishii-Iwamoto. Changes in Energy Metabolism and Antioxidant Defense Systems During Seed Germination of the Weed Species Ipomoea triloba L. and the Responses to Allelochemicals. Journal of Chemical Ecology 37.5 (2011): 500-513.

[156] Singh, Harminder P., Daizy R. Batish, Shalinder Kaur, Komal Arora, Radinder K. Kohli. a-Pinene Inhibits Growth and Induces Oxidative Stress in Roots. Annals of Botany 98.6 (2006): 1261-1269.

[157] Michaelakis, Antonios, Dimitrios Papachristos, Athanasios Kimbaris, George Koliopoulos, Athanasios Giatropoulos, and Moschos Polissiou. "Citrus essential oils and four enantiomeric pinenes against Culex pipiens (Diptera: Culicidae)." Parasitology Research 105.3 (2009): 769-773.

[158] Bester T, Schmitt MH, Shrader AM. The deterrent effects of individual monoterpene odors on the dietary decisions of African elephants. Anim Cogn. 2023 Jun;26(3):1049-1063.

Pinene is found in hundreds of essential oils available in natural preparations, or in blended oils containing other terpene constituents. The cosmetics industry uses essential oils containing pinene to add to cosmetic products, primarily to mask or enhance their aroma.

Used in the food industry as a flavoring agent, pinene is often added to sweets, baked goods, specialty beverages, and other value-added foodstuffs. Interestingly, as the primary constituents of ferula oil, beta-pinene (47.1%) and alpha-pinene (21.36%) exhibited significant antioxidant activity, with the study authors suggesting that the EO of ferula could accordingly "be used as safe and effective natural antioxidants to improve the oxidative stability of fatty foods during storage and to preserve foods against food burn pathogens"[159].

As a highly flammable yet stable liquid, pinene has also garnered the attention of special interest groups, including those in the hydrocarbon fuels industry. One study determined that pinene possesses a gravimetric energy density comparable to that of traditional hydrocarbon fuels (fuels consisting only of hydrogen and carbon atoms: propane, butane, hexane, and methane gasses for example), and found that the monoterpene is suitable for operation in turbocharged SI (spark ignition) engines[160].

POTENTIAL USES IN MEDICINE

Pinene is well-studied and offers promise in several medical fields, and with additional research we will discover more medical uses than detailed below. However, it should be noted that, while this substance is highly bioavailable with 60% pulmonary uptake and rapid metabolism[161], it is also potentially toxic in high concentrations, with as little as half of an ounce proving fatal to a child[162], and fatal doses for adults estimated at about 180 grams[163]. Toxicity figures are estimated based on turpentine consumption or contamination, where pinene is the primary constituent at up to 60%.

Treatment for Stress: To say that stress is a killer is no longer cliché; it is a reality for millions of people. Related to this, the current growth of the aroma therapy industry is buoyed on the concept that aromatic compounds can help relieve, reduce, and even eliminate the negative physical effects of stress. As it turns out, there is scientific evidence to support these practices. In one study, fifteen young men took part in an experiment where they breathed controlled emissions containing alpha-pinene, which lowered the systolic blood pressure of the participants[164], indicating relaxation. A follow-up study was conducted by one of the original study authors, but this time the participants were all young women. In the second study, olfactory stimulation by alpha-pinene significantly decreased the heart rate of the participants, again indicating physiological relaxation[165].

Insect Repellant: Because of the spread of severe mosquito-borne diseases like the Zika virus, Dengue fever, and malaria, new types of natural and easily reproducible insect repellants are needed, and pinene is a good candidate. As a primary constituent of the essential oils of several plant species, alpha-pinene was shown to exhibit strong

---

[159] Evaluation of antioxidant and antimicrobial activities of essential oils from Carum copticum seed and Ferula assafoetida latex. Kavoosi G, Tafsiry A, Ebdam AA, Rowshan V. J Food Sci. 2013 Feb;78(2):T356-61.

[160] Raman, V., Sivasankaralingam, V., Dibble, R., and Sarathy, S., a-Pinene - A High Energy Density Biofuel for SI Engine Applications, SAE Technical Paper 2016-01-2171, 2016.

[161] Russo, E. B (2011). Taming THC: potential cannabis synergy and phytocannabinoid-terpenoid entourage effects. British Journal of Pharmacology. 163 (7): 1344-1364.

[162] Gosselin, R.E., R.P. Smith, H.C. Hodge. Clinical Toxicology of Commercial Products. 5th ed. Baltimore: Williams and Wilkins, 1984., p. III-393.

[163] The Merck Index. 9th ed. Rahway, New Jersey: Merck & Co., Inc., 1976., p. 969.

[164] Tsunetsugu Y, Park BJ, Miyazaki Y (2010) Trends in research related to "Shinrin-yoku" (taking in the forest atmostphere or forest bathing) in Japan. Environ Health Prev Med 15:27-37.

[165] Ikei, Harumi, Chorong Song, and Yoshifumi Miyazaki. Effects of olfactory stimulation by a-pinene on autonomic nervous activity. Journal of Wood Science 62.6 (2016): 568-572.

insect repellant activity[166], but researchers noted that, due to the volatile nature of this bicyclic monoterpene, re-application is needed frequently and thus more study is required before the compound could be a viable option as a commercial insect repellant.

**Antioxidant:* Pinene is a well-known antioxidant and shows potential of being used in the treatment of oxidative stress-induced neurodegenerative disorders. In one study, hydrogen peroxide-induced oxidative stress in rat cells was treated with both 1,8-cineole and alpha-pinene, and the effects were significant, attenuating the loss of cell viability and changes in cell morphology, inhibiting intracellular ROS production, and markedly enhancing the expression of antioxidant enzymes[167]. In a separate study by the same authors, the essential oil of Salvia lavandulifoia (Spanish sage), which featured alpha-pinene and 1,8-cineole as the primary constituents, respectively, inhibited and decreased ROS production, and increased the endogenous antioxidant status, again while attenuating the loss of cell viability and morphology[168].

**Anti-parasitic:* In a blinded, controlled study, an essential oil blend where the primary constituent was alpha-pinene was fed to pigs infected with Ascaris lumbricoides (large roundworms that infect humans), which resulted in significantly reduced total worm counts in the animals, while not causing any adverse reactions[169].

**Anti-inflammatory:* As part of an essential oil that featured alpha-pinene and 1,8-cineole as the primary constituents, these compounds have been shown to greatly reduce inflammation, in one case increasing the rate of skin flap survival in rats[170] (skin flaps in medicine refer to a section of healthy skin that is partly excised and used to cover a nearby wound). In another study, alpha-pinene and beta-pinene were the primary constituents of the essential oil of Fructus Alpiniae zerumbet, which showed significant anti- inflammatory and analgesic effects in mice[171].

**Antibacterial:* The essential oil extracted from the leaves of Myracrodruon urundeuva, a tree native to South America, featured a-pinene and B-pinene as the primary compounds, which exhibited antibacterial activities against some of the most common strains of bacteria, including Staphylococcus aureus, Staphylococcus epidermidis, Escherichia coli, Pseudomonas aeruginosa, and Salmonella Enteritidis[172]. Another study conducted in 2010 echoed these findings regarding E. coli, where a-pinene eliminated the formation of bacterial colonies after just 2 hours of exposure[173].

---

166 Nerio, Luz Stella, Jesus Olivero-Verbel, and Elena Stashenko. Repellent activity of essential oils: A review. Bioresource Technology 101.1 (2010): 372-378.

167 Porres-Martinez, Maria, Elena Gonzalez-Burgos, M. Emilia Carretero, and M. Pilar G6mez-Serranillos.In vitro europrotective potential of the monoterpenes a-pinene and 1,8-cineole against H2O2-induced oxidative stress in PC12 cells. Zeitschrift fir Naturforschung C 71 (2016).

168 Porres-Martinez, M., E. Gonzalez-Burgos, M. E. Carretero, and M. P. G6mez-Serranillos. Major selected monoterpenes a-pinene and 1,8-cineole found in Salvia lavandulifolia (Spanish sage) essential oil as regulators of cellular redox balance. Pharmaceutical Biology 53.6 (2015): 921-929.

169 Kaplan RM, Storey BE, Vidyashankar AN, Bissinger BW, Mitchell SM, Howell SB, Mason ME, Lee MD, Pedroso AA, Akashe A, Skrypec DJ. Antiparasitic efficacy of a novel plant-based functional food using an Ascarissuum model in pigs. Acta Trop. 2014 Nov; 139:15-22.

170 Ince B, Dadacl M, K1l1n; I, Oltulu P, Yarar S, Uyar M. Effect of cineole, alpha-pinene, and camphor on survivability of skin flaps. Turk J Med Sci. 2018 Jun 14;48(3):644-652.

171 Xiao, Rui-Yao, Ling-Jing Wu, Xiao-Xiao Hong, Ling Tao, Peng Luo, and Xiang-Chun Shen. Screening of analgesic and anti-inflammatory active component in Fructus Alpiniae zerumbet based on spectrum-effect relationship and GC-MS. Biomedical Chromatography 32.3 (2018).

172 Rebouas de Araujo TD, Coriolano de Aquino N, Veras de Aguiar Guerra AC, Ferreira de Almeida Junior R, Mendona Araujo R, Fernandes de Araujo Junior R, Silva Farias KJ, Fernandes JV, Sousa Andrade V. Chemical composition and evaluation of the antibacterial and Cytotoxic activities of the essential oil from the leaves of Myracrodruon urundeuva. BMC Complement Altern Med. 2017 Aug 22;17(1):419.

173 de Sousa Eduardo L, Farias TC, Ferreira SB, Ferreira PB, Lima ZN, Ferreira SB. Antibacterial Activity and Time-kill Kinetics of Positive Enantiomer of a-pinene Against Strains of Staphylococcus aureus and Escherichia coli. Curr Top Med Chem. 2018;18(11):917-924.

**Treatment for Liver Cancer:* In a promising study of liver cancer, alpha-pinene obviously inhibited the growth of hepatoma carcinoma cells by 79.3% in vitro and 69.1% in vivo[174], an impressive effect against a deadly type of cancer.

**Protection against UVA Cellular Damage:* alpha-pinene has been shown to prevent UVA-induced cytotoxicity, generation of ROS, lipid peroxidation and DNA strand breaks, and apoptosis in human skin cells[175].

**Prevention of Aspirin Toxicity:* Thanks mostly to its well-established antioxidant properties, alpha-pinene can protect small intestinal epithelial cells from aspirin-induced oxidative stress[176].

**Treatment for Prostate Cancer:* alpha-pinene has a profound impact on human prostate cancer cells, inducing apoptosis and cell cycle arrest, and inhibiting tumor progression in xenograft tumors in mice[177] (xenograft tumors are human tumor cells that have been introduced to immunocompromised mice that will not reject human cells).

**Treatment for Melanoma:* Lending further credibility to the potential power of aromatherapy, an impressive study with mice has shown that inhalation of environmental aromatic alpha-pinene reduced melanoma growth by about 40% over controls[178], which leads to the conclusion that people who have melanoma may benefit from forest bathing as mentioned earlier in this chapter.

**Synergistic Treatment for Lung Cancer:* To add to the already considerable potential for the use of pinene in cancer treatment, alpha-pinene and beta-pinene showed synergistic effects against non-small-cell lung cancer cells when combined with Paclitaxel (Taxol), a common chemotherapy medication[179].

**Treatment for Cervical Cancer:* As an isolated compound, alpha-pinene has been shown to inhibit the growth of HeLa cells (cervical carcinoma), stall the cells in the G0/G1 phase, and induce the cells to apoptosis[180].

**Airborne Anti-tumor Agent:* In a study involving the implantation of tumors in mice, the animals were subjected to a fragrant environment containing alpha-pinene for 5 hours per day for 4 weeks prior to implantation and 3 weeks afterward, which reduced the size of the tumors by around 40%[181]. Researchers theorized that a fragrant environment containing compounds such as alpha-pinene; "may activate the hypothalamus/sympathetic nerve/leptin axis and immune system, thereby retarding tumor growth."

**Prevention of Hepatitis E:* As a primary constituent in the essential oil of Sakhalin spruce, alpha-pinene contributed to a strong reduction in immunoreactivity of hepatitis E virus antigen proteins, with researchers

---

[174] Weiqiang Chen, Ying Liu, Ming Li, Jianwen Mao, Lirong Zhang, Rongbo Huang, Xiaobao Jin, Lianbao Ye. Anti-tumor effect of a-pinene on human hepatoma cell lines through inducing G2/M cell cycle arrest. Journal of Pharmacological Sciences Volume 127, Issue 3, March 2015, Pages 332-338.

[175] Karthikeyan R, Kanimozhi G, Prasad NR, Agilan B, Ganesan M, Srithar G. Alpha pinene modulates UVA- induced oxidative stress, DNA damage and apoptosis in human skin epidermal keratinocytes. Life Sci. 2018 Nov 1; 212:150-158.

[176] Bouzenna H, Hfaiedh N, Giroux-Metges MA, Elfeki A, Talarmin H. Potential protective effects of alpha-pinene against cytotoxicity caused by aspirin in the IEC-6 cells. Biomed Pharmacother. 2017 Sep; 93:961-968.

[177] Zhao Y, Chen R, Wang Y, Yang Y. a-Pinene Inhibits Human Prostate Cancer Growth in a Mouse Xenogra Model. Chemotherapy. 2018;63(1):1-7.

[178] Kusuhara M, Urakami K, Masuda Y, Zangiacomi V, Ishii H, Tai S, Maruyama K, Yamaguchi K. Fragrant environment with a-pinene decreases tumor growth in mice. Biomed Res. 2012 Feb;33(1):57-61.

[179] Zhang Z, Guo S, Liu X, Gao X. Synergistic antitumor effect of a-pinene and B-pinene with paclitaxel against non-small-cell lung carcinoma (NSCLC). Drug Res (Stuttg). 2015 Apr;65(4):214-8.

[180] Huang X, Yan P, Ding W, Zhou C, Xu Q, Li M, Ye L, Chen W. α-Pinene inhibits the growth of cervical cancer cells through its proapoptotic activity by regulating the miR-34a-5p/Bcl-2 signaling axis. Drug Dev Res. 2022 Dec;83(8):1766-1776.

[181] Kusuhara M, Maruyama K, Ishii H, Masuda Y, Sakurai K, Tamai E, Urakami K. A Fragrant Environment Containing α-Pinene Suppresses Tumor Growth in Mice by Modulating the Hypothalamus/Sympathetic Nerve/Leptin Axis and Immune System. Integr Cancer Ther. 2019 Jan-Dec;18:1534735419845139.

proposing that the compound could be used in a blend with other terpenes to develop a virus inactivating agent for meat processing and cooking utensils, such as knives and chopping boards[182].

**Anti-Seizure Agent:* In male Wistar rats induced to temporal lobe epilepsy by kainic acid, pretreatment with the isolated compound alpha-pinene alleviated epileptic seizures, diminished oxidative stress indicators, blocked the mitochondrial apoptotic pathway, and decreased neuronal death in the CA3 and hilus regions[183] (rat hippocampus).

**Treatment for Huntington's Disease:* In an animal model of Huntington's disease where Wistar rats were subjected to 3-Nitropropionic acid (a mycotoxin that acts as a potent mitochondrial inhibitor), alpha-pinene prevented the motor dysfunction induced by the toxin, significantly counteracting the induced fall in antioxidant enzymes, changes in body weight[184], and ameliorating other disease markers.

**Treatment of Neurodegenerative Disease:* Alpha-pinene has been shown to strengthen the antioxidant system and prevent neuroinflammation in the hippocampus, improve spatial learning and memory, and reduce anxiety in mice treated with beta-amyloid (Aβ)1-42[185] (a major component of amyloid plaques that accumulate in neurons of the brains of Alzheimer's disease patients).

**Antifungal Agent:* As the primary constituent in the essential oil of Rosmarinus officinalis L (rosemary), alpha-pinene contributed to the inducement of a "generalized cell wall stress response, causing damage to cellular and organelle membranes, along with elevated chitin production and increased cell surface adhesion and elasticity, leading to complete vacuolar segregation, mitochondrial depolarization, elevated reactive oxygen species, microtubule dysfunction, and cell cycle arrest mainly at the G1/S phase, consequently triggering cell death[186]" against Candida albicans, a fungi that often cause infections in humans. Other work has found that alpha-pinene's antifungal activity involves binding to ergosterol (a sterol in fungi that maintains cell membrane integrity) in the cellular membrane, rupturing 88% of biofilm while remaining non-cytotoxic to human keratinocytes[187] (an epidermal cell).

**Treatment for Meth Addition:* In a murine model of methamphetamine abuse and addiction, pretreatment with alpha-pinene significantly attenuated methamphetamine-induced behavior and appeared to afford restorative effects on the meth-induced dopaminergic toxicity[188], among other effects.

**Treatment for Adenovirus Type 3:* Adenovirus Type 3 causes severe upper respiratory tract infections with significant morbidity in humans. As an isolated compound, beta-pinene significantly inhibited proliferation of the

---

[182] Maeda N, Horochi S, Hasegawa Y, Iwasaki T, Nakatani N, Miyasho T, Hagiwara K, Yokota H, Funatsu Y. Decreased Immunoreactivity of Hepatitis E Virus Antigen Following Treatment with Sakhalin Spruce (Picea glehnii) Essential Oil. Chem Biodivers. 2023 Apr;20(4):e202200924.

[183] Hashemi P, Ahmadi S. Alpha-Pinene Exerts Antiseizure Effects by Preventing Oxidative Stress and Apoptosis in the Hippocampus in a Rat Model of Temporal Lobe Epilepsy Induced by Kainate. Mol Neurobiol. 2023 Jun;60(6):3227-3238.

[184] Rahmani H, Moloudi MR, Hashemi P, Hassanzadeh K, Izadpanah E. Alpha-Pinene Alleviates Motor Activity in Animal Model of Huntington's Disease via Enhancing Antioxidant Capacity. Neurochem Res. 2023 Jun;48(6):1775-1782.

[185] Khan-Mohammadi-Khorrami MK, Asle-Rousta M, Rahnema M, Amini R. Neuroprotective effect of alpha-pinene is mediated by suppression of the TNF-α/NF-κB pathway in Alzheimer's disease rat model. J Biochem Mol Toxicol. 2022 May;36(5):e23006.

[186] Shahina Z, Al Homsi R, Price JDW, Whiteway M, Sultana T, Dahms TES. Rosemary essential oil and its components 1,8-cineole and α-pinene induce ROS-dependent lethality and ROS-independent virulence inhibition in Candida albicans. PLoS One. 2022 Nov 16;17(11):e0277097.

[187] Bomfim de Barros D, de Oliveira E Lima L, Alves da Silva L, Cavalcante Fonseca M, Ferreira RC, Diniz Neto H, da Nóbrega Alves D, da Silva Rocha WP, Scotti L, de Oliveira Lima E, Vieira Sobral M, Cançado Castellano LR, Moura-Mendes J, Queiroga Sarmento Guerra F, da Silva MV. α-Pinene: Docking Study, Cytotoxicity, Mechanism of Action, and Anti-Biofilm Effect against Candida albicans. Antibiotics (Basel). 2023 Feb 28;12(3):480.

[188] Lee C, Jang JH, Park GH. α-Pinene Attenuates Methamphetamine-Induced Conditioned Place Preference in C57BL/6 Mice. Biomol Ther (Seoul). 2023 Jul 1;31(4):411-416.

virus, impeded cell invasion, and improved cell survival, inhibiting the virus more potently than the reference antiviral medication, ribavirin[189].

Treatment for Depression: Intrigued by the traditional medicinal use of essential oils of Litsea glaucescens (Mexican bay leaf) in Mexico, researchers evaluated the properties of the oil on the central nervous systems of mice. The EO showed antidepressant-like activity as well as sedative activity, with beta-pinene as the primary constituent[190].

IMPLICATIONS FOR HUMAN HEALTH & NUTRITION

The easiest method of obtaining pinene is via a healthy diet and time spent outdoors near plants. Consider regularly eating fruits like apples, grapefruit, guava, lemon, lime, mango, blackberries, blueberries, oranges, pomelos, figs, strawberries, tangerines, or plums. Vegetables like bell peppers, carrots, celery, and tomatoes are rich in pinene and should be included as part of a regular diet. When cooking or preparing foods, consider working with spices like allspice, anise, black pepper, bay, ginger, caraway, nutmeg, oregano, parsley, rosemary, saffron, cilantro, coriander, dill, sage, tarragon, or thyme, which all have high concentrations of pinene.

Consider consuming cannabis strains that are rich in pinene such as ACDC, Banana OG, Blue Dream, Bubba Kush, Cannatonic, Chem D, DJ Short Blueberry, Dutch Treat, Girl Scout Cookies, Gorilla Glue #4, Harlequin, Haze Berry, Island Sweet Skunk, Jack Herer, Obama Kush, OG Kush, Purps, Romulan, Royal Jack, Skywalker OG, Strawberry Cough, or Trainwreck, for another way of obtaining biologically available pinene. Smoking or combustion of cannabis is not recommended, but oils and tinctures are good alternative options for consumption of this bicyclic monoterpene.

Use of or experimentation with essential oils containing pinene as the primary constituent should be undertaken with care considering the potentially toxic nature of the purified compound in high doses.

---

[189] Qiu, B., Wei, F., Su, J. et al. The Effects of β-Pinene, a Pine Needle Oil Monoterpene, on Adenovirus Type 3. Bull Exp Biol Med 172, 345–351 (2022).

[190] Guzmán-Gutiérrez SL, Gómez-Cansino R, García-Zebadúa JC, Jiménez-Pérez NC, Reyes-Chilpa R. Antidepressant activity of Litsea glaucescens essential oil: identification of β-pinene and linalool as active principles. J Ethnopharmacol. 2012 Sep 28;143(2):673-9.

Pinene Review

Answer the following questions to test your knowledge of this terpene:

Question #1: What type of terpene is pinene?

a. Acyclic Sesquiterpene
b. Monocyclic Sesquiterpene
c. Bicyclic Monoterpene
d. Linear Monoterpene

Question #2: Which of the following best describes the primary flavor & aroma profile of pinene?

a. Hops, Cloves, Wood, Spicy
b. Floral, Woody, Green, Citrus
c. Spicy, Turpentine, Fuel, Pine
d. Depends on the isomer
e. Turpentine, Pine, Woody, Balsam

Question #3: How common is pinene in cannabis?

a. Top 3
b. Top 5
c. Top 10
d. Top 20

Question #4: What is the chemical formula for pinene?

a. $C_{15}H_{26}$
b. $C_{15}H_{24}$
c. $C_{10}H_{16}$
d. $N_{15}H_{26}$

Question #5: Name two uses of pinene in plants:

1 ______________________________ 2 ______________________________

Question #6: Name two potential medical uses of pinene:

1 ______________________________ 2 ______________________________

For the answer key to all individual compound chapters please go here: http://thebigbookofterps.com/bbot-quiz-answer-key/

BETA-CARYOPHYLLENE

Type: Bicyclic sesquiterpene
Chemical Formula: $C_{15}H_{24}$
Molecular Weight: 204.357 g/mol
Boiling Point: 264 to 266 °F @ 14mmHg[191]
Flash Point: 200.00 °F. TCC (93.33 °C.)[192]
Melting Point: <25 °C (77 °F)
Solubility: Insoluble in water, soluble in ethanol, ether, alcohol, propylene glycol, various oils
Oral LD50: 5,000 mg/kg (mouse)[193]
Biological Role: Antibacterial agent, insect repellant
Therapeutic Role: Anticancer agent, addiction treatment, anti-inflammatory agent
Commercial Use: Fragrance, flavoring agent
Occurrence in Cannabis: Top three
Aroma/Flavor: Spicy, cloves, dry

Occurs in Cannabis Varieties: Beta-caryophyllene occurs in most cannabis strains, but is particularly prominent in varieties like 303, ACDC, Afgoo, Alien Cookies, Banana Kush, Banana Lemonade, Banana OG, Biker OG, Black Mamba, Blueberry Limeade, Blue Cheese, Blue Dream, Bubba Kush, Cannatonic, Cheese, Chemdawg, Chernobyl, Critical Jack, Dark Star, Diablo, Deadhead OG, Deathstar, Gelato, GG#4, Girl Scout Cookies, Gorilla Glue, Green Crack, Gumbo, Harlequin, Jack Herer, Killer Tahoe OG, Kosher Tangie, OG Kush, Miracle Alien Cookies, Purps, Skywalker OG, Sour Diesel, Sour Dream, Star Dawg, Super Sour Diesel, Tosh OG, and Trainwreck

INTRODUCTION

Beta-caryophyllene is one of the most common terpenes found in cannabis, and is prized by many for its spicy, dry, clove-like taste and aroma. Of all the terpenes in cannabis, beta-caryophyllene is among the most fascinating. That is because beta-caryophyllene is considered a dietary cannabinoid. While widely known cannabinoids like THC and CBN bind to the human body's psychoactive CB1 endocannabinoid receptors, beta-caryophyllene selectively binds to the non-psychoactive CB2 receptors.

However, despite the fact that beta-caryophyllene has been classified Generally Recognized as Safe, and appears with U.S. FDA approval in various foods, other non- psychoactive cannabinoids like CBD have been outlawed or severely restricted in the United States as recently as early 2019. This has also occurred in Europe, where CBD has been banned from food products and other consumables.

In a 2019 air emission field sampling study conducted by the Colorado Department of Public Health and Environment Air Pollution Control Division, beta-caryophyllene was found to be the primary VOC constituent in the exhaust of 3 large, high-production commercial cannabis cultivation facilities[194] in the state.

---

[191] National Toxicology Program, Institute of Environmental Health Sciences, National Institutes of Health (NTP). 1992. National Toxicology Program Chemical Repository Database. Research Triangle Park, North Carolina.

[192] The Good Scents Company. Beta-caryophyllene, from: http://www.thegoodscentscompany.com/data/rw1060851.html Accessed July 22, 2023.

[193] Sigma-Aldrich Safety Data Sheet for Beta-Caryophyllene, Version 6.3, 7/22/2023, from: https://www.sigmaaldrich.com/US/en/sds/ALDRICH/W225207. Accessed July 23, 2023.

[194] Urso K, Frazier A, Heald S, Khlystov A. Terpene exhaust emissions and impact ozone modeling from cannabis plants at commercial indoor cultivation facilities in Colorado. J Air Waste Manag Assoc. 2022 Aug;72(8):828-848.

As a dietary cannabinoid, this compound should be of particular interest to scientists, geneticists, growers, and others working in the cannabis industry, as beta-caryophyllene may be able to - with research and development - safely replace some of the same formulations and functions as the recent spate of CBD products that flooded the worldwide cannabis markets over the last 5-8 years.

Interestingly, famed Nobel Prize winning chemist Elias J. Corey made headlines in the scientific community in the 1960's, thanks in large part to his team's first total syntheses of several compounds, one of which was caryophyllene[195]. This total synthesis of caryophyllene was considered a brilliant example of what could be accomplished in organic chemistry, an emerging field at the time.

CHEMICAL STRUCTURE

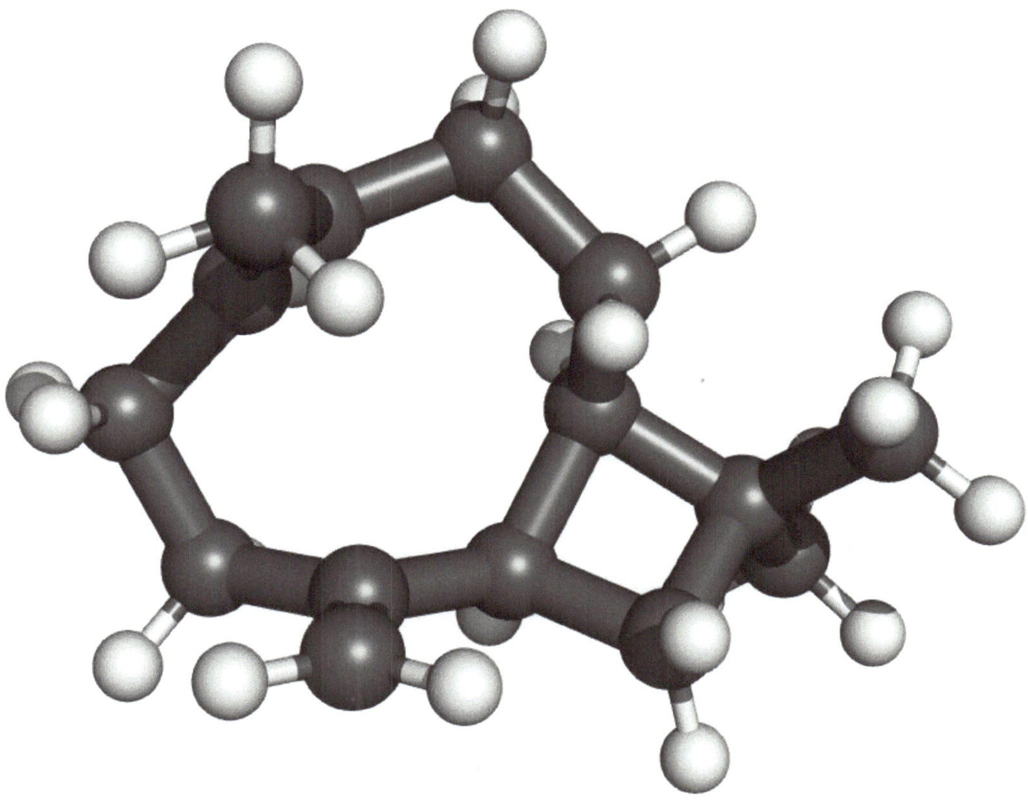

Beta-caryophyllene is formed by three interlinked units of isoprene, making it a sesquiterpene, with an isoprenoid skeleton containing fifteen carbon atoms and twenty- four hydrogen atoms, notated as $C_{15}H_{24}$. This compound has two rings which further classify it as a bicyclic sesquiterpene; one of these rings is comprised of cyclobutene, unlike many of the other terpenes that are common to cannabis. While unique, this cyclobutene ring is unstable because of the angle and torsion of the ring strain, despite the fact that, generally speaking, sesquiterpenes are less volatile than monoterpenes. Beta-caryophyllene is also unique for its second ring, which is a 9-membered ring with a trans-double bond (two carbon groups on opposite sides of the double bond). In total, beta-caryophyllene features two double bonds; one endocyclic in the 9-membered ring, and one exocyclic.

[195] E. J. Corey, Rajat B. Mitra, and Hisashi. Uda. Total Synthesis of d,l-Caryophyllene and d,l-Isocaryophyllene. J. Am. Chem. Soc. 1964, 86, 3, 485-492 Publication Date: February 1, 1964.

This bicyclic sesquiterpene is often found naturally with its cis double bond isomer, isocaryophyllene, where the carbon groups are on the same side as the double bond, and with alpha-humulene (alpha-caryophyllene), which is also common in many cannabis strains.

OCCURRENCE IN PLANTS

While not quite as prevalent in nature as myrcene or pinene, beta-caryophyllene is common in the plant world, and especially common in cannabis, generally occurring in the top three terpenes of many strains.

Abies holophylla	Allspice	Angelica
Anise	Apple	Argulua
Basil	Bay	Bell pepper
Bergamot	Bilberry	Black currant
Black elder	Black pepper	Blueberry
Boronia	Burdock	Cabocla
Cajuput	Calamus	Cananga
Cannabis	Caraway	Cardamom
Carrot	Cassia	Cayenne
Celery	Champaca	Cilantro
Cinnamon	Citronella	Clove
Coleus barbatus	Commophoara gil.	Coriander
Corn	Cubeb	Cumin
Curry	Dill	Elyonurus muticus
Erigeron graveolens	Eucalyptus	Featherfew
Fennel	Fig	Fir
Frankincense	Geranium	Ginger
Grapefruit	Guaiacwood	Guava
Helichrysum	Hinoki	Hops
Horse chestnut	Hyssop	Irish lace
Jambu	Juniper	Labdanum
Laurel	Lavender	Layana
Lemon	Lime	Mace
Mandarin	Manuka	Marigold
Marjoram	Mastic	Melissa
Micromeria fru.	Mint	Myrrh
Neroli	Niaouli	Nutmeg
Oat	Orange	Oregano
Orthosiphon sta.	Palmarosa	Parsley
Patchouli	Pennyroyal	Pennywort
Perilla	Petitgrain	Pimento
Pine	Pokeweed	Pomelo
Prangos	Ptychopetalum ola.	Raspberry
Rose	Rosemary	Rue
Safflower	Sage	Salvia
Savory	Sideritis	Spanish needles
Spearmint	Tagete	Tamarind
Tangerine	Tansy	Taxodium
Tea tree	Thyme	Tomato
Turmeric	Valerian	Vassoura
Verbena	Vitex agnus-castus	Witch hazel
Wormwood	Yarrow	Ylang ylang
Yuzu	Zanthoxylum species	

BIOLOGICAL ACTIVITY IN PLANTS

**Indirect Defense:* Beta-caryophyllene levels in tobacco and cotton plants were found to increase by as much as five times in response to herbivory (feeding) by H. virescens and H. zea (two types of worms), chemically attracting the attention of a species of wasp, C. nigriceps[196], which then parasitized the worms. Interestingly, the emission of beta-caryophyllene can also work to a plant's detriment, in one case attracting an insect, C. picta, which acts as a vector for bacteria that parasitizes plant tissue[197], and in another case, attracting the hungry Egyptian cotton leaf worm[198].

**Antibacterial:* Beta-caryophyllene and caryophyllene oxide have been shown to exhibit significant antibacterial activity against a wide spectrum of gram-positive bacteria[199], with at least one study indicating that the antibacterial activity of beta-caryophyllene against both gram-positive and gram-negative bacteria was stronger in essential oils collected in the summer[200], than at other times in the season - an important consideration for cannabis growers seeking to maximize the amount of this compound in their crops.

**Insect Repellant*: Lima beans respond to attack by spider mites by substantially increasing the amount of a terpene blend where the main constituent is beta-caryophyllene, releasing this through leaf emissions[201]. This is valuable information to indoor and controlled-environment cannabis growers, who often fight losing battles with spider, russet, and other mites. Cannabis strains with higher concentrations of beta- caryophyllene, or insecticides featuring this compound, may be effective methods of treating mites in cannabis sativa. Other research has shown that beta-caryophyllene significantly deterred harvest activities of two genera of fungus-harvesting ants, atta cephalotes and acromyrmex octospinosus[202]. The cultivation of fungi by these ants is often detrimental to host plants.

**Insecticide:* Beta-caryophyllene has been shown to exhibit both fumigant and direct contact toxicity in weevils and red flour beetles[203], while other variations of caryophyllene have been observed to exhibit insecticidal effects against the Angoumois grain moth[204].

---

[196] De Moraes, C. M., W. J. Lewis, P. W. Pare, H. T. Alborn, and J. H. Tumlinson. Herbivore-infested plants selectively attract parasitoids. Nature 393.6685 (1998): 570-573.

[197] Mayer, Christoph, Andreas Vilcinskas, and Jurgen Gross. Pathogen-induced Release of Plant Allomone Manipulates Vector Insect Behavior. Journal of Chemical Ecology 34.12 (2008): 1518-1522.

[198] Anderson, P., B. S. Hansson, and J. Lofqvist. Plant-odour-specific receptor neurones on the antennae of female and male Spodoptera littoralis. Physiological Entomology 20.3 (1995): 189-198.

[199] Xiong, Liang, Cheng Peng, Qin-Mei Zhou, Feng Wan, Xiao-Fang Xie, Li Guo, Xiao-Hong Li, Cheng-Jun He, and Ou Dai. Chemical Composition and Antibacterial Activity of Essential Oils from Different Parts of Leonurus japonicus Houtt. Molecules 18.1 (2013): 963-973.

[200] de Alencar Filho, Jose M. T., Luciana C. da Araujo, Ana P. Oliveira, Amanda L. Guimaraes, Alessandra G. M. Pacheco, Fabricio S. Silva, Leonardo S. Cavalcanti, Angelica M. Lucchese, Jackson R. G. S. da Almeida, and Edigenia C. C. da Araujo. Chemical composition and antibacterial activity of essential oil from leaves of Croton heliotropiifolius in different seasons of the year. Revista Brasileira de Farmacognosia 27.4 (2017): 440-444.

[201] Feng-qi LI, Ning-ning FU, Jing-jiang Zhou, Gui-rong Wang. Functional characterization of (E)-Beta-caryophyllene synthase from lima bean and its up-regulation by spider mites and alamethicin, Journal of Integrative Agriculture, Volume 16, Issue 10, 2017, Pages 2231-2238.

[202] Howard, J.J., Green, T.P. & Wiemer, D.F. Comparative deterrency of two terpenoids to two genera of attine ants. J Chem Ecol 15, 2279-2288 (1989).

[203] Liu, Peng, Xin-Chao Liu, Hui-Wen Dong, Zhi-Long Liu, Shu-Shan Du, and Zhi-Wei Deng. Chemical Composition and Insecticidal Activity of the Essential Oil of Illicium pachyphyllum Fruits against Two Grain Storage Insects. Molecules 17.12 (2012): 14870-14881.

[204] Adjalian, Elvis, Philippe Sessou, Theophile Odjo, Gilles Figueredo, Dansou Kossou, Felicien Avlessi, Chantal Menut, and Dominique Sohounhloue. Chemical Composition and Insecticidal and Repellent Effect of Essential Oi ls of Two Premna Species against Sitotroga cerealella. Journal of Insects 2015.2015 (2015).

**Deterring Ovipositing Insects:* Tobacco plants have been shown to nocturnally release a blend of terpenes, most notably beta-caryophyllene, that deters the tobacco budworm female from laying her eggs on the leaves of the plant[205].

USES IN INDUSTRY

Beta-caryophyllene is often used to add flavor, fragrance, or both to packaged foods, especially baked goods, savory items, chewing gum, condiments, non-alcoholic beverages, hard candy, and frozen dairy products. The compound is also used in cosmetic products to mask odors, and to perfume products directly. Commercially available samples of this bicyclic sesquiterpene generally appear as a light-yellow liquid that smells strongly of cloves, spice, wood, and turpentine.

Thanks to its high flammability yet relative stability, it is possible that one day, we might see Beta-caryophyllene used to manufacture cheap and easily reproducible high-density biofuel. Fascinatingly, scientists have used an engineered strain of bacteria, E. coli, to produce beta-caryophyllene through conversion of acetic acid in the organism[206], with the end goal of bio-fuel mass production.

POTENTIAL USES IN MEDICINE

As mentioned previously, the United States Federal Department of Agriculture has designated beta-caryophyllene Generally Recognized as Safe, and for decades this compound has appeared in food, personal care products, and other commercial items as detailed in the preceding section. Beta-caryophyllene enjoys this status not only because it is non-psychoactive, but also because it exhibits extremely low toxicity. In fact, in one study involving mice, acute dosing for 2 weeks or more with beta-caryophyllene resulted in no deaths, illness, or injuries, nor did it reduce the body weight of the mice, change their eating or drinking habits, or otherwise exhibit any signs of internal damage[207].

Beta-caryophyllene is well-studied and has significant potential for use in medicine and human health. While additional research is ongoing, the current potential medical uses of this sesquiterpene include:

**Addiction Treatment:* Recently there has been a large movement building in the United States and Europe that supports the use of cannabis to treat drug addiction and alcoholism. In the US there is a particular need, where prescription opiate addiction has reached epidemic levels. This movement postulates that cannabis can be used to safely wean addicts from their drugs of choice, based on the concept of harm reduction, and that consuming cannabis is safer than being addicted to heroin, meth, or alcohol. But there is even more to this concept than previously thought, since beta-caryophyllene - a sesquiterpene common to many strains of cannabis - has been shown to decrease voluntary alcohol intake and reduce alcohol-related place preference and sensitivity in mice[208].

**Analgesic Agent:* Already well-known in the cannabis industry for its analgesic properties, beta-caryophyllene has been shown to relieve inflammation and chronic, debilitating pain, especially that involving neuropathic pain, without the development of tolerance, even after prolonged treatment[209]. Because drug tolerance often leads to

---

[205] De Moraes, Consuelo M., Mark C. Mescher, and James H. Tumlinson. Caterpillar-induced nocturnal plant volatiles repel conspecific females. Nature 410.6828 (2001): 577-580.

[206] Jianming Yang, Zhengfeng Li, Lizhong Guo, Juan Du, Hyeun-Jong Bae, Biosynthesis of Beta-caryophyllene, a novel terpene-based high-density biofuel precursor, using engineered Escherichia coli, Renewable Energy, Volume 99, 2016, Pages 216-223.

[207] Oliveira GLDS, Machado KC, Machado KC, da Silva APDSCL, Feitosa CM, de Castro Almeida FR. Non- clinical toxicity of Beta-caryophyllene, a dietary cannabinoid: Absence of adverse effects in female Swiss mice. Regul Toxicol Pharmacol. 2018 Feb; 92:338-346.

[208] Shamma Al Mansouria, Shreesh Ojhab, Elyazia Al Maamaria, Mouza Al America, Syed M Nurulain, Amine Bahia. Pharmacology Biochemistry and Behavior Volume 124, September 2014, Pages 260-268.

[209] Klauke AL, Racz I, Pradier B, Markert A, Zimmer AM, Gertsch J, Zimmer A. The cannabinoid CB2 receptor- selective phytocannabinoid beta-

addiction - especially drugs used in chronic pain management - beta-caryophyllene can likely be used to prevent addiction while still treating pain, and allow addicted patients to transition to a safer substance. In a formalin-induced inflammatory pain model in mice, the combination of cannabidiol (CBD) and beta-caryophyllene produced synergistic analgesic effects[210] related to an unknown anti-inflammatory mechanism.

Chemoprotective & Anti-Cancer Treatments: Convincing studies have shown that beta-caryophyllene might be a valuable tool in the fight against cancer. This double-ringed sesquiterpene exhibits multiple chemoprotective and anti-cancer properties, including inducing apoptosis (natural, programmed cell death) of skin cancer cells[211], strong inhibition of solid tumor growth[212], and inhibition of cellular damage that could lead to malignancies caused by exposure to cigarette smoke[213].

Treatment for Lung Cancer: As a primary constituent in an essential oil made from the leaves of Myrcia splendens (a semi-deciduous tree), beta-caryophyllene contributed to high cytotoxicity against A549 (lung cancer) tumor cells by reducing colony formation and inhibiting the migratory capacity of the cells, while also causing apoptotic morphological changes in the nucleus and cytoplasm[214]. Beta-caryophyllene has also been shown to enhance the anti-tumor activities of cisplatin (a common chemotherapeutic agent) against lung cancer cells[215].

Anti-inflammatory Agent: Because it is likely that many cancers are caused at least in part by prolonged inflammation of cells, beta-caryophyllene might be able to prevent cancer before it starts, thanks to the significant anti-inflammatory properties of this terpene. Beta-caryophyllene has been shown to reduce inflammation by the inhibition of cytokines (proteins secreted by immune system cells) and by increasing the permeability of capillaries to white blood cells (via the histamine pathway), thus allowing them to engage pathogens in the infected tissues[216]. Beta-caryophyllene has also been shown to reduce arthritic inflammation and oxidative damage, without causing hepatoxicity[217] (damage to the liver caused by a chemical or medicine).

Treatment for High Cholesterol & Atherosclerosis: With high cholesterol and heart disease a major risk factor for many adults, natural treatments for these significant medical issues are critically needed. Beta-caryophyllene has high potential as one of these alternatives, having been shown to reduce cardiac lipid content and overall atherogenic and coronary risk indices[218], thanks to the compound's strong antioxidant properties, and by improving the immune response through cellular signaling in the purinergic pathway[219].

---

caryophyllene exerts analgesic effects in mouse models of inflammatory and neuropathic pain. Eur Neuropsychopharmacol. 2014 Apr; 24(4):608-20.

[210] Blanton H, Yin L, Duong J, Benamar K. Cannabidiol and Beta-Caryophyllene in Combination: A Therapeutic Functional Interaction. Int J Mol Sci. 2022 Dec 7;23(24):15470.

[211] Pavithra PS, Mehta A, Verma RS. Synergistic interaction of Beta-caryophyllene with aromadendrene oxide 2 and phytol induces apoptosis on skin epidermoid cancer cells. Phytomedicine. 2018 Aug 1;47:121-134.

[212] Jung, Jae In, Eun Ji Kim, Gyoo Taik Kwon, Yoo Jin Jung, Taesung Park, Yongkang Kim, Rina Yu, Myung-Sook Choi, Hyang Sook Chun, Seung-Hae Kwon, Song Her, Ki Won Lee, and Jung Han Yoon Park.

[213] Di Giacomo S, Abete L, Cocchiola R, Mazzanti G, Eufemi M, Di Sotto A. Caryophyllane sesquiterpenes inhibit DNA-damage by tobacco smoke in bacterial and mammalian cells. Food Chem Toxicol. 2018 Jan;111:393-404.

[214] Montalvão, M.M., Felix, F.B., Propheta dos Santos, E.W. et al. Cytotoxic activity of essential oil from Leaves of Myrcia splendens against A549 Lung Cancer cells. BMC Complement Med Ther 23, 139 (2023).

[215] Ahmed EA, Abu Zahra H, Ammar RB, Mohamed ME, Ibrahim HM. Beta-Caryophyllene Enhances the Anti-Tumor Activity of Cisplatin in Lung Cancer Cell Lines through Regulating Cell Cycle and Apoptosis Signaling Molecules. Molecules. 2022 Nov 30;27(23):8354.

[216] Oliveira-Tintino CDM, Pessoa RT, Fernandes MNM, Alcantara IS, da Silva BAF, de Oliveira MRC, Martins AOBPB, da Silva MDS, Tintino SR, Rodrigues FFG, da Costa JGM, de Lima SG, Kerntopf MR, da Silva TG, de Menezes IRA. Anti-inflammatory and anti-edematogenic action of the Croton campestris A. St.-Hil (Euphorbiaceae) essential oil and the compound Beta-caryophyllene in in vivo models. Phytomedicine. 2018 Mar 1;41:82-95.

[217] Ames-Sibin AP, Barizão CL, Castro-Ghizoni CV, Silva FMS, Sá-Nakanishi AB, Bracht L, Bersani-Amado CA, Marçal-Natali MR, Bracht A, Comar JF. β-Caryophyllene, the major constituent of copaiba oil, reduces systemic inflammation and oxidative stress in arthritic rats. J Cell Biochem. 2018 Dec;119(12):10262-10277.

[218] Baldissera MD, Souza CF, Grando TH, Stefani LM, Monteiro SG. Beta-caryophyllene reduces atherogenic index and coronary risk index in

**Diabetes Treatment:* Beta-caryophyllene shows promise as a treatment for diabetes. This sesquiterpene has been observed to significantly reduce glucose levels while simultaneously increasing insulin levels[220], and is more potent yet has fewer side effects than pioglitazone[221], a common drug used to treat diabetes.

**Treatment for Obesity:* In a diet-induced obesity model in mice, beta-caryophyllene ameliorated metabolic parameters including increased total body weight, fasting glucose levels, oral-glucose tolerance, insulin tolerance, fasting triglycerides, adipocyte hypertrophy, and liver macrovesicular steatosis (fatty liver), while also modulating the levels and expression of several immune response factors[222]. These results improved metabolic and immunological processes in obese mice.

**Treatment for MS (Multiple Sclerosis):* With further research and development, MS patients may soon have a novel natural treatment for their condition. That is because beta-caryophyllene significantly improves the clinical and pathological symptoms of MS by inhibiting microglial cells[223] (in experimental models). These cells normally remove damaged neurons and infections, but in MS they malfunction and attack the nerve linings. Beta-caryophyllene has also been found to greatly reduce inflammatory infiltrates and lessen neurological damage[224] (also in experimental models).

**Treatment of Canine Dental Plaque:* Dental plaque in dogs has been shown to be reduced by the direct administration of a topical solution of beta-caryophyllene, with researchers suggesting that the compound could replace the veterinary industry standard treatment drug, chlorhexidine[225].

**Skin Penetration Enhancer:* As an isolated compound, beta-caryophyllene has been found to exhibit significant penetration enhancement of hydrophilic macromolecules in male Sprague-Dawley rats due to its stratum corneum retention and lipid fluidization ability[226]. This research is useful in the development of transdermal drug delivery products, which are required for certain treatment protocols.

**Treatment for Spinal Cord Injury Chronic Pain:* Chronic neuropathic pain is often associated with spinal cord injury. Researchers seeking new treatments for this debilitating affliction found that co-treatment with cannabidiol

---

hypercholesterolemic rats: The involvement of cardiac oxidative damage. Chem Biol Interact. 2017 May 25;270:9-14.

[219] Baldissera MD, Souza CF, Doleski PH, Leal DBR, Stefani LM, Boligon AA, Monteiro SG. Enzymes that hydrolyze adenine nucleotides in a model of hypercholesterolemia induced by Triton WR-1339: protective effects of Beta-caryophyllene. Mol Cell Biochem. 2017 Oct;434(1-2):127-134.

[220] Basha RH, Sankaranarayanan C. Beta-caryophyllene, a natural sesquiterpene lactone attenuates hyperglycemia mediated oxidative and inflammatory stress in experimental diabetic rats. Chem Biol Interact. 2016 Feb 5;245:50-8.

[221] Youssef DA, El-Fayoumi HM, Mahmoud MF. Beta-caryophyllene protects against diet-induced dyslipidemia and vascular inflammation in rats: Involvement of CB2 and PPAR-y receptors. Chem Biol Interact. 2019 Jan 5; 297:16-24.

[222] Franco-Arroyo NN, Viveros-Paredes JM, Zepeda-Morales ASM, Roldán E, Márquez-Aguirre AL, Zepeda-Nuño JS, Velázquez-Juárez G, Fafutis-Morris M, López-Roa RI. β-Caryophyllene, a Dietary Cannabinoid, Protects Against Metabolic and Immune Dysregulation in a Diet-Induced Obesity Mouse Model. J Med Food. 2022 Oct;25(10):993-1002.

[223] Alberti, Thais Barbosa, Wagner Luiz Ramos Barbosa, Jose Luiz Fernandes Vieira, Nadia Rezende Barbosa Raposo, and Rafael Cypriano Dutra. (-)-Beta-caryophyllene, a CB2 Receptor-Selective Phytocannabinoid, Suppresses Motor Paralysis and Neuroinflammation in a Murine Model of Multiple Sclerosis. International Journal of Molecular Sciences 18.4 (2017).

[224] Livia B.A. Fontes, Debora dos S. Dias, Beatriz J.V. Aarestrup, Fernando M. Aarestrup, Ademar A. Da Silva Filho, Jose Otavio do Amaral Correa. Beta-caryophyllene ameliorates the development of experimental autoimmune encephalomyelitis in C57BL/6 mice. Biomedicine & Pharmacotherapy, Volume 91, 2017, Pages 257-264.

[225] Pieri, F.A., Souza, M.C.d.C., Vermelho, L.L.R. et al. Use of Beta-caryophyllene to combat bacterial dental plaque formation in dogs. BMC Vet Res 12, 216 (2016).

[226] Tang Q, Xu F, Wei X, Gu J, Qiao P, Zhu X, Yin S, Ouyang D, Dong J, Yao J, Wang Y, Chen J. Investigation of β-caryophyllene as terpene penetration enhancer: Role of stratum corneum retention. Eur J Pharm Sci. 2023 Apr 1;183:106401.

(CBD) and beta-caryophyllene reduced tactile and cold hypersensitivity in a rat spinal cord injury model, and produced an enhanced reduction in allodynic responses[227].

Treatment for Ulcerative Colitis: As an isolated compound from the essential oil of Artemisia argyi (silvery wormwood), beta-caryophyllene activated the type 2 cannabinoid receptor in mice, inhibiting macrophage ferroptosis (nonapoptotic cell death), reversing the effects of inflammation, and inhibiting macrophage lipid peroxidation[228].

IMPLICATIONS FOR HUMAN HEALTH & NUTRITION

Normal dietary intake of foods containing beta-caryophyllene is the best and safest method of consumption of this bicyclic sesquiterpene. This can easily be accomplished by adding black pepper to food items, or by eating fruits like apples, mandarins, tangerines, or oranges. Try adding lemon to water and other beverages, or eat berries that are rich in beta-caryophyllene and antioxidants derived from flavonoids such as blueberries and raspberries. For vegetables, consider eating carrots, celery, bell peppers, and spicy cayenne peppers. When preparing, cooking, or serving food, the addition of spices like thyme, parsley, cilantro, cinnamon, sage, or dill may help increase dietary cannabinoid levels of beta-caryophyllene.

Cannabis strains that are high in Beta-caryophyllene like ACDC, Alien Cookies, Banana Kush, Biker OG, Black Mamba, Blue Cheese, Blue Dream, Cannatonic, Cheese, Chemdawg, Dark Star, Deadhead OG, Gelato, GG#4, Girl Scout Cookies, Green Crack, Harlequin, Jack Herer, Killer Tahoe OG, Kosher Tangie, OG Kush, Purps, Skywalker OG, Sour Dream, Star Dawg, Trainwreck, or phenotypes of these or similar strains can be eaten raw or juiced, or smoked or vaporized in water pipes and electronic vaporizers. For medical safety, smoking (combustion) or vaporization of cannabis should be avoided whenever possible. Other options include cannabis extracts that feature intact or added beta-caryophyllene.

Essential oils made from the plants listed in the "Occurrence in Plants" section are likely to be high in beta-caryophyllene. These oils can be used in foods, products, or in aromatherapy via diffusers. As always, caution should be used when working with complex essential oil mixtures, and extreme caution should be used when working with the isolated pure compound of this bicyclic sesquiterpene.

---

[227] Eeswara A, Pacheco-Spiewak A, Jergova S, Sagen J. Combined non-psychoactive Cannabis components cannabidiol and β-caryophyllene reduce chronic pain via CB1 interaction in a rat spinal cord injury model. PLoS One. 2023 Mar 13;18(3):e0282920.

[228] Wu YT, Zhong LS, Huang C, Guo YY, Jin FJ, Hu YZ, Zhao ZB, Ren Z, Wang YF. β-Caryophyllene Acts as a Ferroptosis Inhibitor to Ameliorate Experimental Colitis. Int J Mol Sci. 2022 Dec 16;23(24):16055.

Beta-caryophyllene Review

Answer the following questions to test your knowledge of this terpene:

Question #1: What type of terpene is beta-caryophyllene?

a. Monocyclic monoterpene
b. Linear sesquiterpene
c. Bicyclic sesquiterpene
d. Acyclic diterpene

Question #2: Which of the following best describes the primary flavor & aroma profile of beta-caryophyllene?

a. Citrus, fruity, dry, terpenic
b. Piney, woody, spicy
c. Lemon, sharp, fuel
d. Spicy, cloves, dry

Question #3: How common is beta-caryophyllene in cannabis?

a. Top ten
b. Top three
c. Top fifteen
d. Less common

Question #4: Why is beta-caryophyllene considered a dietary cannabinoid?

a. Beta-caryophyllene regulates uptake of serotonin and dopamine
b. Beta-caryophyllene selectively binds to CB2 receptors
c. Beta-caryophyllene inhibits enzymatic proteins in nerve synapses
d. None of the above

Question #5: Name two uses of beta-caryophyllene in plants:

1 ______________________________ 2 ______________________________

Question #6: Name two potential medical uses of beta-caryophyllene:

1 ______________________________ 2 ______________________________

For the answer key to all individual compound chapters please go here: http://thebigbookofterps.com/bbot-quiz-answer-key/

BETA-MYRCENE

Type: Acyclic monoterpene
Chemical formula: $C_{10}H_{16}$
Molecular Weight: 136.238 g/mol
Boiling Point: 166.00 to 167.00 °C. @ 760.00 mm Hg[229]
Flash Point: 39.44 °C (103 °F)
Melting Point: -10 °C (50 °F)
Solubility: Insoluble in water, soluble in alcohol, chloroform, ether, ethanol, benzene
Oral LD50: 5 g/kg (rat)[230]
Biological Role: Insect repellant, indirect defense agent, antibacterial agent
Therapeutic Role: Sedative, antibacterial agent, anti-tumor agent
Commercial Use: Fragrance/perfume precursor, flavor enhancer, scent additive
Occurrence in cannabis: Top three
Aroma/Flavor: Balsam, pepper, wood, spice

Occurs in Cannabis Varieties: Myrcene likely occurs in most cannabis strains, lending earthy flavors and sedative properties to ACDC, Alien Cookies, Angel Food Cake, Banana Kush, Blackberry Kush, Black Lime, Black Widow, Blue Dream, Cannatonic, Cheese, Chemdawg, Cherry Diesel, Chunkberry, Covelo Sour Diesel, Deathstar, Desert Unicorn, Desert Snow, DJ Short Blueberry, Durban Poison, Dutchie, Fruit Cake, Gelato, Ghost Train Haze, Goo, Island Sweet Skunk, Jack Herer, Kosher Tangie, Lemon, Lemon Kush, Lemon Skunk, Maui Waui, Outer Space, Platinum Dolato, Purple Cadillac, Purple Punch, SFV OG, Sour Jack, Strawberry Short Cookies, Whitaker Blues, and other varieties of cannabis.

INTRODUCTION

Myrcene is one of the most common terpenes in the plant world, and among the most common in today's cannabis chemovars. In the unrefined state, myrcene is a yellow, oily liquid, though industrial preparations of this terpene are typically colorless. Myrcene is found in high concentrations in classic cannabis strains like ACDC, Chemdawg, Jack Herer, Maui Waui, and many others. This monoterpene lends these and other strains some of their characteristic flavors and aromas, which most describe as balsamic, peppery, woody, and spicy.

In a 2019 air emissions field sampling study conducted by the Colorado Department of Public Health and Environment's Air Pollution Control Division, beta-myrcene was found to be the dominant constituent in the emissions of different areas of 4 commercial cannabis cultivation facilities, with the lowest concentration of overall VOC in the vegetative area, and the highest in the post-harvest areas[231].

If you have ever consumed a beer, chances are great that you ingested myrcene, as this monoterpene is a primary constituent in key ingredients used in these products. Found in significant quantity in hops (along with another terpene often found in cannabis, humulene), myrcene is largely responsible for the sedative properties of many types of beer. In fact, on hop farms in the United States, it is well known that just being around the hop vines - especially in the summer - can make a person sleepy, due to the daytime vaporization of myrcene and other terpenes in the hop plants. This author once lived next to a hops farm in Vermont and can personally attest to the sedative power of the volatile emissions of this plant.

---

[229] The Good Scents Company Information System - Myrcene. From: http://www.thegoodscentscompany.com/data/rw1016531.html. Accessed July 23, 2023.

[230] ThermoFisher Scientific Safety Data Sheet for Myrcene, Stabilized, Revision Number 6, May 2011, from: https://www.fishersci.com/store/msds?partNumber=AC128080050&productDescription=MYRCENE. Accessed July 23, 2023.

[231] Urso K, Vizuete W, Moravec R, Khlystov A, Frazier A, Morrison G. Indoor monoterpene emission rates from commercial cannabis cultivation facilities in Colorado. J Air Waste Manag Assoc. 2023 Apr;73(4):321-332.

However, even if you have never consumed beer, you have probably inhaled myrcene at some point in your life, most likely when walking through forests, parks, or streets lined with coniferous trees like pine, spruce, aspen, or fir. Myrcene is found in high concentrations in these trees in particular, and is responsible for the sharp smell of the trees. Like pinene, myrcene is a primary constituent of the emissions of pine and other evergreen forests, and the effects of this and other monoterpenes has led to the practice of forest bathing, as discussed in the individual chapter for pinene.

Interestingly, the well-known sedative properties of myrcene extend to fish, as evidenced by the work of A. Mirghaed et al, which showed that beta-myrcene successfully anesthetized the fish species oncorhynchus mykiss[232], commonly referred to as the rainbow trout. Another unique use of myrcene is by the mountain pine beetle, which uses the volatile emissions of myrcene from pine trees in the production of its own pheromones[233].

CHEMICAL STRUCTURE

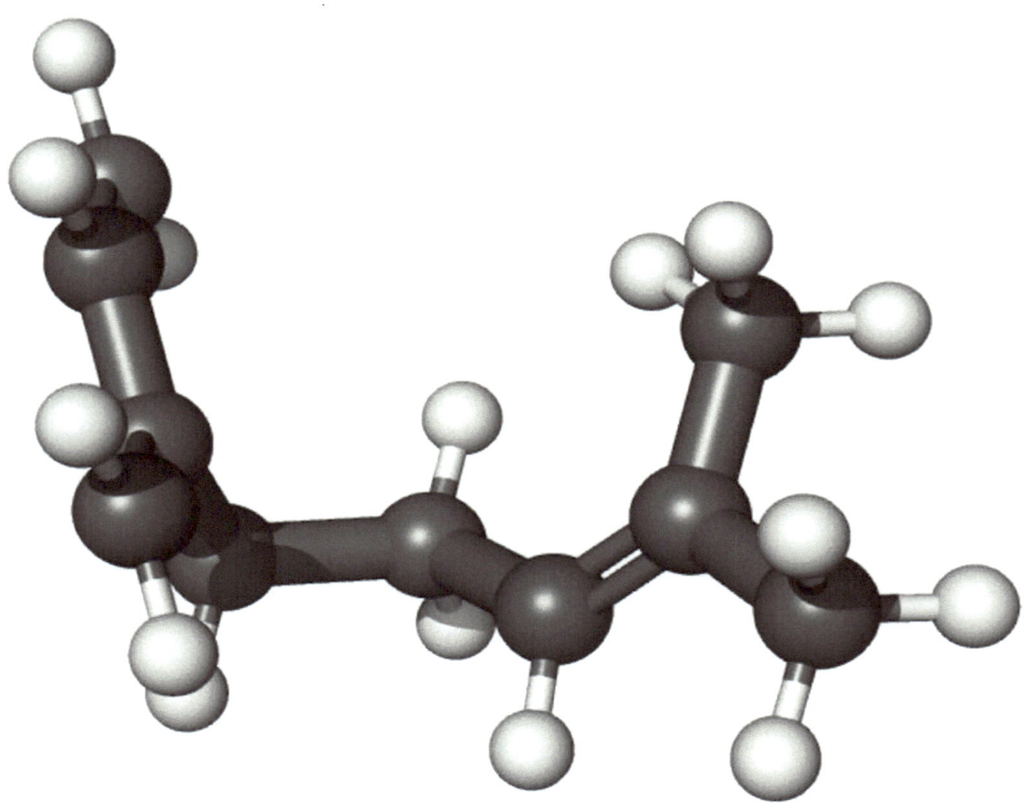

[232] Ali Taheri Mirghaed, Mahyar Yasari, Seyed Saeed Mirzargar, Seyyed Morteza Hoseini. Rainbow trout Oncorhynchus mykiss) anesthesia with myrcene: efficacy and physiological responses in comparison with eugenol. Fish Physiology and Biochemistry June 2018, Volume 44, Issue 3, pp 919-926.

[233] Cale, Jonathan A, Marlena Muskens, Ahmed Najar, Guncha Ishangulyyeva, Altaf Hussain, Sanat S Kanekar, Jennifer G Klutsch, Spencer Taft, and Nadir Erbilgin. Rapid monoterpene induction promotes the susceptibility of a novel host pine to mountain pine beetle colonization but not to beetle-vectored fungi. Tree Physiology 37.12 (2017): 1597-1610.

To be precise, myrcene is actually beta-myrcene. Alpha-myrcene has not been found in nature, though it can be produced in the laboratory. Biosynthesized from geranyl pyrophosphate in the mevalonate pathway, myrcene is a monoterpene based on two units of isoprene, containing ten carbon atoms (two isoprene units, each with five carbon atoms) and sixteen hydrogen atoms, notated as $C_{10}H_{16}$.

This isoprenoid compound contains twenty-five chemical bonds, including three double bonds, and one covalent bond. Because the molecular skeleton of this monoterpene contains no rings, myrcene is classified as an acyclic monoterpene.

OCCURRENCE IN PLANTS

As one of the most widespread monoterpenes, myrcene is found in many plants, although at varying concentrations depending on the species and chemotype. While one could argue that most plants contain at least some myrcene, the following list features plants that have a substantial amount of this isoprenoid compound:

Abies holophylla	Achillea millefolium	Agathosma crenulate
Ajowan	Allspice	Alpinia galanga
Amomum testaceum	Angelica	Anise
Apricot	Artesmisia	Aspen
Baccharis coridifolia	Backhousia cit.	Basil
Bay laurel	Beragmot	Betel
Birch	Black currant	Black elder
Black pepper	Blood orange	Blueberry
Bois de rose	Boldo	Boldus
Buchu	Calamintha nepeta	Calamus
Cajuput	Cananga	Cannabis
Capsicum	Caraway	Cardamom
Carrot	Cascarilla	Cassia
Cassis	Cayenne	Cedar
Celery	Chamchwi	Chamomile
Champaca	Chenopodium amb.	Cherry
Cinnamon	Citronella	Clary sage
Clove	Coriander	Corn
Cornmint	Cotton	Couroupita gui.
Crithmum mar.	Croton	Cubeb
Cumin	Curcuma amada rox.	Curry
Cymbopogon citratus	Cypress	Dill
Echinophora cinerea	Elemi	Eucalyputs
Fennel	Ferula cupularis	Feverfew
Fig	Fir	Fleabane
Frankincense	Galangal	Galbanum
Geranium	Ginger	Glycosmis pen.
Grapefruit	Guava	Helichrysumita.
Hop	Horsemint	Hyacinthus ori.
Hypericum dog.	Hyptis	Hyssop
Jambu	Juniper	Kachur
Kewda	Kiwi	Laurel
Lavandin	Lavandula	Lavender
Layana	Lemon	Lemongrass
Levisticum off.	Lime	Lippia alnifolia sch.
Lovage	Mace	Mandarin
Mango	Manuka	Maple

Marjoram	Mastic	Melissa
Michelia alba	Microstrobos	Mikan
Mugwort	Myristica fragrans	Myrrhinium atr.
Myrtle	Myrcia	Narcissus absolute
Neroli	Nutmeg	Oat
Ocimum gratissimum l.	Orange	Oregano
Palmarosa	Papaya	Parsely
Parsnip	Pennyroyal	Peppermint
Petitgrain	Pimenta	Pimento
Pine	Pittosporum spp.	Plectranhus glandulosus
Pomelo	Poplar	Prangos
Protium heptaphyllum	Pteronia	Rosemary
Rue	Saffron	Sage
Salvia	Santolina	Satureja
Savin	Savory	Sideritis
Skimmia	Snake root	Spearmint
Spruce	Sugandha kokila	Tagete
Tamarind	Tangerine	Tansy
Tarragon	Taxodium distichum	Tea tree
Tejpat	Thyme	Turmeric
Valerian	Vassoura	Verbena
Watermint	Wormseed	Wormwood
West Indian bay tree	Yarrow	Ylang ylang
Yuzu	Xylopia	Zanthoxylum species

BIOLOGICAL ACTIVITY IN PLANTS

Like many terpenes and terpenoids, myrcene serves multiple roles in plants:

**Insect Trap*: A primary constituent of the sap of coniferous trees and other plants including cannabis, myrcene acts as a trap to stop insect attacks and seal bugs out of sensitive or damaged plant tissue. The action of this sap also works to protect the tree against fungal and bacterial infection, acting like a protective wound dressing.

**Insect Repellant:* As the primary constituent of the essential oil of rosemary, myrcene significantly repelled adult E. obliqua (a species of moth), although the EO was stronger than the monoterpene individually[234], suggesting synergy. It should be noted that in the study referenced, rosemary is not a host plant for E. obliqua. As the primary constituent in an essential oil made from spent hops, myrcene contributed to significant repellant activity against R. dominica and S. granaries[235], insects that commonly infect grain stores.

**Allelopathic Agent:* Myrcene has been shown to not only inhibit the growth of rice seedlings by causing oxidative damage and increased ROS production in roots, but the compound also activated the defense responses of the rice plant[236]. As the primary constituent in the essential oil of Artemisia scoparia (a sunflower-like plant), myrcene

[234] Zhang, Zhengqun, Lei Bian, Xiaoling Sun, Zongxiu Luo, Zhaojun Xin, Fengjian Luo, and Zongmao Chen. Electrophysiological and behavioural responses of the tea geometrid Ectropis obliqua (Lepidoptera: Geometridae) to volatiles from a non-host plant, rosemary, Rosmarinus officinalis (Lamiaceae). Pest Management Science 71.1 (2015): 96-104.

[235] Bedini, S., Flamini, G., Girardi, J. et al. Not just for beer: evaluation of spent hops (Humulus lupulus L.) as a source of eco-friendly repellents for insect pests of stored foods. J Pest Sci 88, 583-592 (2015).

[236] Hsiung, Yu-Chywan, Yun-An Chen, Shao-Yen Chen, Wen-Chang Chi, Ruey-Hua Lee, Tzen-Yuh Chiang, and Hao-Jen Huang. Volatilized myrcene inhibits growth and activates defense responses in rice roots. Acta Physiologiae Plantarum 35.8 (2013): 2475-2482.

contributed to a significant reduction in the germination and seedling growth of three weeds; Avena fatua, Cyperus rotundus, and Phalaris minor[237].

**Indirect Defense:* Myrcene is a chemical signal used by wasps to find prey. As part of the volatile emissions of arabidopsis thaliana (thale cress) when under attack by pieris rapae caterpillars, myrcene attracted the parasitic wasp, cotesia rubecula, which used the emissions to locate and prey on the larvae of the caterpillars[238]. In another case, when S. frugiperda (the fall armyworm) feeds on corn, the plant releases volatile emissions where the primary constituents are a-pinene and beta-myrcene, which attract the parasitic wasp, Polybia Fasticiosuscula , to feed on the larvae of the worms[239].

**Antibacterial Agent:* Myrcene has been shown to exhibit significant antibacterial activity against two types of bacteria that are pathogenic to plants: agrobacterium tumefaciens (causes crown gall disease), and erwinia carotovora (affects many agriculturally and scientifically important plants), showing antibacterial activity and inhibitory effects that were stronger than all of the twelve other terpenes tested except thymol, which exhibited similar activity[240].

**Antifungal Agent:* Myrcene caused both type I and type III epigenetic modifications in the genomes of Fusarium graminearum PH-1 and Fusarium culmorum FcUK99 reference standards, causing morphological, genetic, epigenetic, and cellular alterations[241]. This should be of particular interest to cannabis cultivators considering the prevalence of Fusarium infections.

USES IN INDUSTRY

Although myrcene can be obtained from many different plants using various methods, nearly all industrial stocks of this product are produced through the pyrolysis (high-temperature decomposition) of beta-pinene, a terpene that is obtained from industrial-grade turpentine (a chemical precursor and solvent), which is harvested from live coniferous trees, most notably, pine species.

Interestingly, myrcene can also be produced by metabolically engineered Escherichia coli[242] (E. coli) bacteria, joining the list of several other terpenes now being produced by bacteria, including pinene and limonene. Recent work has significantly improved methods for obtaining better yields of myrcene from E. coli by modifying strains that have been engineered to produce geraniol, with researchers achieving de novo biosynthesis of this acyclic monoterpene at 1.25 g/L from glycerol during an 84-hour aerobic-anaerobic two-stage fermentation process[243].

---

[237] Singh, H.P., Kaur, S., Mittal, S. et al. Essential Oil of Artemisia scoparia Inhibits Plant Growth by Generating Reactive Oxygen Species and Causing Oxidative Damage. J Chem Ecol 35, 154-162 (2009).

[238] Van Poecke, Remco, Maarten Posthumus, and Marcel Dicke. Herbivore-Induced Volatile Production by Arabidopsis thaliana Leads to Attraction of the Parasitoid Cotesia rubecula: Chemical, Behavioral, and Gene- Expression Analysis. Journal of Chemical Ecology 27.10 (2004): 1911-1928.

[239] Saraiva, N. B., F. Prezoto, M. das G. Fonseca, M. C. Blassioli-Moraes, M. Borges, R. A. Laumann, and A. M. Auad. The social wasp Polybia fastidiosuscula Saussure (Hymenoptera: Vespidae) uses herbivore-induced maize plant volatiles to locate its prey. Journal of Applied Entomology 141.8 (2017): 620-629.

[240] A Abdel Rasoul, Mona; I Kh, Gehan; Marei Abdelgaleil, Samir. (2012). Evaluation of antibacterial properties and biochemical effects of monoterpenes on plant pathogenic bacteria. African Journal of Microbiology Research Vol. 6(15), pp. 3667-3672, 23 April, 2012.

[241] Albayrak G, Yörük E, Teker T, Sefer Ö. Investigation of antifungal activities of myrcene on Fusarium reference strains. Arch Microbiol. 2023 Feb 4;205(3):82.

[242] Kim EM, Eom JH1 Um Y, Kim Y, Woo HM. Microbial Synthesis of Myrcene by Metabolically Engineered Escherichia coli. J Agric Food Chem. 2015 May 13;63(18):4606-12.

[243] Wang X, Wang J, Zhang X, Zhang J, Zhou Y, Wang F, Li X. Efficient myrcene production using linalool dehydratase isomerase and rational biochemical process in Escherichia coli. J Biotechnol. 2023 Jul 20;371-372:33-40.

Myrcene is regularly used by the perfume and fragrance industries as an intermediate or precursor chemical to produce other more desirable aromatic compounds such as geraniol or linalool. In this industry, recommendations for myrcene levels are just 1-5% in fragrance concentrates, indicating the potent nature of the compound.

Myrcene is used even more widely in the foodservice industries, typically as a flavor enhancer. This monoterpene is used in baked goods, frozen dairy items, fruit ices, hard candy, gum, nonalcoholic beverages, and other food items, although the compound is rarely shown in consumer ingredient lists.

Myrcene is also found in many household products, particularly to add scent to cleaning products, and as a deodorizing agent in some applications. This terpene can also be used to dilute paint, varnish, and dyes, and to remove certain oils. In some cases, myrcene is made into a type of latex.

POTENTIAL USES IN MEDICINE

Myrcene shows excellent potential for use in a variety of medicines and medical treatments, however, it should be noted that some research has indicated that this compound could be genotoxic after beta-myrcene was found to cause significant DNA damage and chromosomal abnormalities including nucleoplasmic bridges and nuclear buds in HepG2/C3A (hepatocellular carcinoma or liver cancer) cells[244]. Researchers posited that the compound could be carcinogenic, however, the substantial body of existing research on this compound has led it to be regarded as generally safe by most scientists.

Anti-ulcer: As a primary constituent in the essential oil of C. aurantium (commonly known as bitter orange), myrcene was found to exhibit significant anti-ulcer activity by greatly decreasing gastric and duodenal (the first part of the small intestine) lesions, and increasing gastric mucous production[245].

Antibacterial: Myrcene successfully eliminated 100% of Staphylococcus aureus and listeria monocytogenes (pathogenic bacteria that causes listeriosis) bacteria by significantly damaging the cell membranes, resulting in complete annihilation of the bacteria after 12 hours of exposure to the essential oil of schinus terebinthifolius raddi[246] (pink pepper), where myrcene was the primary constituent.

Treatment for Breast Cancer: As a top constituent in the essential oil of pinus koraiensis (a species of pine tree), myrcene showed potent inhibition of tumor necrosis factor, and anti-metastatic activity in breast cancer cells[247].

Antioxidant: While studying the antibacterial properties of pinene as extracted from the essential oil of rosemary, scientists observed that myrcene exhibited significant free radical scavenging capacity[248], an important antioxidant function.

---

[244] Orlando JB, Silva BO, Pires-Cunha CL, Hiruma-Lima CA, Gaivão IOM, Maistro EL. Genotoxic effects induced by beta-myrcene following metabolism by liver HepG2/C3A human cells. J Toxicol Environ Health A. 2019;82(3):176-185.

[245] Bonamin, Flavia, Thiago M. Moraes, Raquel C. dos Santos, Helio Kushima, Felipe M. Faria, Marcos A. Silva, Ivan V. Junior, Leonardo Nogueira, Tais M. Bauab, Alba R.M. Souza Brito, Lucia R.M. da Rocha, and Clelia A. Hiruma-Lima. The effect of a minor constituent of essential oil from Citrus aurantium: The role of B-myrcene in preventing peptic ulcer disease. Chemico-Biological Interactions 212 (2014): 11-19.

[246] Dannenberg, Guilherme da Silva, Graciele Daiana Funck, Wladimir Padilha da Silva, and Angela Maria Fiorentini. Essential oil from pink pepper (Schinus terebinthifolius Raddi): Chemical composition, antibacterial activity and mechanism of action. Food Control 95 (2019): 115-120.

[247] Lee, Jeong-Ho, Kicheol Lee, Da Lee, Soon Shin, Yeonjoong Yong, and Young Lee. Anti-invasive effect of B- myrcene, a component of the essential oil from Pinus koraiensis cones, in metastatic MDA-MB-231 human breast cancer cells. Journal of the Korean Society for Applied Biological Chemistry 58.4 (2015): 563-569.

[248] Ojeda-Sana, Adriana M., Catalina M. van Baren, Miguel A. Elechosa, Miguel A. Juarez, and Silvia Moreno. New insights into antibacterial and antioxidant activities of rosemary essential oils and their main components. Food Control 31.1 (2013): 189-195.

Treatment of Ischemic Stroke: In a murine model of stroke, researchers found that myrcene attenuated the negative effects of histological damage in brain tissue caused by global cerebral ischemia, thanks to the monoterpene's strong antioxidant and radical scavenging properties[249].

Sedative: Already a well-known sedative, myrcene presented sedative and motor relaxant effects, and at higher levels produced a potentiation of the pentobarbital-induced sleeping time in mice[250].

Treatment for Osteoarthritis: Myrcene has exhibited significant anti-inflammatory and anti-catabolic (prevents or minimizes the breakdown of muscle tissue) properties, decreasing or stopping cartilage destruction and osteoarthritis progression[251].

Antileishmanial Agent: A troubling disease spread by a species of sandfly, leishmaniasis can be severe. As the primary constituent in the essential oil of Myracrodruon urundeuv (a species of tree), myrcene has been shown to exhibit significant antileishmanial activity by decreasing the rate of parasite infection in blood cells[252], while maintaining low toxicity to host cells.

Treatment for Organophosphate Toxicity: Another interesting medical application for myrcene comes in the form of a treatment for organophosphate or carbamate toxicity - overexposure to or poisoning by commercial insecticides. In a patent filed with the United States patent office, a treatment for organophosphate and carbamate toxicity makes use of cannabinoids and terpenes, most notably, myrcene[253].

Treatment for Colon Inflammation (related to cancer): As an isolated compound, beta-myrcene reduced inflammation in human colorectal adenocarcinoma cells by inhibiting proinflammatory cytokine release, mitogen-activated protein kinases, and Nf-κB (nuclear factor kappa-light-chain-enhancer of activated B cells, a protein complex that controls cytokine production, among other roles) signaling pathways that fuel inflammation[254].

IMPLICATIONS FOR HUMAN HEALTH & NUTRITION

Dietary consumption of myrcene can be achieved by regularly eating fruits like apricots, blood oranges, blueberries, figs, guava, grapefruit, kiwi, lemons, limes, mandarins, mangos, oranges, tangerines, and yuzu (a fruit like a lemon). Vegetables that are high in this acyclic monoterpene include carrots, celery, and corn. When cooking or preparing foods, consider using spices like allspice, anise, basil, black pepper, cayenne, cinnamon, clove, curry, ginger, dill, oregano, parsley, rosemary, saffron, sage, and savory.

---

[249] Ciftci, Osman, M. Oztanir, and Aslh Cetin. Neuroprotective Effects of B-Myrcene Following Global Cerebral Ischemia/Reperfusion-Mediated Oxidative and Neuronal Damage in a C57BL/J6 Mouse. Neurochemical Research 39.9 (2014): 1717-1723.

[250] Gurgel do Vale, T., E. Couto Furtado, J.G. Santos, and G.S.B. Viana. Central effects of citral, myrcene and limonene, constituents of essential oil chemotypes from Lippia alba (Mill.) N.E. Brown. Phytomedicine 9.8 (2002): 709-714.

[251] Rufino, Ana Teresa, Madalena Ribeiro, Catia Sousa, Fernando Judas, Ligia Salgueiro, Carlos Cavaleiro, and Alexandrina Ferreira Mendes. Evaluation of the anti-inflammatory, anti-catabolic and pro-anabolic effects of E-caryophyllene, myrcene and limonene in a cell model of osteoarthritis. European Journal of Pharmacology 7 50 (2015): 141-150.

[252] C.E.S. Carvalho, E.P.C. Sobrinho-Junior, L.M. Brito, L.A.D. Nicolau, T.P. Carvalho, A.K.S. Moura, K.A.F. Rodrigues, S.M.P. Carneiro, D.D.R. Arcanjo, A.M.G.L. Cito, F.A.A. Carvalho. Anti-Leishmania activity of essential oil of Myracrodruon urundeuva (Engl.) Fr. All.: Composition, cytotoxity and possible mechanisms of action. Experimental Parasitology Volume 175, April 2017, Pages 59-67.

[253] Use Of Cannabinoids And Terpenes For Treatment Of Organophosphate And Carbamate Toxicity; US Patent US2015313868. From; https://pubchem.ncbi.nlm.nih.gov/patent/US2015313868.

[254] Almarzooqi S, Venkataraman B, Raj V, Alkuwaiti SAA, Das KM, Collin PD, Adrian TE, Subramanya SB. β-Myrcene Mitigates Colon Inflammation by Inhibiting MAP Kinase and NF-κB Signaling Pathways. Molecules. 2022 Dec 9;27(24):8744.

Teas or essential oils made from the plants listed in the "Occurrences in Plants" section may contain significant amounts of myrcene, however, extreme care should be taken when working with complex EO or isolated compounds, as these can be toxic in some situations.

Cannabis strains that are known to be high in myrcene include ACDC, Angel Food Cake, Blackberry Kush, Black Lime, Cannatonic, Chemdawg, Deathstar, DJ Short Blueberry, Fruit Cake, Island Sweet Skunk, Jack Herer, Kosher Tangie, Lemon Skunk, Maui Waui, SFV OG, Sour Jack, and Whitaker Blues. However, it is not likely that any bioavailable myrcene survives the smoke-stream of cannabis when combusted or vaporized, unless done so at extremely low temperatures. The safest and most efficient way of obtaining this terpene from cannabis is through the use of essential oils, tinctures, or by eating parts of the plant raw, particularly the flowering tops.

Myrcene Review

Answer the following questions to test your knowledge of this terpene:

Question #1: What type of terpene is myrcene?

a. Linear monoterpene
b. Monocyclic sesquiterpene
c. Bicyclic monoterpene
d. Acyclic monoterpene

Question #2: Which of the following best describes the primary flavor & aroma profile of myrcene?

a. Hops, cloves, wood, spicy
b. Balsam, pepper, wood, spice
c. Spicy, turpentine, fuel, pine
d. Turpentine, pine, woody, balsam

Question #3: How common is myrcene in cannabis?

a. Top three
b. Top five
c. Top ten
d. Top twenty

Question #4: What is the chemical formula for myrcene?

a. $C_{15}H_{26}$
b. C15H24
c. $C_{10}H_{16}$
d. $N_{15}H_{26}$

Question #5: Name two uses of myrcene in plants:

1 ______________________________ 2 ______________________________

Question #6: Name two potential medical uses of myrcene:

1 ______________________________ 2 ______________________________

For the answer key to all individual compound chapters please go here: http://thebigbookofterps.com/bbot-quiz-answer-key/

LIMONENE

Type: Monocyclic monoterpene
Chemical Formula: $C_{10}H_{16}$
Molecular Weight: 136.2375 g/mol
Boiling Point: *D-limonene:* 175.00 to 177.00 °C. @ 760.00 mm Hg[255]
L-limonene: 176.00 to 177.00 °C. @ 760.00 mm Hg[256]
Flash Point: *D-limonene:* 117.00 °F. TCC (47.22 °C.)
L-limonene: 127.00 °F. TCC (52.78 °C.)
Melting Point: *D-limonene:* -96.00 °C. @ 760.00 mm Hg
L-limonene: −74.35 °C.
Solubility: *D-limonene:* Soluble in alcohol, fixed oils, kerosene, paraffin oil, insoluble in water, propylene glycol
L-limonene: Soluble in alcohol, fixed oils, insoluble in water
Oral LD50: *D-limonene:* 5,200 mg/kg (rat)[257]
Biological Role: Insecticide, antifeedant, allelopathic agent, chemical signaling agent
Therapeutic Role: Anticancer agent, antifungal agent, UV radiation protection
Commercial Use: Food additive, fragrance agent, solvent
Occurrence in Cannabis: Top five
Aroma/Flavor *D-limonene*: Citrus, orange, fresh, sweet, terpenic
L-limonene: Pine, herbal, peppery, terpenic

Occurs in Cannabis Varieties: Limonene probably occurs in most cannabis varieties at differing concentrations, depending on the chemovar. At this time, most laboratory testing does not distinguish between the optical isomers of this monoterpene, therefore, although the following cannabis varieties are known to contain limonene, the list is general and does not identify the isomers: 1991 OG, Afgoo, Alien Cookies, Angel Food Cake, Banana Kush, Banana OG, Blood Drive, Blue Cheese, Cheese, Chemdawg, Deadhead OG, Desert Snow, Diablo, Durban Poison, Fruit Cake, Ghost Train Haze, Gelato, Gorilla Glue, Killer Tahoe OG, King Tut, Kosher Kush, Miracle Alien Cookies, Miss X, Obama Kush, Outer Space, Pink Lemonade, Platinum Dolato, Purple Punch, SFV OG, Shiatsu Kush, Skunkdawg, Skywalker OG, Sour Dream, Sour Cherry Sherbet, Sour Jack, Star Dawg, Star Pie, Strawberry Short Cookies, Sugar Cookies, Sunset Sherbet, Sunset Sour, Super Sour Diesel, Tosh OG, Trainwreck, Ultimate Purple

INTRODUCTION

Limonene is one of the primary constituents in the essential oil of cannabis, and it is also one of the most widely encountered terpenes in the natural world. Primarily derived from citrus fruit - particularly oranges - limonene is used for a multitude of industrial and agricultural purposes, although the name is derived from the word 'limon', or, lemon. As one of the most studied monoterpenes, there is substantial evidence in the literature of this terpene's medical potential, which is covered below in the "Potential Uses in Medicine" section.

Most strains of cannabis contain limonene. While the majority of cannabis strains exhibit a large percentage of this terpene, some varieties contain limonene in smaller but not insignificant proportions of the essential oil. Strains that are rich in citrus scents or flavors are likely to have a high concentration of limonene, but because there are two isomers of this molecule, cannabis strains that do not have citrusy features may still contain a variation of limonene in high concentration. In these cases, the flavor or aroma profile is likely to be piney instead of citrusy.

---

[255] The Good Scents Company, Data Sheet for D-Limonene, from: http://www.thegoodscentscompany.com/data/rw1013772.html. Accessed July 24, 2023.

[256] The Good Scents Company Data Sheet for L-limonene, from: http://www.thegoodscentscompany.com/data/rw1007831.html. Accessed July 24, 2023.

[257] ThermoFisher Scientific Safety Data Sheet for Limonene. Oct. 2010. From: https://www.fishersci.com/store/msds?partNumber=AC203735000&countryCode=US&language=en. Accessed July 24, 2023.

CHEMICAL STRUCTURE

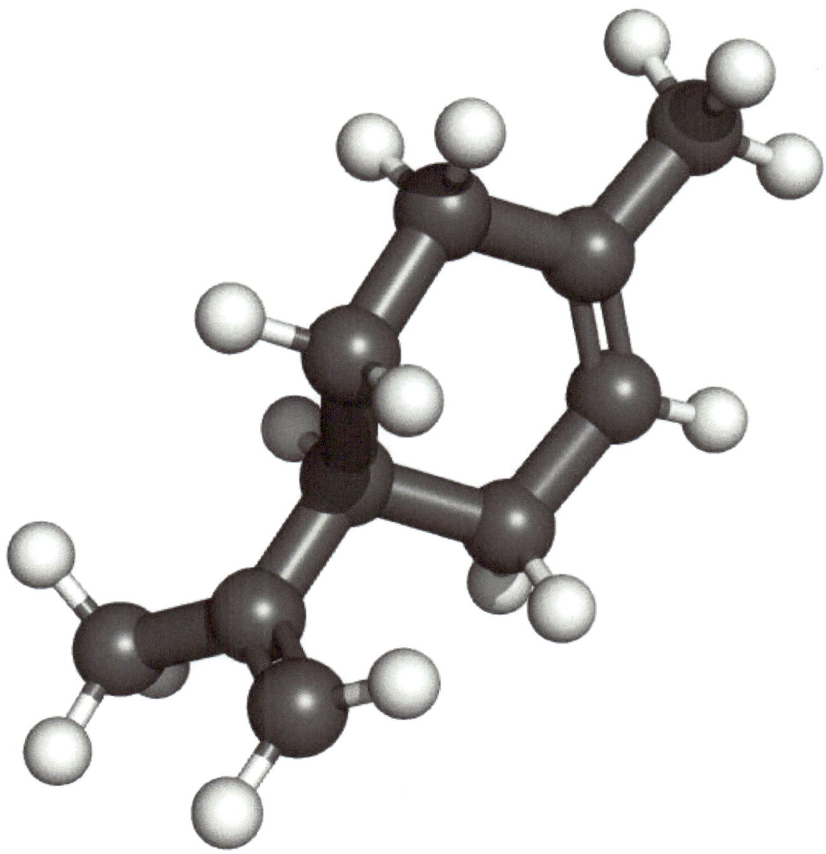

Like all the monoterpenes, limonene contains ten carbon atoms and sixteen hydrogen atoms, scientifically notated as $C_{10}H_{16}$. Because of its hexagonal shape, limonene is categorized as a cyclic monoterpene. Limonene contains two double bonds, one of which is endocyclic (occurring inside of the molecule's ring).

An isoprenoid compound biosynthesized from geranyl pyrophosphate, limonene exists in two isomeric forms: l-limonene, a primary constituent of coniferous tree essential oils, especially pines, and d-limonene, the primary terpene constituent of the essential oils of citrus fruits. As explained in the preceding section, this means that cannabis strains containing high amounts of limonene will likely feature aroma and taste profiles that are dependent on which isomer is present. Additionally, some plants – including cannabis – produce limonene in racemic mixtures.

Limonene is a precursor or intermediate in the biosynthesis of other monoterpenes, most notably 1,8-cineole (eucalyptol), pulegone, and alpha-terpineol[258], all of which commonly occur in many varieties of cannabis. Industrial use of this terpene also includes limonene as a precursor in the production of other terpenes, similar to how both myrcene and pinene are used by the fragrance industries to produce more desirable compounds.

258 Kjonaas R, Croteau R. (1983) Demonstration that limonene is the first cycle intermediate in the biosynthesis of oxygenated p-menthane monoterpenes in Mentha piperita and other Mentha species. Archives of Biochemistry and Biophysics, 22O, 79-89.

OCCURRENCE IN PLANTS

Although limonene is commonly known as the primary constituent in the essential oil of citrus fruits, this terpene is found in a wide variety of plants, and, because of the differences between l-limonene and d-limonene, this terpene can exhibit dramatically different properties. Limonene is found in moderate to high concentration among the following plants:

D-limonene:

Anise	Bay laurel	Bergamot
Black pepper	Caraway	Cardamom
Celery	Dill	Fennel
Grapefruit	Guava	Hyssop
Lemon	Lime	Nutmeg
Orange	Parsley	Peppermint
Rosemary	Tea tree	

L-limonene:

Anise	Black pepper	Cajuput
Caraway	Copal congo	Corn mint
Cotton	Dill	Fennel
Horsemint	Peppermint	Pine
Sage	Spearmint	Tea tree
Water mint	Wormseed	

BIOLOGICAL ACTIVITY IN PLANTS

Insecticidal Agent: Limonene is an effective natural insecticide, particularly against weevils[259], mealybugs, and whiteflies[260] - a common pest of cannabis. In the case of whiteflies, a solution of 1% limonene, used at just half strength, killed 99% of the pests.

Ovicidal/Larvicidal Agent: Ovicidal and larvicidal agents work by inhibiting or outright killing the ova/eggs or larvae of an insect. D-limonene has long been known to inhibit the egg hatching rate of western corn rootworm and exhibits toxicity in the larvae[261]. Corn rootworm species have been known to attack both fiber and drug varieties of cannabis sativa.

Oviposition Deterrent: As emitted from a variety of seeds, volatile d-limonene was shown to change the behavior of seed beetles, in particular repelling egg-laying females[262].

[259] Haag KH. (1986) Effects of herbicide application on mortality and dispersive behavior of the water hyacinth weevils, Neochetina eichhorniae and Neochetina bruchi (Coleoptera: Curculionidae). Environmental Entomology, 15, 1192-1198.

[260] Hollingsworth, Robert G. Limonene, a Citrus Extract, for Control of Mealybugs and Scale Insects. Journal of Economic Entomology 98.3 (2005): 772-779.

[261] L. Karr, Laura & Coats, Joel. (1988). Insecticidal Properties of d-Limonene. Journal of Pesticide Science. 13. 287-290.

[262] Hudaib, Taghread, William Hayes, Sarah Brown, and Paul E. Eady. Effect of seed moisture content and d-limonene on oviposition decisions of the seed beetle Callosobruchus maculatus. Entomologia Experimentalis Et Applicata 137.2 (2010): 120-125.

**Herbivore Repellant / Antifeedant*: All the bitter terpenes, including limonene, likely act as herbivore repellants, especially those that occur in the leaves of plants, including cannabis.

**Nematocidal Agent:* Of thirty-four phytoterpenoids tested, limonene was one of just two monoterpenes (the other was alpha-terpinene) that exhibited toxicity against caenorhabditis elegans (parasitic roundworms), killing more than 50% of nematode populations at a dose of about 80 µg/mL[263]. Based on the study referenced here, higher doses of limonene would likely prove even more fatal.

**Chemical Communication:* As discussed in the chapter "Terpenes 101", terpenes are the most-used communication medium in the world. This includes plant-to-plant communication. Corn plants have been shown to emit volatile limonene when eggs are laid on the plant by the parasitic stemborer Chilo partellus (the spotted stalk borer, a moth). These emissions of limonene induce a defense response in neighboring corn plants, which activate their defenses without first being attacked or infested with eggs, indicating the ability of the plants to send and receive advance warning signals[264]. In another case of plant-to-plant use of limonene as a communication medium, Japanese mustard spinach (Brassica rapa var. perviridis) infected with the pathogenic plant fungus Colletotrichum higginsianum releases gaseous limonene, which activates the defenses of nearby, uninfected conspecific plants[265].

**Allelopathic Agent:* Limonene is a potent allelopathic agent. As the major constituent in the essential oil of shinus molle L. (an evergreen tree native to Peru), limonene inhibited the germination and radicle elongation (a major biological process of seedling growth) of nearby wheat[266] (Triticum aestivum L.). In a direct test of limonene, the monoterpene was significantly phytotoxic to white cabbage and carrot plants, primarily through inhibition of chlorophyll fluorescence[267] (an important process in photosynthesis). Interestingly, limonene as produced and emitted by cyanobacteria (bacteria that derives energy via photosynthesis) acts allelopathically against nearby algae by inhibiting cell growth and killing cells[268].

Interestingly, some plants are being bred to overexpress, under-express, or eliminate the expression of terpenes from their volatile emissions for the purpose of developing disease or pest resistance, among other traits. For instance, a rice plant was bred that featured much less limonene, which made it far less attractive to Nilaparvata lugens (the Brown planthopper), a common pest of rice.

USES IN INDUSTRY

Nearly all the commercially produced limonene on the market today is derived from the byproducts of the orange industry; especially via waste products generated in the production of orange juice. This monoterpene is obtained from the essential oil of discarded orange peels, typically through steam distillation or centrifugal separation.

While limonene is used extensively in the food industries, mainly as an additive or flavoring agent, or as a dietary supplement, the number of commercial and household non-food products that contain limonene is astounding. This

[263] Abdel-Rahman FH, Alaniz NM, Saleh MA. Nematicidal activity of terpenoids. J Environ Sci Health B. 2013;48(1):16-22.

[264] Mutyambai, Daniel M., Toby J. A. Bruce, Johnnie van den Berg, Charles A. O. Midega, John A. Pickett, and Zeyaur R. Khan. An Indirect Defence Trait Mediated through Egg-Induced Maize Volatiles from Neighbouring Plants. PLoS ONE 11.7 (2016).

[265] Fujioka, Kayoko, Haruko Gotoh, Taku Noumi, Ami Yoshida, Yoshiteru Noutoshi, Yoshishige Inagaki, Mikihir Yamamoto, Yuki Ichinose, Tomonori Shiraishi, and Kazuhiro Toyoda. Protection induced by volatile limonene against anthracnose disease in Arabidopsis thaliana. Journal of General Plant Pathology 81.6 (2015): 415-419.

[266] Zahed, Nesrine, Karim Hosni, Nadia Brahim, Monem Kallel, and Houcine Sebei. Allelopathic effect of Schinus molle essential oils on wheat germination. Acta Physiologiae Plantarum 32.6 (2010): 1221-1227.

[267] Ibrahim, MA, EJ Oksanen, and JK Holopainen. Effects of limonene on the growth and physiology of cabbage (Brassica oleracea L) and carrot (Daucus carota L) plants. Journal of the Science of Food and Agriculture 84.11 (2004): 1319-1326.

[268] Zhao, Jingxian, Lin Yang, Lv Zhou, Yan Bai, Bin Wang, Ping Hou, Qinghuan Xu, Wangting Yang, and Zhaojiang Zuo. Inhibitory effects of eucalyptol and limonene on the photosynthetic abilities in Chlorella vulgaris (Chlorophyceae). Phycologia 55.6 (2016): 696-702.

includes detergents, aftershave lotions, adhesive removers, odor control agents, soaps, perfumes, bath products, hand cleansers, hair products, inks, dishwashing liquid, household cleaners, institutional cleaners, electronics cleaners, air fresheners, and other similar personal care products and household or commercial cleaners.

Limonene is known for its ability to dissolve oils and is often used as a solvent in a variety of applications, including in glue and paint solvents, sewage treatment solvents, and as a solvent in the removal of oil from industrial or agricultural equipment.

Examples of common consumer products containing limonene include the Axe brand of hair gels, pastes, and waxes, which typically contain terpenes including limonene, linalool, and eugenol. Then there is the Adams brand of flea and tick shampoo for dogs and cats, which features d-limonene as its primary active ingredient. Finally, the active ingredient in the Avenger brand of organic herbicide is also d-limonene, about which the company wrote; "The active ingredient, d-limonene (citrus oil) naturally strips away the waxy plant cuticle, causing it to dry up and die[269];" which is also another testament, albeit a commercial one, to the potent herbicidal effects of limonene. The pesticidal effects of this monoterpene are echoed almost identically in the literature for an all-natural insecticide called D'Bug, made by the Total Solutions company; "The d'Limonene [sic] instantly strips away the protective wax coating from the exoskeletons of insects, causing them to suffocate and die on contact[270]."

Apart from the extensive medical benefits of limonene as discussed below, perhaps one of the most interesting uses of this monoterpene is as a novel fuel source. One company, called Cyclone, has developed an external combustion engine, in which d-limonene was successfully used as a fuel source[271].

Other interesting work includes the development of a limonene nanoemulsion that researchers developed to protect and prolong the shelf life of bananas. In this case, the nanoemulsion was effective against several types of common bacteria, and reduced water loss and weight loss in the bananas[272].

Limonene generally occurs as a combustible, slippery, colorless liquid in industrial samples.

POTENTIAL USES IN MEDICINE

Although limonene shows promise in medicine, it should be noted that this monocyclic monoterpene can cause allergies in many people. This is especially true for people who have allergies where the cause is unknown or uncertain, as patch tests for limonene are uncommon. In a large study conducted in the United Kingdom and echoed by other European research, scientists found significant rates of contact allergies to both limonene and linalool[273] - common terpenes used heavily in fragrances, cleaners, and household products, as discussed in the "Uses in Industry" section above.

Treatment for Prostate Cancer: d-limonene has enhanced the antitumor effect of Docetaxel (a common chemotherapy agent) against prostate cancer cells without being toxic to normal prostate epithelial cells. This occurs

---

[269] Avenger Organic Weed Killer - From the Frequently Asked Questions web page; http://avengerorganics.com/organic-weed-killer-faqs.html, accessed January 15, 2019.

[270] Valtec Industries - TOTAL SOLUTIONS D-Bug All-Natural D-Limonene Insecticide. From: https://valtecindustries.com/products/dbug-all-natural-d-limonene-insecticide Accessed March 11, 2019.

[271] Cyclone Power to Showcase External Combustion Engine at SAE Event, Green Car Congress, 20 September 2007. From: https://www.greencarcongress.com/2007/09/cyclone-power-t.html. Accessed January 10, 2019.

[272] Hou CY, Hazeena SH, Hsieh SL, Li BH, Chen MH, Wang PY, Zheng BQ, Liang YS. Effect of D-Limonene Nanoemulsion Edible Film on Banana (Musa sapientum Linn.) Post-Harvest Preservation. Molecules. 2022 Sep 20;27(19):6157.

[273] Audrain, H., C. Kenward, C.R. Lovell, C. Green, A.D. Ormerod, J. Sansom, M.M.U. Chowdhury, S.M. Cooper, G.A. Johnston, M. Wilkinson, C. King, N. Stone, H.L. Horne, C.R. Holden, S. Wakelin, and D.A. Buckley. Allergy to oxidized limonene and linalool is frequent in the U.K. British Journal of Dermatology 171.2 (2014): 292-297.

via increased ROS generation and caspase (enzymes that play a role in apoptosis or programmed cell death) cascades resulting in cell death[274].

Antimicrobial Agent: Limonene offers the potential for a powerful weapon against microbial infection. In fact, while preparations of the essential oils of citrus fruits have been shown to inhibit the growth of common and sometimes dangerous strains of bacteria such as Escherichia coli (widely known as E. coli) by up to 58%, a direct application of pure limonene inhibited the growth of the same strain of bacteria by 100%[275]. This is interesting to note considering that essential oils are often more potent than their individual constituents.

UV Radiation Protection: Because of the environmental risks posed by the extensive use of sunscreens, a more natural type of UV protection is needed. D-limonene is a likely option, as it has been shown to protect against UVB-induced DNA photodamage and sunburn in mice[276], and is generally considered non-toxic.

Antifungal Agent: Limonene has been shown to exhibit potent antifungal effects against trichophyton rubrum[277] - a dermatophytic fungus that is the most common cause of athlete's foot, jock itch, and ringworm. In the case of candida albicans, which causes vaginal yeast infections, oral thrush, and other fungal infections of the skin and nails, limonene was shown to inhibit the growth of the fungus by damaging the cell membrane and inducing oxidative stress, resulting in DNA damage[278].

Antifibrotic Agent: Fibrosis is the thickening and scarring of connective tissue, usually caused by injury. Limonene is a strong antifibrotic, acting by boosting antioxidants, regulating collagen content, and inhibiting inflammation[279], thanks to this monoterpene's well-known anti-inflammatory properties.

Breast Cancer Prevention/Treatment: D-limonene exhibits both chemotherapeutic and chemopreventive effects[280], and has been shown to preferentially concentrate in human breast tissue and may lead to cell-cycle arrest and reduced cell proliferation[281]. When combined with the chemotherapeutic agent tamoxifen, d-limonene significantly improved the efficacy of the cancer drug, increasing apoptosis and cell viability in MCF 7 breast cancer cells[282].

---

[274] Rabi, Thangaiyan and Anupam Bishayee. d-Limonene sensitizes docetaxel-induced cytotoxicity in human prostate cancer cells: Generation of reactive oxygen species and induction of apoptosis. Journal of carcinogenesis vol. 8 (2009): 9.

[275] Dabbah R, Edwards VM, Moats WA. (1970) Antimicrobial action of some citrus fruit oils on selected food-borne bacteria. Applied Microbiology, 19, 27-31.

[276] Ahmed N Uddin, Feng Wu, Ivica Labuda, Kam-Meng Tchou-Wong, Fredric J Burns. d-limonene prevents ultraviolet irradiation: Induced cyclobutane pyrimidine dimers in Skh1 mouse skin. World J Dermatol. Aug 2, 2014; 3(3): 64-72.

[277] Chee, Hee Youn, Hoon Kim, and Min Hee Lee. In vitro Antifungal Activity of Limonene against Trichophyton rubrum. Mycobiology 37.3 (2009): 243-246.

[278] Thakre, Archana, Gajanan Zore, Santosh Kodgire, Rubina Kazi, Shradha Mulange, Rajendra Patil, Amruta Shelar, Bayitigeri Santhakumari, Mahesh Kulkarni, Kiran Kharat, and Sankunny Mohan Karuppayil. Limonene inhibits Candida albicans growth by inducing apoptosis. Medical Mycology Advance Article.5 (2017): 1-578.

[279] Ahmad, Sheikh Bilal, Muneeb U. Rehman, Bilques Fatima, Bilal Ahmad, Ishraq Hussain, Sheikh Pervaiz Ahmad, Adil Farooq, Showkeen Muzamil, Rahil Razzaq, Shahzada Mudasir Rashid, Showkat Ahmad Bhat, and Manzoor Ur Rahman Mir. Antifibrotic effects of D limonene (5(1 methyl 4 1 methylethenyl) cyclohexane) in CCl 4 induced liver toxicity in Wistar rats. Environmental Toxicology 33.3 (2018): 361-369.

[280] Miller, Jessica, Patricia Thompson, Iman Hakim, H.-H. Chow, and Cynthia Thomson. d-Limonene: a bioactive food component from citrus and evidence for a potential role in breast cancer prevention and treatment. Oncology Reviews 5.1 (2010): 31-42.

[281] Miller JA, Lang JE, Ley M, Nagle R, Hsu CH, Thompson PA, Cordova C, Waer A, Chow HH.. Human breast tissue disposition and bioactivity of limonene in women with early-stage breast cancer. Cancer prevention research (Philadelphia, Pa.) vol. 6,6 (2013): 577-84.

[282] Mandal D, Sahu BR, Parija T. Combination of tamoxifen and D-limonene enhances therapeutic efficacy in breast cancer cells. Med Oncol. 2023 Jun 30;40(8):216.

**Treatment for Herpes:* In an in vitro study of the Herpes Simplex Virus type 1 (HSV- 1), limonene and beta-pinene were shown to reduce viral infectivity by 100%, by directly interacting with free virus particles[283]. These monoterpenes are promising as an early treatment option for herpes, for which there is currently no cure.

**Treatment for Neuroblastoma:* SH-SY5Y neuroblastomas are a rare type of cancer that primarily occurs in children. As the major constituents in the essential oil of bergamot (a type of citrus fruit), limonene and linalyl acetate exhibited significant cytotoxicity against neuroblastoma cells, resulting in DNA fragmentation, cell shrinkage, cytoskeletal alterations, and necrotic and apoptotic cell death[284].

**Treatment for Lung Injury:* In a study that investigated the protective role of limonene in lipopolysaccharide (LPS)-induced acute lung injury in mice, limonene inhibited inflammatory cells and proinflammatory cytokines[285], indicating a significant protective effect.

**Treatment for Parkinson's Disease:* In a rotenone-induced model of Parkinson's disease in rats, limonene preserved dopaminergic neurons, attenuated lipid peroxidation, enhanced the activities and concentrations of enzymatic and non-enzymatic antioxidant status in the midbrain, attenuated proinflammatory cytokines and the activation of microglia and astrocytes, inhibited expression of inflammatory mediators and NF-κB/IκB activation, and several other effects that indicate limonene can be useful in the treatment of neurodegeneration[286].

**Treatment for Obesity:* As an isolated compound, d-limonene reduced body weight, total fat tissue weight, adipose cell size, serum triglycerides, and low-density lipoprotein cholesterol in high-calorie diet-induced rats[287]. Additionally, researchers seeking a treatment option for neurodegeneration caused by obesity-related inflammation and oxidative stress found that treatment with d-limonene counteracted the neurodegeneration caused by high fat diet (HFD)-induced obesity in mice, and that these effects were even stronger when combined with honey[288].

**Treatment for Pigeon Feather Lice:* D-limonene has been shown to completely eliminate pigeon lice within 24 hours, without apparent toxicity in kidney or liver functions in pigeons[289].

**Protective Agent Against Heart Attack:* In an isoproterenol-induced myocardial infarction model in Wistar rats, s-limonene attenuated isoproterenol-induced injury in the animals via several mechanisms related to mitochondrial reactive oxygen species[290].

---

[283] Astani, Akram and Paul Schnitzler. Antiviral activity of monoterpenes beta-pinene and limonene against herpes simplex virus in vitro. Iranian journal of microbiology vol. 6,3 (2014): 149-55.

[284] Russo R, Ciociaro A, Berliocchi L, Cassiano MG, Rombola L, Ragusa S, Bagetta G, Blandini F, Corasaniti MT. Implication of limonene and linalyl acetate in cytotoxicity induced by bergamot essential oil in human neuroblastoma cells. Fitoterapia. 2013 Sep;89:48-57.

[285] Chi, Gefu, Miaomiao Wei, Xianxing Xie, L. Soromou, Fang Liu, and Shuhua Zhao. Erratum to: Suppression of MAPK and NF-KB Pathways by Limonene Contributes to Attenuation of Lipopolysaccharide-Induced Inflammatory Responses in Acute Lung Injury. Inflammation 36.4 (2013): 982-982.

[286] Eddin LB, Azimullah S, Jha NK, Nagoor Meeran MF, Beiram R, Ojha S. Limonene, a Monoterpene, Mitigates Rotenone-Induced Dopaminergic Neurodegeneration by Modulating Neuroinflammation, Hippo Signaling and Apoptosis in Rats. Int J Mol Sci. 2023 Mar 9;24(6):5222.

[287] Liao JT, Huang YW, Hou CY, Wang JJ, Wu CC, Hsieh SL. D-Limonene Promotes Anti-Obesity in 3T3-L1 Adipocytes and High-Calorie Diet-Induced Obese Rats by Activating the AMPK Signaling Pathway. Nutrients. 2023 Jan 4;15(2):267.

[288] Terzo S, Calvi P, Nuzzo D, Picone P, Allegra M, Mulè F, Amato A. Long-Term Ingestion of Sicilian Black Bee Chestnut Honey and/or D-Limonene Counteracts Brain Damage Induced by High Fat-Diet in Obese Mice. Int J Mol Sci. 2023 Feb 9;24(4):3467.

[289] Gadelhaq SM, Aboelhadid SM, Abdel-Baki AS, Hassan KM, Arafa WM, Ibrahium SM, Al-Quraishy S, Hassan AO, Abd El-Kareem SG. Safety and Efficacy of Pure and a Nanosuspension of D-limonene for Controlling Pigeon Lice. J Med Entomol. 2023 Jan 12;60(1):148-158.

[290] Rhana P, Barros GM, Santos VCO, Costa AD, Santos DMD, Fernandes-Braga W, Durço AO, Santos MRV, Roman-Campos D, Vasconcelos CML, Cruz JS, Souza DS. S-limonene protects the heart in an experimental model of myocardial infarction induced by isoproterenol: Possible involvement of mitochondrial reactive oxygen species. Eur J Pharmacol. 2022 Sep 5;930:175134.

**Protective Agent Against Photodamage and Photoaging:* Pretreatment with limonene has been shown to increase cell viability and inhibit intracellular ROS generation in UVB-irradiated HaCaT cells (immortalized, human keratinocyte line) , with researchers attributing the antioxidant activity of limonene to an increase in expression of endogenous antioxidants[291].

IMPLICATIONS FOR HUMAN HEALTH & NUTRITION

Dietary limonene is easy to obtain by eating fruits like bergamot, grapefruit, guava, lemons, limes, mandarins, or oranges, although the most potent source of limonene is in the essential oils in the peels of these fruits. Many people supplement their diet with limonene by adding a drop or two of the essential oil to beverages or food items. Other sources of this terpene include celery, black walnuts, or canarium nuts. When preparing or cooking foods, consider using spices that are high in limonene like black pepper, dill, fennel, marjoram, oregano, parsley, rosemary, or sage. Valerian root and tea tree oil - both of which are common at health food stores - are both high in limonene.

Recent research in sheep showed that limonene as extracted from grapefruit peels (Exentiae srl) was readily absorbed and transferred to the animals' mammary glands, but that inhalation of the monoterpene was more efficacious in the transfer process than oral administration[292]. This could indicate that, while oral consumption of this compound is effective, inhalation appears to be more so, which means that diffusers and aromatherapy involving limonene may be a viable option to acquire this compound in addition to a regular diet based on foods that are high in this terpene.

Cannabis strains that contain moderate to high limonene profiles include Afgoo, Alien Cookies, Banana Kush, Banana OG, Blood Drive, Blue Cheese, Cheese, Deadhead OG, Desert Snow, Diablo, Fruit Cake, Ghost Train Haze, Gorilla Glue, Killer Tahoe OG, Kosher Kush, Purple Punch, SFV OG, Skywalker OG, Sour Jack, Star Dawg, and Tosh OG. The healthiest way to consume these strains is by juicing or blending the leaves and flowers of cannabis directly. Smoking or vaporizing these varieties are common methods of ingestion that we recommend against for obvious health reasons, as well as the fact that monoterpenes are easily lost when vaporized or combusted.

Essential oils made from the plants listed in the "Occurrences in Plants" section are likely to contain high amounts of limonene, including in the racemic mixtures of both the l- and d- isomers. Extreme care should be taken when working with these complex mixtures or with isolated compounds, as they could be toxic in some situations.

[291] Kumar KJS, Vani MG, Wang SY. Limonene protects human skin keratinocytes against UVB-induced photodamage and photoaging by activating the Nrf2-dependent antioxidant defense system. Environ Toxicol. 2022 Dec;37(12):2897-2909.

[292] Faccia M, Maggiolino A, Natrella G, Zizzadoro C, Mazzone A, Poulopoulou I, Bragaglio A, De Palo P. Ingested versus inhaled limonene in sheep: A pilot study to explore potential different transfer to the mammary gland and effects on milk and Caciotta cheese aroma. J Dairy Sci. 2022 Oct;105(10):8143-8157.

Limonene Review

Answer the following questions to test your knowledge of this terpene:

Question #1: What type of terpene is limonene?

a. Linear monoterpene
b. Monocyclic sesquiterpene
c. Bicyclic monoterpene
d. Monocyclic monoterpene

Question #2: Which of the following best describes the primary flavor & aroma profile of limonene?

a. Hops, cloves, wood, spicy
b. Balsam, pepper, wood, spice
c. Depends on the isomer
d. Turpentine, pine, woody, balsam

Question #3: How common is limonene in cannabis?

a. Top three
b. Top five
c. Top ten
d. Top twenty

Question #4: What is the chemical formula for limonene?

a. $C_{15}H_{26}$
b. C15H24
c. $C_{10}H_{16}$
d. $N_{15}H_{26}$

Question #5: Name two uses of limonene in plants:

1 ______________________________ 2 ______________________________

Question #6: Name two potential medical uses of limonene:

1 ______________________________ 2 ______________________________

For the answer key to all individual compound chapters' please go here: http://thebigbookofterps.com/bbot-quiz-answer-key/

HUMULENE

Type: Monocyclic sesquiterpene
Chemical Formula: $C_{15}H_{24}$
Molecular Weight: 204.357 g/mol
Boiling Point: 166.00 to 168.00 °C. @ 760.00 mm Hg[293]
Flash Point: 90 °C (194 °F)
Melting Point: 25 °C (77 °F)
Solubility: Soluble in alcohol, weakly soluble in water
Oral LD50: 4,710 mg/kg (rat)[294]
Biological Role: Insecticide, fungicide, allelopathic agent
Therapeutic Role: Insecticide, anticancer agent, antibacterial
Commercial Use: Flavoring agent
Occurrence in Cannabis: Top ten
Aroma/Flavor: Hops, woody, spicy, clove, bitter

Occurs in Cannabis Strains: Biker OG, Blueberry Peaches, Blue Dream, Bubba Kush, Blue Cheese, Cheese, Chemdawg, Deadhead OG, Gelato, Girl Scout Cookies, Goldmeyer, Gorilla Glue, Green Crack, Jack Herer, Killer Tahoe OG, Lemon, Lemon Kush, Lemon Meringue, Lemon Skunk, Miracle Alien Cookies, OG Kush, Pink Lemonade, Platinum Dolato, Skunkdawg, Sour Diesel, Sour Dream, Sour OG, Star Dawg, Strawberry Short Cookies, Sunset Sherbet, Super Sour Diesel, Sugar Cookies, Trainwreck, Tosh OG, Ultimate Purple

INTRODUCTION

Humulene, generally referred to as alpha-humulene, is a monocyclic sesquiterpene that is closely related to beta-caryophyllene. In fact, these two compounds typically occur together in cannabis and other plants, and at one time alpha-humulene was even mistakenly categorized as alpha-caryophyllene.

Like beta-caryophyllene, alpha-humulene imparts plants including cannabis with the rich flavor and aromas of cloves, wood, spice, and bitterness, but this compound differs significantly from beta-caryophyllene in that the primary scent and flavor profile of alpha-humulene is that of hops (Humulus lupulus). Hop products are the main ingredient in most types of beers, and are responsible for the sedative effects of beer, due to humulene and the high concentration of the monoterpene myrcene found in this plant. However, despite the high prevalence of myrcene in hops, the primary terpene found in that plant is alpha-humulene, with Humulus lupulus lending this compound its name.

---

[293] The Good Scents Company; Data Sheet for Alpha-Humulene, from: http://www.thegoodscentscompany.com/data/rw1027221.html. Accessed July 26, 2023.

[294] Agilent Technologies - Safety Data Sheet for Alpha-humulene, from: https://www.agilent.com/cs/library/msds/TRP-135-1_NAEnglish.pdf. Accessed July 26, 2023.

CHEMICAL STRUCTURE

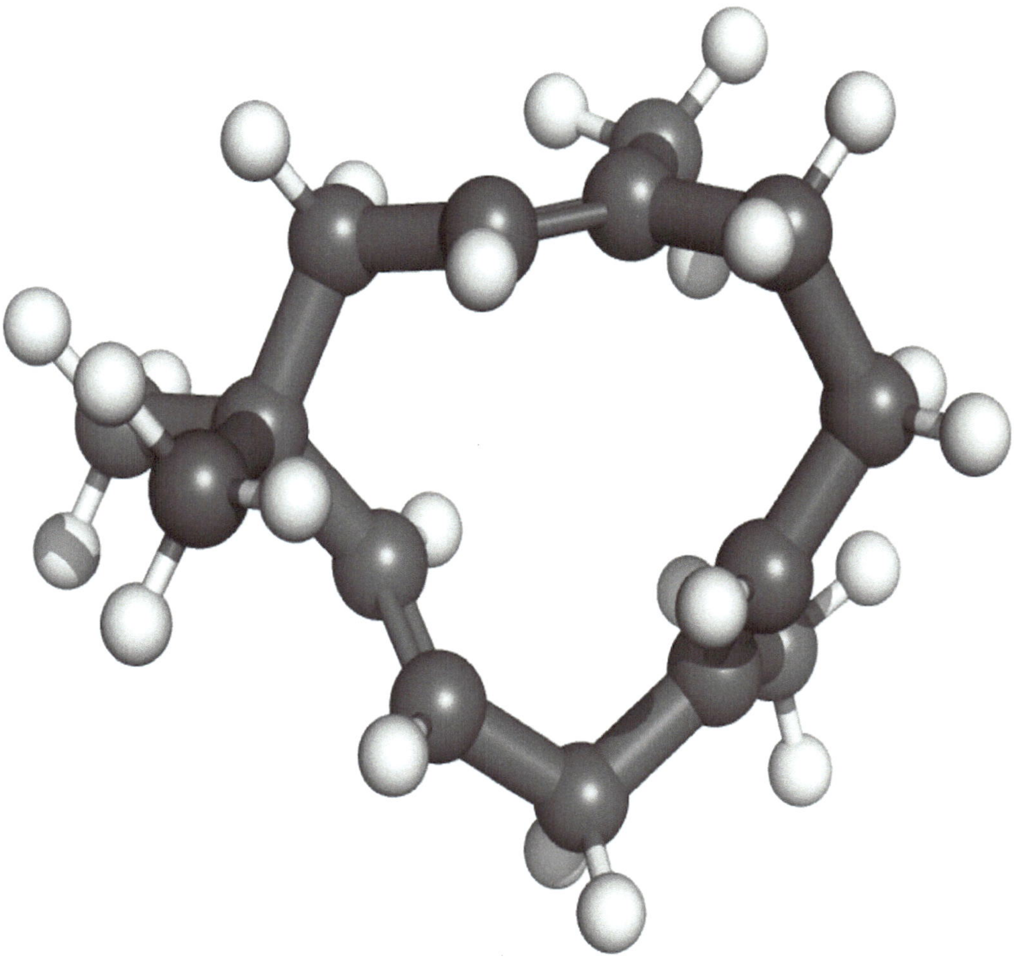

Alpha-humulene is a ring-opened isomer of beta-caryophyllene, which explains why these compounds often co-occur. Like most sesquiterpenes, alpha-humulene is derived primarily from farnesyl diphosphate. This compound consists of three isoprene units with one 11-membered ring, which classifies it as a monocyclic sesquiterpene. This ring contains 3 endocyclic double bonds, which afford this compound modified properties. With fifteen carbon atoms and twenty-four hydrogen atoms, the molecule is notated as $C_{15}H_{24}$.

Although alpha-humulene is an isomer of beta-caryophyllene, the most striking and important difference between the two compounds is that, unlike beta-caryophyllene, alpha-humulene is not considered a dietary cannabinoid because it does not exhibit activity at CB2 endocannabinoid receptor sites.

OCCURRENCE IN PLANTS

While seemingly not quite as prevalent as beta-caryophyllene, alpha-humulene occurs in many of the same plants. In cannabis, alpha-humulene usually occurs together with beta-caryophyllene, although in some varieties like Blue Dream, alpha-humulene can be the main constituent.

Allamanda cathartica	Allspice	Alpinia galanga
Amyris wood	Angelica	Anise
Apple custard plant	Armoise	Artemisia

Artichoke
Bay
Black pepper
Cajuput
Cannabis
Cassia
Cinnamon
Clausena anisata
Croton
Cypress
Eucalyptus
Fir
Geranium
Guava
Hop
Jambu
Laurel
Lemon
Lippia aff.
Mandarin
Marjoram
Mayur pankh
Mikan
Niaouli
Orange
Palmarosa
Pepper tree
Phlomis fruticosa l.
Pimento
Plectranthus australis
Rose
Safflower
Savory
Tagete
Tea tree
Valerian
Witch hazel
Ylang ylang
Zingiber zerumbet

Baccharis coridifolia
Bergamot
Blood orange
Calamus
Cardamom
Celery
Citronella
Clove
Cubeb
Elyonurus muticus
Featherfew
Fleabane
Ginger
Helichrysum
Hyptis suaveolens
Juniper
Lavandin
Lemongrass
Gracillis
Mango
Marsypianthes cha.
Melissa
Myrtle
Nutmeg
Oregano
Parsley
Peppermint
Phoebe
Pine phlomis
Prangos
Rosemary
Sage
Spearmint
Tamarind
Thyme
Vassoura
Wormwood
Yuzu
Zinnia elegans jac.

Basil
Black currant
Cabocla
Canaga
Carrot
Chamomile
Clary Sage
Cotton
Curry
Erigeron graveolens l.
Fennel
Galangal
Grapefruit
Hinoki
Hyssop
Labdanum
Lavender
Lime
Mace
Manuka
Mastic
Mentha longifolia
Neroli
Ocimum graissimum l.
Orthodon dianthera max.
Pennyroyal
Petitgrain
Pimenta
Pittosporum spp.
Raspberry
Rue
Salvia
Sugandha kokila
Taxodium distichum
Turmeric
Verbena
Yarrow
Zanthoxylum species

BIOLOGICAL ACTIVITY IN PLANTS

Although the biological roles of alpha-humulene are not well-studied, it is clear that this sesquiterpene serves several functions:

****Allelopathic Agent:*** Alpha-humulene exhibits allelopathic properties and has been shown to inhibit the germination of seeds and root growth in nearby plants[295].

[295] Bouajaj S, Benyamna A, Bouamama H, Romane A, Falconieri D, Piras A, Marongiu B. Antibacterial, allelopathic and antioxidant activities of essential oil of Salvia officinalis L. growing wild in the Atlas Mountains of Morocco. Nat Prod Res. 2013;27(18):1673-6.

Antifungal Agent: As a major component of the essential oils of clove, lemongrass, mint, and eucalyptus, alpha-humulene exhibited significant antifungal activity against Fusarium oxysporum[296], a common fungal infection in cannabis, especially when grown outdoors. This antifungal activity is likely synergistic, especially in combination with beta-caryophyllene. Other studies have shown that alpha-humulene exhibits strong antifungal activity against a number of yeasts, including candida species[297].

Insecticidal Agent: Alpha-humulene - again in conjunction with beta-caryophyllene - has exhibited potent contact toxicity against Liposcelis bostrychophila (common booklouse), and actively repels this bug, along with the red flour beetle[298]. As a primary constituent of the essential oil of varronia curassavica (black sage), alpha-humulene was shown to kill, repel, and reduce the locomotor activity of the ant, dorymyrmex thoracicus[299].

Larvicidal Agent: Alpha-humulene has demonstrated excellent larvicidal activities against the cotton bollworm[300], one of the most prolific agricultural insect pests in the world. This monocyclic sesquiterpene has also exhibited a significant toxic effect in larvae of mosquitoes that carry malaria and dengue fever, with few effects on non-target organisms[301], making alpha-humulene an excellent candidate as a bio-friendly larvicide and mosquito control agent.

USES IN INDUSTRY

While not as common in industry and in commercial applications as other terpenes like linalool, limonene, pinene, and myrcene, alpha-humulene is used primarily in the processed foods industry, where it serves as a flavoring agent. This sesquiterpene can be found in dairy products, some fats and oils, dessert items like sherbet and sorbet, processed fruit and vegetables, confectionary items, cereals, baked goods, processed meats including fish and crustaceans, spices, soups, non-alcoholic beverages, alcoholic beverages, packaged savory items, casseroles, pies, etcetera.

Research into the metabolic production of humulene by microorganisms is ongoing in various scientific circles. For instance, once group has engineered the yeast Yarrowia lipolytica to produce alpha-humulene from waste cooking oil[302]. Other work showed that alpha-humulene can be produced by Candida tropicalis by co-overexpression of the codon-optimized alpha-humulene synthase gene and the entire endogenous farnesyl diphosphate synthesis

---

[296] Abhishek Sharma, Sasireka Rajendran, Ankit Srivastava, Satyawati Sharma, Bishwajit Kundu. Antifungal activities of selected essential oils against Fusarium oxysporum f. sp. lycopersici 1322, with emphasis on Syzygium aromaticum essential oil. Journal of Bioscience and Bioengineering, Volume 123, Issue 3, 2017, Pages 308-313.

[297] Pereira FG, Marquete R, Domingos LT, Rocha MEN, Ferreira-Pereira A, Mansur E, Moreira DL. Antifungal activities of the essential oil and its fractions rich in sesquiterpenes from leaves of Casearia sylvestris Sw. An Acad Bras Cienc. 2017 Oct-Dec;89(4):2817-2824.

[298] Zhang, Hai, Shan Guo, Bin Fan, Shu Du, Yong Wang, and Zhi Deng. Evaluation of efficacy of the essential oil from Ostericum viridiflorum (Turcz.) Kitagawa in control of stored product insects. Environmental Science and Pollution Research 26.2 (2018): 1406-1413.

[299] Oliveira, Bruna, Carlisson Melo, Ane Santos, Luis Nascimento, Daniela Nizio, Paulo Cristaldo, Arie Blank, and Leandro Bacci. Essential oils from Varronia curassavica (Cordiaceae) accessions and their compounds (E)- caryophyllene and a-humulene as an alternative to control Dorymyrmex thoracius (Formicidae: Dolichoderinae). Environmental Science and Pollution Research 26.7 (2019): 6602-6612.

[300] Benelli, Giovanni & Govindarajan, Marimuthu & Rajeswary, Mohan & Vaseeharan, Baskaralingam & Alyahya, Sami & Alharbi, Naiyf & Km, Shine & Khaled, Jamal & Maggi, Filippo. (2017). Insecticidal activity of camphene, zerumbone and a-humulene from Cheilocostus speciosus rhizome essential oil against the Old-World bollworm, Helicoverpa armigera. Ecotoxicology and environmental safety. 148. 781-786.

[301] Govindarajan, Marimuthu, and Giovanni Benelli. a-Humulene and B-elemene from Syzygium zeylanicum (Myrtaceae) essential oil: highly effective and eco-friendly larvicides against Anopheles subpictus, Aedes albopictus, and Culex tritaeniorhynchus (Diptera: Culicidae). Parasitology Research 115.7 (2016): 2771-2778.

[302] Guo Q, Peng QQ, Chen YY, Song P, Ji XJ, Huang H, Shi TQ. High-yield α-humulene production in Yarrowia lipolytica from waste cooking oil based on transcriptome analysis and metabolic engineering. Microb Cell Fact. 2022 Dec 24;21(1):271.

pathway[303], while yeast peroxisomes have been engineered to produce alpha-humulene using glucose as the sole carbon source[304].

Commercial preparations of alpha-humulene generally appear as a clear liquid tinged with yellow or green.

POTENTIAL USES IN MEDICINE

Although the United States Federal Department of Agriculture has classified alpha-humulene as Generally Recognized as Safe, recent research has demonstrated that this compound (as well as beta-caryophyllene and caryophyllene oxide) significantly inhibits cytochrome enzymes that metabolize different drugs[305], leading the study authors to recommend a drug interaction warning. After taking all concurrent medications into account, alpha-humulene has considerable potential for uses in medicine:

Anti-inflammatory Agent: Alpha-humulene exhibits potent anti-inflammatory activity in several different experimental models, with results showing that the compound is comparable to that of dexamethasone[306], a common prescription-strength (steroidal) anti-inflammatory drug.

Mosquito-borne Disease Prevention: Alpha-humulene is an excellent natural larvicide, capable of inhibiting or killing insect larvae. This larvicidal activity has the potential to help millions of people living in regions affected by mosquito-borne illnesses such as dengue fever, because alpha-humulene has also been shown to deter egg-laying female Aedes aegypti mosquitos, the carrier of dengue[307] (in the study referenced here, the authors concluded that alpha-humulene exhibits some of the most potent oviposition deterrent effects reported thus far). We have previously mentioned and referenced work showing that alpha-humulene has toxic effects on larvae of mosquitos that carry malaria, a disease that killed around 429,000 people in 2015[308]. Dengue fever infects an estimated 390 million people per year[309].

Anti-Cancer Agent: Alpha-humulene could be the next important anti-cancer compound, having been shown to inhibit cancer cell growth by approximately 50% as a single compound, and by as much as 75% when blended with beta-caryophyllene[310].

Anti-Tumor Agent: Alpha-humulene has been shown to inhibit the growth of both cancer cells and tumors[311], and to decrease cellular GSH (glutathione, an antioxidant) content while increasing ROS (reactive oxygen species) production[312].

---

303 Zhang, L., Yang, H., Xia, Y. et al. Engineering the oleaginous yeast Candida tropicalis for α-humulene overproduction. Biotechnol Biofuels 15, 59 (2022).

304 Zhang C, Li M, Zhao GR, Lu W. Harnessing Yeast Peroxisomes and Cytosol Acetyl-CoA for Sesquiterpene α-Humulene Production. J Agric Food Chem. 2020 Feb 5;68(5):1382-1389.

305 Nguyen LT, Mysliveckova Z, Szotakova B, Spicakova A, Lnenickova K, Ambroz M, Kubicek V, Krasulova K, Anzenbacher P, Skalova L. The inhibitory effects of Beta-caryophyllene, Beta-caryophyllene oxide and a-humulene on the activities of the main drug-metabolizing enzymes in rat and human liver in vitro. Chem Biol Interact. 2017 Dec 25; 278:123-128.

306 Fernandes ES, Passos GF, Medeiros R, da Cunha FM, Ferreira J, Campos MM, Pianowski LF, Calixto JB. Anti- inflammatory effects of compounds alpha-humulene and (-)-trans-caryophyllene isolated from the essential oil of Cordia verbenacea. Eur J Pharmacol. 2007 Aug 27;569(3):228-36.

307 da Silva, Rayane Cristine Santos, Paulo Milet-Pinheiro, Patricia Cristina Bezerra da Silva, Alexandre Gomes da Silva, Marcia Vanusa da Silva, Daniela Maria do Amaral Ferraz Navarro, and Nicacio Henrique da Silva. (E)- Caryophyllene and a-Humulene: Aedes aegypti Oviposition Deterrents Elucidated by Gas Chromatography- Electrophysiological Assay of Commiphora leptophloeos Leaf Oil. PLoS ONE 10.12 (2015).

308 10 Facts on Malaria World Health Organization From: https://www.who.int/features/factfiles/malaria/en/ Accessed February 19, 2019.

309 Bhatt, Samir, Peter W. Gething, Oliver J. Brady, Jane P. Messina, Andrew W. Farlow, Catherine L. Moyes. The global distribution and burden of dengue. Nature 496.7446 (2013): 504-507.

310 Legault, Jean, and Andre Pichette. Potentiating effect of Beta-caryophyllene on anticancer activity of a-humulene, isocaryophyllene and paclitaxel. Journal of Pharmacy and Pharmacology: An International Journal of Pharmaceutical Science 59.12 (2007).

311 Costa EV, Menezes LR, Rocha SL, Baliza IR, Dias RB, Rocha CA, Soares MB, Bezerra DP. Antitumor Properties of the leaf essential oil of Zornia

Chemotherapeutic Agent Precursor: Zerumbone, a sesquiterpenoid and cyclic ketone, has strong potential as an anti-cancer or chemotherapeutic agent, but it is difficult and expensive to produce. But because it is so closely related to alpha-humulene, scientists have devised a way to produce zerumbone from alpha-humulene, using E. coli bacteria as an intermediary[313], which means that zerumbone may be more widely available as a cancer treatment or preventative in the near future.

Treatment for Prostate Cancer: In research that investigated the cytotoxicity of the essential oils of several plants, the highest activity was found in alpha-humulene, which inhibited cell growth of LNCaP human prostate cancer cells[314].

Antibacterial Agent: With drug-resistant bacteria rapidly becoming a significant threat to human health, novel new treatments are needed, and alpha-humulene shows potential to become an excellent tool in this regard. This compound has been shown to exhibit strong antibacterial activity against a range of organisms, including E. coli, sarcina lutea, bacillus subtilis, pseudomonas sp., klebsiella pneumoniae, and xanthomonas campestris[315]. Alpha-humulene also exhibits strong antibacterial properties in the centella plant, where, as the primary constituent of the essential oil of the plant, the sesquiterpene showed a variety of antibacterial activities against both gram-positive bacteria like bacillus subtilis and staphylococcus aureus, and gram-negative bacteria like E. coli[316].

Post-Surgery Periodontal Treatment: Researchers have shown that alpha-humulene can be added to periodontal dressings, where the compound acts as a post-surgery antimicrobial agent with efficacy similar to the commercial gold standards Barricaid and Periobond[317].

IMPLICATIONS FOR HUMAN HEALTH & NUTRITION

Although alpha-humulene is not a dietary cannabinoid like beta-caryophyllene, it still has significant value for human health and nutrition. To include this terpene in your diet, consider regularly eating fruits like grapefruits, mangos, oranges, guava, or raspberries. Vegetables that are high in this monocyclic sesquiterpene include artichokes, celery, and carrots. When preparing, cooking, or serving foods, consider using spices and herbs like cloves, bay, black pepper, oregano, cinnamon, fennel, ginger, nutmeg, or parsley. Responsibly consuming beer is also a good way to obtain alpha-humulene.

Although we do not recommend habitual consumption of cannabis by smoking or combustion of the plant material, eating, juicing, or using other preparations of raw cannabis will allow you to safely include alpha-humulene in your regular diet. Varieties that feature this terpene prominently include Blue Dream, Bubba Kush, Blue Cheese, Cheese, Chemdawg, Gelato, Girl Scout Cookies, Gorilla Glue, Green Crack, Jack Herer, Lemon Meringue, Lemon Skunk,

---

brasiliensis. Planta Med. 2015 May;81(7):563-7.

312 Legault J, Dahl W, Debiton E, Pichette A, Madelmont JC. Antitumor activity of balsam fir oil: production of reactive oxygen species induced by alpha-humulene as possible mechanism of action. Planta Med. 2003 May; 69(5):402-7.

313 Alemdar, Semra, Jan C. Konig, Steffen Hartwig, Thore Frister, Thomas Scheper, and Sascha Beutel. Bioproduction of a-humulene in metabolically engineered Escherichia coli and application in zerumbone synthesis. Engineering in Life Sciences 17.8 (2017): 900-907.

314 Loizzo MR, Tundis R, Menichini F, Saab AM, Statti GA, Menichini F. Cytotoxic activity of essential oils from labiatae and lauraceae families against in vitro human tumor models. Anticancer Res. 2007 Sep-Oct;27(5A):3293-9.

315 Atiqur Rahman, Zakia Sultana Shanta, M.A. Rashid, Tanzima Parvin, Shajia Afrin, Mst Khodeza Khatun, M.A. Sattar. In vitro antibacterial properties of essential oil and organic extracts of Premna integrifolia Linn. Arabian Journal of Chemistry, Volume 9, Supplement 1, 2016, Pages S475-S479.

316 O.A. Oyedeji & A.J. Afolayan (2005) Chemical Composition and Antibacterial Activity of the Essential Oil of Centella asiatica. Growing in South Africa, Pharmaceutical Biology, 43:3, 249-252.

317 Rossato TCA, Alves T, Cuevas-Suárez CE, Rosa WLOD, Silva AFD, Piva E, ZANCHi CH, Lund RG. Effect of alpha-humulene incorporation on the properties of experimental light-cured periodontal dressings. Braz Oral Res. 2022 Jul 11;36:e091.

Miracle Alien Cookies, OG Kush, Pink Lemonade, Sour Diesel, Sour Dream, Super Sour Diesel, Sugar Cookies, and Trainwreck. Some cannabis extracts that have retained or added alpha-humulene may be suitable as well.

Another effective way to consume alpha-humulene is to purchase essential oil (EO) with this terpene as the primary constituent. Essential oils made from the plants listed in the "Occurrences in Plants" section above are likely to be high in this compound. These EO can be added to recipes, used in tinctures, taken or used directly, or blended with other essential oils and added to salves or creams. As always, extreme care should be exercised when working with complex oil mixtures or isolated compounds, as these can be toxic in some situations.

Humulene Review

Answer the following questions to test your knowledge of this terpene:

Question #1: What type of terpene is humulene?

a. Monocyclic monoterpene
b. Linear sesquiterpene
c. Bicyclic monoterpene
d. Monocyclic sesquiterpene

Question #2: Which of the following best describes the primary flavor & aroma profile of humulene?

a. Spicy, turpentine, dry
b. Hops, woody, spicy, clove, bitter
c. Lemon, sharp, fuel
d. Citrus, fruity, dry, terpenic

Question #3: How common is humulene in cannabis?

a. Top three
b. Top ten
c. Top twenty
d. Less common

Question #4: Which plant is humulene named after?

a. Hops
b. Hyacinth
c. Hyssop
d. Hinoki

Question #5: Name two uses of humulene in plants:

1 ____ 2 ____

Question #6: Name two potential medical uses of humulene:

1 ____ 2 ____

For the answer key to all individual compound chapters please go here: http://thebigbookofterps.com/bbot-quiz-answer-key/

LINALOOL

Type: Acyclic monoterpenoid
Chemical Formula: $C_{10}H_{18}O$
Molecular Weight: 154.2526 g/mol
Boiling Point: 194.00 to 197.00 °C. @ 760.00 mm Hg[318]
Flash Point: 78 °C (172 °F)
Melting Point: 25 °C (77 °F)
Solubility: Soluble in water, alcohol, ethanol, ether, propylene glycol, fixed oils, paraffin oil, Insoluble in glycerin
Oral LD50: 2,790 mg/kg (rat)[319]
Biological Role: Allelopathic agent, insecticide, antifeedant
Therapeutic Role: Anticancer agent, anti-inflammatory, neurological agent
Commercial Use: Fragrance, food additive, chemical precursor
Occurrence in Cannabis: Top ten
Aroma/Flavor: *R-Linalool:* Lavender, citrus, rose, woody
S-Linalool: Floral, sweet, tropical, coriander

Occurs in Varieties: Linalool occurs at some concentration in most cannabis strains, including 1991 OG, Amnesia Haze, Biker OG, Desert Snow, Girl Scout Cookies, Island Sweet Skunk, King Tut, LA Confidential, Lavender, Master Kush, Miracle Alien Cookies, Miss X, Obama Kush, OG Kush, OG Shark, Pink Kush, Purple Punch, Purple Hindu Kush, Sour Kush, Star Pie.

INTRODUCTION

Linalool is common to hundreds of plants including cannabis, in which it typically occurs in the top ten terpenes for many varieties. This compound's unique aroma and taste profile is easy for most people to discern, and consists of a lavender, floral, or rose scent, and a woody, tropical, coriander-like taste. In fact, Linalool's aroma is so unique that its primary commercial use is to add scent to products like perfumes and hair care products.

---

[318] The Good Scents Company Data Sheet for Linalool, from: http://www.thegoodscentscompany.com/data/rw1007872.html. Accessed July 27, 2023.

[319] Agilent Technologies Safety Data Sheet for Linalool, from: https://www.agilent.com/cs/library/msds/TRP-145-1_NAEnglish.pdf. Accessed July 27, 2023.

CHEMICAL STRUCTURE

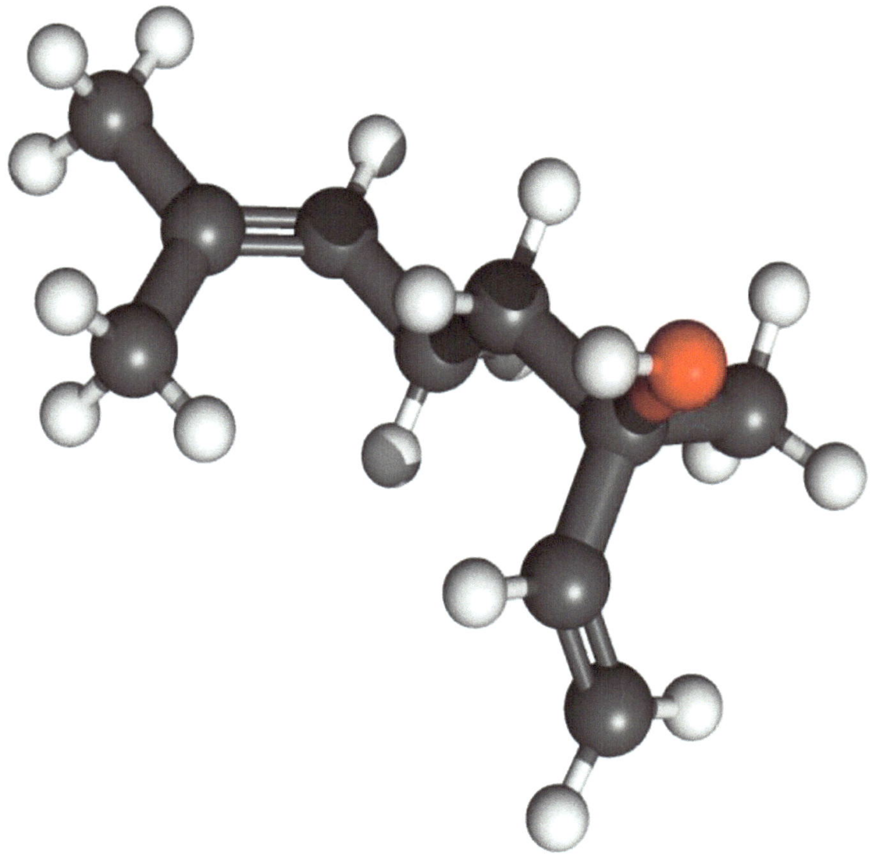

Like many of the top ten terpenes in cannabis, linalool is a monoterpene, but instead of the traditional $C_{10}H_{16}$ arrangement, this compound is comprised of ten carbon atoms, eighteen hydrogen atoms, and an oxygen atom, notated as $C_{10}H_{18}O$. Linalool's chemical skeleton has seven bonds (including two double bonds); two hydrogen, four rotatable, and one covalent, and typically features methyl groups at positions 3 and 7, and a hydroxy group at position 3 (methyl and hydroxyl groups add functionality to the compound, see "Terpenes 101.") Because linalool does not contain rings, it is considered an acyclic monoterpene.

Produced in the mevalonate pathway from isopentenyl pyrophosphate and geranyl pyrophosphate, linalool has two stereoisomers: r-linalool and s-linalool, where the former is called licareol, and the latter coriandrol. Similar to how d-limonene is associated with citrus, while l-limonene is associated with pine, each stereoisomer of linalool has a distinct scent and flavor profile; r-linalool is most often described as lavender and woody, while s-linalool is described as floral and sweet. Some plants contain the racemic mixture of both stereoisomers.

OCCURRENCE IN PLANTS

While many plants are known to produce linalool, recent work has shown that cinnamon might be the world's powerhouse linalool producer, with the leaf of one variety containing more than 92% linalool in the essential oil[320]. Linalool is also found in significant quantity in the following plants:

Allamanda cathartica	Allspice	Alpinia galangal

[320] Aromatic plants containing essential oil component-linalool, eugenol and methyl chavicol reported from North-East India Barnali Dutta1, Kongkona Borborah and S. K. Borthakur J. Nat. Prod. Plant Resource., 2015, 5 (1):6-10.

Angelica	Anise	Apple
Apricot	Artemisia	Ayou wood
Barbatus	Basil	Basswood
Bay	Bergamot	Betel leaf
Birch	Black currant	Black elder
Black pepper	Blood orange	Blueberry
Bois de rose	Cabocla	Cacao
Cajuput	Calamus	Cananga
Cannabis	Caraway	Cardamom
Carnation	Carrot	Cascarilla
Cassia	Celery	Cilantro
Chamomile	Champaca	Cherry
Cinnamon	Citronella	Clary sage
Clove	Coleus	Common tea
Coriander	Corn	Cranberry
Croton	Cubeb	Cumin
Curry	Cypress	Dill
Eucalyptus	Fennel	Feverfew
Fig	Fleabane	Frankincense
Galangal root	Garlic	Geranium
Ginger	Goldenrod	Grape
Grapefruit	Guava	Gum
Helichrysum	Ho leaf	Hop
Horsemint	Hyacinthus orientalis	Hyssop
Jambu	Jasmine	Juniper
Kewda	Kohlrabi	Laurel
Lavandin	Lavender	Lemon
Lemongrass	Lime	Lovage
Mace	Mandarin	Marjoram
Mastic	Melissa	Michelia
Mint	Mugwort	Myrtle
Nectarine	Neroli	Nutmeg
Orange	Oregano	Palmarosa
Papaya	Parsley	Peach
Pennyroyal	Peppermint	Perilla
Peumus boldus	Pimento	Pine
Pineapple	Plum	Rose
Rosemary	Rosewood	Saffron
Sage	Salvia	Sandalwood
Savory	Sour orange	Spearmint
Strawberry	Sunflower	Tagete
Tamarind	Tarragon	Tea tree
Thyme	Tomato	Turmeric
Valerian	Verbena	Witch hazel
Wormseed	Wormwood	Yarrow
Ylang ylang	Yuzu	

BIOLOGICAL ACTIVITY IN PLANTS

Like most terpenes, linalool serves several roles in plants, depending on the species and environment, and likely serves multiple roles in a given plant. Some include:

**Antifeedant & Oviposition Deterrent:* Many monoterpenes are considered bitter antifeedants; especially those that are concentrated in the broad leaves of plants. In one study that reviewed essential oils that were high in linalool, cotton and tobacco plant volatiles containing this terpene as the primary constituent were shown to significantly repel aphids, and deterred the cotton bollworm from depositing eggs on the plants[321]. This should be of particular importance to cannabis growers, who often battle sap- sucking aphids.

**Allelopathic Agent:* Linalool exhibits strong allelopathic properties and can influence the growth of numerous competitors, including restricting the growth of lettuce[322], inhibiting germination rates and seedling growth of the noxious weed parthenium hysterophorus[323] (Santa-Maria feverfew), and, as a major constituent of the essential oil of tomato plants, inhibiting seed germination and seedling growth of several other plants[324]. For outdoor cannabis growers, the development of strains that are high in linalool, or localized treatment of plants using linalool, may be an effective and natural method of controlling weeds.

**Nematocidal:* As the major constituent in the essential oil of several different plants, linalool exhibited nematocidal activity against seed-gall, citrus, root-knot, and pigeon-pea nematodes[325].

**Indirect Defense:* When Mountain birch trees come under attack by autumnal moths, the trees release volatile compounds consisting mostly of three terpenes, including ocimene and linalool, which attract insectivorous birds that feed on the moths[326].

**Insecticidal Agent*: Linalool is a well-known insecticidal agent, capable of repelling numerous types of insects. Used in the veterinary industry to repel ticks and fleas[327], this monoterpene is also highly repellant to mites[328] - a significant pest of cannabis crops.

**Antibacterial Agent:* In tests that examined bacterial growth on the petals of flowers from penstemon digitalis, researchers were able to significantly inhibit the growth of bacteria after treating the petals with linalool, which showed greater antibacterial activity than methyl nicotinate treatment[329], another natural antibacterial agent. Linalool has also been shown to in reduce biofilm formation, hamper motility, decrease acyl-homoserine lactone production, and reduce secretion of plant cell wall-degrading enzymes in Pseudomonas syringae pv; a damaging

---

[321] Huang XZ, Xiao YT, Kollner TG, Jing WX, Kou JF, Chen JY, Liu DF, Gu SH, Wu JX, Zhang YJ, Guo YY. The terpene synthase gene family in Gossypium hirsutum harbors a linalool synthase GhTPS12 implicated in direct defence responses against herbivores. Plant Cell Environ. 2018 Jan;41(1):261-274.

[322] El Ayeb-Zakhama, Asma, Lamia Sakka-Rouis, Afifa Bergaoui, Guido Flamini, Hichem Ben Jannet, and Fethia Harzallah-Skhiri. Chemical Composition and Allelopathic Potential of Essential Oils Obtained from Acacia cyanophylla Lindl . Cultivated in Tunisia." Chemistry & Biodiversity 12.4 (2015): 615-626.

[323] Vaid, Supriya & Batish, Daizy & P. Singh, H & Kohli, Ravinder. (2005). Allelopathich Effect of Linalool against Parthenium hysterophorus L. International Journal Bioscience Reporter. 3. 76 - 81.

[324] Young Sik Kim & Bong-Seop Kil. Allelopathic Effects of Some Volatile Substances from the Tomato Plant. Journal of Crop Production Volume 4, 2001 - Issue 2, Pages 313-321.

[325] Naresh K. Sangwan Braham S. Verma Kailash K. Verma Kuldip S. Dhindsa. Nematicidal activity of some essential plant oils. Pest Management Science Volume 28, Issue3 1990 Pages 331-335.

[326] Elina, Mantyla, Alessio Giorgio A., Blande James D., Heijari Juha, Holopainen Jarmo K., Laaksonen Toni, Piirtola Panu, and Klemola Tero. From Plants to Birds: Higher Avian Predation Rates in Trees Responding to Insect Herbivory. PLoS ONE 3.7 (2008)e2832.

[327] US Patent: Controlling ticks and fleas with linalool. PATENT# US4933371A. W. Fredric Hink, Thomas E. Duffey From: https://patents.google.com/patent/US4933371A/en. Accessed March 22, 2019.

[328] Sanchez-Ramos I I1, Castafiera P. Acaricidal activity of natural monoterpenes on Tyrophagus putrescentiae (Schrank), a mite of stored food. J Stored Prod Res. 2000 Jan 15;37(1):93-101.

[329] Burdon RCF, Junker RR, Scofield DG, Parachnowitsch AL. Bacteria colonising Penstemon digitalis show volatile and tissue-specific responses to a natural concentration range of the floral volatile linalool. Chemoecology. 2018;28(1):11-19.

pathogen of tomato plants[330]. Deepa et al also found that linalool affected the phenotypic and genotypic expressions associated with quorum sensing, thereby significantly reducing the infection rate of the bacteria.

Antifungal Agent: As an isolated compound, linalool exhibited strong inhibitive effects on mycelial growth of Botrytis cinerea, inhibiting spore germination, inducing oxidative damage, destroying the cell membrane integrity, and significantly decreasing activities of superoxide dismutase, catalase, and peroxidase[331].

USES IN INDUSTRY

The primary commercial use of linalool is to add scent to a wide variety of products, particularly in the perfume and fragrance industries. This monoterpene is used as an aroma-enhancer in consumer personal care products, toiletries, and hygiene products like shampoos, conditioners, deodorants, perfumes, aftershaves, colognes, hair gels and sprays, deodorizers, and some cleaning products. In fact, linalool is added to 60% to 80% of scented personal care products and cleaners, with the oxidation of this compound causing or exacerbating allergies for a significant number of consumers[332]. In many cases, consumers are unaware that linalool can cause allergies.
Because of its uniquely woody, floral, tropical, and somewhat spicy taste, linalool is added to food products, especially sweets like gum, hard candy, soft drinks, pudding, and baked goods, among others.

Linalool is also used as a common precursor in the production of Vitamin E; an important industrial, medical, nutritional, and commercial compound, which is added to many of the products that already contain linalool, like hand lotions and hair care products.

Linalool is an acyclic monoterpene that appears as a clear to light yellow liquid exhibiting the scent of lavender in most commercial preparations. Linalool is produced for commercial and industrial use by several methods, including chemical synthesis from geraniol or myrcene, by pyrolysis, or fractional distillation. There are also many small companies offering this terpene as derived from specific plants, either in pure form or as formulations with other similar compounds.

Interestingly, scientists have produced linalool oxidation products by harnessing the biotransformation power of corynespora cassiicola, a species of sac fungus that is well-known as a plant pathogen. Essentially, researchers fed the fungus nearly pure linalool, which the fungus consumed and released back into the environment as linalool oxides[333].

Wine consumers might be interested to know that linalool can have a significant impact on the aroma of Pinot Gris wines. Researchers found that the addition of linalool to the wine did not alter aroma quality when limonene content in the wine was at low concentrations, but it did change aroma perception at high limonene levels, and at high concentrations, linalool and α-terpineol added to the wine presented tropical fruit aromas with some floral notes, irrespective of limonene levels[334].

---

[330] Deepa N, Chauhan S, Kumari P, Rai AK, Tandon S, Singh A. Linalool reduces the virulence of Pseudomonas syringae pv. tomato DC 3000 by modulating the PsyI/PsyR quorum-sensing system. Microb Pathog. 2022 Dec;173(Pt A):105884.

[331] Wang QF, Wang XY, Li HS, Yang XY, Zhang RM, Gong B, Li XM, Shi QH. Effects of linalool on Botrytis cinerea growth and control of tomato gray mold. Ying Yong Sheng Tai Xue Bao. 2023 Jan;34(1):213-220. English.

[332] Widely Used Fragrance Ingredients In Shampoos And Conditioners Are Frequent Causes Of Eczema. Medical News Today. Published Saturday 28 March 2009 from: https://www.medicalnewstoday.com/releases/144041.php. Accessed January 18, 2019.

[333] Bormann, Sebastian, Maria Etschmann, Marco-Antonio Mirata, and Jens Schrader. Integrated bioprocess for the stereospecific production of linalool oxides from linalool with Corynespora cassiicola DSM 62475. Journal of Industrial Microbiology. Biotechnology 39.12 (2012): 1761-1769.

[334] Chigo-Hernandez MM, Tomasino E. Aroma Perception of Limonene, Linalool and α-Terpineol Combinations in Pinot Gris Wine. Foods. 2023 Jun 16;12(12):2389.

Finally, as a primary constituent in the extract and powder made from Cinnamomum verum (true cinnamon tree), linalool contributed to contact toxicity and repellent activity against the stored grain pests Cryptolestes ferrugineus, Rhyzopertha dominica, and Sitophilus granaries, with researchers suggesting us of the extract as an insecticidal agent to control pests in grain stores[335]. It should be noted that the study in question here is a preprint study as of late July 2023, and has not been published by a peer-reviewed journal.

POTENTIAL USES IN MEDICINE

As discussed in the "Uses in Industry" section above, linalool is a potential allergen for some people. In fact, in a European patch test of 821 people, more than 9% reacted positively to hydroperoxides of limonene and linalool[336], with scientists calling for these compounds to be included in European baseline series patch tests as recently as 2019. However, it should be noted that these allergies are primarily to oxidation products of linalool, and not the acyclic monoterpene itself. While some people might be allergic to linalool oxides as found in fragrances, cleaners, or other products, these same people might not be allergic to, for instance, the purified form of linalool, or an essential oil where this terpene is the primary constituent. Caution should be exercised if allergies are suspected or present.

Antibacterial Agent and Synergist: As the primary constituent in the essential oil made from Citrus aurantium L. var. amara Engl (sour orange), linalool exhibited strong antibacterial activity, influencing the synthesis and modification of lipopolysaccharide, significantly inhibiting ribosomal assembly and protein synthesis, and altering the expression and uptake of several proteins[337]. As an isolated compound, linalool damaged cell structural integrity leading to the leakage of intracellular components, exhibited a scavenging effect on biofilm, disrupted intracellular metabolism and caused energy metabolism disorder, obstructed nucleic acid synthesis, downregulated amino acids, and upregulated saturated fatty acids in Streptococcus agalactiae[338]. Linalool has also been shown to work synergistically to enhance the antibacterial properties of essential oils, showing efficacy against common bacterial strains such as gram-positive Staphylococcus aureus, and gram-negative bacteria like Escherichia coli[339]. Linalool has also been shown to potentiate the antibacterial activity of existing clinical antibiotics through synergistic interactions[340]. These findings indicate that it might not be new drugs that are needed to combat bacterial resistance, but the application of new drug synergies.

Anti-inflammatory Agent: Linalool is well known for its anti-inflammatory properties and has been shown to inhibit tumor necrosis factor alpha-induced cell adhesion molecules in both murine and human endothelial cells[341], and attenuate pathogen-mediated inflammation in human lung cells[342] through several mechanisms.

---

[335] Hend A. Mahmoud, Mohamed M. Azab, Fatma Mohamed Amin Sleem et al. Bioactivity of Cinnamomum verum powder and extract against Cryptolestes ferrugineus S., Rhyzopertha dominica F. and Sitophilus granarius L. (Coleoptera), 08 March 2023, PREPRINT (Version 1)

[336] Dittmar, Daan, and Marie L. A. Schuttelaar. Contact sensitization to hydroperoxides of limonene and linalool: Results of consecutive patch testing and clinical relevance. Contact Dermatitis 80.2 (2019): 101-109.

[337] Gao Z, Jiang S, Zhong W, Liu T, Guo J. Linalool controls the viability of Escherichia coli by regulating the synthesis and modification of lipopolysaccharide, the assembly of ribosome, and the expression of substrate transporting proteins. Food Res Int. 2023 Feb;164:112337.

[338] Liang T, Huo G, Chen L, Ding L, Wu J, Zhang J, Wang R. Antibacterial activity and metabolomic analysis of linalool against bovine mastitis pathogen Streptococcus agalactiae. Life Sci. 2023 Jan 15;313:121299.

[339] Herman, Anna, Krzysztof Tambor, and Andrzej Herman. Linalool Affects the Antimicrobial Efficacy of Essentia l Oils. Current Microbiology 72.2 (2015): 165-172.

[340] Silva V.A., Sousa, J.P., Guerra F.Q.S., Pessoa H.L.F., Freitas A.F.R., Coutinho, H.D.M., Alves L.B.N, Lima E.O. Antibacterial Activity of the Monoterpene Linalool: Alone and in Association with Antibiotics Against Bacteria of Clinical Importance. International Journal of Pharmacognosy and Phytochemical Research 2015; 7(5); 1 022-1026.

[341] Aoe M1, Ueno-Iio T, Shibakura M, Shinohata R, Usui S, Arao Y, Ikeda S, Miyahara N, Tanimoto M, Kataoka M. Lavender Essential Oil and Its Main Constituents Inhibit the Expression of TNF-a-induced Cell Adhesion Molecule in Endothelial Cells. Acta Med Okayama. 2017 Dec;71(6):493-503.

[342] Wu Q, Yu L, Qiu J, Shen B, Wang D, Soromou LW, Feng H. Linalool attenuates lung inflammation induced by Pasteurella multocida via activating Nrf-2 signaling pathway. Int Immunopharmacol. 2014 Aug;21(2):456-63.

Treatment of Neurodegenerative Disease: Lipopolysaccharide (LPS) - normally found within the cell walls of gram-negative bacteria - causes an acute inflammatory reaction when released into the blood stream. Linalool has been shown to inhibit LPS-induced inflammation in microglial cells by activating the Nrf2 signaling pathway[343] - a protein that regulates the expression of protective antioxidants. Because of these antioxidant properties, linalool is a potential tool in the treatment of neurodegenerative diseases.

Treatment for Carpal Tunnel Syndrome: In a study that set out to evaluate linalool's antioxidant properties in people with Carpal Tunnel Syndrome (CTS), researchers found that simply inhaling linalool increased antioxidant levels in all patients - even the healthy control group - and reduced the blood pressure and pulse rate in CTS patients[344]. This study was another testament to the potent antioxidant properties of linalool, and a strong nod to the aromatherapy industry.

Treatment of Liver, Lung Cancer: Linalool inhibited cell proliferation in liver and lung cancer cell lines and synergistically inhibited cholesterol synthesis[345] - an important consideration for people with these cancers, but also those with cardiovascular disease. Interestingly, to overcome linalool's poor aqueous solubility, scientists have successfully used nanoparticles to deliver linalool intracellularly[346] in recent research that involved the same liver and lung cancer cell lines mentioned previously. Linalool has also been shown to exhibit excellent binding affinity with selected targets, and modulated biomarkers associated with cancer progression in lung cancer cells[347].

Treatment of Colon Cancer: In a study involving xenografted mice, treatment with linalool reduced tumor weight by up to 55% and induced apoptosis of human colon cancer cells via oxidative stress[348], impressively only targeting cancer cells and leaving healthy tissue and organs seemingly unaffected.

Anticancer Agent: Because of its ability to induce oxidative stress in cells, linalool offers potential as an anti-cancer tool, having exhibited potent antitumor activity both in vitro and in vivo[349]. The study referenced here was inspired by the long well-known folk medicine effects of coriander, where linalool is the primary constituent of the essential oil. Linalool has also been shown to induce apoptosis in MCF-7 and MDA-MB-231 human breast cancer cells[350].

---

[343] Li, Yang, Ou Lv, Fenggang Zhou, Qingsong Li, Zhichao Wu, and Yongri Zheng. Linalool Inhibits LPS-Induced Inflammation in BV2 Microglia Cells by Activating Nrf2. Neurochemical Research 40.7 (2015): 1520-1525.

[344] Seol, Geun-Hye, Purum Kang, Hui Lee, and Geun Seol. Antioxidant activity of linalool in patients with carpal tunnel syndrome. BMC Neurology 16.1 (2016): 1-6.

[345] Rodenak Kladniew, Boris, Monica Polo, Sandra Montero Villegas, Marianela Galle, Rosana Crespo, and Margarita Garcia de Bravo. Synergistic antiproliferative and anticholesterogenic effects of linalool, 1,8-cineole, and simvastatin on human cell lines. Chemico-Biological Interactions 214 (2014): 57-68.

[346] Rodenak-Kladniew, Boris, German A. Islan, Margarita G. de Bravo, Nelson Duran, and Guillermo R. Castro. Design, characterization and in vitro evaluation of linalool-loaded solid lipid nanoparticles as potent tool in cancer therapy. Colloids and Surfaces B: Biointerfaces 154 (2017): 123-132.

[347] Singh J, Meena A. Exploration of Chemopreventive Potential of Linalool in Targeting Lung Cancer Biomarkers. Endocr Metab Immune Disord Drug Targets. 2022;22(14):1416-1424.

[348] Kenichi Iwasaki, Yun-Wen Zheng, Soichiro Murata, Hiromu Ito, Ken Nakayama, Tomohiro Kurokawa, Naoki Sano, Takeshi Nowatari, Myra O Villareal, Yumiko N Nagano, Hiroko Isoda, Hirofumi Matsui, and Nobuhiro Ohkohchi. Anticancer effect of linalool via cancer-specific hydroxyl radical generation in human colon cancer. World J Gastroenterol. 2016 Nov 28; 22(44): 9765-9774.

[349] Jana S, Patra K, Sarkar S, Jana J, Mukherjee G, Bhattacharjee S, Mandal DP. Antitumorigenic potential of linalool is accompanied by modulation of oxidative stress: an in vivo study in sarcoma-180 solid tumor model. Nut Cancer. 2014;66(5):835-48.

[350] Elbe H, Ozturk F, Yigitturk G, Baygar T, Cavusoglu T. Anticancer activity of linalool: comparative investigation of ultrastructural changes and apoptosis in breast cancer cells. Ultrastruct Pathol. 2022 Jul 4;46(4):348-358.

**Sleep Therapy for Cancer Patients:* As a primary constituent in the essential oil of pelargonium roseum (rose geranium), linalool contributed to significantly positive effects in an aromatherapy application designed to help cancer patients achieve better sleep[351].

**Prevention of Mosquito-borne Illnesses:* As the major constituents of the essential oil of pelargonium roseum (rose geranium), the terpenes citronellol, geraniol, and linalool were shown to exhibit a significant toxic effect against the larvae and egg rafts of the mosquito culex pipiens[352], which is known to carry West Nile and other dangerous diseases.

**Treatment for Schistosomiasis:* Linalool has been shown to exhibit strong molluscicidal and larvicidal activity, damaging the gills and hepatopancreas of the intermediate host species of schistosomiasis japonicum[353], which causes a persistent parasitic infection called schistosomiasis that is common in China, Indonesia, and the Philippines.

**Antifungal Agent:* Linalool markedly inhibited biofilm formation in the common pathogenic fungus, Candida albicans, and prevented the yeast's transition to the hyphal stage[354], offering a promising treatment for this sometimes serious and invasive fungal infection.

**Treatment for Lice:* Malathion-resistant and permethrin-resistant strains of human head lice appeared to be no match for linalool, which exhibited toxic fumigant activity against adults and inhibited egg hatching by 100% at 0.25 mg/cm after 24 hours of exposure[355] to the monoterpene.

**Treatment for Alzheimer's Disease*: Thanks once again to linalool's well-established anti-inflammatory effects, this compound may one day become an effective treatment for Alzheimer's disease, after linalool was shown to reverse histopathological hallmarks of the disease in older mice, restoring their cognitive and emotional functions[356]. As mentioned earlier in this section, the potential to use linalool to treat various neurodegenerative disorders is promising.

**Inhalation Treatment for Dental Anxiety and Pain:* As the primary constituent in the essential oil made from Lavandula angustifolia (lavender), inhaled linalool reduced dental anxiety by three to four times and pain by two times while significantly reducing fear of dental procedures in patients[357].

**Treatment for Non-alcoholic Fatty Liver Disease:* Oral administration of linalool to rats with high-fat diet–induced non-alcoholic fatty liver disease for 45 days suppressed the production of inflammatory cytokines,

---

[351] Cheng H, Lin L, Wang S, Zhang Y, Liu T, Yuan Y, Chen Q, Tian L. Aromatherapy with single essential oils can significantly improve the sleep quality of cancer patients: a meta-analysis. BMC Complement Med Ther. 2022 Jul 14;22(1):187.

[352] Tabari MA, Youssefi MR, Esfandiari A, Benelli G. Toxicity of B-citronellol, geraniol and linalool from Pelargonium roseum essential oil against the West Nile and filariasis vector Culex pipiens (Diptera: Culicidae). Res Vet Sci. 2017 Oct;114:36-40.

[353] Yang F, Long E, Wen J, Cao L, Zhu C, Hu H, Ruan Y, Okanurak K, Hu H, Wei X, Yang X, Wang C, Zhang L, Wang X, Ji P, Zheng H, Wu Z, Lv Z. Linalool, derived from Cinnamomum camphora (L.) Presl leaf extracts, possesses molluscicidal activity against Oncomelania hupensis and inhibits infection of Schistosoma japonicum.

[354] Manoharan RK, Lee JH, Kim YG, Kim SI, Lee J. Inhibitory effects of the essential oils a-longipinene and linalool on biofilm formation and hyphal growth of Candida albicans. Biofouling. 2017 Feb;33(2):143-155.

[355] Yang YC, Lee SH, Clark JM, Ahn YJ. Ovicidal and adulticidal activities of Origanum majorana essential oil constituents against insecticide-susceptible and pyrethroid/malathion-resistant Pediculus humanus capitis (Anoplura: Pediculidae). J Agric Food Chem. 2009 Mar 25;57(6):2282-7.

[356] Sabogal-Guaqueta AM, Osorio E, Cardona-Gomez GP. Linalool reverses neuropathological and behavioral impairments in old triple transgenic Alzheimer's mice. Neuropharmacology. 2015;102:111-20.

[357] Alkanan SAM, Alhaweri HS, Khalifa GA, Ata SMS. Dental pain perception and emotional changes: on the relationship between dental anxiety and olfaction. BMC Oral Health. 2023 Mar 26;23(1):175.

increased secreted IgA levels, and enhanced the mRNA expressions of occludin and zonula occludens-1 in the intestines, alleviating insulin resistance and anomalies in lipid homeostasis[358].

**Treatment for Osteoarthritis:* As an isolated compound, linalool reduced the catabolism of the extracellular matrix, upregulated the expression of type II collagen and aggrecan, and inhibited the signal transduction of nuclear factor kappa B in chondrocytes in a murine model of osteoarthritis[359].

**Acaricidal and Anthelmintic Agent:* As a primary constituent in the essential oil of Ocimum basilicum (basil), linalool contributed to significant acaricidal activity against Hyalomma scupense (an endophilic domestic tick), and substantial inhibition of the enzyme acetylcholinesterase in Heligmosomoides polygyrus[360] (an intestinal nematode).

**Hepatoprotective Agent:* As an isolated compound, linalool significantly increased interleukin-10 (an anti-inflammatory cytokine) while decreasing toxicity in other cytokine levels, thereby inhibiting carbon tetrachloride-induced hepatotoxicity and oxidative damage in mice[361].

**Treatment for Benzene Cytotoxicity:* Linalool has been shown to reverse benzene-induced increases in cytotoxicity, ROS formation, mitochondrial membrane potential collapse, lipid peroxidation, and oxidative stress in human lymphocytes[362]. This is important information, as benzene exposure comes in the form of tobacco smoke, automobile service stations and gas stations, exhaust from motor vehicles, industrial emissions, and off-gassing from petroleum-based products in homes such as paints, varnishes, certain furniture, carpeting, etc.

IMPLICATIONS FOR HUMAN HEALTH & NUTRITION

Dietary consumption of linalool should be relatively simple for most people considering the wide variety of fruits, vegetables, and spices that contain linalool. For instance, there are nearly two-dozen fruits that contain significant amounts of linalool, including apples, apricots, black currants, blood oranges, blueberries, cherries, cranberries, figs, grapes, grapefruit, guava, lemons, limes, mandarins, nectarines, oranges, papaya, peaches, pineapples, plums, strawberries, and yuzu. Vegetables that are high in linalool include carrots, celery, corn, and tomatoes.

When preparing and cooking foods, consider spices and herbs like allspice, anise, basil, black pepper, cilantro, cinnamon, cloves, coriander, dill, fennel, garlic, ginger, jasmine, lemongrass, mint, oregano, parsley, rosemary, saffron, sage, and turmeric, which all contain linalool and a number of other beneficial terpenes and flavonoids.

Cannabis varieties that have moderate to high concentrations of linalool include Amnesia Haze, Biker OG, Girl Scout Cookies, Island Sweet Skunk, King Tut, LA Confidential, Lavender, Master Kush, Miss X, Obama Kush, OG Shark, Pink Kush, Purple Punch, Sour Kush, and Star Pie. Juicing, blending, or eating these strains raw is an excellent, healthy way to unlock some of the nutritional and other value of linalool and other mono- and sesquiterpenes. Smoking or vaporizing cannabis is not as effective a method of consumption, and in fact we do not recommend this type of cannabis consumption. Instead, careful experimentation with essential oils may be a better, healthier option.

---

[358] Periyasamy, T., Sathibabu Uddandrao, V.V., Ponnusamy, C. et al. Linalool Mitigated High-Fat Diet–Induced Non-alcoholic Fatty Liver Disease by Regulating the Intestinal-Hepatic Axis via TGF-β/NF-kB/TLR4/ZO-1 Pathway. Rev. Bras. Farmacogn. 33, 617–628 (2023).

[359] Miao Z, Dong M, Wang Z, Ma J, Lin Y, Wu Y. Linalool inhibits the progression of osteoarthritis via the Nrf2/HO-1 signal pathway both in vitro and in vivo. Int Immunopharmacol. 2022 Dec;113(Pt A):109338.

[360] Alimi D, Hajri A, Jallouli S, Sebai H. Acaricidal and anthelmintic efficacy of Ocimum basilicum essential oil and its major constituents estragole and linalool, with insights on acetylcholinesterase inhibition. Vet Parasitol. 2022 Sep;309:109743.

[361] Hsouna AB, Sadaka C, Beyrouthy ME, Hfaiedh M, Dhifi W, Brini F, Saad RB, Mnif W. Immunomodulatory effect of Linalool (Lin) against CCl4 -induced hepatotoxicity and oxidative damage in rats. Biotechnol Appl Biochem. 2023 Feb;70(1):469-477.

[362] Salimi A, Khodaparast F, Bohlooli S, Hashemidanesh N, Baghal E, Rezagholizadeh L. Linalool reverses benzene-induced cytotoxicity, oxidative stress and lysosomal/mitochondrial damages in human lymphocytes. Drug Chem Toxicol. 2022 Nov;45(6):2454-2462.

Finally, teas and essential oils made from plants listed in the "Occurrences in Plants" section are likely to contain high amounts of linalool. Caution should be used when working with these complex EO mixtures or isolated compounds, as they can be toxic in some situations.

Linalool Review

Answer the following questions to test your knowledge of this terpene:

Question #1: What type of terpene is linalool?

a. Monoterpene
b. Monocyclic monoterpene
c. Acyclic monoterpenoid
d. Linear monoterpene

Question #2: Which of the following best describes the primary flavor & aroma profile of linalool?

a. Citrus, sweet, tropical, spicy
b. Clove, nutmeg, wood, terpenic
c. Pine, earth, dry, woody
d. Depends on the isomer

Question #3: How common is linalool in cannabis?

a. Top three
b. Top five
c. Top ten
d. Top twenty

Question #4: What is the chemical formula for linalool?

a. $C_{15}H_{26}O$
b. C15H24
c. $C_{10}H_{16}O$
d. $C_{10}H_{18}O$

Question #5: Name two uses of linalool in plants:

1 ____ 2 ____

Question #6: Name two potential medical uses of linalool:

1 ____ 2 ____

For the answer key to all individual compound chapters please go here: http://thebigbookofterps.com/bbot-quiz-answer-key/

BISABOLOL

Type: Monocyclic sesquiterpenoid
Chemical Formula: $C_{15}H_{26}O$
Molecular Weight: 222.3714 g/mol
Boiling Point: *Alpha:* 314.00 to 315.00 °C. @ 760.00 mm Hg[363]
Beta: 305.00 to 307.00 °C. @ 760.00 mm Hg[364]
Flash Point: *Alpha:* 275.00 °F. TCC (135.00 °C.)
Beta: 267.00 °F. TCC (130.56 °C.)
Solubility: *Alpha:* Soluble in alcohol, slightly soluble in ethanol, weakly soluble in water
Beta: Soluble in alcohol, insoluble in water
Oral LD50: *Alpha:* 4,710 mg/kg (rat)[365]
Beta: Unknown
Biological Role: Antiphytoviral, antifeedant, allelopathic agent
Therapeutic Role: Antitumor agent, anticancer agent, anti-inflammatory
Commercial Use: Flavoring agent, fragrance, insecticide
Occurrence in Cannabis: *Alpha:* Top ten
Beta: Unknown
Aroma/flavor: *Alpha:* Floral, peppery, balsamic, clean
Beta: Citrus, floral, tart, sweet, herbal

Occurs in Cannabis Strains: 303, Agent Orange, Banana Kush, Banana Lemonade, Bast 3, Black Lime, Blueberry Peaches, Blue Cheese, Blue Dragon, Blue Dream, Bubba Kush, Chemdawg, Dark Star, Deadhead OG, Deathstar, Diablo, GG #4, Girl Scout Cookies, God's Green Crack, Goo, Goldmeyer, Gorilla Glue, Green Crack, Island Sweet Skunk, Kosher Kush, Mickey Kush, Miracle Alien Cookies, Obama Kush, OG Kush, Purple Nepal, Purple Cadillac, Purple Punch, Purple Punch 2.0, SFV OG, Skunkdawg #9, Skywalker OG, Speakeasy, Sour Diesel, Sour Dream, Super Sour Diesel, Tosh OG, Ultimate Purple.

INTRODUCTION

Bisabolol, generally known as alpha-bisabolol since the beta- version is rarely reported, is one of the most-used terpenes in industry, and one of the most well-studied of the sesquiterpenes. Occurring generally within the top ten terpenes in cannabis, alpha-bisabolol imparts a floral, peppery, balsamic, and clean flavor and aroma, while beta-bisabolol imparts citrus, floral, tart, sweet, and herbal characteristics. Appearing in moderate to high concentrations in many common cannabis strains, alpha-bisabolol also contributes to the citrus profile of the cyclic monoterpene, limonene[366], although it is unclear if there is a positive association between alpha-bisabolol and limonene.

Originally isolated via extraction of the essential oil of Matricaria chamomilla[367], alpha-bisabolol is a well-known anti-inflammatory and antimicrobial agent, in use for centuries in various herbal, natural, and folk remedies, and

[363] The Good Scents Company Data Sheet for Alpha-bisabolol, from: http://www.thegoodscentscompany.com/data/rw1042821.html. Accessed July 30, 2023.

[364] The Good Scents Company Data Sheet for Beta-bisabolol, from: http://www.thegoodscentscompany.com/data/rw1012461.html. Accessed July 30, 2023.

[365] Agilent Technologies Safety Data Sheet for Alpha-bisabolol, from: https://www.agilent.com/cs/library/msds/TRP-100-1_NAEnglish.pdf. Accessed July 30, 2023.

[366] Bisabolol Naturhaus http://www.omikron-online.de/naturhaus/angebote/info/bisab.htm From: https://web.archive.org/web/20071010110347/. Accessed February 20, 2019.

[367] Kamatou, Guy P. P., and Alvaro M. Viljoen. A Review of the Application and Pharmacological Properties of a- Bisabolol and a-Bisabolol-Rich Oils. Journal of the American Oil Chemists' Society 87.1 (2010): 1-7.

found in dozens and likely hundreds of skin care and personal hygiene products today. Beta-bisabolol on the other hand is rarely reported in the literature. Interestingly, researchers studying cotton gin trash (stalks, seeds, leaves, burrs, and husks discarded mechanically during the cotton separating and sorting process) found that beta-bisabolol was the primary constituent in several different extracts made from these plant parts[368]. The researchers also found another rarely reported version of bisabolol named 6-epi-β-bisabolol.

CHEMICAL STRUCTURE

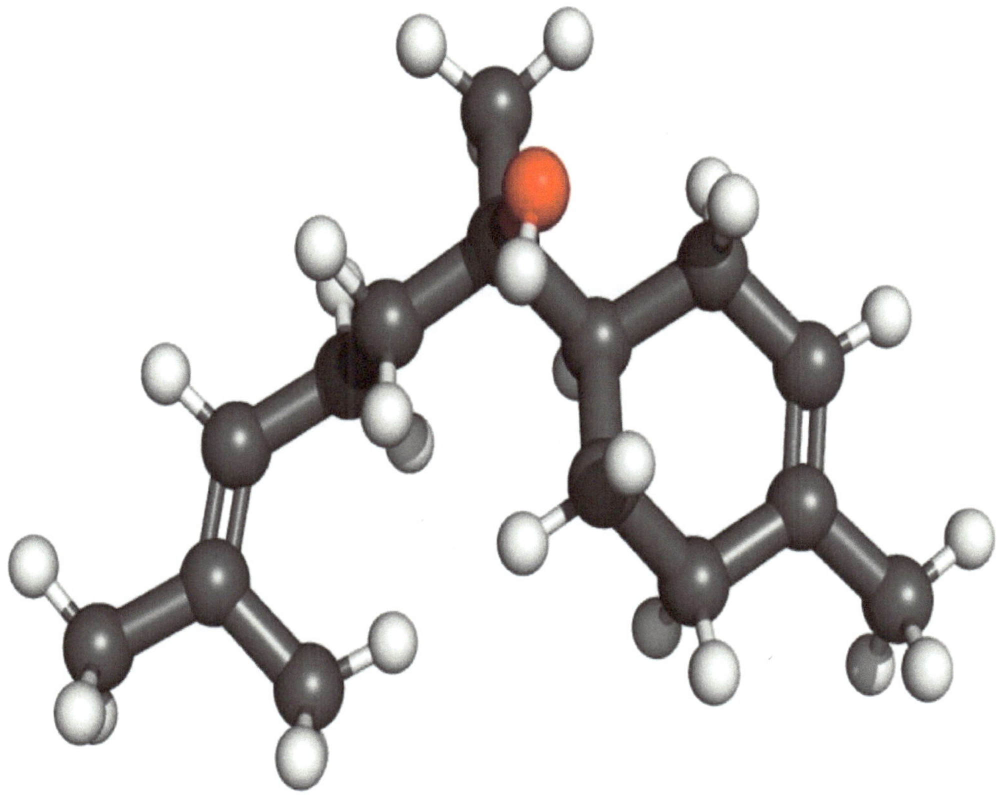

Bisabolol is formed from three linked units of isoprene consisting of fifteen carbon atoms, twenty-six hydrogen atoms, and one oxygen atom, classifying it as a sesquiterpene. Formed from two molecules of isopentenyl diphosphate in the mevalonate pathway, bisabolol is derived primarily from farnesyl diphosphate.

Bisabolol consists of two isomers differing in the position of the oxygen atom. The hydrocarbon skeleton of alpha-bisabolol contains 4 rotatable bonds and 1 covalent bond, and 1 exocyclic and 1 endocyclic double bond structured around 1 ring, which further classifies this compound as a *monocyclic* sesquiterpene. Beta-bisabolol also has 2 double bonds - one exocyclic and one endocyclic - and one oxygen atom attached to the ring, whereas in the alpha-version, the oxygen atom is attached to the center of the molecule.

OCCURRENCE IN PLANTS

Bisabolol appears in many types of plants and is common in hemp oil and many varieties of cannabis. Plants containing bisabolol include:

[368] Egbuta MA, McIntosh S, Waters DLE, Vancov T, Liu L. Chemical volatiles present in cotton gin trash: A by-product of cotton processing. PLoS One. 2019 Sep 18;14(9):e0222146.

Alpha:

Abies holophylla
Artemisia species
Basil
Bergamot
Carrot
Cascarilla
Cedarwood
Chamomile
Corn
Croton cajucara
Cubeb
Fleabane
Ginger
Guava
Juniper
Lavender
Lime
Nepeta
Orange
Phoebe
Pine
Plinia cerrocampanensis
Prangos
Pteronia
Rosemary
Sandalwood
Tansy
Wormwood
Yarrow
Ylang ylang
Zingiber

Beta:

Ayou wood
Bergamot
Black pepper
Bombycilaena erecta
Cascarilla
Cedrela
Cotton
Curcuma amada
Ginger
Guava
Lavandin
Lavebder
Lemon
Lemongrass
Phlomis fruticosa
Phoebe
Sandalwood

BIOLOGICAL ACTIVITY IN PLANTS

Bisabolol serves numerous biological functions in plants, but although the compound seems to be quite effective by itself, many researchers have observed synergistic effects when blended with other terpenes, or when tested as part of a plant's complete essential oil. These biological uses include:

Antiphytoviral Agent: Alpha-bisabolol has been shown to be an excellent antiphytoviral agent (chemical agent that exhibits antiviral activity against plant viruses), in one case reducing the number of phytovirus-caused lesions and inhibiting the development of new lesions in quinoa plants by nearly 60%[369].

Antifeedant: Some plants use this sesquiterpene to deter herbivory not only by insects, but also by mammals, as evidenced in the case of the balsam poplar tree, which deters winter feeding by snowshoe hares (a type of rabbit) through volatile emissions of terpenes including alpha-bisabolol[370]. In the US Midwest, where the Colorado potato beetle is a significant pest, alpha-bisabolol has been observed to inhibit herbivory or feeding activity of the potato beetle by more than 96%, with the study authors concluding that "alpha-bisabolol is the best option to develop natural antifeedant formulations against L. decemlineata on the basis of its high antifeedant and low phytotoxic activities[371]."

[369] Yuko, Elma, Yalerija Dunkic, Nada Bezic, Mirko Ruscic, and Dario Kremer. Chemical Composition and Antiphytoviral Activity of Essential Oil of Micromeria graeca. Natural Product Communications 7.9 (2012)1.

[370] Reichardt, P., J. Bryant, B. Mattes, T. Clausen, F. Chapin, and M. Meyer. Winter chemical defense of Alaskan balsam poplar against snowshoe hares. Journal of Chemical Ecology 16.6 (2005): 1941-1959.

[371] Ortiz De Elguea-Culebras, Gonzalo et al. Antifeedant effects of common terpenes from Mediterranean aromatic plants on Leptinotarsa decemlineata. J. Soil Sci. Plant Nutr. 2017, vol.17, n.2 pp.475-485.

**Allelopathic Agent:* Alpha-bisabolol acts as an effective allelopathic agent (a chemical that influences other plants in the environment). As the primary constituent of the essential oil of chamomile, alpha-bisabolol clearly inhibited the growth of nearby plants[372], and as a major constituent of the EO from the leaves of Prangos pabularia Lindl. (a perennial herb occurring from Russia to western Europe), exhibited high allelopathic activity against competing plant species[373].

**Insecticidal Agent:* Alpha-bisabolol (as isolated from heartwood) showed pronounced insecticidal effects against bemisia argentifolii[374] (the silverleaf white fly), a common pest of cannabis and many other plants. This suggests that sprays made with essential oils where the primary constituent is alpha-bisabolol could be effective against white flies in outdoor cannabis gardens.

**Larvicidal & Oviposition Deterrent:* Bisabolol has been shown to act as both a larvicidal and oviposition (egg-laying) deterrent agent against mosquitos (including those that spread malaria, chikungunya, and St. Louis encephalitis), as evidenced by studies examining the essential oil of hedychium larsenii, where a variation of bisabolol was one of the major constituents[375].

Interestingly, some fungi such as Schizophyllum (split gill mushroom) have been found to significantly increase emissions of beta-bisabolol after mating has been completed[376].

USES IN INDUSTRY

Alpha-bisabolol is used in many commercial and industrial applications, so much so that the production of this monocyclic sesquiterpene is pushing one tree species to the brink of extinction. Eremanthus erythropappus, (Candeia tree) is native to Brazil, where the trees have been harvested in huge numbers for the production of alpha-bisabolol, which has shrunk the population of these trees to near-extinction levels[377]. This has spurred scientists to develop an alternative source of alpha-bisabolol, with many experimenting with using engineered strains of the common bacteria, E. coli, to produce the compound[378].

While not as common in the prepared foods industry as terpenes like limonene or linalool, alpha-bisabolol is commonly used as a flavoring agent in baked goods, alcoholic and non-alcoholic beverages, breakfast cereal, chewing gum, confectionary, fats and oils, frozen dairy, fruit ices, gelatins, hard candy, herbs and spices, tea, imitation dairy, instant coffee, seasonings and flavorings, soft candy, and soups, among other food items.

Alpha-bisabolol is also used heavily in fragrances, cosmetics, and personal hygiene products, where it is recommended at not more than 8% in the fragrance concentrate. This sesquiterpene can be found in lipstick, facial cleansers, hand and body lotions, conditioners, moisturizers, sunscreen, soothing agents, etc. While alpha-bisabolol

---

[372] Demarque, Daniel & Sab6ia, J.F. & Fabri, Joao & Alexandre Carollo, Carlos. (2012). Allelopathic activity of Matricaria chamomilla essential oil in bioautography test. Allelopathy Journal. 29. 171-176.

[373] Razavi SM. Chemical and allelopathic analyses of essential oils of Prangos pabularia Lindl. from Tran. Nat Prod Res. 2012;26(22):2148-51.

[374] de Andrade TL, Bezerra JN, Lima MA, de Faria RA, Lima MA, Andrade-Neto M, Cavalcanti FS, Mesquita AL. Chemical composition and insecticidal activity of essential oils from Vanillosmopsis pohlii baker against Bemisia argentifolii. J Agric Food Chem. 2004 Sep 22;52(19):5879-81.

[375] Al Shebly MM, AlQahtani FS, Govindarajan M, Gopinath K, Vijayan P, Benelli G. Toxicity of ar-curcumene and epi-beta-bisabolol from Hedychium larsenii (Zingiberaceae) essential oil on malaria, chikungunya and St. Louis encephalitis mosquito vectors. Ecotoxicol Environ Saf. 2017 Mar;137:149-157.

[376] Freihorst D, Brunsch M, Wirth S, Krause K, Kniemeyer O, Linde J, Kunert M, Boland W, Kothe E. Smelling the difference: Transcriptome, proteome and volatilome changes after mating. Fungal Genet Biol. 2018 Mar;112:2-11.

[377] Alves Gomes Albertti, Leticia, Thierry L. Delatte, Katyuce Souza de Farias, Amanda Galdi Boaretto, Francel Verstappen, Adele van Houwelingen, Katarina Cankar, Carlos Alexandre Carollo, Harro J. Bouwmeester, and Jules Beekwilder. Identification of the Bisabolol Synthase in the Endangered Candeia Tree (Eremanthus erythropappus McLeisch. Frontiers in Plant Science 9 (2018).

[378] Han G.H., Kim, S.K., Yoon. Fermentative production and direct extraction of bisabolo in metabolically engineered E.coli. Microb Cell Fact 15, 185 (2016).

has long been known for its anti-inflammatory properties, many product formulations take advantage of other characteristics of the compound, including its ability to smooth and soothe skin, and because this terpene penetrates and absorbs readily through skin[379]. alpha-bisabolol also significantly lightens skin pigmentation[380] in some formulations.

Other industrial and commercial uses of alpha-bisabolol include all-natural insect repellant lotions for children[381], and mixed insecticides patented in part for their uniquely low toxicity levels[382]. Other work underway includes a project that seeks to produce hernadulcin - a safe, low-calorie sugar substitute - from alpha-bisabolol[383].

Commercial formulations of alpha-bisabolol generally appear as a colorless or pale-yellow, slightly viscous oil.

POTENTIAL USES IN MEDICINE

Bisabolol shows strong potential in medicine, and overall this substance has been reviewed extensively, although the current literature suggests that much more research is needed, as there is almost certainly even more medical potential than suspected in this isoprenoid molecule:

**Antifungal Agent:* Alpha-bisabolol exhibits potential as a new, potent, non-toxic antifungal agent; a growing need as fungal resistances increase worldwide. Fortunately, alpha-bisabolol has been shown to interfere with fungi-specific ergosterol biosynthesis[384]; a critical first development step in fungi.

**Anti-inflammatory Agent:* As mentioned previously in the "Uses in Industry" section, alpha-bisabolol is easily absorbed through skin. Topical application of this compound has also been observed to significantly inhibit skin inflammation in mice by reducing cytokine production[385], confirming its already well-known status as an anti-inflammatory agent.

**Anti-tumor Agent:* Alpha-bisabolol exhibits strong anti-tumor properties, inhibiting tumor growth through enhancement of cytokine production. Interestingly, this was observed in a water-soluble extract from the mushroom, Agrocybe aegerita, which contained primarily triethylene glycol, and alpha-bisabolol[386], which induced apoptosis (programmed cell death) of malignant pancreatic carcinoma cells[387].

---

[379] Kerry Bone, Simon Mills, Principles and Practice of Phytotherapy. 2nd Edition 2013. Pgs 475-580.

[380] Lee, J., H. Jun, E Jung, J. Ha, and D. Park. Whitening effect of bisabolol in Asian women subjects. International Journal of Cosmetic Science 32.4 (2010).

[381] Amerta All Natural Insect Repellent Lotion for Kids. DEET FREE. Ingredients - Lemongrass distilled hydrolate, Lemongrass essential oil, Lavender essentail oil, Rosemary essential oil, Peppermint essential oil, Jojoba oil, Bisabolol. (February 21, 2019).

[382] US Patent US 20100040705A1. (February 2019).

[383] Attia M, Kim SU, Ro DK. Molecular cloning and characterization of (+)-epi-a-bisabolol synthase, catalyzing the first step in the biosynthesis of the natural sweetener, hernandulcin, in Lippia dulcis. Arch Biochem Biophys. 2012 Nov 1;527(1):37-44.

[384] Alexander Pauli, a-Bisabolol from Chamomile - A specific ergosterol biosynthesis inhibitor? International Journal of Aromatherapy, Volume 16, Issue 1, 2006, Pages 21-25.

[385] Maurya AK, Singh M, Dubey V, Srivastava S, Luqman S, Bawankule DU. a-(-)-bisabolol reduces pro- inflammatory cytokine production and ameliorates skin inflammation. Curr Pharm Biotechnol. 2014;15(2):173-81.

[386] Liang Y, Liu HH, Chen YJ, Sun H. Antitumor activity of the protein and small molecule component fractions from Agrocybe aegerita through enhancement of cytokine production. J Med Food. 2014 Apr; 17(4):439-46.

[387] Darra E, Abdel-Azeim S, Manara A, Shoji K, Maréchal JD, Mariotto S, Cavalieri E, Perbellini L, Pizza C, Perahia D, Crimi M, Suzuki H. Insight into the apoptosis-inducing action of alpha-bisabolol towards malignant tumor cells: involvement of lipid rafts and Bid. Arch Biochem Biophys. 2008 Aug 15; 476(2):113-23.

**Precursor for Anti-tumor Agent:* In relation to the anti-tumor properties of alpha-bisabolol, this compound also serves as a precursor for thiosemicarbazones (a class of ligands that show a diverse range of biological activity and are useful in the treatment of cancer), which have been shown to exhibit inhibitory effects on a wide range of cancer cells, especially myeloid leukemia cells[388].

**Treatment for Brain & Spinal Cord Cancer:* Alpha-bisabolol has been shown to have a strong effect on glioma cells (cancerous cells in the brain and spinal cord), reducing cell viability by 50% with moderate doses of the monocyclic sesquiterpene, and resulting in 100% cell death with high doses[389].

**Gastroprotective Agent:* This sesquiterpene shows potential as a gastroprotective agent, and may be useful in the treatment of alcoholism. That's because alpha-bisabolol reduces gastric lesions caused by absolute ethanol, attenuating these lesions by as much as 96%[390].

**Antinociceptive Agent:* Scientists in the Balkans have demonstrated that alpha-bisabolol exhibits both antinociceptive (blocks pain perception) and anti-edematous (stops or reduces swelling or edema) properties[391], suggesting that this compound may prove effective as a pain medicine, and to treat inflammation, as mentioned previously.

**Pro-Apoptotic Agent for Leukemia Cells:* In the first study of its kind, scientists in Italy have demonstrated that alpha-bisabolol is a pro-apoptotic (promoter of programmed cell death) agent for human leukemia cells[392].

**Treatment of Leishmaniasis:* Visceral Leishmaniasis, a parasitic infection spread by sandflies, is a severe form of black fever, that often results in death if not treated. Alpha-bisabolol has been shown to reduce leishmaniasis parasite load in the spleen and liver by more than 89%, and was more effective than meglumine antimoniate (a medicine used to treat leishmaniasis) at reducing spleen parasite load, and comparably effective as meglumine at reducing parasite load in the liver, but without toxicity[393].

**Attenuation of Chemotherapeutic Agent Toxicity:* Although many chemotherapeutic agents such as Doxorubicin are very effective at destroying or inhibiting cancer cells, healthy cells and organs often suffer collateral cytotoxic damage. Researchers working to solve this problem found that alpha-bisabolol prevented testicular weight loss and oxidative stress, activated Nrf2 signaling and triggered upregulation of antioxidant defenses, attenuated the levels and expression levels of pro-inflammatory cytokines, protected the testicular architecture, attenuated the expression levels of inflammatory mediators and downregulated NF-κB/MAPK signaling pathway, and attenuate ER-stress-mediated testicular apoptosis in DOX-induced testicular damage in rats[394]. The same researchers also found that

---

[388] da Silva AP, Martini MV, de Oliveira CM, Cunha S, de Carvalho JE, Ruiz AL, da Silva CC. Antitumor activity of (-)-alpha-bisabolol-based thiosemicarbazones against human tumor cell lines. Eur J Med Chem. 2010 Jul; 45(7):2987-93.

[389] Elisabetta Cavalieri, Sofia Mariotto, Cinzia Fabrizi, Alessandra Carcereri de Prati, Rossella Gottardo, Stefano Leone, Luigi Valentino Berra, Giuliana Maria Lauro, Anna Rosa Ciampa, Hisanori Suzuki. a-Bisabolol, a nontoxic natural compound, strongly induces apoptosis in glioma cells. Biochemical and Biophysical Research Communications, Volume 315, Issue 3, 2004, Pages 589-594.

[390] Bezerra SB, Leal LK, Nogueira NA, Campos AR. Bisabolol-induced gastroprotection against acute gastric lesions: role of prostaglandins, nitric oxide, and KATP+ channels. J Med Food. 2009 Dec; 12(6):1403-6.

[391] Popovic, Visnja, Silvana Petrovic, Maja Tomic, Radica Stepanovic-Petrovic, Ana Micov, Milica Pavlovic- Drobac, Maria Couladis, and Marjan Niketic. Antinociceptive and Anti-edematous Activities of the Essential Oils of Two Balkan EndemicLaserpitium Species. Natural Product Communications 9.1 (2014)1.

[392] Cavalieri, Elisabetta, Antonella Rigo, Massimiliano Bonifacio, Alessandra Carcereri de Prati, Emanuele Guardalben, Christian Bergamini, Romana Fato, Giovanni Pizzolo, Hisanori Suzuki, and Fabrizio Vinante. Pro- apoptotic activity of a-bisabolol in preclinical models of primary human acute leukemia cells. Journal of Translational Medicine 9 (2011): 45-45.

[393] Corpas-L6pez, Victoriano & Morillas-Marquez, Francisco & Navarro-Moll, M & Merino Espinosa, Gemma & Diaz-Saez, Victoriano & Martin-Sanchez, Joaquina. (2015). (-)-a-Bisabolol, a Promising Oral Compound for the Treatment of Visceral Leishmaniasis. Journal of natural products. 78. 10.1021/np5008697.

[394] Arunachalam S, Nagoor Meeran MF, Azimullah S, Kumar Jha N, Saraswathiamma D, Albawardi A, Beiram R, Ojha S. α-Bisabolol Attenuates NF-κB/MAPK Signaling Activation and ER-Stress-Mediated Apoptosis by Invoking Nrf2-Mediated Antioxidant Defense Systems against Doxorubicin-Induced

alpha-bisabolol prevented body weight loss, restored antioxidants, mitigated lipid peroxidation, inhibited the rise in proinflammatory cytokines, and favorably modulated the expressions of NF-κB/MAPK signaling and apoptosis markers in doxorubicin-induced nephrotoxicity in male Wistar albino rats[395].

Treatment for Arthritis: In a complete Freund's adjuvant (CFA)-induced arthritis model in Male albino Swiss mice, researchers found that treatment with alpha-bisabolol significantly attenuated spontaneous "pain-like" behaviors, mechanical hyperalgesia, grip strength deficit, and knee edema, reducing joint pain and functional disability associated with arthritis while also inhibiting the generation of inflammatory and oxidative markers in the knee, and blocked mitogen-activated protein kinases in the spinal cord [396].

Treatment for Dermatitis: As an isolated compound, alpha-bisabolol reduced atopic dermatitis-like symptoms and the release of interleukin (IL)-4 without immunoglobulin (Ig)-E production, reduced epidermal thickness and inhibited mast cells in the atopic dermatitis-like skin lesions, and suppressed mast-cell-mediated passive cutaneous anaphylaxis, among several other effects in 2,4-dinitrochlorobenzene-induced atopic dermatitis in BALB/c mice[397] (a very common mouse bred for laboratory research).

Neuroprotective Agent: Alpha-bisabolol has been shown to cause a significant increase in cell viability in undifferentiated cells while also demonstrating a significant neuroprotective effect against amyloid beta (Aβ1-42) protein exposure, with researchers writing, "[the] bioactivity of selected terpenes should be a consideration in the emergent use of medicinal cannabis formulations for the treatment of neurodegenerative diseases[398].

Hepatoprotective Agent: As an isolated compound, alpha-bisabolol exerted a hepatoprotective effect against Cadmium (a heavy metal) toxicity by reducing the hepatic content of the toxin as well as antagonizing oxidative insults, inflammation, and apoptosis in cadmium-induced liver injury in ducks[399].

IMPLICATIONS FOR HUMAN HEALTH & NUTRITION

Dietary bisabolol can be acquired by regularly consuming fruits like guava, oranges, lemons, and limes. When cooking, consider using spices like rosemary, basil, black pepper, or ginger. Chamomile, lemongrass, and ginger teas typically contain this terpene, and vegetables like corn and carrots may also contain bisabolol.

Cannabis strains that are reportedly high in a-bisabolol like 303, Agent Orange, Chemdawg, Dark Star, Gorilla Glue, Mickey Kush, Purple Punch 2.0, Speakeasy, Goldmeyer, Skunkdawg, Sour Diesel, or Super Sour Diesel can be eaten raw, juiced, or included in teas. These same strains or others like them (or any strain that tests high in a-bisabolol) can also be smoked or vaporized, however, from a good health standpoint, inhaling the smoke of any substance is not recommended. Cannabis extracts that have retained or otherwise have added alpha-bisabolol are alternative options, provided dosing and usage recommendations are followed.

Testicular Toxicity in Rats. Nutrients. 2022 Nov 3;14(21):4648.

[395] Arunachalam S, Nagoor Meeran MF, Azimullah S, Jha NK, Saraswathiamma D, Subramanya S, Albawardi A, Ojha S. α-Bisabolol Attenuates Doxorubicin Induced Renal Toxicity by Modulating NF-κB/MAPK Signaling and Caspase-Dependent Apoptosis in Rats. Int J Mol Sci. 2022 Sep 10;23(18):10528.

[396] Heimfarth L, Rezende MM, Pereira EWM, Passos FRS, Monteiro BS, Santos TKB, Lima NT, Souza ICL, de Albuquerque Junior RLC, de Souza Siqueira Lima P, de Souza Araújo AA, Quintans Júnior LJ, Kim B, Coutinho HDM, de Souza Siqueira Quintans J. Pharmacological effects of a complex α-bisabolol/β-cyclodextrin in a mice arthritis model with involvement of IL-1β, IL-6 and MAPK. Biomed Pharmacother. 2022 Jul;151:113142.

[397] Li G, Wu H, Sun L, Cheng K, Lv Z, Chen K, Qian F, Li Y. (-)-α-Bisabolol Alleviates Atopic Dermatitis by Inhibiting MAPK and NF-κB Signaling in Mast Cell. Molecules. 2022 Jun 21;27(13):3985.

[398] Laws Iii JS, Shrestha S, Smid SD. Cannabis terpenes display variable protective and anti-aggregatory actions against neurotoxic β amyloid in vitro: highlighting the protective bioactivity of α-bisabolol in motorneuronal-like NSC-34 cells. Neurotoxicology. 2022 May;90:81-87.

[399] Elazab ST, Hsu WH. Antagonism of cadmium-induced liver injury in ducks by α-bisabolol. Front Vet Sci. 2022 Nov 7;9:1024549.

Supplementation or experimentation with bisabolol is also possible by purchasing or making essential oils from plants that are high in this compound such as those listed in the "Occurrences in Plants" section above, however, the best way to consume this monocyclic sesquiterpene is as part of a regular healthy diet. Extreme care should be exercised when working with complex essential oil mixtures or pure compounds, and they can be toxic in some situations.

Bisabolol Review

Answer the following questions to test your knowledge of this terpene:

Question #1: What type of terpene is bisabolol?

a. Monocyclic monoterpene
b. Bicyclic monoterpene
c. Monocyclic sesquiterpene
d. Bicyclic sesquiterpene

Question #2: Which of the following best describes the primary flavor & aroma profile of bisabolol?

a. Citrus, fruity, dry, terpenic
b. Sweet, floral, balsamic, tangy
c. Spicy, turpentine, dry
d. Depends on the isomer
e. Hops, cloves, wood, spicy

Question #3: How common is bisabolol in cannabis?

a. Top three
b. Top five
c. Top ten
d. Top twenty

Question #4: What is the chemical formula for bisabolol?

a. $C_{10}H_{16}C$
b. $H_{24}CO_{20}$
c. $C_{20}H_{32}$
d. $C_{15}H_{26}O$

Question #5: Name two uses of bisabolol in plants:

1 ________________________ 2 ________________________

Question #6: Name two potential medical uses of bisabolol:

1 ________________________ 2 ________________________

For the answer key to all individual compound chapters please go here: http://thebigbookofterps.com/bbot-quiz-answer-key/

NEROLIDOL

Type: Acyclic sesquiterpenoid
Chemical Formula: $C_{15}H_{26}O$
Molecular Weight: 222.372 g/mol
Boiling Point: *Cis:* 114.00 °C. @ 1.00 mm Hg[400]
Trans ((E)-nerolidol): 276.00 to 277.00 °C. @ 760.00 mm Hg[401]
D ((+)-nerolidol): 290.00 to 293.00 °C. @ 760.00 mm Hg[402]
Z ((Z)-nerolidol): 276.00 °C. @ 760.00 mm Hg (estimated)[403]
Flash Point: *Cis:* 205.00 °F. TCC (96.11 °C.)
Trans ((E)-nerolidol): 230.00 °F. TCC (> 110.00 °C.)
D ((+)-nerolidol): 229.00 °F. TCC (109.44 °C.)
Z ((Z)-nerolidol): 230.00 °F. TCC (109.90 °C.)
Melting Point: *Cis and Trans:* -75 °C (-103 °F)[404]
D ((+)-nerolidol): 19.00 to 21.00 °C. @ 760.00 mm Hg
Solubility: *Cis:* Soluble in alcohol, ethanol, dipropylene glycol, fixed oils, propylene glycol, weakly soluble in water, insoluble in glycerin
Trans ((E)-nerolidol): Soluble in alcohol, fixed oils, propylene glycol, weakly soluble in water, insoluble in glycerol
D ((+)-nerolidol): Soluble in alcohol, weakly soluble in water
Z ((Z)-nerolidol): Soluble in alcohol, weakly soluble in water
S-(-)-(E)-nerolidol: Weakly soluble in water
Oral LD50: *Cis:* 5628 mg/kg (rat)[405]
Mixture of cis- and trans-nerolidol: >8,000 mg/kg (mouse)[406]
Biological Role: Insect repellant, allelopathic agent
Therapeutic Role: Antimalarial, trypanocidal, antifungal, sedative
Commercial Use: Flavoring agent, fragrance
Occurrence in Cannabis: Top ten for cis- and trans-nerolidol, unknown for other isomers
Aroma/Flavor: *Cis:* Floral, green, waxy, citrus, woody
Trans ((E)-nerolidol): Floral, green, citrus, woody, waxy
D ((+)-nerolidol): Floral
Z ((Z)-nerolidol): Waxy, floral
S-(-)-(E)-nerolidol: Sweet, mild, soft, flowery[407]

Occurs in Cannabis Strains: Afgoo, Agent 4, Agent Orange, Black Lime, Blueberry Peaches, Blueberry Peaches, Blue Cheese, Blue Dream, Bubba Kush, Cannatonic, Chemdawg, Critical Jack, Dark Star, Deadhead OG, Diablo, Elvis, GG #4, Ghost Train Haze, Gorilla Glue, Green Crack, Kosher Kush, Mickey Kush, Obama Kush, OG Kush,

---

[400] The Good Scents Company Data Sheet for Nerolidol, from: http://www.thegoodscentscompany.com/data/rw1009032.html. Accessed July 31, 2023.

[401] The Good Scents Company Data Sheet for Trans-nerolidol ((E)-nerolidol), from: http://www.thegoodscentscompany.com/data/rw1053991.html. Accessed July 31, 2023.

[402] The Good Scents Company Data Sheet for D-nerolidol ((+)-nerolidol), from: http://www.thegoodscentscompany.com/data/rw1384621.html. Accessed July 31, 2023.

[403] The Good Scents Company Data Sheet for Z-nerolidol, from: http://www.thegoodscentscompany.com/data/rw1102331.html. Accessed July 31, 2023.

[404] Chemical Book Product Description for Nerolidol, from: https://www.chemicalbook.com/ChemicalProductProperty_US_CB2206877.aspx. Accessed July 31, 2023.

[405] Spex OSHA Safety Data Sheet for Cis-nerolidol, from: https://www.spex.com/MSDS/S-7608.pdf. Accessed July 31, 2023.

[406] The John D Walsh Company Safety Data Sheet for Racemic Nerolidol, from: https://johndwalsh.com/wp-content/uploads/2016/09/NEROLIDOL-GHS-SDS.pdf. Accessed July 31, 2023.

[407] The Good Scents Company Data Sheet for S-(-)-(E)-nerolidol, from: http://www.thegoodscentscompany.com/data/rw1556301.html. Accessed July 31, 2023.

Pink Lemonade, Purple Cadillac, Purple Nepal, Purps, Sour Cherry Sherbet, Sour Diesel, Sour Jack, Sunset Sherbet, Super Sour Diesel, The Gift, Trainwreck.

INTRODUCTION

Nerolidol is a sesquiterpene produced by many types of plants, most notably found in the oil of the bitter orange flower blossom, and from ginger, tea tree, and lemongrass. This compound is known for its subtle aroma and flavor profile, which is floral, woody, green, waxy, and citrus for the trans- and cis- isomers. Often, the overall scent of this sesquiterpene is described as 'clean.'

Nerolidol occurs commonly in cannabis, typically within the top 10 terpenes by concentration for many varieties. Although this hydrocarbon appears often in cannabis varieties, it is usually present in moderate to low concentrations, although strains like those listed above - particularly Black Lime, Agent Orange, Bubba Kush, and Jack Herer, feature nerolidol as a primary terpene.

Interestingly, although scientists are working on commercial production of nerolidol by synthesis from certain bacteria, and from brewer's yeast (read more about this in the "Uses in Industry" section below), a natural, non-plant source of this sesquiterpene has been discovered; the compound is also produced in the pheromones of the fruit spotting bug[408], and by one of the most feared pests of cannabis; the spider mite (see more on this in the "Biological Activity In Plants" section).

One of the most prolific natural users of nerolidol is termites in the Syntermitinae family, which emits (3R,6E)-nerolidol in the development of queens, and to suppress females from developing into queens, and is proposed as the key regulatory element in the breeding system of the termites[409]. In Embiratermes neotenicus termites, queens emit significant amounts of the R- version the sesquiterpene alcohol trans-nerolidol, which is found on the surface of the abdomen and in internal tissues in queens, as well as on the surface of eggs, but completely absent in kings, workers, and soldiers[410].

---

[408] Le Thanh C, Chauhan KR. Simple and short synthesis of trans-(R)-nerolidol, a pheromone component of fruit spotting bug. Nat Prod Commun. 2014 Mar;9(3):297-8.

[409] Dolejšová K, Křivánek J, Štáfková J, Horáček N, Havlíčková J, Roy V, Kalinová B, Roy A, Kyjaková P, Hanus R. Identification of a queen primer pheromone in higher termites. Commun Biol. 2022 Nov 2;5(1):1165.

[410] Havlíčková J, Dolejšová K, Tichý M, Vrkoslav V, Kalinová B, Kyjaková P, Hanus R. (3R,6E)-nerolidol, a fertility-related volatile secreted by the queens of higher termites (Termitidae: Syntermitinae). Z Naturforsch C J Biosci. 2019 Sep 25;74(9-10):251-264.

CHEMICAL STRUCTURE

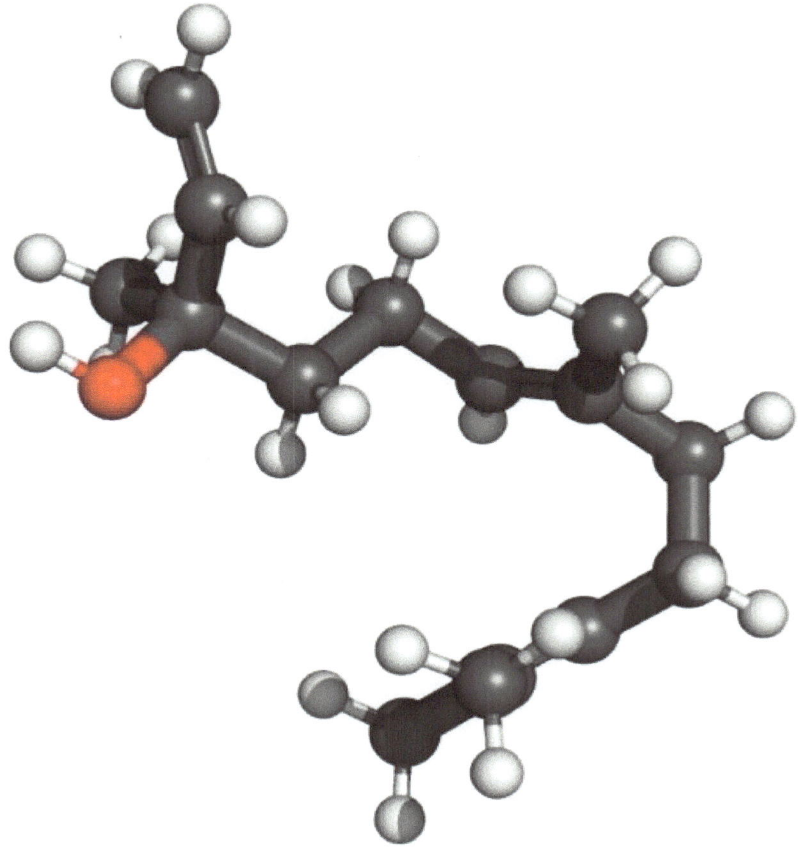

Nerolidol is derived from farnesyl diphosphate in the cytosol of plant cells. This compound is comprised of three units of isoprene, which classifies it as a sesquiterpene. Nerolidol has fifteen carbon atoms, twenty-six hydrogen atoms, and one oxygen atom, notated as $C_{15}H_{26}O$.

This isoprenoid skeleton contains seven rotatable and one covalent bond, and is organized without a ring, further classifying this molecule as an acyclic sesquiterpene. The molecule features three double bonds.

Nerolidol occurs as two primary isomers, cis- and trans-, which are differentiated by the geometry of the central double bond and the position of the functional group. These isomers are generally found in plants as a racemic mixture, and several other isomers have been found as well. In this chapter we discuss the following, although there are at least 2 other isomers that are mostly unreported in the literature and not discussed here:

Cis-nerolidol
Trans ((E)-nerolidol)
D ((+)-nerolidol)
Z ((Z)-nerolidol)
S-(-)-(E)-nerolidol

One interesting chemical characteristic of nerolidol, as observed in kiwifruit, is that the terpene farnesol, a variation of which occurs occasionally in cannabis (typically in the top thirty terpenes of some strains), competes for the available pool of farnesyl diphosphate used in sesquiterpene biosynthesis, and ultimately influences the production of nerolidol[411]. For cannabis breeders, geneticists, or growers seeking to produce strains that are high in nerolidol, it may be critical to ensure that the genetics of the target strain not contain high amounts of farnesol or farnesene.

OCCURRENCE IN PLANTS

Although not as common as monoterpenes like linalool, limonene, or myrcene, nerolidol is found in a wide variety of plant species, including the following:

Cis:

Ambrette seed	Anise	Basil
Basswood	Bergamot	Black pepper
Cabreuva	Cananga	Cardamom
Carrot seed	Cascarilla	Chamomile
Cinnamon	Clary sage	Coriander
Cornmint	Galangal	Ginger
Gingergrass	Grapefruit	Guava
Jasmin	Lemon	Mandarin
Marjoram	Melon	Michelia champaca
Neroli	Nutmeg	Orange
Osmanthus absolute	Parsley	Peppermint
Plectranthus glandulosus	Plum	Plumcot
Pomelo	Sideritis species	Skimmia laureola
Stachys recta	Strawberry (wild)	Tarragon
Tea	Thyme	Water mint
Ylang ylang	Yuzu	

Trans:

Allamanda cathartica	Ambrette	Amyris
Artemisia species	Ayou	Bergamot
Black pepper	Blood orange	Cedrela wood
Champaca	Cinnamon	Cistus
Couroupita guianensis	Croton species	Cubeb
Curry	Fleabane	Ginger
Grapefruit	Guava	Hyssop
Jambu	Kewda	Laurel
Lavender	Layana	Lime
Mandarin	Mangrove	Manuka species
Microstrobos fitzgeraldii	Mikania cordata	Myrrhinium atropurpureum
Myrtle	Nepeta species	Neroli
Orange	Orthodon dianthera	Palmarosa
Persea americana	Phoebe	Pteronia
Rue oil	Salvia	Sandalwood
Satureja viminea	Strawberry (wild)	Vassoura
Verbena	Witch hazel	Wormwood
Yarrow	Ylang ylang	

[411] Green, Sol A et al. Identification, functional characterization, and regulation of the enzyme responsible for floral (E)-nerolidol biosynthesis in kiwifruit (Actinidia chinensis). Journal of experimental botany vol. 63,5 (2012): 1951-67

Z:

Clary sage	Helichrysum italicum	Ho leaf
Lavender	Neroli	Peppermint

D:

Black pepper	Ginger	Hyssop
Peru balsam	Phlox paniculata	Tobacco jasmine

BIOLOGICAL ACTIVITY IN PLANTS

Nerolidol appears to perform an impressive number of biological functions in plants, including antibacterial, indirect defense, allelopathic, and insecticidal actions, among others.

Antibacterial Agent: Trans-nerolidol as produced by and extracted from rice has been shown to exhibit significant antibacterial activity against Xanthomonas oryzae pv.[412], a common bacterial pathogen of rice plants.

Indirect Defense: When attacked by the Egyptian cotton leafworm, corn plants release a mixture of mono- and sesquiterpenes - notably trans-nerolidol - which is used as a chemical signal by parasitic wasps to locate the larvae of the Egyptian cotton leafworm, which the wasp then parasitizes[413].

Allelopathic Agent: Nerolidol has been shown to exhibit potent allelopathic properties. In the case of cleome droserifolia - a plant that features capitate trichomes that are strikingly like those found in cannabis - the dominant sesquiterpene in the essential oil of the plants was cis-nerolidol, which inhibited the growth of seeds, shoots, and roots of nearby clover and other weeds[414]. In weeds like Bidens pilosa and Urospermum picroides, emissions of z-nerolidol from the nearby cullen plicata plant *completely* inhibited their growth[415]. This compound and its variations have significant implications as a natural, non-toxic pesticide, especially considering that chemical pesticides are dangerously impacting the environment and human health.

Insect Repellant: The essential oil of Siam-wood and amyris, a primary component of which is trans-nerolidol, showed repellant activity against both the 24-hour house fly, and the brown dog tick, which were observed to repeatedly avoid and move away from surfaces that had been treated with a spray containing the nerolidol-rich essential oil[416]. Other studies have shown that nerolidol successfully repels aphids[417], a common pest of cannabis. Perhaps most interestingly, cis-nerolidol has been observed to effectively repel and confuse the male spider mite.

---

[412] Kiryu, M., Hamanaka, M., Yoshitomi, K. et al. Rice terpene synthase 18 (OsTPS18) encodes a sesquiterpene synthase that produces an antibacterial (E)-nerolidol against a bacterial pathogen of rice. J Gen Plant Pathol 84, 221-229 (2018).

[413] Schnee, Christiane et al. The maize gene terpene synthase 1 encodes a sesquiterpene synthase catalyzing the formation of (E)-beta-farnesene, (E)-nerolidol, and (E,E)-farnesol after herbivore damage. Plant physiology vol. 130,4 (2002): 2049-60.

[414] Abd El-Gawad AM, El-Amier YA, Bonanomi G. Essential Oil Composition, Antioxidant and Allelopathic Activities of Cleome droserifolia (Forssk.) Delile. Chem Biodivers. 2018 Dec;15(12):e1800392.

[415] Ahmed M. Abd El-Gawad. Chemical constituents, antioxidant and potential allelopathic effect of the essential oil from the aerial parts of Cullen plicata. Industrial Crops and Products, Volume 80, 2016, Pages 36-41.

[416] Paluch, Gretchen & Zhu, Junwei & Bartholomay, Lyric & Coats, Joel. (2010). Amyris and Siam-wood Essential Oils: Insect Activity of Sesquiterpenes.

[417] M. Itkin and A. Aharoni, with Anne E. Osbourn, Virginia Lanzotti Editors. Plant-derived Natural Products: Synthesis, Function, and Application; Production of Medical Agents in Plants, Page 456 Jul 7, 2009.

Researchers observed that the female spider mite produces both farnesol and cis-nerolidol as chemical signals to attract male spider mites for mating. Subsequently, a spray was developed that contained farnesol, cis-nerolidol, and a miticide. Male mites were attracted to the spray and exhausted their energies attempting to find the non-existent females, instead finding only spray-on terpenes and perishing in the miticide[418]. This is important information for cannabis growers, and provides the potential for a natural, non-toxic method of controlling mites. For instance, a spray like the one mentioned here could be developed without the chemical miticide.

Growers could co-plant a minor crop near cannabis plants, and, upon signs of mite infestation, spray the minor crop with the terpene blend, ultimately inhibiting the mating process as male spider mites expend their energies on the plants where no female mites are signaling, causing the reproduction cycle of those female mites using cannabis as a host to be broken.

Larvicidal Agent: Trans-nerolidol has been found to be toxic to the larvae of Aedes aegypti[419] - the mosquito responsible for spreading Zika virus, dengue fever, yellow fever, and other diseases.

Interestingly, the beetle Cyclocephala paraguayensis was shown to be significantly attracted to blends of methyl benzoate and nerolidol – two of the primary compounds found in the essential oil of Brugmansia suaveolens (trumpet flowers)[420], although it's unknown what purpose this might serve.

USES IN INDUSTRY

Nerolidol is frequently used in the prepared and processed foods industry, usually in teas in the form of raspberry, peach, or other floral-fruity flavors. This acyclic sesquiterpene is used in baked goods, hard candy, dairy products, ices, sherbet, sorbet, processed fruit, cereal, processed vegetables, egg products, sweeteners, spices, sauces and soups, prepared salads, and alcoholic beverages.

Nerolidol is found as a fragrance and masking agent in detergents and cleaners, perfumes, shampoos, conditioners, body lotions, etc. Although currently produced as an extract from a variety of plants, scientists are working on the production of commercial nerolidol from other, more efficient sources, including brewer's yeast, wherein one group of researchers sought to build a cell factory[421] to produce this terpene.

Because nerolidol is difficult and inefficient to produce from plant sources, other efforts are underway to produce this compound using several other methods. In one case, researchers engineered Escherichia coli bacteria via optimization of its biosynthetic pathways to produce trans-nerolidol from sugar and glycerol[422]. Other work focused on protein engineering and pathway compartmentation in Yarrowia lipolytica (an ascomycetous yeast) to enhance production of trans-nerolidol[423].

---

[418] Office of Pesticide Programs; Environmental Protection Agency. 3,7,11-Trimethyl-2,6,10-dodecatriene-1-ol (Farnesol) (128910) 3,7,11-Trimethyl-1,6,10-dodecatriene-3-ol (Nerolidol) (128911) Fact Sheet. From: https://www3.epa.gov/pesticides/chem_search/reg_actions/registration/fs_G-40_01-Apr-01.pdf Accessed February 22, 2019.

[419] Chantraine, Jean-Marie, Dominique Laurent, Carolt Ballivian, Gloria Saavedra, Ruben Ibanez, and L. Antonio Vilaseca. Insecticidal activity of essential oils on Aedes aegypti larvae. Phytotherapy Research 12.5 (1998):350-354.

[420] Favaris AP, Túler AC, Silva WD, Pec M, Rodrigues SR, Maia ACD, Bento JMS. Methyl benzoate and nerolidol attract the cyclocephaline beetle Cyclocephala paraguayensis to trumpet flowers. Naturwissenschaften. 2023 Jan 26;110(1):3.

[421] Zhang LL, Ma XL, Wang D, Yu P, Huang LQ, Zhang XL, Dai ZB. [Construction of cell factories for high production of nerolidol in Saccharomyces cerevisiae]. Zhongguo Zhong Yao Za Zhi. 2017 Aug;42(15):2962-2968.

[422] Tan N, Ong L, Shukal S, Chen X, Zhang C. High-Yield Biosynthesis of trans-Nerolidol from Sugar and Glycerol. J Agric Food Chem. 2023 Jun 7;71(22):8479-8487.

[423] Liu F, Liu SC, Qi YK, Liu Z, Chen J, Wei LJ, Hua Q. Enhancing Trans-Nerolidol Productivity in Yarrowia lipolytica by Improving Precursor Supply and Optimizing Nerolidol Synthase Activity. J Agric Food Chem. 2022 Dec 7;70(48):15157-15165.

POTENTIAL USES IN MEDICINE

Nerolidol shows great promise in medicine, thanks to its antifungal, antibacterial, anti-tumor, antimalarial, antioxidant, and other properties, for which there is promising evidence as detailed below.

**Sensitize Bacteria to Antibiotics:* Perhaps one of the most fascinating and exciting things about nerolidol is its ability to weaken and make bacteria more susceptible to antibiotics. Researchers demonstrated that nerolidol sensitized Staphylococcus aureus and Escherichia coli bacteria to a range of traditional chemical antibiotics, including ciprofloxacin, clindamycin, erythromycin, gentamicin, tetracycline, and vancomycin[424]. This has important implications in medicine considering the growing antibiotic resistance of many strains of bacteria, which has led to worldwide policy changes in the way antibiotics are administered.

**Antifungal Agent:* Nerolidol isolated from Japanese cypress trees attenuated skin lesions infected with microsporum gypseum (a pathogenic fungus that commonly lives in soil and infects the upper dead skin layer of animals including humans) and showed clinical effectiveness in healing infections of Microsporum gypseum in guinea pigs[425].

**Anti-tumor Agent:* The essential oil of zornia brasiliensis, which included the isomer trans-nerolidol as a major constituent, showed cytotoxicity against tumor cells and inhibited tumor growth rates by as much as 38%[426].

**Antimalarial:* Nerolidol shows astounding promise as an antimalarial agent, inhibiting infection with the parasite in mice by 99% as long as two weeks post-infection, resulting in survival rates of 90%, compared to just 16% in controls[427].

**Trypanocidal:* Trypanosoma evansi is a dangerous, highly fatal zoonotic parasite that often leads to brain damage and death in mammals including humans. Nerolidol may prove to be the next effective medicine in the treatment of this disease, as the acyclic sesquiterpene increased the survival rate of mice infected with T. evansi by 66% and was more effective than traditional chemotherapy in one study[428], and was shown to prevent DNA damage and cell mortality in another thanks to the compound's anti-inflammatory properties[429].

**Treatment of Parkinson's Disease:* Nerolidol has been shown to be an effective neuroprotectant because of its antioxidant and anti-inflammatory properties and shows therapeutic potential as a treatment for Parkinson's Disease[430], a complicated neurological disorder.

---

[424] Brehm-Stecher, Byron F, and Eric A Johnson. Sensitization of Staphylococcus aureus and Escherichia coli to antibiotics by the sesquiterpenoids nerolidol, farnesol, bisabolol, and apritone. Antimicrobial agents and chemotherapy vol. 47,10 (2003): 3357-60.

[425] Lee SJ, Han JI, Lee GS, Park MJ, Choi IG, Na KJ, Jeung EB. Antifungal effect of eugenol and nerolidol against Microsporum gypseum in a guinea pig model. Biol Pharm Bull. 2007 Jan;30(1):184-8.

[426] Costa EV, Menezes LR, Rocha SL, Baliza IR, Dias RB, Rocha CA, Soares MB, Bezerra DP. Antitumor Properties of the leaf essential oil of Zornia brasiliensis. Planta Med. 2015 May;81(7):563-7.

[427] Alexandre Y. Saito, Adriana A. Marin Rodriguez, Danielle S. Menchaca Vega, Rodrigo A.C. Sussmann, Emflia, A. Kimura, Alejandro M. Katzin. Antimalarial activity of the terpene nerolidol. International Journal of Antimicrobial Agents, Volume 48, Issue 6, 2016, Pages 641-646.

[428] Baldissera MD, Grando TH, Souza CF, Cossetin LF, Sagrillo MR, Nascimento K, da Silva AP, Dalla Lana DF,Da Silva AS, Stefani LM, Monteiro SG. Nerolidol nanospheres increases its trypanocidal efficacy against Trypanosoma evansi: New approach against diminazene aceturate resistance and toxicity. Exp Parasitol. 2016 Jul;166:144-9.

[429] Baldissera, Matheus, Carine Souza, Roberta Rieffel, Maiara Velho, Andiara Ramos, Katia Nascimento, Michele Sagrillo, Aline Ourique, Aleksandro Silva, Lenita Stefani, and Silvia Monteiro. Protective effect of nerolidol-loaded in nanospheres against cerebral damage caused by Trypanosoma evansi. Naunyn-Schmiedeberg's Archives of Pharmacology 391.7 (2018): 753-759.

[430] Javed, Hayate, Sheikh Azimullah, Salema Abul Khair, Shreesh Ojha, and M. Haque. Neuroprotective effect of nerolidol against neuroinflammation and oxidative stress induced by rotenone. BMC Neuroscience 17.1 (2016): 1-12.

**Sedative:* As mentioned previously, nerolidol is a well-known antioxidant, and studies have shown the sesquiterpene to exhibit neuroprotective activity. One study suggests that these antioxidant and neuroprotective activities also contribute to the sedative properties of nerolidol as demonstrated in mice[431], and, somewhat surprisingly, as already well-known in the cannabis industry.

**Treatment for Kindling:* Kindling refers to the progressive worsening of withdrawal and detoxification symptoms after repeated exposure to a drug or substance. For instance, in alcoholics, each relapse after a period of abstinence results in a worsening series of symptoms, and longer recovery time. Nerolidol has been shown to significantly inhibit the progression of kindling and decreases associated depression and memory impairment in a murine model[432], suggesting that this compound may be a novel tool in the fight against not only addiction and alcoholism, but other conditions where kindling is developed.

**Treatment of Schistosomiasis:* Schistosomiasis, also called snail fever, is a disease of the urinary tract or intestines caused by parasitic flatworms. Nerolidol was shown to morphologically alter the worms including sloughing and disintegration of worm bodies, reducing the activity of the worms in smaller doses, and killing all the schistosomes in higher doses[433]. The potential of using nerolidol as a new treatment for schistosomiasis is an urgent need for hundreds of millions of people infected with this disease, especially considering there is currently only one drug to treat this condition.

**Treatment for Leishmaniasis:* Nerolidol has been shown to inhibit the growth of Leishmania braziliensis (a protozoan that causes mucocutaneous leishmaniasis) promastigotes by inducing cell death[434], offering a promising new treatment for this disease that affects more than 12 million people.

**Treatment for Polycystic Ovary Syndrome:* Nerolidol has been shown to significantly reduce increases of lipid peroxidation, oxidative stress, inflammation, histopathological damage, and apoptosis in dehydroepiandrosterone-induced polycystic ovary syndrome in pre-pubertal female Sprague-Dawley rats[435].

**Treatment for Renal Ischemia–reperfusion Injury:* In a rat model of renal ischemia–reperfusion injury, nerolidol was shown to significantly attenuate alterations in serum creatinine and urea, creatinine clearance, urinary albumin, urinary albumin–creatinine ratio, and markers of kidney injury including proinflammatory, profibrotic and apoptotic cytokines, oxidative stress markers, and histological changes[436].

**Alzheimer's Disease Prevention:* In an amyloid-beta (Aβ)-induced model of Alzheimer's disease in Wistar rats, researchers found that administration of nerolidol improved the deleterious effects of this condition by decreasing Aβ and increasing brain-derived neurotrophic factor (serves several important roles in neurons and neurotransmitters) and CREB-1 (a protein that stimulates gene transcription) expression and latency time[437]. Taheri

---

[431] Nogueira Neto, Jose, Antonia Almeida, Johanssy Silva Oliveira, Pauline Santos, Damiao Sousa, and Rivelilson Freitas. Antioxidant Effects of Nerolidol in Mice Hippocampus After Open Field Test. Neurochemical Research 38.9 (2013): 1861-1870.

[432] Kaur, Dilpreet, Priyanka Pahwa, and Rajesh Goel. Protective Effect of Nerolidol Against Pentylenetetrazol- Induced Kindling, Oxidative Stress and Associated Behavioral Comorbidities in Mice. Neurochemical Research 41.11 (2016): 2859-2867.

[433] Silva, Marcos P.N., George L.S. Oliveira, Rusbene B.F. de Carvalho, Damiao P. de Sousa, Rivelilson M. Freitas, Pedro L.S. Pinto, and Josue de Moraes. Antischistosomal Activity of the Terpene Nerolidol. Molecules 19.3 (2014): 3793-3803.

[434] Ceole LF, Cardoso MDG, Soares MJ. Nerolidol, the main constituent of Piper aduncum essential oil, has anti-Leishmania braziliensis activity. Parasitology. 2017 Aug;144(9):1179-1190.

[435] Türkmen NB, Yüce H, Aydın M, Taşlıdere A, Doğan A, Özek DA, Hayal TB, Yaşar Ş, Çiftçi O, Ünüvar S. Nerolidol attenuates dehydroepiandrosterone-induced polycystic ovary syndrome in rats by regulating oxidative stress and decreasing apoptosis. Life Sci. 2023 Feb 15;315:121380.

[436] Hammad FT, Al-Salam S, Ahmad R, Yasin J, Hammad AF, Rasheed JA, Lubbad L. The Effect of Nerolidol Renal Dysfunction following Ischemia-Reperfusion Injury in the Rat. Nutrients. 2023 Jan 15;15(2):455.

[437] Taheri P, Yaghmaei P, Hajebrahimi Z, Parivar K. Neuroprotective effects of nerolidol against Alzheimer's disease in Wistar rats. Drug Dev Res. 2022 Dec;83(8):1858-1866.

et al summarized by explaining that; "it seems that the administration of nerolidol before the onset of the disease will be more effective than after."

Chemotherapeutic Toxicity Adjuvant: In combination with the anticancer drug Cisplatin, nerolidol inhibited cell viability, increased reactive oxygen species generation, apoptotic induction, and DNA damage in human laryngeal carcinoma Hep 2 cells[438].

Treatment for Bone Cancer: As an isolated compound, nerolidol induced oxidative stress and morphological changes, regulated the protein PI3K/AKT mechanistically via induction of apoptosis, and suppressed osteosarcoma MG-63 cells by PI3K/AKT by cell cycle arrest at early phase of G0/G1[439].

Prevention of Oral Cancer: Nerolidol has been shown to protect against oral carcinogenesis by modulating biochemical profiles including antioxidants, detoxification agents, lactoperoxidase, and renal function markers, significantly inhibiting tumor development in 7,12-dimethylbenz(a)anthracene-induced buccal pouch carcinogenesis in hamsters[440].

Treatment for Colon Cancer: Recent work by Zhao et al has shown that nerolidol substantially inhibited cell viability, boosted ROS levels, induced apoptosis, and inhibited cell cycle at the G0/G1 phase in colon cancer HCT-116 cells[441].

Treatment for Rheumatoid Arthritis: In a Freund's complete adjuvant rheumatoid model in rats, nerolidol prevented body weight loss, stabilized biochemical and haematological homeostasis, significantly reduced paw inflammation volume, and improved joint structure in the subject animals while downregulating proinflammatory cytokines and upregulating anti-inflammatory cytokines[442].

Treatment for Heart Attack Damage: In an isoproterenol-induced myocardial infarction model in rats, nerolidol attenuated deleterious effects of the treatment by reducing myocardial infarct size, and by suppression of oxidative stress[443].

Treatment for Diabetes: In rats induced to diabetes via a high-fat diet and treatment with streptozotocin, nerolidol significantly reduced the level of elevated glucose and glycosylated hemoglobin, improved body weight and insulin levels, considerably improved carbohydrate metabolic enzyme activities, increased glycogen storage in the liver, increased serum triglycerides, significantly lowered total cholesterol, low-density lipoproteins-C and very low-density lipoproteins-C levels, attenuated oxidative stress markers by significantly increasing the levels of superoxide dismutase and catalase, reduced glutathione, and lowered the level of thiobarbituric acid reactive substances and lipid hydroperoxide, among several other antidiabetic affects[444].

---

[438] Balakrishnan V, Ganapathy S, Veerasamy V, Duraisamy R, Jawaharlal S, Lakshmanan V. Nerolidol assists Cisplatin to induce early apoptosis in human laryngeal carcinoma Hep 2 cells through ROS and mitochondrial-mediated pathway: An in vitro and in silico view. J Food Biochem. 2022 Dec;46(12):e14465.

[439] Yu Y, Velu P, Ma Y, Vijayalakshmi A. Nerolidol induced apoptosis via PI3K/JNK regulation through cell cycle arrest in MG-63 osteosarcoma cells. Environ Toxicol. 2022 Jul;37(7):1750-1758.

[440] Balakrishnan V, Ganapathy S, Veerasamy V, Duraisamy R, Sathiavakoo VA, Krishnamoorthy V, Lakshmanan V. Anticancer and antioxidant profiling effects of Nerolidol against DMBA induced oral experimental carcinogenesis. J Biochem Mol Toxicol. 2022 Jun;36(6):e23029.

[441]Zhao, X., Chinnathambi, A., Alharbi, S.A. et al. Nerolidol, Bioactive Compound Suppress Growth of HCT-115 Colorectal Cancer Cells Through Cell Cycle Arrest and Induction of Apoptosis. Appl Biochem Biotechnol (2023).

[442] Akhter S, Irfan HM, Alamgeer, Jahan S, Shahzad M, Latif MB. Nerolidol: a potential approach in rheumatoid arthritis through reduction of TNF-α, IL-1β, IL-6, NF-kB, COX-2 and antioxidant effect in CFA-induced arthritic model. Inflammopharmacology. 2022 Apr;30(2):537-548.

[443] Gonçalves MSS, Silva EAP, Santos DM, Santana IR, Souza DS, Araujo AM, Heimfarth L, Vasconcelos CML, Santos VCO, Santos MRV, de S S Barreto R, Quintans Júnior LJ, Barreto AS. Nerolidol attenuates isoproterenol-induced acute myocardial infarction in rats. Naunyn Schmiedebergs Arch Pharmacol. 2022 Mar;395(3):353-363.

[444] Jiang N, Zhang Y. Antidiabetic effects of nerolidol through promoting insulin receptor signaling in high-fat diet and low dose streptozotocin-induced type 2 diabetic rats. Hum Exp Toxicol. 2022 Jan-Dec;41:9603271221126487.

IMPLICATIONS FOR HEALTH & HUMAN NUTRITION

The safest and most efficient method of consuming nerolidol is by eating fruits and herbs as part of your normal meals and snacks that contain this acyclic sesquiterpene. Fruits that are high in nerolidol include strawberries, kiwi, guava, grapefruit, mandarin, and oranges. Spices and herbs that contain nerolidol include anise, basil, cinnamon, ginger, nutmeg, parsley, tarragon, and thyme, giving most people a wide variety of aromas and flavors to work with in the kitchen.

Because not many vegetables are high in nerolidol, consider adding black pepper to food, or lemon to water or other beverages. You can also obtain nerolidol from teas, especially flavored teas like raspberry, peach, or other fruity/floral flavors.

Cannabis strains that feature cis- or trans-nerolidol in moderate amounts include Agent Orange, Black Lime, Blueberry Peaches, Blue Dream, Bubba Kush, Cannatonic, Deadhead OG, Elvis, Green Crack, Purple Cadillac, and Sour Jack. Inhalation of smoke or combustion of plant materials isn't recommended, and vaporizing cannabis strains that are high in nerolidol may not provide a safer alternative. Oils, edibles, and tinctures are acceptable forms of consumption.

Essential oils made from the plants listed in the "Occurrence in Plants" section above are likely to contain variations or racemic mixtures of nerolidol isomers. However, extreme care should be taken when working with these complex mixtures, or with pure compounds, as they can be dangerous in some situations.

Nerolidol Review

Answer the following questions to test your knowledge of this terpene:

Question #1: What type of terpene is nerolidol?

a. Acyclic sesquiterpene
b. Linear sesquiterpene
c. Monocyclic sesquiterpene
d. Bicyclic sesquiterpene

Question #2: Which of the following best describes the primary flavor & aroma profile of nerolidol?

a. Hops, cloves, wood, spice
b. Depends on the isomer
c. Floral, woody, green, citrus
d. Spicy, turpentine, fuel, pine
e. Citrus, fruity, dry, terpenic

Question #3: How common are cis- and trans-nerolidol in cannabis?

a. Top three
b. Top five
c. Top ten
d. Top twenty

Question #4: What is the chemical formula for nerolidol?

a. $C_{15}H_{32}0$
b. $H_{24}C0_{20}$
c. $C_{15}H_{26}B$
d. $C_{15}H_{26}0$

Question #5: Name two uses of nerolidol in plants:

1 ____ 2 ____

Question #6: Name two potential medical uses of nerolidol:

1 ____ 2 ____

For the answer key to all individual compound chapters please go here: http://thebigbookofterps.com/bbot-quiz-answer-key/

OCIMENE

Type: Acyclic monoterpene
Chemical Formula: $C_{10}H_{16}$
Molecular Weight: 136.237 g/mol
Boiling Point: *Alpha:* 169.00 to 171.00 °C. @ 760.00 mm Hg (estimated)[445]
Beta: 176.00 to 178.00 °C. @ 760.00 mm Hg[446]
Z-beta: 177.00 °C. @ 760.00 mm Hg[447]
E-beta: 174.00 to 175.00 °C. @ 760.00 mm Hg[448]
Flash Point: *Alpha:* 112.00 °F. TCC (44.40 °C.) (estimated)
Beta: 143.00 °F. TCC (61.67 °C.)
Z-beta: 135.00 °F. TCC (57.22 °C.)
E-beta: 16.00 °F. TCC (46.67 °C.)
Solubility: *Alpha:* Soluble in alcohol, slightly soluble in water
Beta: Soluble in alcohol, oils, ethanol, weakly soluble in water
Z-beta: Soluble in alcohol, weakly soluble in water
E-beta: Soluble in alcohol, oils, ethanol, slightly soluble in water
Oral LD50: *Alpha:* Unreported
Beta: 5,000 mg/kg (rat)[449]
Z-beta: Unreported
E-beta: 5,000 mg/kg (rat)[450]
Biological Role: Indirect defense, antifungal, allelopathic agent
Therapeutic Role: Anticancer agent, antibacterial, anti-inflammatory
Commercial Use: Fragrance, flavoring agent
Occurrence in Cannabis: Top ten
Aroma/Flavor: *Alpha:* Fruity, floral, wet, cloth
Beta: Citrus, tropical, green, terpenic, woody
Z-beta: Warm, floral, herb, sweet, green
E-beta: Sweet, herbal

Occurs in Cannabis Strains: Beta-ocimene is the most likely ocimene isomer to occur in cannabis, generally within the top ten terpenes by concentration for many varieties. However, as always, it's important to note that strain or variety variability is extremely diverse, and few laboratories have standards for the other isomers of ocimene. The following varieties are known to contain ocimene, although laboratory reports used to generate this list rarely specified an isomer: Agent Orange, Amnesia, Black Widow, Chernobyl, Chocolope, Critical Jack, Deathstar, Durban Poison, Dutchie, Dutch Treat, Gelato, Ghost Train Haze, Golden Goat, Goo, Green Mountain, Jack Herer, OG Kush, Sour Diesel, Space Queen, Super Lemon Haze, Super Silver Haze, Strawberry Cough, Whitaker Blues, White Fire OG.

---

[445] The Good Scents Company Data Sheet for Alpha-ocimene, from: http://www.thegoodscentscompany.com/data/rw1423541.html. Accessed August 2, 2023.

[446] The Good Scents Company Data Sheet for Beta-ocimene, from: http://www.thegoodscentscompany.com/data/rw1017311.html, Accessed August 2, 2023.

[447] The Good Scents Company Data Sheet for Z-beta-ocimene, from: http://www.thegoodscentscompany.com/data/rw1010541.html. Accessed August 2, 2023.

[448] The Good Scents Company Data Sheet for E-beta-ocimene, from: http://www.thegoodscentscompany.com/data/rw1423741.html. Accessed August 2, 2023.

[449] Echemi Safety Data Sheet for Beta-ocimene, from: https://www.echemi.com/sds/ocimene-pd180521128683.html. Accessed August 3, 2023.

[450] Sigma-Aldrich Safety Data Sheet for E-beta-ocimene, from; https://www.sigmaaldrich.com/US/en/sds/ALDRICH/W353901. Accessed August 3, 2023.

INTRODUCTION

Ocimene is an acyclic monoterpene with a unique aroma that is best described as floral-tropical regarding the racemic mixture of the primary isomers. Occurring among the top ten terpenes in most strains of cannabis, ocimene is also common in many other plants, where it is an important primary phytochemical performing a variety of roles, depending on the organ of origin and environmental and other factors. The word 'ocimene' comes from the ancient Greek word for basil, a plant that produces this terpene in large proportions.

Insects also produce ocimene, including the queen honeybee, which releases ocimene in a blend of pheromones to chemically prevent the development of ovaries in other honeybees[451]. Starving honeybee larvae also use a variation of ocimene as a method of signaling worker bees to bring food[452].

Male mealworm beetles are another insect that emits ocimene as part of its pheromones, which causes both males and females to form lines while walking up to the source of the odor[453].

Finally, the male butterfly Heliconius Melpomene rubs complex pheromones containing (E)-beta-ocimene on the bellies of females during copulation, which then renders the females unattractive to other males[454]. Researchers confirmed this when they treated un-mated females with (E)-beta-ocimene, which repelled what would otherwise be interested males.

Ocimene should be of particular interest to the cannabis industry, as it has potential as a treatment for spider mites, powdery mildew, and botrytis - three of the most difficult problems faced by cultivators of cannabis. See more in the "Biological Activity In Plants" section below.

[451] Maisonnasse, Alban, Jean-Christophe Lenoir, Guy Costagliola, Dominique Beslay, Fanny Choteau, Didier Crauser, Jean-Marc Becard, Erika Plettner, and Yves Le Conte. A scientific note on E-B-ocimene, a new volatile primer pheromone that inhibits worker ovary development in honey bees. "Apidologie" 40.5 (2009): 562-564.

[452] Xu Jiang He, Xue Chuan Zhang, Wu Jun Jiang, Andrew B. Barron, Jian Hui Zhang & Zhi Jiang Zeng Starving honey bee (Apis mellifera) larvae signal pheromonally to worker bees Scientific Reports volume 6, Article number: 22359 Published: 29 February 2016.

[453] Cosse, Allard, and Bruce Zilkowski. Behavioral Responses of Lesser Mealworm Beetles, Alphitobius diaperinus, (Coleoptera: Tenebrionidae) to Pheromone Components Using a Wind Tunnel Dual Choice Walking Bioassay. Journal of Insect Behavior 28.2 (2015): 202-210.

[454] Schulz, Stefan, Catalina Estrada, Selma Yildizhan, Michael Boppre, and Lawrence Gilbert. An Antiaphrodisiac in Heliconius melpomene Butterflies. Journal of Chemical Ecology 34.1 (2007): 82-93.

CHEMICAL STRUCTURE

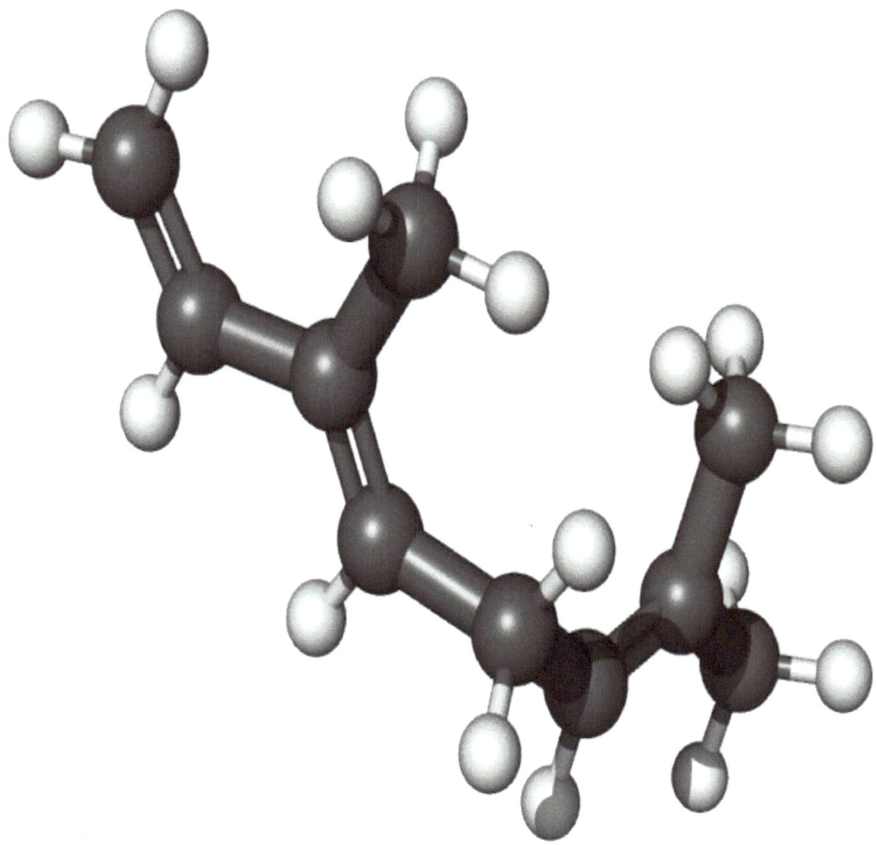

Produced from geranyl pyrophosphate and arranged as the typical isoprenoid skeleton, ocimene has ten carbon atoms and sixteen hydrogen atoms, scientifically notated as $C_{10}H_{16}$. This molecule has three double bonds, three rotatable bonds, and one covalent bond.

There are two primary versions of this acyclic (does not contain rings) monoterpene; alpha-ocimene, and beta-ocimene, with the latter splitting into cis- (Z) and trans- (E) stereoisomeric forms. Alpha-ocimene and beta-ocimene are differentiated by the position of the isolated double bond (two other double bonds are located close together), which is terminal in the alpha- version, meaning it occurs at the tail end of the molecule. Plants that produce ocimene generally produce a mixture or racemic presentation of these four variations.

Although the isomers of this compound smell and taste different and serve different purposes, ocimene is closely related to myrcene, one of the top three terpenes in most strains of cannabis.

OCCURRENCE IN PLANTS

Ocimene is produced by hundreds of different plants. The emission of this terpene has been shown to be controlled by a circadian clock in the snapdragon plant[455], and is likely used similarly by other plant species. Ocimene is known to occur in the following plants:

---

[455] N Dudareva, D Martin, CM Kisha, N Kolosova, N Gorensteina, J Faldt, B Millerb and J Bohlmann, (E)-b-Ocimene and Myrcene Synthase Genes of Floral Scent Biosynthesis in Snapdragon: Function and Expression of Three Terpene Synthase Genes of a New Terpene Synthase Subfamily The Plant Cell, 15(5) 122 7-1241 (2003).

Alpha:

Calycanthus floridus	Celery seed	Fennel
Melissa	Ocimum gratissimum	Papaya

Beta:

Basil	Black currant	Black elder
Black pepper	Cannabis	Coriander
Cotton	Ferula cupularis	Ho leaf
Hop	Kumquat	Lemon
Mandarin	Mango	Marjoram
Mint	Neroli	Oregano
Parsley	Peppermint	Pine
Rosemary	Spearmint	Tarragon
Thyme		

Z-beta:

Allspice	Angelica	Anise
Anise star	Artemisia species	Baccharis coridifolia
Basil	Bay laurel	Bay leaf
Bergamot	Betel	Black currant
Black pepper	Blood orange	Bois de rose
Cajuput	Calamintha nepeta	Calamus
Cananga	Caraway	Carrot
Cascarilla	Cassis	Celery
Chamomile	Campaca	Cinnamon
Citronella	Clary sage	Cocoa
Coriander	Cornmint	Crithmum maritimum
Croton species	Cubeb	Curry
Cypress	Dill	Eschweilera coriacea
Eucalyptus species	Fennel	Ferula
Geranium	Grapefruit	Guava
Helichrysum italicum	Horsemint	Hyacinthus orientalis
Hyptis	Hyssop	Lavandin
Lavandula	Lavender	Layana
Lemon	Lemongrass	Levisticum officinale
Lime	Lovage	Mandarin
Mango	Manuka	Marjoram
Mastic	Narcissus	Nectarine
Nepeta	Neroli	Ocimum gratissimum
Orange	Oregano	Palmarosa
Papaya	Parsley	Parsnip
Passiflora	Pennyroyal	Peppermint
Phlomis fruticosa	Pimenta	Pimento
Prangos species	Ravintsara	Rosemary
Sage	Salvia	Savory
Spearmint	Stachys recta	Sugandha kokila
Tagete	Tarragon	Tea

Thyme
Water mint
Wormwood

E-beta:

Allspice
Angelica
Apple
Artemisia species
Basil
Bay
Bergamot
Betel
Black currant
Black pepper
Blood orange
Bois de rose
Boronia
Buchu
Cajuput
Calamus
Cananga
Caraway
Carrot
Cascarilla
Cassis
Celery
Chamomile
Champaca
Cinnamon
Citronella
Clary sage
Coriander
Courourpita guianensis
Curry
Cypress
Dill
Eschweilera coriacea
Eucalyptus
Fennel
Fig
Fleabane
Frankincense
Geranium
Grapefruit
Guava
Horsemint
Hyptis
Hyssop
Jambu
Juniper
Lallemantia peltata
Laurel
Lavandin
Lavender
Layana
Lemon
Lemongrass
Levisticum
Lime
Lovage
Mace
Mandarin
Manuka
Marjoram
Mastic
Myrtle
Neroli
Ocimum gratissimum
Orange
Oregano
Palmarosa
Papaya
Parsley
Parsnip
Peucedanum petiolare
Pine
Pomelo
Prangos species
Ravintsara
Rosemary
Sage
Salvia
Skimmia laureola
Soybean
Spearmint
Tagete
Tarragon
Tea
Thyme
Wormwood
Ylang ylang
Yuzu
Zanthoxylum species

BIOLOGICAL ACTIVITY IN PLANTS

Ocimene seems to serve many different roles in plants, and - as is generally the case with terpenes - often serves multiple roles in the same plant. Well-documented in the literature, the following are just some of the ways that plants may use this acyclic monoterpene:

Allelopathic Agent & Insect Growth Inhibitor: In some plants, beta-ocimene is used to retard or inhibit the growth of insects, including influencing the weight gain, reproduction, and feeding behavior of aphids[456], a common pest of cannabis.

---

[456] Kang, Zhi-Wei, Fang-Hua Liu, Zhan-Feng Zhang, Hong-Gang Tian, and Tong-Xian Liu. Volatile B-Ocimene Can Regulate Developmental Performance of Peach Aphid Myzus persicae Through Activation of Defense Responses in Chinese Cabbage Brassica pekinensis. Frontiers in Plant Science 9 (2018).

Oviposition Deterrent: As a major constituent in the essential oil of basil plants, E-beta-ocimene was shown to significantly deter female leaf miners from laying eggs on tomato plants treated with the EO[457].

Attract Pollinators: A review study carried out in 2017 found that the high concentration and temporal-spatial patterns of emission from beta-ocimene in various flowers and plant organs were typical of emissions that are involved in pollinator attraction, suggesting that this compound may play generalist and specialist roles in the attraction of pollinators[458].

Indirect Defense: Ocimene has been shown to be released by plants in response to attack by herbivores[459], and is one of the most common terpenes found in the volatile emissions of plants responding to herbivory or mechanical wounding[460]. Ocimene has been shown to attract predatory lady beetles[461], which are commonly used by cannabis growers to stem aphid infestations, as lady beetles are voracious predators of aphids. Additionally, (E)-beta-ocimene is released by Royal Gala apple trees when under attack by larvae of the light brown apple moth, which the parasitoid dolichogenidia tasmanica then uses to find and consume the larvae[462]. Perhaps most interestingly for the cannabis industry, ocimene has been shown to attract predatory mites to plants infested with spider mites[463] - one of the most persistent and challenging pests encountered by cannabis growers. Pear plants have been shown to increase emissions of a blend of terpenes where beta-ocimene was a major component in response to herbivory by Cacopsylla bidens (pear psyllid), which is used by chrysopid larvae (green lacewing) as a chemical cue to locate and prey upon the psyllids[464]. Tomato plants responding to emissions of beta-ocimene from nearby tobacco plants were less attractive to aphids, and more attractive to the predator of aphids - the parasitoid Aphidius ervi[465], a wasp which lays its eggs in aphids, where wasp larvae then eat the aphid from the inside.

Antioxidant: As the primary constituent in the essential oil of three species of Lavandula, (E)-beta-ocimene displayed significant antioxidant activity that varied according to the extract and species[466].

Antifungal Agent: Beta-ocimene had a primary role in decreasing the severity of powdery mildew infections in cultivated tobacco plants, which significantly increased the emission of (E)-beta-ocimene in response to the pathogenic fungus[467]. This is potentially critical information for cannabis growers, as entire crops are sometimes

---

457 Yarou, Boni, Thomas Bawin, Antoine Boullis, Stephanie Heukin, Georges Lognay, Francois Verheggen, and Frederic Francis. Oviposition deterrent activity of basil plants and their essentials oils against Tuta absoluta (Lepidoptera: Gelechiidae). Environmental Science and Pollution Research 25.30 (2017): 29880-29888.

458 Farre-Armengol, Gerard, Iolanda Filella, Joan Llusia, and Josep Peñuelas. B-Ocimene, a Key Floral and Foliar Volatile Involved in Multiple Interactions between Plants and Other Organisms. Molecules: A Journal of Synthetic Chemistry and Natural Product Chemistry 22.7 (2017).

459 O. Borges, Eliane, Camila Martins, Rodolfo R. da Silva, and Paulo Zarbin. Terpenoids dominate the bouquet of volatile organic compounds produced by Passiflora edulis in response to herbivory by Heliconius erato phyllis (Lepidoptera: Nymphalidae). Arthropod-Plant Interactions 12.1 (2017): 123-131.

460 Faldt, Jenny, Gen-ichiro Arimura, Jonathan Gershenzon, Junji Takabayashi, and Jorg Bohlmann. Functional identification of AtTPS03 as (E)-B-ocimene synthase: a monoterpene synthase catalyzing jasmonate- and wound- induced volatile formation in Arabidopsis thaliana. Planta 216.5 (2003): 745-751.

461 Laznik, Ziga, and Stanislav Trdan. Are synthetic volatiles, typically emitted by insect-damaged peach cultivars, navigation signals for two-spotted lady beetle (Adalia bipunctata L.) and green lacewing (Chrysoperla carnea [Stephens]) larvae? Journal of Plant Diseases and Protection 125.6 (2018): 529-538.

462 Suckling DM, Twidle AM, Gibb AR, Manning LM, Mitchell VJ, Sullivan TE, Wee SL, El-Sayed AM. Volatiles from apple trees infested with light brown apple moth larvae attract the parasitoid Dolichogenidia tasmanica. J Agric Food Chem. 2012 Sep 26;60(38):9562-6.

463 Kappers, Iris, Francel Verstappen, Ludo Luckerhoff, Harro Bouwmeester, and Marcel Dicke. Genetic Variation in Jasmonic Acid- and Spider Mite-Induced Plant Volatile Emission of Cucumber Accessions and Attraction of the Predator Phytoseiulus persimilis. Journal of Chemical Ecology 36.5 (2010): 500-512.

464 Valle, D., Mujica, V. & Gonzalez, A. Herbivore-Dependent Induced Volatiles in Pear Plants Cause Differential Attractive Response by Lacewing Larvae. J Chem Ecol (2023).

465 Cascone P, Iodice L, Maffei ME, Bossi S, Arimura G, Guerrieri E. Tobacco overexpressing β-ocimene induces direct and indirect responses against aphids in receiver tomato plants. J Plant Physiol. 2015 Jan 15;173:28-32.

466 Messaoud C, Chograni H, Boussaid M. Chemical composition and antioxidant activities of essential oils and methanol extracts of three wild Lavandula L. species. Nat Prod Res. 2012 Nov;26(21):1976-84.

467 Quaglia M, Fabrizi M, Zazzerini A, Zadra C. Role of pathogen-induced volatiles in the Nicotiana tabacum- Golovinomyces cichoracearum interaction. Plant

lost to powdery mildew (PM). Furthermore, a variation of ocimene (allo-ocimene) has been shown to reduce the occurrence and severity of botrytis cinerea infections in thale cress, retarding the development of disease symptoms[468]. Botrytis cinerea is commonly known as "bud rot" in the cannabis industry, and is one of the most feared complications of cultivating cannabis. Ocimene might provide a novel approach to treating both powdery mildew and botrytis in a natural, non-toxic, and sustainable manner.

Insect Repellant: Tea plants responding to herbivory by Ectropis obliqua (a species of moth) caused neighboring, non-infested plants to increase emissions of a blend of terpenes where beta-ocimene was a principal component, causing direct repellency of mated E. obliqua females[469].

USES IN INDUSTRY

The primary industrial or commercial use of ocimene is to add fragrance to products. These typically include detergents, household and industrial cleaners, deodorizers, antiperspirants, shampoos, conditioners, soaps, fabric softeners, and lotions, among many other products. Ocimene is also used by perfumeries, where the maximum recommended content of this terpene in a fragrance is just 3%. Ocimene is extremely volatile in the air, which makes it an excellent medium for perfumes and colognes.

In the prepared foods industry, ocimene is used as a flavoring agent for desserts and sweet items like ice cream and frozen yogurt, pudding, gelato, sorbet, soft candy, and baked goods. Ocimene is also readily available naturally in the plants listed previously, especially basil, mangos, apples, limes, lemons, and oranges. (Z)-beta-ocimene is likely the most commonly used isomer in food products.

Ocimene is also used in the food industry as a preservative; table grapes are often sprayed with the essential oil of thyme, with ocimene as the primary constituent[470]. This prevents or delays the development of fungal decay, without changing the flavor of the grapes.

Interestingly, some researchers have suggested the use of beta-ocimene as a lure for the predator insect Nesidiocoris tenuis (predatorial insect already used as a biological control against pests), which uses the compound to locate its prey, Tuta absoluta (a species of moth), and Trialeurodes vaporariorum (glasshouse whitefly), two significant pests of tomato plants[471].

POTENTIAL USES IN MEDICINE

Medical potential of ocimene and its variations include the following:

Anti-tumor Agent: Ocimene has been shown to exhibit significant anti-tumor properties. In one case, in-vivo tumor growth was inhibited by more than 34% after treatment with the essential oil of annona vepretorum, where (E)-beta-ocimene was a primary constituent[472]. As the chief constituent of the essential oils of Laurus nobilis, beta-ocimene had a central role in the inhibition of the K562 human chronic myelogenous leukaemia tumour cell line[473].

Physiol Biochem. 2012 Mar; 52:9-20.

[468] Kishimoto, Kyutaro, Kenji Matsui, Rika Ozawa, and Junji Takabayashi. Volatile C6-aldehydes and Allo-ocimene Activate Defense Genes and Induce Resistance against Botrytis cinerea in Arabidopsis thaliana. Plant and Cell Physiology 46.7 (2005): 1093-1102.

[469] Jing T, Qian X, Du W, Gao T, Li D, Guo D, He F, Yu G, Li S, Schwab W, Wan X, Sun X, Song C. Herbivore-induced volatiles influence moth preference by increasing the β-Ocimene emission of neighbouring tea plants. Plant Cell Environ. 2021 Nov;44(11):3667-3680.

[470] Abdollahi A, Bassani A, Ghosta Y, Bernousi I, Meshkatalsadat MB, Shabani R, Ziaee SM. Evaluation of essential oils for maintaining postharvest quality of Thompson seedless table grape. Nat Prod Res. 2012;26(1):77-83.

[471] Ayelo PM, Yusuf AA, Pirk CW, Chailleux A, Mohamed SA, Deletre E. Terpenes from herbivore-induced tomato plant volatiles attract Nesidiocoris tenuis (Hemiptera: Miridae), a predator of major tomato pests. Pest Manag Sci. 2021 Nov;77(11):5255-5267.

[472] Bomfim, Larissa M., Leociley R. A. Menezes, Ana Carolina B. C. Rodrigues, Rosane B. Dias, Clarissa A. Gurgel Rocha, Milena B. P. Soares, Albertino F.

**Anti-cancer Agent:* The essential oil of neolitsea variabillima - of which (E)-beta-ocimene was the primary constituent - exhibited in-vitro cytotoxicity against human oral, liver, lung, colon, melanoma, and leukemic cancer cells[474]. Beta-caryophyllene, another top ten terpene in cannabis, was also a primary constituent of the EO of N. variabillima.

**Mosquitocidal Agent and Mosquito Repellant:* As part of the essential oil of the leaves of chloroxylon swietenia, (E)-beta-ocimene displayed remarkable fumigant toxicity against anopheles gambiae, culex quinquefasciatus, and aedes aegypti species of mosquitos[475]. In other research, beta-ocimene exhibited strong repellant activity against female host-seeking mosquitos, with efficacy comparable to DEET[476], one of the world's most common commercial mosquito repellants.

**Anti-schistosomiasis and Anti-leischmaniasis:* The essential oil of syzygium cumini leaves - where (Z)-beta-ocimene and (E)-beta-ocimene were the primary components - exhibited strong activity against leishmania amazonensis, and overall promising activity as a leishmanicidal and molluscicidal agent[477]. As an isolated compound, beta-ocimene was shown to inhibit promastigotes and axenic amastigotes by inducing changes in membrane permeability and abnormalities in the parasitic cell shape, with a higher selectivity against intra-macrophagic amastigote forms at 469.52 times more selective than meglumine antimoniate, and 42.88 times more selective than amphotericin B, with low levels of cytotoxicity to host cells[478].

**Antibacterial Agent:* The essential oil of echinophora platyloba – where ocimene was the primary constituent - showed potent antimicrobial activity against several bacteria including Staphylococcus aureus, Listeria monocytogenes, Salmonella thyphimurium, and Escherichia coli [479]. With two variations of ocimene - (E)-beta-ocimene and (Z)-beta-Ocimene - as the primary constituents, the essential oil of artemisia dracunculusas exhibited strong antimicrobial activity against multiple bacteria and yeasts, including Streptococcus pyogenes, staphylococcus aureus, and candida albicans[480].

S. Neto, Magaly P. Nascimento, Adriana F. Campos, Lidercia C. R. C. e. Silva, Emmanoel V. Costa, and Daniel P. Bezerra. Antitumour Activity of the Microencapsulation of Annona vepretorum Essential Oil. Basic and Clinical Pharmacology & Toxicology 118.3 (2016): 208-213.

473 Saab AM, Tundis R, Loizzo MR, Lampronti I, Borgatti M, Gambari R, Menichini F, Esseily F, Menichini F. Antioxidant and antiproliferative activity of Laurus nobilis L. (Lauraceae) leaves and seeds essential oils against K562 human chronic myelogenous leukaemia cells. Nat Prod Res. 2012;26(18):1741-5.

474 Su YC, Hsu KP, Wang EI, Ho CL. Composition and in vitro anticancer activities of the leaf essential oil of Neolitsea variabillima from Taiwan. Nat Prod Commun. 2013 Apr;8(4):531-2.

475 Ravi Kiran, S., and P. Sita Devi. Evaluation of mosquitocidal activity of essential oil and sesquiterpenes from leaves of Chloroxylon swietenia DC. Parasitology Research 101.2 (2007): 413-418.

476 Dube, Fitsum Fikru et al. Fresh, dried or smoked? Repellent properties of volatiles emitted from ethnomedicinal plant leaves against malaria and yellow fever vectors in Ethiopia. Malaria journal vol. 10 375. 19 Dec. 2011.

477 Dias, Clarice N., Klinger A. F. Rodrigues, Fernando A. A. Carvalho, Sabrina M. P. Carneiro, Jose G. S. Maia, Eloisa H. A. Andrade, and Denise F. C. Moraes. Molluscicidal and Leishmanicidal Activity of the Leaf Essential Oil of Syzygium cumini (L.) Skeels from Brazil. Chemistry & Biodiversity 10.6 (2013): 1133-1141.

478 Sousa JMS, Nunes TAL, Rodrigues RRL, Sousa JPA, Val MDCA, Coelho FADR, Santos ALSD, Maciel NB, Souza VMR, Machado YAA, Sousa PSA, Araújo AR, Rocha JA, Sousa DP, Silva MVD, Arcanjo DDR, Rodrigues KADF. Cytotoxic and Antileishmanial Effects of the Monoterpene β-Ocimene. Pharmaceuticals (Basel). 2023 Jan 26;16(2):183.

479 Hashemi, Mohammad, Ali Ehsani, Nima Hosseini Jazani, Javad Aliakbarlu, and Razzaqh Mahmoudi. Chemical composition and in vitro antibacterial activity of essential oil and methanol extract of Echinophora platyloba D.C against some of food-borne pathogenic bacteria. Veterinary Research Forum 4.2 (2013): 123-127.

480 Behbahani, Behrooz, Fakhri Shahidi, Farideh Yazdi, Seyed Mortazavi, and Mohebbat Mohebbi. Antioxidant activity and antimicrobial effect of tarragon (Artemisia dracunculus) extract and chemical composition of its essential oil. Journal of Food Measurement and Characterization 11.2 (2017): 847-863.

**Antifungal Agent:* As a primary component in the essential oil of angelica major, (Z)-beta-ocimene displayed low minimum inhibitory concentrations (MIC) against dermatophytes and cryptococcus neoformans, and inhibited yeast-mycelium transition at sub-MIC levels in candida albicans[481].

**Antitrypanosomal Agent:* As isolated from apiaceae essential oils, beta-ocimene exhibited significant inhibitory effects against trypanosoma[482] - the cause of sleeping sickness.

**Anti-inflammatory:* The essential oil of neolitsea sericea leaves - of which (E)-beta-ocimene was a primary constituent - showed significant anti-inflammatory activity via inhibition of NF-kappa beta activation and MAPK phosphorylation[483], while the essential oils from C. unshiu flower - where beta-ocimene was a primary constituent - suppressed the production of inflammatory cytokines including interleukin (IL)-1beta, tumor necrosis factor (TNF)-alpha, and IL-6[484].

IMPLICATIONS FOR HUMAN HEALTH & NUTRITION

To obtain ocimene as part of a healthy diet, consider regularly eating fruits like apples, blood oranges, currants, figs, grapefruit, guava, kumquats, lemons, limes, mangos, mandarins, oranges, papaya, or yuzu. Vegetables that likely contain this acyclic monoterpene include celery, soybeans, carrots, and parsnip. When preparing or cooking foods, consider using spices like basil, black pepper, cinnamon, coriander, dill, fennel, marjoram, oregano, parsley, rosemary, sage, or thyme. Teas made from green tea leaves, lemongrass, eucalyptus, chamomile, lavender, anise, mint, spearmint are also an excellent way to obtain dietary ocimene.

Cannabis varieties that are known to have moderate to high amounts of ocimene include Agent Orange, Amnesia, Black Widow, Chernobyl, Chocolope, Critical Jack, Deathstar, Durban Poison, Dutchie, Dutch Treat, Gelato, Ghost Train Haze, Golden Goat, Goo, Green Mountain, Jack Herer, OG Kush, Sour Diesel, Space Queen, Super Lemon Haze, Super Silver Haze, Strawberry Cough, Whitaker Blues, and White Fire OG. For health reasons, we do not recommend smoking or vaporizing cannabis. However, eating, blending, or juicing these varieties raw are excellent alternatives, as is experimenting with essential oils that contain ocimene as the primary constituent, whether derived from cannabis or not. Essential oils made from the plants listed in the "Occurrences in Plants" section are likely to feature ocimene as a primary constituent, however, extreme care should be used when working with complex EO or isolated compounds, as these can be toxic in some situations.

---

[481] Cavaleiro, Carlos, Ligia Salgueiro, Maria-Jose Goncalves, Karnjana Hrimpeng, Jessica Pinto, and Eugenia Pinto. Antifungal activity of the essential oil of Angelica major against Candida, Cryptococcus, Aspergillus, and dermatophyte species. Journal of Natural Medicines 69.2 (2015): 241-248.

[482] Ngahang Kamte, Stephane L., Farahnaz Ranjbarian, Kevin Cianfaglione, Stefania Sut, Stefano Dall'Acqua, Maurizio Bruno, Fariba Heshmati Afshar, Romilde Iannarelli, Giovanni Benelli, Loredana Cappellacci, Anders Hofer, Filippo Maggi, and Riccardo Petrelli. Identification of highly effective antitrypanosomal compounds in essential oils from the Apiaceae family. Ecotoxicology and Environmental Safety 156 (2018): 154-165.

[483] Yoon WJ, Moon JY, Kang JY, Kim GO, Lee NH, Hyun CG. Neolitsea sericea essential oil attenuates LPS- induced inflammation in RAW 264.7 macrophages by suppressing NF-kappaB and MAPK activation. Nat Prod Commun. 2010 Aug;5(8):1311-6.

[484] Kim MJ, Yang KW, Kim SS, Park SM, Park KJ, Kim KS, Choi YH, Cho KK, Hyun CG. Chemical composition and anti-inflammation activity of essential oils from Citrus unshiu flower. Nat Prod Commun. 2014 May;9(5):727- 30.

Ocimene Review

Answer the following questions to test your knowledge of this terpene:

Question #1: What type of terpene is ocimene?

a. Monocyclic monoterpene
b. Linear sesquiterpene
c. Linear monoterpene
d. Acyclic monoterpene

Question #2: Which of the following best describes the primary flavor & aroma profile of ocimene?

a. Lemon, sharp, fuel, sweet
b. Hops, cloves, wood, spicy
c. Lavender, green, woody, tropical
d. Depends on the isomer
e. Citrus, fruity, dry, turpentine

Question #3: How common is ocimene in cannabis?

a. Top three
b. Top ten
c. Top twenty
d. Less common

Question #4: Which word is ocimene named after?

a. Ancient Sumerian word for Thyme
b. Ancient Greek word for Basil
c. Ancient Greek word for Hops
d. Ancient Roman word for Bay Laurel

Question #5: Name two uses of ocimene in plants:

1 ______________________________ 2 ______________________________

Question #6: Name two potential medical uses of ocimene

1 ______________________________ 2 ______________________________

For the answer key to all individual compound chapters please go here: http://thebigbookofterps.com/bbot-quiz-answer-key

TERPINOLENE

Type: Monocyclic monoterpene
Chemical Formula: $C_{10}H_{16}$
Molecular Weight: 136.238 g/mol
Boiling Point: 183.00 to 185.00 °C. @ 760.00 mm Hg[485]
Flash Point: 148.00 °F. TCC (64.44 °C.)
Solubility: Soluble in benzene, carbon tetrachloride, alcohol, ether, glycol, oils, ethanol, insoluble in water[486]
Oral LD50: 3,740 mg/kg (rat)[487]
Biological Role: Antifungal, insecticide, antibacterial
Therapeutic Role: Antibacterial agent, sedative, anti-inflammatory
Commercial Use: Flavoring agent, fragrance, plastics manufacture, solvents
Occurrence in Cannabis: Top ten
Aroma/Flavor: Fresh, pine, woody, citrus, sweet, terpenic

Occurs in Cannabis Strains: Terpinolene is often quantified in many different cannabis varieties, and is a standard that most laboratories carry for cannabis testing purposes. While variations between each strain likely mean that each variety appears as a distinct chemovar, terpinolene is theorized to be responsible for the "haze" characteristics of strains featuring the name. Banana Lemonade, Black Widow, Blueberry Limeade, Blue Dream, Cherry Diesel, Critical Jack, Durban Poison, Dutchie, Dutch Treat, Ghost Train Haze, Island Sweet Skunk, Jack Herer, Pineapple Kush, Royal Jack Automatic, Sensi Star, Shipwreck, Snowcap, Sour Tangie, Sour Tsunami, Super Lemon Haze, Super Silver Haze.

INTRODUCTION

Sourced industrially for decades by extraction from wood turpentine, terpinolene is common to cannabis and many other plants. Some in the cannabis industry refer to terpinolene as the *'least-common common terpene'*, due to the fact that even though this terpene is recorded in laboratory test results (typically via gas chromatography) in many strains, it often occurs in insignificant amounts. Nevertheless, some cannabis varieties do feature terpinolene prominently, particularly those noted above. Interestingly, research published late in 2022 found a positive correlation between the cannabinoid cannabigerol (CBG) and terpinolene[488].

Terpinolene has a piney flavor, and an aroma that is strongly of turpentine, blended with citrus and woody notes. It is used commonly as a flavoring in a wide variety of foods, especially processed foods, and to add scent to household and industrial cleaners and other products.

Finally, terpinolene is also produced by some insects, including as part of the venom of one species of African ant in the myrmicaria genus, in combination with the monoterpene limonene[489]. Additionally, the mountain pine beetle

[485] The Good Scents Company Data Sheet for Terpinolene, from: http://www.thegoodscentscompany.com/data/rw1007171.html. Accessed August 6, 2023.

[486] PubChem, National Library of Medicine Data Sheet for Terpinolene, from: https://pubchem.ncbi.nlm.nih.gov/compound/Terpinolene#section=Solubility. Accessed August 6, 2023.

[487] Sigma-Aldrich Safety Data Sheet for Terpinolene, from: https://www.sigmaaldrich.com/US/en/sds/aldrich/w304603. Accessed August 6, 2023.

[488] Li D, Ilnytskyy Y, Ghasemi Gojani E, Kovalchuk O, Kovalchuk I. Analysis of Anti-Cancer and Anti-Inflammatory Properties of 25 High-THC Cannabis Extracts. Molecules. 2022 Sep 16;27(18):6057.

[489] Matthew Gronquist, Frank C. Schroeder. Insect Natural Products, Reference Module in Chemistry, Molecular Sciences and Chemical Engineering, Comprehensive Natural Products II, Chemistry and Biology, Volume 2, 2010, Pages 67-108.

uses terpinolene and other terpenes emitted by at least one type of pine tree as a synergist in the production of their own pheromones[490].

CHEMICAL STRUCTURE

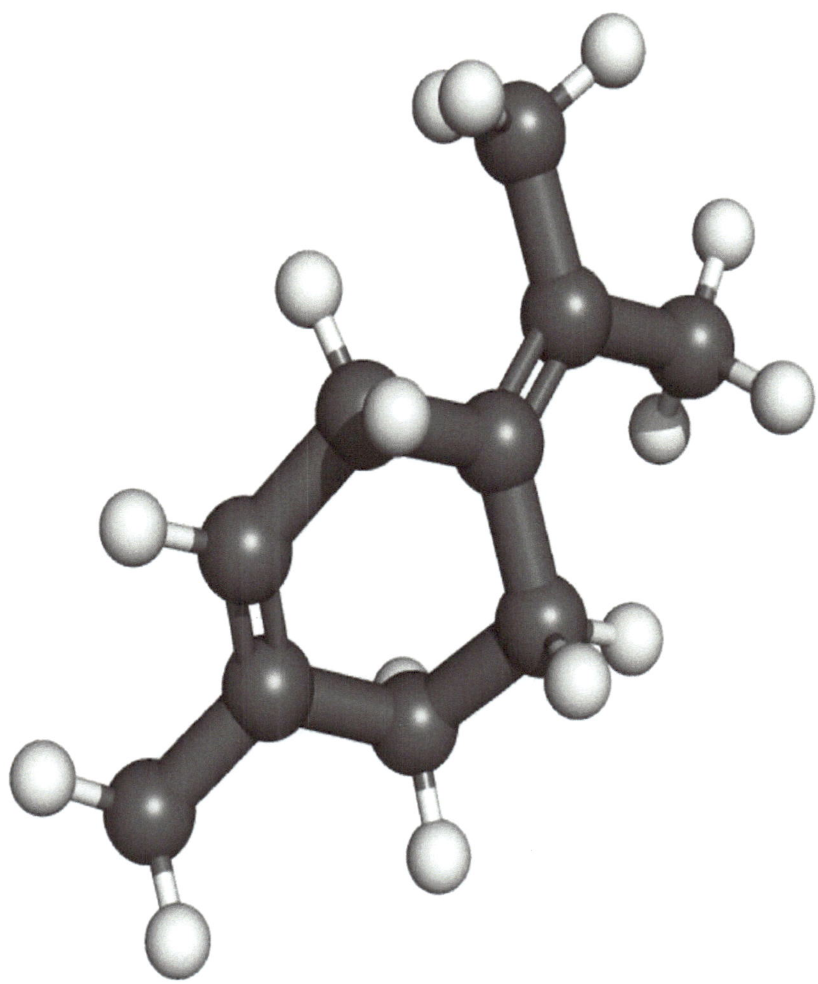

Like most of the monoterpenes common to cannabis, terpinolene is comprised of two units of isoprene with ten carbon atoms and sixteen hydrogen atoms, notated as $C_{10}H_{16}$. Produced in the mevalonate pathway from geranyl pyrophosphate, the terpinolene molecule features twenty-six bonds, with double bonds at positions 1 and 4(8), ten non-H bonds, two multiple bonds, and one covalent bond. Because the carbon skeleton of terpinolene has one ring, this compound is classified as a monocyclic monoterpene. Of the double bonds, one is exocyclic, and one is endocyclic.

Terpinolene is the delta- isomer of terpinene, although chemically, terpinolene is closely related to the terpene thymol, which, interestingly, does not generally occur in cannabis (based on hundreds of laboratory tests examined for this book.) Terpinolene is also likely an intermediate in the biosynthesis of other important plant compounds[491].

OCCURRENCE IN PLANTS

490 John H Borden, Deepa S Pureswaran, Jean Pierre Lafontaine. Synergistic blends of monoterpenes for aggregation pheromones of the mountain pine beetle (Coleoptera: Curculionidae). Journal of economic entomology 2008 Aug; 101(4):1266-75.

491 Takako Harada, Eriko Harada, Ryoko Sakamoto, Tatsuya Ashitani, Koki Fujita, Ken-ichi Kuroda Regio- and Substrate-Specific Oxidative Metabolism of Terpinolene by Cytochrome P450. Monooxygenases in Cupressus lusitanica Cultured Cells American Journal of Plant Sciences, 2012, 3, 268-275.

Terpinolene is produced by many different plants including;

Ajowan
Angelica
Artemisis species
Betel
Blood orange
Cabocla
Canaga
Cascarilla
Cedar
Champaca
Clary sage
Couroupita guianensis
Cubeb
Cypress
Elemi
Fennel
Foeniculum vulgare
Geranium
Guava
Ho
Hyptis
Kewda
Ladvandin
Lemon
Lovage
Mandarin
Mastic
Myristica fragrans
Nutmeg
Papaya
Passion fruit
Phoebe
Pine
Pteronia
Rosemary
Salvia
Savory
Sugandha kokila
Tarragon
Turmeric
Water mint
Yuzu
Allspice
Anise
Basil
Black currant
Buchu
Cajuput
Cannabis
Cassia
Celery
Cinnamon
Coriander
Crithmum maritimum
Cumin
Dill
Eucalyptus species
Ferula
Frankincense
Ginger
Helichrysum italicum
Horsemint
Hyssop
Labdanum
Lavender
Lemongrass
Lychee
Manuka varieties
Microstrobos species
Myrtle
Orange
Parsley
Pepper tree
Pimenta
Prangos
Raspberry
Saffron
Santolina
Snake root
Tagete
Tea tree
Valerian
Wormwood
Alpinia
Apricot
Bay
Black pepper
Cabbage
Calamintha nepeta
Carrot
Cassis
Chamomile
Citronella
Costmary
Croton
Curry
Echinophora cinerea
Featherfew
Fir
Galangal
Grapefruit
Hinoki
Hypericum dogonbadanicum
Juniper
Laurel
Layana
Lime
Mace
Marjoram
Mint
Niaouli
Oregano
Parsnip
Peppermint
Pimento
Protium heptaphyllum
Ravintsara
Sage
Satureja species
Spearmint
Tansy
Thyme
Vassoura
Yarrow

BIOLOGICAL ACTIVITY IN PLANTS

Terpinolene serves many roles in plants, the most prominent of which is its antifungal properties - potent activity against pathogenic fungi that cannabis growers should take note of.

Antifungal Agent: During an investigation into the antifungal properties of tea tree oil, it was found that terpinolene, among other terpenes, had significant antifungal activity against Botrytis cinerea[492], which is one of the most common and significant species of fungi that attack cannabis[493]. In another study, the essential oil of the leaves of gypsophila bicolor - where terpinolene was a primary constituent - exhibited substantial antifungal activity against aspergillus niger, penicillium sp., and botrytis cinerea[494] - again, the much-feared bud rot that afflicts cannabis. Italian cypress trees have been shown to strongly increase production of several terpenes including terpinolene in response to infestation by the aggressive fungus seiridium cardinale [495], while garlic was shown to release terpinolene as an antifungal agent against sclerotium cepivorum[496], which causes allium white rot disease. In the latter study, it was suggested that the antifungal action of terpinolene is via disruption of fungal membrane integrity.

Antifeedant: Terpinolene has been shown to function as an effective antifeedant agent in plants, discouraging, and, in some cases, completely stopping the feeding activity of the cotton bollworm[497].

Larvicidal Agent: Terpinolene was found to be the most effective pure terpene compound out of numerous compounds tested, showing strong larvicidal and repellant activity against the Asian tiger mosquito[498]. In Douglas fir trees, terpinolene contributed to the inhibition of larval growth rate and pupal weight in Choristoneura occidentalis (the western spruce budworm)[499].

Fumigant: This monoterpene has been shown to exhibit toxic fumigant activity against the red flour beetle, and obvious fumigant toxicity against the common booklouse[500].

Antibacterial Agent: In a 2005 study, terpinolene was shown to have the highest antimicrobial activity of ten other terpenes occurring in the essential oil of diplotaenia damavandica (a plant that is native to Iran), with terpinolene showing significant antibacterial activity against all test organisms in the study[501].

---

[492] Yu D, Wang J, Shao X, Xu F, Wang H Antifungal modes of action of tea tree oil and its two characteristic components against Botrytis cinerea. J Appl Microbiol. 2015 Nov;119(5):1253-62.

[493] McPartland, J. M., 1996. A review of Cannabis diseases. Journal of the International Hemp Association 3(1): 19-23 From: http://druglibrary.org/olsen/hemp/iha/iha03111.html Accessed January 27, 2019.

[494] Shafaghat, Ali & Shafaghat Lonbar, Masoud. (2011). Antimicrobial Activity and Chemical Constituents of the Essential Oils from Flower, Leaf and Stem of Gypsophila bicolor from Iran. Natural product communications. 6. 275-6. 10.1177/1934578X1100600230.

[495] Achotegui-Castells, Ander, Roberto Danti, Joan Llusia, Gianni Rocca, Sara Barberini, and Josep Pefiuelas. Strong Induction of Minor Terpenes in Italian Cypress, Cupressus sempervirens, in Response to Infection by the Fungus Seiridium cardinale. Journal of Chemical Ecology 41.3 (2015): 224-243.

[496] Pontin M, Bottini R, Burba JL, Piccoli P. Allium sativum produces terpenes with fungistatic properties in response to infection with Sclerotium cepivorum. Phytochemistry. 2015 Jul; 115:152-60.

[497] M. Liao, J.-J. Xiao, L.-J. Zhou, X. Yao, F. Tang, R.-M. Hua, X.-W. Wu, H.-Q. Cao Chemical composition, insecticidal and biochemical effects of Melaleuca alternifolia essential oil on the Helicoverpa armigera. Journal of Applied Entomology, Volume 141, Issue 9 November 2017 Pages 721-728.

[498] Conti B, Benelli G, Flamini G, Cioni PL, Profeti R, Ceccarini L, Macchia M, Canale A. Larvicidal and repellent activity of Hyptis suaveolens (Lamiaceae) essential oil against the mosquito Aedes albopictus Skuse (Diptera: Culicidae). Parasitology Research 09 Dec 2011, 110(5):2013-2021.

[499] Zou, Jiping, and Rex Cates. Effects of Terpenes and Phenolic and Flavonoid Glycosides from Douglas Fir on Western Spruce Budworm Larval Growth, Pupal Weight, and Adult Weight. Journal of Chemical Ecology 23.10 (2004): 2313-2326.

[500] Jun-Yu Liang, Shan-Shan Guo, Wen-Juan Zhang, Zhu-Feng Geng, Zhi-Wei Deng, Shu-Shan Du & Ji Zhang Fumigant and repellent activities of essential oil extracted from Artemisia dubia and its main compounds against two stored product pests Natural Product Research, Volume 32, 2018 - Issue 10 Pages 1234-1238.

[501] Fereshteh Eftekhar, Morteza Yousefzadi, Dina Azizian, Ali Sonboli, and Peyman Salehi Essential Oil Composition and Antimicrobial Activity of Diplotaenia Z. Naturforsch. 60c, 8210825 (2005); received March 21 - April 27, 2005.

**UV-B Protection:* Apparently usurping a role normally reserved for flavonoids, emissions of terpinolene in vitis vinifera (common grape vine) significantly increased after high doses of UV-B radiation, suggesting the compound is used as a mechanism of defense[502] against radiation.

**Attract Pollinators:* In addition to their role as pests in households all over the world, the common housefly is also a pollinator of plants like caralluma europaea, which attracts the flies by emitting several terpenoid volatiles; primarily terpinolene, alpha-terpinene, and linalool[503].

USES IN INDUSTRY

Like many of the top ten terpenes in cannabis, terpinolene is used both to flavor foods, and to add scent to cleaners, detergents, and personal care products. Examples of food products containing added terpinolene include hard candy, baked goods, ice cream, and fruit ices. Examples of cleaning and personal products containing terpinolene include many pine fragrances, air care products, surface cleaners, liquid detergents, shampoos, aftershaves, deodorants, fabric softeners, soaps, and furniture polish. When added as a fragrance, this terpene is recommended at not more than 10%, which is significantly higher than the fragrance industry recommendations for most other terpenes, where 1% to 3% is more common.

In heavy industry, terpinolene is used to manufacture plastics and resins, and it is used as a solvent in the production of other synthetic products, including synthetic food flavorings.

For much of the last century, terpinolene was derived from fractionation of wood turpentine. Today, this terpene is made in the laboratory, using alpha-pinene as a precursor. Limonene can also be used to make terpinolene. In commercial preparations, terpinolene is a whitish to light amber or light straw-colored liquid.

Interestingly, recent work found that a sweet and fetid chemical blend high in terpinolene was significantly attractive to musca domestica[504] (the common house fly), with researchers suggesting its potential use in lures or traps designed for this pest.

POTENTIAL USES IN MEDICINE

Terpinolene exhibits therapeutic and medicinal traits that could be developed into effective, low-toxicity medicines and treatments for a variety of common conditions. This potential includes but is not limited to:

**Antibacterial Agent:* Perhaps one of the most valuable traits of terpinolene is its antibacterial properties, especially important considering the ever-increasing bacterial resistance to current antibiotic medications. As a major constituent in the essential oil of Astronium fraxinifolium (a tree native to Brazil), terpinolene showed high antibacterial activity against staphylococcus aureus, bacillus cereus, and escherichia coli[505] - some of the most prolific types of bacteria. As a primary constituent in the root steam distillate of paeonia lactiflora (Chinese peony), terpinolene was shown to strongly inhibit growth of harmful human gut bacteria[506], but the compound also inhibited

---

[502] Gil M, Pontin M, Berli F, Bottini R, Piccoli P. Metabolism of terpenes in the response of grape (Vitis vinifera L.) leaf tissues to UV-B radiation. Phytochemistry. 2012 May; 77:89-98.

[503] Zito, Pietro, Salvatore Guarino, Ezio Peri, Maurizio Sajeva, and Stefano Colazza. Electrophysiological and behavioural responses of the housefly to "sweet" volatiles of the flowers of Caralluma europaea (Guss.) N.E. Br. Arthropod-Plant Interactions 7.5 (2013): 485-489.

[504] Guarino, S., Caimi, M., Abdulsattar Arif, M. et al. Use of lures with a mix of sweet and fetid odors for catching Musca domestica L. in domestic environments. Int J Trop Insect Sci 42, 2709–2715 (2022).

[505] Montanari, Ricardo M et al. Exposure to Anacardiaceae volatile oils and their constituents induces lipid peroxidation within food-borne bacteria cells. Molecules (Basel, Switzerland) vol. 17,8 9728-40. 14 Aug. 2012.

[506] Ngan, Luong, Joon-Kwan Moon, Jeong-Han Kim, Takayuki Shibamoto, and Young-Joon Ahn. Growth inhibiting effects of Paeonia lactiflora root steam distillate constituents and structurally related compounds on human intestinal bacteria. World Journal of Microbiology and Biotechnology 28.4 (2011): 1575-1583.

the growth of all beneficial gut bacteria, a common problem of antibiotic drugs. Finally, terpinolene combined with a traditional physico-chemical method of acoustic cavitation delivered efficient disinfection of the secondary effluent of field scale hospital wastewater, completely disinfecting and eliminating various multi-drug-resistant coliforms within 25 minutes of application[507].

**Anticancer Agent:* Terpinolene was shown to inhibit cell proliferation and markedly reduce the expression of protein kinase AKT in human leukemia cells by more than 95%[508]. This is important work considering the activation of AKT or protein-kinase-B is associated with numerous cancers.

**Insecticidal Agent:* Terpinolene - as extracted via steam distillation of the roots of cyperus rotundus - exhibited high contact and fumigant toxicity against German cockroaches[509].

**Antifungal Agent:* As a primary constituent in the essential oil from the leaves of turmeric, terpinolene inhibited growth of the fungus aspergillus flavus by 100% at just 1.5% concentration[510].

**Antiviral Agent:* The H1N1 virus - also known as 'Swine Flu' - is highly contagious. Fortunately, there may be a novel new medication on the horizon to treat this disease, as the essential oil of Melaleuca alternifolia or tea tree oil - where terpinolene is a primary constituent - has been shown to exhibit an inhibitory effect on the replication of the H1N1 virus[511].

**Sedative:* The essential oil of the perennial herb microtoena patchouli - where terpinolene was a primary compound - almost completely immobilized mice that inhaled vapors from the oil for just 35 minutes, with no adverse effects, even after caffeine-induced excitation[512]. In a follow-up study, the authors found that terpinolene's sedative activities work via a double-bond in the side chain or pi bonds in the 6-membered ring[513].

**Antioxidant:* In cultured human blood cells, terpinolene increases total antioxidant capacity without altering total oxidative stress levels, and with no genotoxicity on lymphocytes [514].

**Anti-inflammatory:* The essential oil of the leaves of Curcuma longa - where terpinolene was the primary constituent - inhibited the production of pro-inflammatory cytokines, regulated the overproduction of oxidative markers, and restored histopathological damage[515], soundly demonstrating this monoterpene's anti-inflammatory activity.

---

[507] Mukherjee A, Ahn YH. Terpinolene as an enhancer for ultrasonic disinfection of multi-drug-resistant bacteria in hospital wastewater. Environ Sci Pollut Res Int. 2022 May;29(23):34500-34514.

[508] Okumura, Naoko et al. Terpinolene, a component of herbal sage, downregulates AKT1 expression in K562 cells. Oncology letters vol. 3,2 (2011): 321-324.

[509] Contact and Fumigant Toxicity of Cyperus rotundus Steam Distillate Constituents and Related Compounds to Insecticide-Susceptible and -Resistant Blattella germanica. Journal of Medical Entomology 49.3 (2012): 631-639.

[510] Sindhu S, Chempakam B, Leela NK, Suseela Bhai R. Chemoprevention by essential oil of turmeric leaves (Curcuma longa L.) on the growth of Aspergillus flavus and aflatoxin production. Food Chem Toxicol. 2011 May 49(5):1188-92.

[511] Garozzo, A., R. Timpanaro, B. Bisignano, P.M. Furneri, G. Bisignano, and A. Castro. In vitro antiviral activity o f Melaleuca alternifolia essential oil. Letters in Applied Microbiology 49.6 (2009).

[512] Ito, Ken, and Michiho Ito. Sedative effects of vapor inhalation of the essential oil of Microtoena patchoulii and its related compounds. Journal of Natural Medicines 65.2 (2011): 336-343.

[513] Ito, Ken, and Michiho Ito. The sedative effect of inhaled terpinolene in mice and its structure-activity relationships. Journal of Natural Medicines 67.4 (2013): 833-837.

[514] Turkez, Hasan, Elanur Ayd n, Fatime Geyikoglu, and Damla Cetin. Genotoxic and oxidative damage potentials in human lymphocytes after exposure to terpinolene in vitro. Cytotechnology 67.3 (2014): 409-418.

[515] Kumar, Anant, Karishma Agarwal, Monika Singh, Archana Saxena, Pankaj Yadav, Anil Maurya, Anju Yadav, Sudeep Tandon, Debabrata Chanda, and Dnyaneshwar Bawankule. Essential oil from waste leaves of Curcuma longa L. alleviates skin inflammation. Inflammopharmacology 26.5 (2018): 1245-1255.

Accelerated Wound Healing: As the primary constituent in the essential oil of Canarium schweinfurthii, terpinolene contributed to significant accelerated wound healing in a dose dependent manner in human keratinocytes, with a theorized mechanism of action being modulation of pro-inflammatory cytokines[516].

IMPLICATIONS FOR HUMAN HEALTH & NUTRITION

To include terpinolene as part of a healthy diet, consider regularly eating fruits that contain this compound like apricots, black currants, blood oranges, grapefruit, guava, mandarin, papaya, passion fruit, raspberries, and yuzu. Vegetables that contain terpinolene include bell peppers, cabbage, carrots, celery, and chickpeas. When preparing and cooking foods, consider working with spices such as allspice, anise, basil, bay, black pepper, coriander, cumin, dill, ginger, mace, mint, nutmeg, oregano, saffron, or thyme. Other options for dietary inclusion of terpinolene include many types of herb-based teas, especially tea with valerian root, lemongrass, or chamomile.

Cannabis varieties that are known to feature moderate to high concentrations of terpinolene include Banana Lemonade, Black Widow, Blueberry Limeade, Blue Dream, Cherry Diesel, Critical Jack, Durban Poison, Dutchie, Dutch Treat, Ghost Train Haze, Island Sweet Skunk, Jack Herer, Pineapple Kush, Royal Jack Automatic, Sensi Star, Shipwreck, Snowcap, Sour Tangie, Sour Tsunami, Super Lemon Haze, and Super Silver Haze. While we do not recommend smoking or vaporizing cannabis as an ideal or healthy manner of consumption, it is possible to acquire trace terpinolene via either method. A better option is to juice, blend or otherwise consume these strains raw.

Essential oils made from the plants listed in the "Occurrences in Plants" section above are likely to contain high amounts of terpinolene, however, we urge extreme caution when working with these complex oils or isolated compounds, as they can be toxic in some situations.

---

[516] Bonnard M, Martin E, Parrot I. Wound Healing Potential of an Oleoresin Essential Oil Chemotype from Canarium schweinfurthii Engl. Molecules. 2022 Nov 17;27(22):7966.

Terpinolene Review

Answer the following questions to test your knowledge of this terpene:

Question #1: What type of terpene is terpinolene?

a. Linear monoterpene
b. Monoterpenic sesquiterpene
c. Monoterpene alcohol
d. Monocyclic monoterpene

Question #2: Which of the following best describes the primary flavor & aroma profile of terpinolene?

a. Spicy, turpentine, fuel, pine
b. Fresh, pine, woody, citrus
c. Hops, cloves, woods, spicy
d. Rose, floral, woody, tea

Question #3: How common is terpinolene in cannabis?

a. Top three
b. Top five
c. Top ten
d. Top twenty

Question #4: What is the chemical formula for terpinolene?

a. $C_{15}H_{26}$
b. $C_{10}N_{16}O$
c. $C_{10}H_{16}$
d. $C_{15}H_{26}O$

Question #5: Name two uses of terpinolene in plants:

1 ____________________ 2 ____________________

Question #6: Name two potential medical uses of terpinolene:

1 ____________________ 2 ____________________

For the answer key to all individual compound chapters please go here: http://thebigbookofterps.com/bbot-quiz-answer-key

FENCHOL

Type: Bicyclic monoterpenoid
Chemical Formula: $C_{10}H_{18}O$
Molecular Weight: 154.25 g/mol
Boiling Point: *Fenchol alcohol:* 201.00 to 202.00 °C. @ 760.00 mm Hg[517]
(-)-alpha-fenchol: 202.00 to 203.00 °C. @ 760.00 mm Hg[518]
(-)-beta-fenchol: 202.00 to 203.00 °C. @ 760.00 mm Hg[519]
Flash Point: *Fenchol alcohol:* 165.00 °F. TCC (73.89 °C.)
(-)-alpha-fenchol: 167.00 °F. TCC (75.00 °C.)
(-)-beta-fenchol: 165.00 °F. TCC (73.89 °C.)
Melting Point: 39 to 45 °C (102 to 113 °F)
Solubility: *All isomers:* Soluble in alcohol, water
Oral LD50: 2,000 mg/kg (rat)[520]
Biological Role: Insecticidal, antifungal, antifeedant
Therapeutic Role: Antibacterial, antifungal, acaricidal
Commercial Use: Fragrance, flavoring agent
Occurrence in Cannabis: Top twenty
Aroma/Flavor: *Fenchol alcohol:* Camphor, pine, woody, citrus, mint
(-)-alpha-fenchol: Earthy, woody, clean, camphoreous
(-)-beta-fenchol: No aroma or flavor descriptors reported

Occurs in Cannabis Strains: Fenchol likely occurs in many more cannabis varieties than currently reported, however, often laboratories do not test for and/or have standards for this compound. As more laboratories start testing for this compound, it's likely that its overall position may change, becoming a top ten terpene in cannabis. Current laboratory testing worldwide shows this compound appearing in the following varieties: Angel Food Cake, Banana Kush, Blue Cheese, Blue Dream, Cannatonic, Cheese, Chemdawg, Covelo Sour Diesel, Diablo, Girl Scout Cookies, GG#4, Kosher Kush, Lemon Skunk, God's Green Crack, Grape Soda OG, Kosher Kush, OG Kush, Pink Lemonade, SFV OG, Sour Cherry Sherbet, Shiatsu Kush, Sour Jack, Sunset Sherbet, Tosh OG.

Of particular interest to cannabis consumers is a 2022 study where researchers consistently found significant oxidized fenchol in the vapors of 12 different cannabis vaporization products sold in California[521], with scientists noting that the medical or other effects of oxidized compounds like fenchol are unknown and require urgent study. Users of cannabis vape products should use extreme caution or avoid these products.

INTRODUCTION

Fenchol is a bicyclic monoterpene that generally occurs within the top fifteen terpenes by concentration in many cannabis varieties, although it also occurs within the top ten for a large number of strains. With a unique flavor and aroma profile that depends on the isomer but generally consists of camphor and woody notes. This compound is widely used in the perfume and fragrance industries, and as a flavoring ingredient in some foods.

---

[517] The Good Scents Company Data Sheet for Fenchol, from: http://www.thegoodscentscompany.com/data/rw1011531.html. Accessed August 9, 2023.

[518] The Good Scents Company Data Sheet for (-)-alpha-fenchol, from: http://www.thegoodscentscompany.com/data/rw1054341.html. Accessed August 8, 2023.

[519] The Good Scents Company Data Sheet for (-)-beta-fenchol, from: http://www.thegoodscentscompany.com/data/rw1046201.html. Accessed August 9, 2023.

[520] Santa Cruz Biotechnology Data Sheet for Beta-eudesmol, from: https://datasheets.scbt.com/sc-235107.pdf. Accessed August 16, 2023.

[521] Zhu J, Niu J, Das D, Cabecinha A, Abramovici H. In-situ TD-GCMS measurements of oxidative products of monoterpenes at typical vaping temperatures: implications for inhalation exposure to vaping products. Sci Rep. 2022 Jun 30;12(1):11019.

Interestingly, fenchol is emitted by larvae of the bark beetle, dendroctonus valens, which a predatory beetle, rhizophagus grandis, uses to locate the larvae and lay its eggs nearby[522]. Additionally, the Southern Pine beetle uses fenchol as a chemical signal to let other beetles know that a particular tree has no further vacancy, once a significant mass of mating pair-offs have occurred[523].

Though not as prevalent in the literature as the top ten terpenes in cannabis, fenchol serves multiple biological roles in plants, and exhibits activity that could be important for use in medicine.

CHEMICAL STRUCTURE

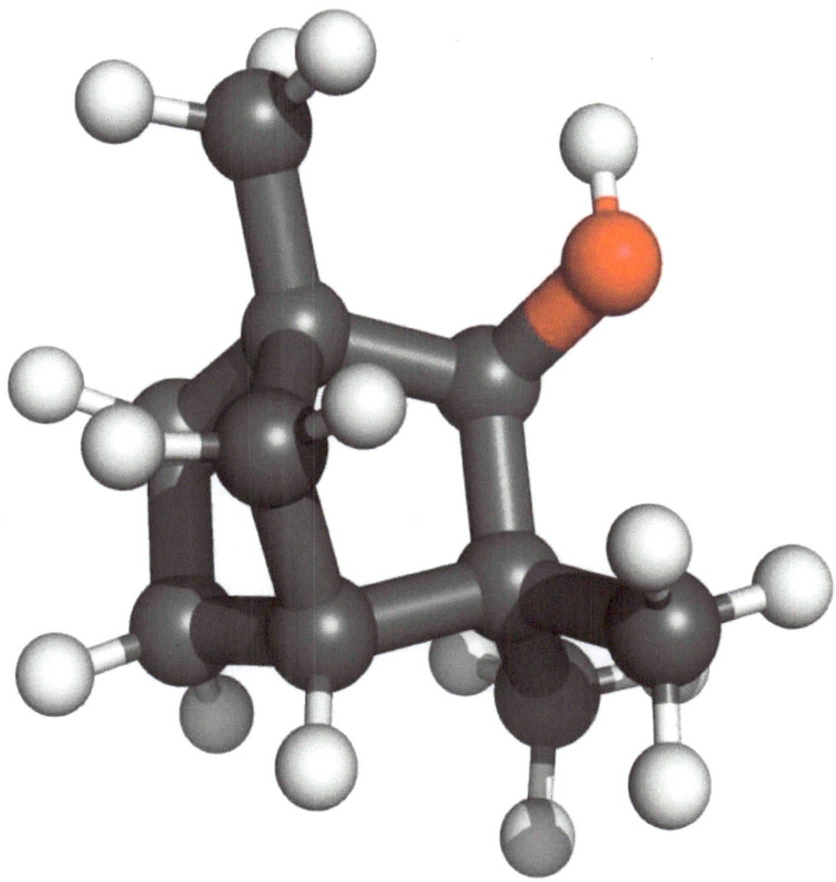

With its name derived from that of the fennel plant, fenchol is a monoterpene biosynthesized in the mevalonate pathway from geranyl pyrophosphate. This compound is based on the typical isoprene skeleton and contains two units of isoprene totaling ten carbon atoms, eighteen hydrogen atoms, and one oxygen atom, notated as $C_{10}H_{18}O$. It is primarily the oxygenation of this compound that classifies fenchol as a monoterpenoid, and because the compound has two rings fused together, it is further classified as bicyclic.

Fenchol is also called fenchyl alcohol, and is sometimes referred to in the literature by this term. There are several known isomers of this compound, although not every isomer has data reported for it. In this chapter we discuss

---

[522] Gregoire JC, Baisier M, Drumont A, Dahlsten DL, Meyer H, Francke W. Volatile compounds in the larval frass of Dendroctonus valens and Dendroctonus micans (Coleoptera: Scolytidae) in relation to oviposition by the predator Rhizophagus grandis (Coleoptera: Rhizophagidae). J Chem Ecol. 1991 Oct;17(10):2003-19.

[523] Sullivan, Brian T. Electrophysiological and Behavioral Responses of Dendroctonus frontalis (Coleoptera: Curculionidae) to Volatiles Isolated from Conspecifics. Journal of Economic Entomology 98.6 (2005): 2067-2078.

fenchol alcohol, and its laevorotatory isomer forms, l-beta-fenchol and l-alpha-fenchol. However, variations include the following (note that The Good Scents Company indicates that the racemic version of alpha-fenchol has not been found in nature):

Fenchol alcohol
L-alpha-fenchol
D-alpha-fenchol
L-beta-fenchol
D-beta-fenchol
Alpha-isofenchol
Fenchone
Racemic: (±)-alpha-fenchol
Racemic: (±)-beta-fenchol

Interestingly, the torsion of the strained ring in the fenchol molecule has led researchers studying quantum tunneling in alpha-fenchol to conclude that "the presented analysis highlights the importance of considering nuclear quantum effects for comprehension of the structure and dynamics of matter, even for cases such as asymmetric biomolecules with their (seemingly) well-defined geometries[524]."

OCCURRENCE IN PLANTS

Fenchol is moderately distributed in nature and occurs in many plant species, including:

Fenchyl alcohol:

Artemisia
Basil
Celery
Fennel
Grape
Hop
Lime
Nutmeg
Picea alba
Pine
Rosemary

L-alpha-fenchol:

Basil
Boldo
Boldus
Carrot
Cascarilla
Cypress
Eucalyptus species
Hinoki
Hypericum dogonbadanicum
Lemongrass
Lime
Lovage
Marjoram
Mastic
Pine
Plectranthus glandulosus
Pteronia
Rosemary
Thyme

L-beta-fenchol:

Artemisia haussknechtii
Carrot weed
Hyptis suaveolens
Lemon
Lime
Tetradenia riparia

---

[524] Medel R, Springborn JR, Crittenden DL, Suhm MA. Hydrogen Delocalization in an Asymmetric Biomolecule: The Curious Case of Alpha-Fenchol. Molecules. 2021 Dec 24;27(1):101.

BIOLOGICAL ACTIVITY IN PLANTS

Although more study is needed to understand how plants use fenchol, this compound likely has important biological roles in plants, most notably for its insecticidal and antifungal properties.

Insecticidal: Fenchol was shown to kill 100% of adult maize weevils (sitophilus zeamais motschulsky) after 96 hours of exposure[525]. As a primary constituent in the essential oil of cardamom (elettaria cardamomum L.), fenchyl alcohol contributed to toxic fumigation activity against the red flour beetle, the bruchid beetle, and the flour moth[526].

Antifungal: As a primary constituent in the essential oil of eucalyptus teretecornis, beta-fenchol exhibited significant antifungal activity against the pathogen, alternaria alternate[527], and also exhibited free-radical scavenging and other antioxidant properties. In a separate study, the same authors demonstrated that, as a primary constituent in the essential oil of zanthoxylum alatum, beta-fenchol exhibited potent antifungal activity against the same fungus researched previously, alternaria alternata, and two others: alternaria brassicae, and curvularia lunata[528]. While studying the EO of zanthoxylum alatum, the authors also discovered that beta-fenchol exhibited strong antibacterial activity against several common microbes.

Antifeedant: In sheep and deer that had never previously had access to Douglas fir trees and shrubs, the essential oil of the needles of this species strongly inhibited microbial activity in the guts of the animals, whereas there was no effect for species of the same animals that had previous access to Douglas fir[529]. Among the constituents of the EO of Douglas fir, fenchyl alcohol contributed to inhibition of the rumen microbial activity, suggesting an antifeedant effect against some animals.

USES IN INDUSTRY

Fenchol is used mostly in the fragrance and perfume industries, with a large amount going toward adding scent to household and commercial products such as air fresheners, dishwashing liquids, shampoos, detergents, etcetera. This bicyclic monoterpene is also used as a bitter flavoring ingredient in food items like baked goods, hard candy, and frozen dairy products.

Interestingly, scientists have discovered new variations of fenchol by feeding cutworm larvae a substrate diet high in this compound, and then measuring what came out in the frass[530] (feces).

POTENTIAL USES IN MEDICINE

Research into the medical potential of fenchol has provided some tantalizing possibilities, including the following:

---

[525] Erol Yildirim, Saban Kordali, Bugrahan Emsen. Insecticidal effects of monoterpenes on Sitophilus zeamais Motschulsky (Coleoptera: Curculionidae). Journal of Applied Botany and Food Quality 86, 198 - 204 (2013).

[526] Abbasipour, Habib et al. Fumigant toxicity and oviposition deterrency of the essential oil from cardamom, Elettaria cardamomum, against three stored-product insects. Journal of insect science (Online) vol. 11 (2011): 165.

[527] Guleria, Sanjay & K. Tiku, A & Gupta, Sahil & Singh, Gurjinder & Koul, Apurva & K. Razdan, V. (2011). Chemical composition, antioxidant activity and inhibitory effects of essential oil of Eucalyptus teretecornis grown in north-western Himalaya against Alternaria alternata. Journal of Plant Biochemistry and Biotechnology. 21.44-50. 10.1007/s13562-011-0073-2.

[528] Guleria, Sanjay, A. K. Tiku, Apurva Koul, Sahil Gupta, Gurjinder Singh, and V. K. Razdan. Antioxidant and Antimicrobial Properties of the Essential Oil and Extracts of Zanthoxylum alatum Grown in North-Western Himalaya. The Scientific World JOURNAL 2013.2013 (2013).

[529] Oh, H K et al. Effect of various essential oils isolated from Douglas fir needles upon sheep and deer rumen microbial activity. Applied microbiology vol. 15,4 (1967): 777-84.

[530] Miyazawa, Mitsuo, and Yohei Miyamoto. Biotransformation of (+)-(1 R ,2 S)-fenchol by the larvae of common cutworm (Spodoptera litura). Tetrahedron 60.13 (2004): 3091-3096.

**Antibacterial:* Although the effects were not as strong as those of penicillin, fenchol exhibited broad-spectrum antibacterial activity[531] when researchers examined the active properties of twenty-one oxygenated monoterpenes. As the primary constituent in the essential oil of the camphor bush (tarchonanthus camphoratus), fenchol exhibited pronounced antibacterial activity against both gram positive (staphylococcus aureus, bacillus ssp.) and gram negative (escherichia coli, pseudomonas aeruginosa, salmonella typhi, klebsiella pneumoniae, proteus mirabilis) bacteria[532]. A Swiss study echoed these findings, demonstrating that, as a primary constituent in the essential oil of C. ornifolia, fenchol exhibited potent antibacterial activity against the same bacterial strains previously mentioned[533], and showed strong antifungal activity against candida albicans.

**Antifungal:* As a primary constituent in the essential oil of cedar leaf, fenchyl alcohol (and a closely related compound, fenchone, which is the oxidized version of fenchol) significantly reduced biofilm formation of candida albicans - a pathogenic fungus that commonly infects humans - by downregulating hypha-specific and biofilm related genes[534].

**Acaricidal:* With tick populations and related diseases on the rise, fenchol might become an important insecticidal tool after it was shown to exhibit acaricidal activity against engorged female rhipicephalus (boophilus) microplus ticks[535].

IMPLICATIONS FOR HUMAN HEALTH & NUTRITION

Dietary intake of fenchol is not as easy as most of the top ten terpenes in cannabis, but it is possible. Fruits containing fenchol include lime - which can be included in foods or added to beverages, and some varieties of grapes. For vegetables, celery is known to contain fenchol, although it is likely that other vegetables do produce this bicyclic monoterpene. When cooking, preparing, or serving foods, consider using spices like basil, fennel, nutmeg, or rosemary. Because the hop plant contains fenchol, it may also be possible to obtain trace amounts of this compound by drinking some types of beer.

For people interested in the Japanese practice of forest bathing, visiting a wooded area containing pine or white spruce trees may allow you to obtain fenchol in the form of volatile emissions from these trees.

Cannabis varieties that are relatively easy to obtain and that are reported to include fenchol include Angel Food Cake, Banana Kush, Blue Cheese, Cheese, Chemdawg, Covelo Sour Diesel, Diablo, Girl Scout Cookies, GG#4, Kosher Kush, Lemon Skunk, Grape Soda OG, OG Kush, Pink Lemonade, SFV OG, Sour Cherry Sherbet, Shiatsu Kush, Sour Jack, Sunset Sherbet, and Tosh OG.

While it is unlikely but possible to smoke or vaporize these strains to obtain trace or secondary fenchol, we recommend instead to consume juiced, blended, or otherwise raw cannabis, or to use fully extracted whole cannabis oil made from one of these strains.

---

531 Kotan R, Kordali S, Cakir A. Screening of antibacterial activities of twenty-one oxygenated monoterpenes. Z Naturforsch C. 2007 Jul-Aug;62(7-8):507-13.

532 Matasyoh, Josphat C., Joyce J. Kiplimo, Nicholas M. Karubiu, and Tiffany P. Hailstorks. Chemical composition and antimicrobial activity of essential oil of Tarchonanthus camphoratus. Food Chemistry 101.3 (2007): 1183-1187.

533 Mothana, Ramzi A et al. Chemical analysis and biological activity of the essential oils of two endemic Soqotri Commiphora species. Molecules (Basel, Switzerland) vol. 15,2 689-98. 1 Feb. 2010.

534 Manoharan, Ranjith Kumar et al. Antibiofilm and Antihyphal Activities of Cedar Leaf Essential Oil, Camphor, and Fenchone Derivatives against Candida albicans. Frontiers in microbiology vol. 8 1476. 3 Aug. 2017.

535 Karina Neoob de Carvalho Castro, Kirley Marques Canuto, Edy de Sousa Brito, Livio Martins Costa-Junior, Ivanilza Moreira de Andrade, Joao Avelar Magalhaes, Dhiessica Morgana Alves Barros. In vitro efficacy of essential oils with different concentrations of 1,8-cineole against Rhipicephalus (Boophilus) microplus Rev. Bras. Parasitol. Vet. vol.27 no.2 Jaboticabal April-June 2018.

Experimentation with pure or near-pure extractions of fenchol or fenchyl alcohol, or use of essential oils made from plants listed in the "Occurrences in Plants" section above should be carried out with extreme care, as these compounds can be toxic in some situations.

Fenchol Review

Answer the following questions to test your knowledge of this terpene:

Question #1: What type of terpene is fenchol?

a. Monocyclic monoterpene
b. Monocyclic sesquiterpene
c. Bicyclic sesquiterpene
d. Bicyclic monoterpenoid

Question #2: Which of the following best describes the primary flavor & aroma profile of fenchol?

a. Lemon, sharp, fuel, terpenic
b. Piney, wood, spicy, dry
c. Camphor, citrus, wood, pine
d. Spicy, cloves, dry, pepper
e. Depends on the isomer

Question #3: How common is fenchol in cannabis?

a. Top ten
b. Top three
c. Top twenty
d. Less common

Question #4: What is the chemical formula for fenchol?

a. $C_{15}H_{32}0$
b. $C_5H_{24}0$
c. $C_{10}H_{24}0$
d. $C_{10}H_{18}0$

Question #5: Name two uses of fenchol in plants:

1 ____________________________________ 2 ____________________________________

Question #6: Name two potential medical uses of fenchol:

1 ____________________________________ 2 ____________________________________

For the answer key to all individual compound chapters please go here: http://thebigbookofterps.com/bbot-quiz-answer-key/

TERPINEOL

Type: Monocyclic monoterpenoid
Chemical Formula: $C_{10}H_{18}O$
Molecular Weight: 154.253 g/mol
Boiling Point: *Alpha-terpineol:* 214.00 to 218.00 °C. @ 760.00 mm Hg[536]
D-alpha-terpineol: 217.50 °C. @ 760.00 mm Hg (estimated)[537]
L-alpha-terpineol: 217.00 to 218.00 °C. @ 760.00 mm Hg[538]
Beta-terpineol: 209.00 to 210.00 °C. @ 760.00 mm Hg[539]
Delta-terpineol: 218.00 to 219.00 °C. @ 760.00 mm Hg[540]
Gamma-terpineol: 218.00 to 219.00 °C. @ 760.00 mm Hg[541]
Terpineol acetate: 232.88 °C. @ 760.00 mm Hg (estimated)[542]
Terpinen-4-ol: 211.00 to 213.00 °C. @ 760.00 mm Hg[543]
Flash Point: *Alpha-terpineol:* 191.00 °F. TCC (88.33 °C.)
D-alpha-terpineol: 194.00 °F. TCC (90.00 °C.)
L-alpha-terpineol: 194.00 °F. TCC (90.00 °C.)
Beta-terpineol: 189.00 °F. TCC (87.22 °C.)
Delta-terpineol: 183.00 °F. TCC (83.89 °C.)
Gamma-terpineol: 191.00 °F. TCC (88.33 °C.)
Terpineol acetate: 212.00 °F. TCC (100.00 °C.)
Terpinen-4-ol: 174.00 °F. TCC (79.00 °C.)
Melting Point: *Alpha-terpineol:* 40.00 to 41.00 °C. @ 760.00 mm Hg
D-alpha-terpineol: 30.00 to 31.00 °C. @ 760.00 mm Hg
L-alpha-terpineol: 31.00 to 35.00 °C. @ 760.00 mm Hg
Beta-terpineol: 32.00 to 33.00 °C. @ 760.00 mm Hg
Gamma-terpineol: 68.00 to 70.00 °C. @ 760.00 mm Hg
Solubility: *Alpha-terpineol:* Soluble in alcohol, paraffin oil, kerosene, benzene, acetone, ether, methanol, ethanol, propylene glycol, weakly soluble in water
D-alpha-terpineol: Soluble in alcohol
L-alpha-terpineol: Soluble in alcohol
Beta-terpineol: Soluble in alcohol, various oils
Trans-beta-terpineol: Weakly soluble in water[544]
Cis-beta-terpineol: Weakly soluble in water[545]

---

[536] The Good Scents Company Data Sheet for Alpha-terpineol, from: http://www.thegoodscentscompany.com/data/rw1011252.html. Accessed August 29, 2023.

[537] The Good Scents Company Data Sheet for D-alpha-terpineol, from: http://www.thegoodscentscompany.com/data/rw1508221.html. Accessed August 29, 2023.

[538] The Good Scents Company Data Sheet for L-alpha-terpineol, from: http://www.thegoodscentscompany.com/data/rw1218601.html. Accessed August 29, 2023.

[539] The Good Scents Company Data Sheet for beta-terpineol, from: http://www.thegoodscentscompany.com/data/rw1031211.html. Accessed August 29, 2023.

[540] The Good Scents Company Data Sheet for Delta-terpineol, from: http://www.thegoodscentscompany.com/data/rw1531781.html. Accessed August 29, 2023.

[541] The Good Scents Company Data Sheet for Gamma-terpineol, from: http://www.thegoodscentscompany.com/data/rw1056011.html. Accessed August 29, 2023.

[542] The Good Scents Company Data Sheet for Terpineol Acetate, from: http://www.thegoodscentscompany.com/data/rw1375491.html. Accessed August 29, 2023.

[543] The Good Scents Company Data Sheet for Terpinen-4-ol, from: http://www.thegoodscentscompany.com/data/rw1618181.html. Accessed August 29, 2023.

[544] The Good Scents Company Data Sheet for Trans-beta-terpineol, from: http://www.thegoodscentscompany.com/data/rw1559421.html. Accessed August 29, 2023.

[545] The Good Scents Company Data Sheet for Cis-beta-terpineol, from: http://www.thegoodscentscompany.com/data/rw1126511.html. Accessed August 29,

Delta-terpineol: Soluble in alcohol, mineral oil, paraffin oil, weakly soluble in water
Gamma-terpineol: Soluble in alcohol, weakly soluble in water
Terpineol acetate: Soluble in alcohol
Terpinen-4-ol: Soluble in alcohol, weakly soluble in water
Oral LD50: *Alpha-terpineol:* 2,830 mg/kg (rat)
Mixture of alpha- and beta-terpineol: 4,300 mg/kg (rat)[546]
Biological Role: Insecticidal, antifungal, antibacterial
Therapeutic Role: Major antibacterial agent**,** antinociceptive, antifungal
Commercial Use: Fragrance, food additive, solvent, chemical precursor
Occurrence in Cannabis: Top twenty
Aroma/Flavor: *Alpha-terpineol:* Pine, floral, woody, citrus
D-alpha-terpineol: Floral
L-alpha-terpineol: Floral, terpenic
Beta-terpineol: Earthy, woody
Gamma-terpineol: Pine, floral
Terpineol acetate: Sweet, herbal, floral
Terpinen-4-ol: Musty

Occurs in Cannabis Strains: Most laboratories do not differentiate between the isomers of terpineol, and it's likely that few of them have standards to do so. Most or all the laboratory reports indicating the presence of terpineol in the following varieties are likely referring to alpha-terpineol; Banana Kush, Cannatonic, Ghost Train Haze, Girl Scout Cookies, God's Green Crack, Jack Herer, Kosher Kush, OG Kush, Shiatsu Kush, Skywalker OG, Sour Cherry Sherbet, Sour Jack, Strawberry Short Cookies, Sunset Sherbet, Super Silver Haze, White Widow

INTRODUCTION

Terpineol is a monocyclic monoterpenoid with a highly prized aroma and flavor profile that is primarily lilac and floral, with hints of woodiness, lime, citrus, and anise. Occurring generally within the top twelve to fifteen terpenes in cannabis, terpineol is found in moderate concentrations in classic cannabis strains like Jack Herer, Ghost Train Haze, Skywalker OG, Super Silver Haze, and White Widow.

Terpineol is produced by hundreds of plants, but commercial samples are usually manufactured in the laboratory by conversion of alpha-pinene. Terpineol is mainly used in the perfume and fragrance industries, where its lilac scent is celebrated.

2023.

[546] Federal Register for Alpha-terpineol, from: https://www.federalregister.gov/documents/2021/12/16/2021-27179/a-terpineol-cas-no-98-55-5-tolerance-exemption. Accessed August 29, 2023.

CHEMICAL STRUCTURE

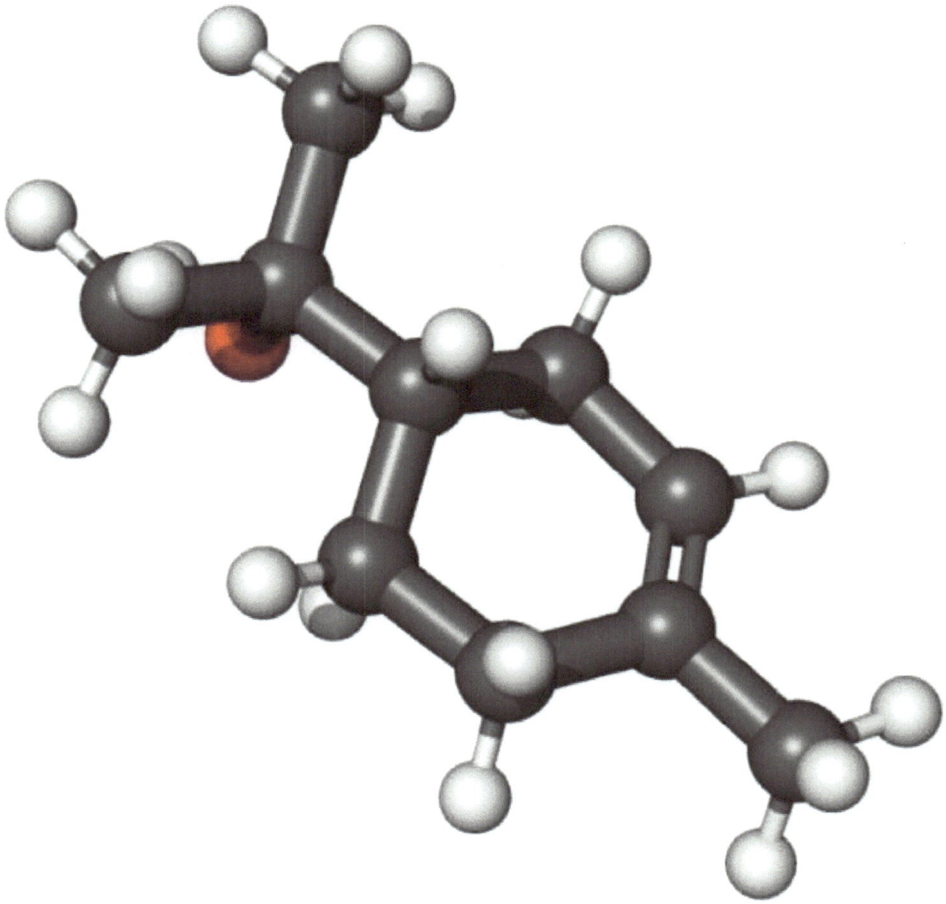

Biosynthesized in the cytosol of plants via the mevalonate pathway, and based on the typical linked isoprene skeleton, terpineol has ten carbon atoms, eighteen hydrogen atoms, and one oxygen atom, notated as $C_{10}H_{18}O$. This compound features one ring, making it monocyclic, and the oxygen atom further classifies it as a monoterpenoid. Terpineol's molecular skeleton contains one double bond, one rotatable bond, and one covalent bond.

Terpineol has a surprising number of isomers, including D-alpha, L-alpha, beta, trans-beta, cis-beta, delta, gamma, acetate, and terpinene-4-ol. Terpineol generally consists of a mixture of some or all of these isomers, with alpha-terpineol being the primary constituent. Beta, delta, gamma, and other versions of this monoterpenoid are less common in nature.

OCCURRENCE IN PLANTS

Like many of the top terpenes in cannabis, terpineol and it variations are produced by hundreds of plant species, including:

Alpha:

Achillea tenuifolia
Ajowan
Allmanda cathartica
Allspice
Alpinia galanga
Ammomum testaceum
Angelica
Anise
Apple

Apricot
Armoise
Artemisia species
Ayou
Basil
Basswood
Bay
Bergamot
Betel
Bilberry
Black currant
Black elder
Black pepper
Blood orange
Blueberry
Bois de rose
Boldo
Boldus
Buchu
Cabbage
Cajuput
Calamus
Camel grass
Cananga
Cannabis
Caraway
Cardamom
Carnation
Carrot
Cascarilla
Cassia
Cassinia
Cassis
Cayenne
Celery
Chamomile
Champaca
Cherry
Cinnamon
Citronella
Clary sage
Clove
Coconut
Coriander
Corn silk
Cornmint
Cranberry
Croton species
Cubeb
Cumin
Curcuma mada
Curry
Cypress
Dill
Echinophora cinerea
Eschweilera coriacea
Eucalyptus species
Featherfew
Fennel
Ferula
Fig
Fir
Fleabane
Frankincense
Galangal
Gardenia
Geranium
Ginger
Grape leaf
Grapefruit
Guava
Hinoki
Ho leaf
Horsemint
Hyacinthus orientalis
Hyptis species
Hyssop
Jambu
Juniper
Kachur
Kewda
Labdanum
Laurel
Lavandin
Lavandula officinalis
Lavender
Layana
Lemon
Lemongrass
Levisticum
Lime
Lovage
Mace
Magnolia
Mandarin
Mangrove
Manuka species
Marjoram
Mastic
Melissa
Melon
Mentha longifolia
Microstrobos species
Mikan
Mugwort
Myrrhinium atropurpureum
Myrtle
Nepeta denudata
Neroli
Nutmeg
Orange
Oregano
Papaya
Parsley
Peach
Pennyroyal
Peppermint
Pimento
Pine
Pineapple
Plum
Plumcot
Pteronia
Raisin
Raspberry
Ravintsara
Rose
Rosemary
Sage
Salvia species
Santolina
Savin
Savory
Skimmia laureola
Snake root
Spearmint
Stachys recta
Star fruit
Strawberry
Sugandha kokila
Sunflower
Tagete
Tamarind
Tangerine
Tansy
Taxodium distichum
Tea
Thevetia peruviana
Thyme
Tomato
Turmeric
Valerian
Verbena
Witch hazel
Wormwood
Yarrow

Ylang ylang
Yuzu

D-alpha:

Cardamom
Cinnamon
Lovage
Marjoram
Tea tree

L-alpha:

Bay
Cinnamon
Mint
Tea tree

Beta:

Boldo
Black pepper
Cardamom
Cascarilla
Clove
Cornmint
Dill
Lavandin
Lemongrass
Lime
Mace
Mandarin
Nutmeg
Orange
Oregano
Pine
Sage
Salvia
Thyme

Trans-beta:

Artemisia species
Caraway
Citrus aurantifolia
Clove
Lime
Mandarin
Rosemary
Sage

Cis-beta:

Artemisia
Citrus aurantifolia
Lavandula
Ledum
Lime
Mandarin
Murraya koenigii
Rosemary

Delta:

Artemisia hypsacea
Buchu
Cardamom
Cypress
Eucalyptus species
Galangal
Laurel
Layana
Mentha longifolia
Myrtle
Pteronia
Rosemary
Sage
Salvia
Spearmint
Sugandha kokila
Thyme

Gamma:

Celery	Cinnamon	Lavandin
Lavender	Lime	Mandarin
Marjoram	Niaouli	Salvia
Thyme		

Terpineol acetate:

Ferula cupularis stem oil (otherwise reported by The Good Scents Company as not found in nature)

Terpinen-4-ol:

Tea tree oil (reported by The Good Scents Company at 24.73% in the EO)

BIOLOGICAL ACTIVITY IN PLANTS

Antifungal Agent: Terpineol has excellent antifungal activity, including against the common plant pathogen penicillium italicum (post-harvest mold), and via the inhibition of the mycelial growth of Penicillium digitatum[547] (another common pathogenic fungus), which is required for the fungus to propagate.

Insect Repellant: Research has shown that terpineol possesses insect repellant properties, especially when acting synergistically with other terpenes and plant phenolics[548]. In fact, one U.S. patent for insect repellant lotions and sprays features terpineol as the primary active ingredient[549].

Antimicrobial Agent: In one study that examined the effectiveness of the essential oil of cinnamon against common bacteria like E. coli, S. aureus, and salmonella, terpineol was found to possess the best antibacterial activity of all the other terpenes measured in the EO[550].

Insecticidal Agent: Alpha-terpineol has been shown to exhibit both fumigant activity (airborne toxicity) and contact toxicity against booklice[551], and is likely used by plants against a wide range of other insects. As a primary constituent of the essential oil of Eucalyptus, the vapors of alpha-terpineol and terpinen-4-ol were found to be toxic to the horn fly[552], a parasite of cattle.

---

[547] Jing, Guo-xing, Neng-guo Tao, Lei Jia, and Hai-en Zhou. Influence of a-terpineol on the growth and morphogenesis of Penicillium digitatum. Botanical Studies 56.1 (2015): 1-6.

[548] Guo, Shanshan, Wenjuan Zhang, Junyu Liang, Chunxue You, Zhufeng Geng, Chengfang Wang, and Shushan Du. Contact and Repellent Activities of the Essential Oil from Juniperus formosana against Two Stored Product Insects. Molecules 21.4 (2016).

[549] Donald T. Beldock, John A. Beldock. Insect repellent lotions and sprays. US Patent US5227406A. Primavera Laboratories Inc. From: https://patents.google.com/patent/US5227406. Accessed January 28, 2019.

[550] Li L, Li ZW, Yin ZQ, Wei Q, Jia RY, Zhou LJ, Xu J, Song X, Zhou Y, Du YH, Peng LC, Kang S, Yu W. Antibacterial activity of leaf essential oil and its constituents from Cinnamomum longepaniculatum. Int J Clin Exp Med. 2014 Jul 15;7(7):1721-7.

[551] Liu, Xin Chao, Yin Ping Li, He Qin Li, Zhi Wei Deng, Ligang Zhou, Zhi Long Liu, and Shu Shan Du. Identification of Repellent and Insecticidal Constituents of the Essential Oil of Artemisia rupestris L. Aerial Parts against Liposcelis bostrychophila Badonnel. Molecules 18.9 (2013): 10733-10746.

[552] Juan LW, Lucia A, Zerba EN, Harrand L, Marco M, Masuh HM. Chemical composition and fumigant toxicity o f the essential oils from 16 species of Eucalyptus against Haematobia irritans (Diptera: Muscidae) adults. J Econ Entomol. 2011 Jun 104(3):1087-92.

Larvicidal Agent: When alpha-terpineol was used synergistically with linalool, 1,8-cineole, and thymol, significant contact toxicity, inhibition of feeding, and inhibition of the growth of chil partellus (a type of moth) larvae occurred[553].

USES IN INDUSTRY

Terpineol is one of the most important and commonly used perfume chemicals and is specially produced in highly pure forms for use in fine perfumery. This terpene is the primary constituent of lilac, lily of the valley, apple blossom, and fougere types of perfumes, and is used in various formulations in many fragrance-added products.

Terpineol is also common in the prepared foods industry as a flavor additive, in addition to the fact that many of the foods we normally eat already naturally contain this compound. Terpineol is found in chicken products like fried chicken, chicken flavoring, chicken broth, and in hot dogs, beer, wine, mezcal & tequila, baked goods, chewing gum, condiments like relish and mustard, frozen dairy items, fruit ices, gelatins and puddings, hard candy, and other sweets.

Terpineol is also added as a fragrance or solvent in many industrial, commercial, agricultural, and consumer products, including lotions, deodorants, fabric softeners, hair sprays and gels, surface cleaners, shampoos & conditioners, soaps, detergents, air care products, furniture polish, paper products, adhesives, sealants, solvents, paints, and more. This compound is also used as a precursor or intermediate in the production of plastic and rubber products, and in some electronic products.

Interestingly, terpineol has potential as a gasoline octane booster, thanks to its high energy density. Researchers have successfully used terpineol to increase the octane of a gasoline fuel blend, improving brake thermal efficiency, reducing total fuel consumption, and reducing emissions by up to 36.8%[554].

Most commercial samples of terpineol are produced in a laboratory, and are derived from alpha-pinene. Terpineol is available as a liquid, a crystalline powder, or a colorless solid, depending on the method of manufacture and the intended use of the finished product.

Concerning the specific isomers of terpineol, the following uses are reported:

Alpha: Used in many deodorants, soaps, detergents, other household products, as well as baked goods, beverages, snacks, gum, and candy.

L-alpha: Used as a perfuming agent.

Beta: Used as a fragrance and in some cosmetics.

Terpineol acetate: Used in lotions, deodorants, detergents, fabric softeners, shampoos, and soaps.

POTENTIAL USES IN MEDICINE

Terpineol shows great promise as a medical or therapeutic agent, especially for its proven antibacterial properties, and for its potential as a tool in pain management, and much more.

---

553 Singh, Rajwinder, Opender Koul, Pushpinder Rup, and Jawala Jindal. Toxicity of some essential oil constituents and their binary mixtures against Chilo pattellas (Lepidoptera: Pyralidae). International Journal of Tropical Insect Science 29.2 (2009): 93-101.

554 Vallinayagam, R., S. Vedharaj, William L. Roberts, Robert W. Dibble, and S. Mani Sarathy. Performance and emissions of gasoline blended with terpineol as an octane booster. Renewable Energy 101 (2017): 1087-1093.

Pain Management: Alpha-terpineol has been shown to significantly reduce both pain and inflammation in a murine model[555], while an earlier study found the monocyclic monoterpenoid exhibited strong analgesic properties[556]. A 2018 study echoed these findings and further determined that alpha-terpineol increased tissue antioxidant capacity, and reduced inducible nitric oxide synthase (iNOS) immunocontent in tumors, making it an excellent candidate as a treatment for pain in cancer patients[557].

Treatment for Fibromyalgia: A blend of alpha-terpineol and beta-cyclodextrin were studied in an animal model of fibromyalgia, with researchers finding that the combination produced a significant decrease in the mechanical hyperalgesia (induced pain or inflammation) via interactions with opioid and serotonin receptors[558].

Antibacterial Agent: Terpineol exhibits strong broad-spectrum antibacterial properties, including against periodontopathic and cariogenic bacteria[559] that can cause cavities and gum disease, and against the drug-resistant helicobacter pylori[560]. As the primary constituent in the essential oil from the flowers of anthemis altissima, alpha-terpineol showed broad-spectrum antibacterial activity against gram-positive staphylococcus aureus, bacillus subtilis, and staphylococcus epidermidis bacteria, and against gram-negative Escherichia coli, Pseudomonas aeruginosa, and Klebsiella pneumoniae bacteria[561]. Alpha-terpineol has also been shown to synergize the antibacterial effects of gentamicin by attenuating quorum sensing-regulated virulence[562], and exhibits synergistic antibacterial effects when combined with Bascillus coagulans against enterotoxigenic escherichia coli[563], while other research has demonstrated that alpha-terpineol and terpinene-4-ol together are synergistic against ESKAPE pathogens[564] (the top 6 most common bacterial pathogens in humans). Finally, terpinen-4-ol has been shown to inhibit the growth, biofilm, spore germination, swarming motility, and protease activity of Bacillus cereus, while also reducing extracellular matrix synthesis[565].

---

[555] Oliveira, Makson G. B., Rosemarie B. Marques, Michele F. Santana, Amanda B. D. Santos, Fabiola A. Brito, Emiliano O. Barreto, Damiao P. Sousa, Fernanda R. C. Almeida, Daniel Badaue-Passos, Angelo R. Antoniolli, and Lucindo J. Quintans-Junior. a-Terpineol Reduces Mechanical Hypernociception and Inflammatory Response.

[556] Quintans-Juunior, Lucindo J., Makson G.B. Oliveira, Michele F. Santana, Marilia T. Santana, Adriana G. Guimaraaes, Jullyana S. Siqueira, Damiaao P. De Sousa, and Reinaldo N. Almeida. a-Terpineol reduces nociceptive behavior in mice. Pharmaceutical Biology 49.6 (2011): 583-586.

[557] Gouveia DN, Costa JS, Oliveira MA, Rabelo TK, Silva AMOE, Carvalho AA, Miguel-Dos-Santos R, Lauton- Santos S, Scotti L, Scotti MT, Santos MRVD, Quintans-Junior LJ, Albuquerque Junior RLC, Guimaraes AG. a-Terpineol reduces cancer pain via modulation of oxidative stress and inhibition of iNOS. Biomed Pharmacother. 2018 Sep;105:652-661.

[558] Oliveira, Makson G.B., Renan G. Brito, Priscila L. Santos, Heitor G. Arat'.jo-Filho, Jullyana S.S. Quintans, Paula P. Menezes, Mairim R. Serafini, Yasmim M.B.G. Carvalho, Juliane C. Silva, Jackson R.G.S. Almeida, Luciana Scotti, Marcus T. Scotti, Saravanan Shanmugam, Parimelazhagan Thangaraj, Adriano A.S. Arat'.jo, and Lu cindo J. Quintans-Jt'.nior. a-Terpineol, a monoterpene alcohol, complexed with beta-cyclodextrin exerts antihyperalges ic effect in animal model for fibromyalgia aided with docking study. Chemico-Biological Tnteractions 254 (2016): 54-62.

[559] Park SN, Lim YK, Freire MO, Cho E, Jin D, Kook JK. Antimicrobial effect of linalool and a-terpineol against periodontopathic and cariogenic bacteria. Anaerobe. 2012 Jun;18(3):369-72.

[560] Miyamoto, Tomofumi, Tadayoshi Okimoto, and Michihiko Kuwano. Chemical Composition of the Essential Oil of Mastic Gum and their Antibacterial Activity Against Drug-Resistant Helicobacter pylori. Natural Products and Bioprospecting 4 (2014): 227-231.

[561] Samadi N, Manayi A, Vazirian M, Samadi M, Zeinalzadeh Z, Saghari Z, Abadian N, Mozaffarian VO, Khanavi M. Chemical composition and antimicrobial activity of the essential oil of Anthemis altissima L. var. altissima. Nat Prod Res. 2012;26(20):1931-4.

[562] Chadha J, Ravi, Singh J, Harjai K. α-Terpineol synergizes with gentamicin to rescue Caenorhabditis elegans from Pseudomonas aeruginosa infection by attenuating quorum sensing-regulated virulence. Life Sci. 2023 Jan 15;313:121267.

[563] Wu T, Zhang Q, Xu H, Li P, Zhao D, Wang L, Yi D, Hou Y. Protective effects of α-terpineol and Bacillus coagulans on intestinal function in weaned piglets infected with a recombinant Escherichia coli expressing heat-stable enterotoxin STa. Front Vet Sci. 2023 Feb 10;10:1118957.

[564] Johansen B, Duval RE, Sergere JC. First Evidence of a Combination of Terpinen-4-ol and α-Terpineol as a Promising Tool against ESKAPE Pathogens. Molecules. 2022 Nov 2;27(21):7472.

[565] Zhao L, Duan F, Gong M, Tian X, Guo Y, Jia L, Deng S. (+)-Terpinen-4-ol Inhibits Bacillus cereus Biofilm Formation by Upregulating the Interspecies Quorum Sensing Signals Diketopiperazines and Diffusing Signaling Factors. J Agric Food Chem. 2021 Mar 24;69(11):3496-3510.

**Antifungal Agent:* As the primary constituents of tea tree oil, terpinen-4-ol and alpha-terpineol displayed potent activity against 69 biofilm-forming strains of Candida albicans, reducing biofilm viability by 94% within 60 minutes[566]. These biofilms eventually lead to diseases of teeth and gums. As a primary constituent in the essential oil of artemisia princeps, the antifungal properties of alpha-terpineol were found to be comparable to clotrimazole[567], a common medication for the treatment of bacterial vaginosis and vulvovaginal candidiasis.

**Anti-tumor:* As a primary constituent in the essential oil of salvia libanotica, alpha-terpineol has been suggested as a potent anti-cancer medication after it was found to inhibit the growth of lung cancer cells by inhibiting NF-kappaB pathway signaling[568].

**Treatment for Liver Cancer:* Gamma-terpineol has been shown to dose-dependently suppress proliferation in BEL-7402 liver cancer cells, exhibiting characteristic changes associated with apoptosis including cell shrinkage, deformation and vacuolization of mitochondria, nuclear chromatin condensation and fragmentation, and formation of apoptotic bodies, thereby suppressing tumor growth[569].

**Treatment for Hypertension:* Oral administration of alpha-terpineol reduced mean arterial pressure, induced vascular endothelium independent vasodilatation, and restored enzymatic antioxidants in hypertensive rats[570], indicating that this monoterpenoid has value as a potential therapeutic agent in the treatment of hypertension - a significant problem for many adults in the United States and Europe.

**Treatment/Prevention of Heart Attack:* Alpha-terpineol has been shown to significantly reduce mortality rates, cardiac marker enzymes, tachycardia, hypertrophy, myocardium necrosis, edema, hemorrhagic areas, infiltration of inflammatory cells, and myocardial infarct size in a Wistar rat model of myocardial infarction[571].

IMPLICATIONS FOR HUMAN HEALTH & NUTRITION

Terpineol is easy to obtain via dietary consumption because it is found in many natural foods. To acquire this monocyclic monoterpene in your diet, consider regularly eating fruits like cherries, coconuts, grapes, grapefruit, guava, lemons, limes, melons, oranges, papaya, peaches, pineapple, raspberries, plums, star fruit, strawberries, tamarinds, tangerines, or yuzu. Vegetables that are high in this terpene include bell peppers, carrots, celery, and tomatoes. When preparing, cooking, or serving foods, consider working with spices like allspice, anise, basil, bay, cayenne, cinnamon, clove, coriander, dill, fennel, ginger, nutmeg, parsley, rosemary, sage, savory, or thyme.

Extracts of valerian or turmeric typically contain high amounts of terpineol; these can be consumed with food, drink, or taken directly. Adding black pepper to finished foods and adding lemon or lime to beverages is another option for terpineol consumption.

---

566 Ramage G, Milligan S, Lappin DF, Sherry L, Sweeney P, Williams C, Bagg J, Culshaw S. Antifungal, cytotoxic, and immunomodulatory properties of tea tree oil and its derivative components: potential role in management of oral candidosis in cancer patients. Front Microbiol. 2012 Jun 18; 3:220.

567 Trinh HT, Lee TA, Hyun YJ, Kim DH. Artemisia princeps Pamp. Essential oil and its constituents eucalyptol and a-terpineol ameliorate bacterial vaginosis and vulvovaginal candidiasis in mice by inhibiting bacterial growth and NF-KB activation. Planta Med. 2011 Dec;77(18):1996-2002.

568 Hassan SB, Gali-Muhtasib H, Goransson H, Larsson R. Alpha terpineol: a potential anticancer agent which acts through suppressing NF-kappaB signaling. Anticancer Res. 2010 Jun;30(6):1911-9.

569 Wu ZL, Yin ZQ, Du YH, Feng RZ, Ye KC, Wei Q, Hu Y, He L, Liao L, Wang Y. γ-terpineol inhibits cell growth and induces apoptosis in human liver cancer BEL-7402 cells in vitro. Int J Clin Exp Pathol. 2014 Sep 15;7(10):6524-33.

570 Sabino, Carla & Ferreira-Filho, Edson & Mendes, Marcelo & Silva-Filho, Jose & Ponte, Marco & Moura, Lucas & Oliveira, Elisangela & Quintans-Junior, Lucindo & Santos, M.rcio & Oliveira, Rita De C.ssia & Oliveira, Aldeidia. (2013). Cardiovascular effects induced by a-terpineol in hypertensive rats. Flavour and Fragrance Journal. 28. 10.1002/ffj.3159.

571 Paulino ET, Rodrigues AKBF, Machado MLDP, de Oliveira KRV, Bernardino AC, Quintans-Júnior LJ, Oliveira AP, Ribeiro ÊAN. Alpha-terpineol prevents myocardial damage against isoproterenol-MI induced in Wistar-Kyoto rats: new possible to promote cardiovascular integrity. Life Sci. 2022 Feb 1;290:120087.

Cannabis strains that typically contain terpineol include Banana Kush, Cannatonic, Ghost Train Haze, Girl Scout Cookies, God's Green Crack, Jack Herer, Kosher Kush, OG Kush, Shiatsu Kush, Skywalker OG, Sour Cherry Sherbet, Sour Jack, Strawberry Short Cookies, Sunset Sherbet, Super Silver Haze, and White Widow, among others. While we do not recommend smoking or vaporizing cannabis due to potential health concerns, juicing, blending, or eating these strains raw is a healthier option. Oils that have retained or had terpineol added can be used directly, or in foods.

Experimentation with purified terpineol or with essential oils made from the plants listed in the "Occurrences in Plants" section above should only be undertaken with extreme caution, as these compounds can be toxic in some situations.

Interestingly, research has shown that volatiles released by sake and by associated wooden drinking cups or bowls – volatiles primarily consisting of several ethyl compounds, as well as delta-cadinene and alpha terpineol – contribute to feelings of relaxation and improve the taste of the sake[572].

[572]Matsubara, E., Kusumoto, N., Hashida, K. et al. Combined effects of a wooden sake vessel and alcohol on subjective and physiological relaxation. Wood Sci Technol 56, 989–1005 (2022).

Terpineol Review

Answer the following questions to test your knowledge of this terpene:

Question #1: What type of terpene is terpineol?

a. Linear monoterpene
b. Monoterpenic sesquiterpene
c. Monoterpene alcohol
d. Monocyclic monoterpenoid

Question #2: Which of the following best describes the primary flavor & aroma profile of terpineol?

a. Spicy, turpentine, fuel, pine
b. Lilac, floral, woody, lime
c. Depends on the isomer
d. Hops, cloves, wood, spicy
e. Rose, floral, woody, tea

Question #3: How common is terpineol in cannabis?

a. Top three
b. Top five
c. Top ten
d. Top twenty

Question #4: What is the chemical formula for terpineol?

a. $C_{15}H_{26}$
b. $C_{10}N_{16}O$
c. $C_{10}H_{18}O$
d. $C_{15}H_{26}O$

Question #5: Name two uses of terpineol in plants:

1 ______________________________ 2 ______________________________

Question #6: Name two potential medical uses of terpineol:

1 ______________________________ 2 ______________________________

For the answer key to all individual compound chapters please go here: http://thebigbookofterps.com/bbot-quiz-answer-key/

GUAIOL

Type: Bicyclic sesquiterpenoid
Chemical Formula: $C_{15}H_{26}0$
Molecular Weight: 222.372 g/mol
Boiling Point: *Guaiol:* 309.00 to 310.00 °C. @ 760.00 mm Hg[573]
Guaiol acetate: 268.00 to 270.00 °C. @ 760.00 mm Hg[574]
Flash Point: *Guaiol:* 236.00 °F. TCC (113.33 °C.)
Guaiol acetate: > 212.00 °F. TCC (> 100.00 °C.)
Melting Point: *Guaiol:* 91.00 to 93.00 °C. @ 760.00 mm Hg
Solubility: *Guaiol:* alcohol, chloroform, dichloromethane, ethyl acetate, acetone
Guaiol acetate: Soluble in alcohol
Oral LD50: 520 mg/kg (rat)[575]
Biological Role: Insecticidal, antibacterial
Therapeutic Role: Antifungal, anticancer agent, antibacterial, anti-leishmaniasis
Commercial Use: Fragrance, food additive, surfactant, emulsifier
Occurrence in Cannabis: Top fifteen
Aroma/Flavor: *Guaiol:* Woody, tea, rose
Guaiol acetate: Balsamic, tea, rose, woody, spicy

Occurs in Cannabis Strains: Guaiol has been shown to occur in several cannabis varieties, and it's likely that this compound will be found more widely as testing standards improve at laboratories. Guaiol has been found in cannabis samples including; Barbara Bud, Bast 3, Blue Kush, Blueberry Peaches, Bubba Kush, Cannatonic, Chocolope, Deadhead OG, Elvis, Goldmeyer, Goo, Jean Guy, Liberty Haze, Mickey Kush, Pineapple Express, Purple Punch, Shiatsu Kush, Skunkdawg #9, Sour Diesel, Super Silver Haze, Whitaker Blues, White Widow.

INTRODUCTION

Guaiol is a bicyclic sesquiterpene that is generally found within the top fifteen terpenes in cannabis varieties by concentration. This isoprenoid compound is known for its aroma and flavor profile, which is uniquely rose, floral, woody, and tea-like. Guaiol occurs in moderate concentrations in some common strains of cannabis like White Widow, Liberty Haze, Chocolope, Sour Diesel, and some Kush varieties such as Blue Kush and Bubba Kush.

In addition to cannabis, guaiol is also produced by many other plants, most notably a species of tree, bulnesia sarmienti, which is the primary industrial and commercial source of this terpene. Guaiol is also produced by the soft coral sinularia sp., commonly referred to as the finger leather coral or soft finger coral[576].

Interestingly, guaiol has been shown to exhibit sedative properties in the fish species rhamdia quelen[577], commonly called the silver catfish. The compound has also puzzled scientists for more than half a century, after it was observed to yield a deep purple color after treatment with bromine reagents[578]. This is particularly fascinating

[573] The Good Scents Company Data Sheet for L-guaiol, from: http://www.thegoodscentscompany.com/data/rw1015101.html/ Accessed August 31, 2023.

[574] The Good Scents Company Data Sheet for Guaiol Acetate, from: http://www.thegoodscentscompany.com/data/rw1005771.html. Accessed August 31, 2023.

[575] Auro Chemicals Safety Data Sheet for Guaiol, from: https://www.aurochemicals.com/wp-content/uploads/2021/07/GUAIOL-NATURAL-SDS.pdf. Accessed August 31, 2023.

[576] Guang-Wen Zhang, Xiang-Quan Ma, Jing-Yu Su, Kun Zhang, Hiroshi Kurihara, Xin-Sheng Yao & Long-Mei Zeng (2006) Two new bioactive sesquiterpenes from the soft coral Sinularia sp., Natural Product Research, 20:7, 659-664.

[577] Benovit SC, Silva LL, Salbego J, Loro VL, Mallmann CA, Baldisserotto B, Flores EM, Heinzmann BM. Anesthetic activity and bio-guided fractionation of the essential oil of Aloysia gratissima (Gillies & Hook.) Tronc. in silver catfish Rhamdia quelen. An Acad Bras Cienc. 2015 Sep;87(3):1675-89.

[578] Waddell TG, Arp NW, Bodine KD, Pagni RM. The guaiol color reaction. Planta Med. 2002 Oct; 68(10):949-50.

considering that - under normal conditions - alkenes (guaiol is an alkene) treated with bromine lose their double bond and instead take on a bromine atom, which then loses its original reddish-brown color and becomes colorless[579].

CHEMICAL STRUCTURE

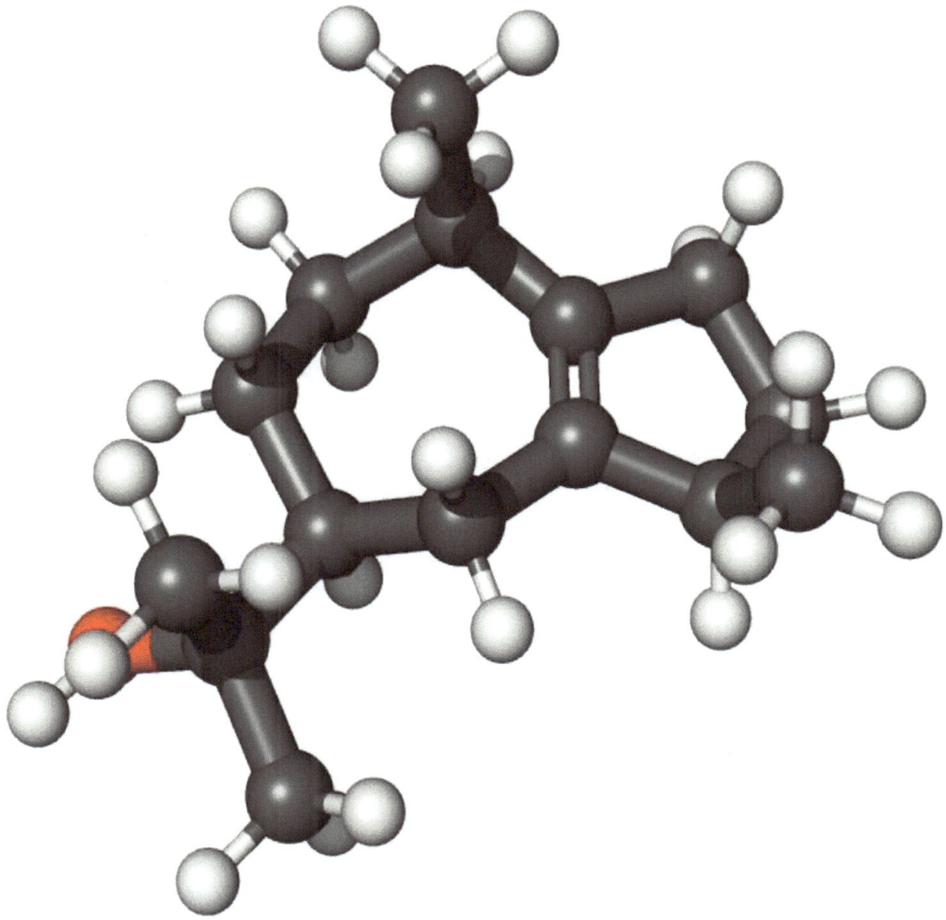

Guaiol is produced in the mevalonate pathway of plants, and is comprised of three linked units of isoprene, which classifies this compound as a sesquiterpene. This molecule has one rotatable bond, one endocyclic double bond, and one covalent bond among the fifteen carbon atoms (five carbon atoms from each of the isoprene units), twenty-six hydrogen atoms, and one oxygen atom, notated as $C_{15}H_{26}O$. The molecular skeleton of guaiol contains two rings, which further classifies this terpene as a bicyclic sesquiterpenoid.

Unlike other sesquiterpenes, guaiol does not appear as an oil. Instead, this compound is technically a sesquiterpenoid alkene alcohol, although in industrial and commercial applications, guaiol is produced as a crystalline solid, or as a yellow, amber, or orange semi-solid.

[579] Jim Clark. Reactions of Alkenes with Bromine. Chemistry Libretexts May 2017.

OCCURRENCE IN PLANTS

To date, guaiol has not been found to occur in many plants. In addition to cannabis, this compound has been found in the following plants, spices, and herbs, and it's likely that further research will find this terpene in many more species:

Guaiol:

Ayou	Black pepper	Cajuput
Callitris species	Cascarilla	Cassinia
Cinnamon	Cubeb	Ferula
Guaiacwood	Helichrysum italicum	Ho leaf
Parsley	Ravintsara	Sage
Tagete	Zanthoxylum bracchyacanthum	

Guaiol acetate:

Eucalyptus camaldulensis	Myrtle

BIOLOGICAL ACTIVITY IN PLANTS

Some of the most important biological uses of guaiol by plants are extremely relevant to cannabis. This includes the ability to potently fight mites and aphids; two of the biggest and most challenging pests for growers of cannabis to control.

Acaricidal Agent: Guaiol has been shown to exhibit potent fumigant toxicity against tetranychus urticae, commonly referred to as the two-spotted spider mite[580]. This mite is a significant pest of cannabis and other plants, especially in indoor or even greenhouse grow operations, where infestations of this arachnid can devastate entire crops. Development of a guaiol-based fumigant for mites could be an effective yet non-toxic method of controlling and eradicating spider mites.

Insecticidal Agent: Guaiol exhibited significant contact toxicity against aphids - one of the most common pests of cannabis - and good contact toxicity against mythimna separata (commonly called the northern armyworm), and the diamondback moth[581]. Taken together with its acaricidal properties, guaiol shows excellent promise as an all-around pesticide for cannabis and other plants.

Antibacterial Agent: Tea tree oil, a popular and well-studied antimicrobial agent, showed less antimicrobial activity against gram-positive bacteria than the essential oil of Guaiacwood, where the primary constituents were bulnesol (36-44%), and guaiol (27- 34%)[582].

USES IN INDUSTRY

[580] Neves, Ilzenayde Araujo, and Claudio Augusto Gomes da Camara. Acaricidal Activity against Tetranychus Urticae and Essential Oil Composition of Four Croton Species from Caatinga Biome in Northeastern Brazil. Natural Product Communications 6.6 (2011).

[581] Liu, Tao, Chun-Juan Wang, Hui-Qin Xie, and Qing Mu. Guaiol - A Naturally Occurring Insecticidal Sesquiterpene. Natural Product Communications 8.10 (2013).

[582] Pegard, A. Antimicrobial activity of essential oil of Bulnesia sarmienti Lorenz (Gaiac Wood) and of a rectified fraction of this oil. Comparison with Melaleuca alternifolia L. (Tea Tree) essential oil activity. Phytoth_rapie 13.6 (2015): 384-388.

Most commercial or industrial samples of guaiol are produced by steam distillation of wood from bulnesia sarmienti, a tree that is indigenous to Paraguay and Argentina. This produces champaca wood oil, where the primary constituent is guaiol. Commercial samples of this sesquiterpenoid appear as a crystalline solid, or a thick, semi-solid yellowish-orange or amber mass.

While guaiol is FDA-approved for use in food, and is generally known as a flavoring agent, few food products on the market as of late summer 2023 featured this ingredient either directly or as a processing agent. Guaiol is also used in the fragrance industries, where it is recommended at not more than 5% of fragrance concentrate.

Guaiol is used in industry as a surfactant to lower the surface tension of water for use in chemical processing, and as an emulsifier to promote the stability of liquid suspensions. This bicyclic sesquiterpenoid is also used as a cleaning agent, foam booster, and solubility agent.

Guaiol's insecticidal properties have been harnessed by at least one company, which holds a patent for a botanical insecticide based on the compound, which the company testifies has been shown to possess antifeedant, contact toxicity, and fumigation properties[583], among others.

POTENTIAL USES IN MEDICINE

Guaiol shows promise in medicine, offering a wide variety of therapeutic possibilities, including being particularly effective at destroying lung cancer cells, among other potential uses as detailed below:

Antifungal Agent: Guaiol exhibits strong antifungal properties. In one study, guaiol exhibited similar antifungal activities against four dermatophytic fungi and six yeasts as itraconazole and fluconazole[584], two of the most common antifungal medications currently on the market. Another study observed that the essential oil of aloysia lycioides cham. (a genus of flowering plants in the verbena family), where the main constituents were beta-myrcene, pinocophone, and guaiol, not only inhibited the growth of all twelve species of candida tested - and with no cytotoxicity - it also showed synergistic effects with fluconazole for eleven of the candida species[585]. The antifungal activity of guaiol is of critical importance considering the increasing resistance of fungi to current antifungal medications.

Dental Treatment: Streptococcus mutans is an anerobic, gram-positive bacteria that is a significant contributor to tooth decay. As part of the essential oil of Aloysia gratissima (whitebrush), guaiol was shown to exhibit strong antibacterial activity against Streptococcus mutans, greatly inhibiting the production of bacterial biofilm that leads to the buildup of plaque[586]. Guaiol has also been observed to exhibit inhibitory activities against Streptococcus sobrinus; bacteria that works in conjunction with S. mutans in the formation of dental caries[587] (cavities).

---

[583] Chinese Insecticide Patent CN102795969A: Natural sesquiterpene guaiol compound and medicinal application thereof. From: https://patents.google.com/patent/CN102795969A/en. Accessed February 26, 2019.

[584] Waikedre, Jean, Carmen I. Vitturo, Ana Molina, Phellipe Norato Estrela Terra Theodoro, Maria do Rosario Rodrigues Silva, Laila Salmen Espindola, Alexandre Maciuk, and Alain Fournet. Antifungal Activity of the Essential Oils of Callitris neocaledonica and C. sulcata Heartwood (Cupressaceae). Chemistry & Biodiversity 9.3 (2012): 644-653.

[585] Mobin, Mitra & Porto, Jhonatas & Almeida, Lorena & Andrade, Laise & Szeszs, Maria & Melhem, Marcia & Brito, Lucas & Carvalho, Fernando & Martins, Galberto & de Lima, Sidney. (2017). Chemical composition and evaluation of the antifungal, cytotoxic and modulatory activity from essential oil of Aloysia Lyciodes Cham. International Journal of Development Research.

[586] Galvao, Livia Camara de Carvalho, Vivian Fernandes Furletti, Salete Meyre Fernandes Bersan, Marcos Guilherme da Cunha, Ana Lucia Tasca Ruiz, Joao Ernesto de Carvalho, Adilson Sartoratto, Vera Lucia Garcia Rehder, Glyn Mara Figueira, Marta Cristina Teixeira Duarte, Masarahu Ikegaki, Severino Matias de Alencar, and Pedro Luiz Rosalen. Antimicrobial Activity of Essential Oils against Streptococcus mutans and their Antiproliferative Effects. Evidence-Based Complementary and Alternative Medicine 2012.2012 (2012).

[587] Wang X, Mitsunaga T, Yamauchi K (2016) Antibacterial Activity of Selected Essential oils against Streptococcus sobrinus and Porphyromonas gingivalis. J Pharm Sci Emerg Drugs 4:2.

**Anti-Leishmaniasis:* The essential oil of Bulnesia sarmientoi - the primary source of industrial guaiol - showed guaiol concentration at 48.29%, and exhibited moderate toxicity against Leishmania amazonensis[588], a parasite spread by sandflies that causes the chronic disease called leishmaniasis. As a major constituent of the essential oil of aloysia gratissima (commonly known as whitebrush or bee-brush), guaiol actively killed promastigotes and intracellular amastigotes[589] (different life cycles of the protozoan responsible for the disease), suggesting that guaiol acts with direct contact toxicity against the parasites that cause leishmaniasis.

**Lung Cancer Treatment:* Researchers in China have shown that guaiol induces autophagy (a process in cells that recycles redundant or dysfunctional cell components) in non-small cell lung cancer by blocking cell signaling[590], thereby inhibiting tumor growth. The same researchers also found that guaiol is involved in cell autophagy that regulates the expression of RAD51 (a eukaryotic gene that assists in repair of double-strand breaks in DNA), leading to cell apoptosis (programmed death), and further chemo-sensitizing non-small cell lung cancer cells to guaiol[591]. Guaiol has also been shown to inhibit cell growth and induce apoptosis via the PI3K/Akt signaling pathway in lung adenocarcinoma[592], while other research found that guaiol induces immunogenic cell death in non-small cell lung cancer[593].

**Mosquito Control:* Because mosquitos are vectors for serious diseases like Zika, Dengue Fever, and malaria, controlling these insects is a critical function of health organizations around the world, and tools are limited. Guaiol may prove to be an excellent method of localized protection, since it has been shown to have mosquito biting deterrent activity similar to that of DEET[594], one of the most common forms of personal mosquito protection.

**Treatment for Klebsiella Pneumoniae:* This condition is caused by a gram-negative bacterium that normally lives on human and animal skin and in the intestines, but that causes serious internal damage if it infects the lungs through aspiration. The sesquiterpenes polygodial, guaiol, and caryophyllene oxide showed inhibitory activity against K. pneumoniae, without toxicity[595], offering scientists a strong lead in the search for a preventative and treatment for this disease.

**Anti-inflammatory:* As the primary constituent of the essential oil of C. sellowianus (a flowering plant in the myrtle family), guaiol significantly reduced inflammation in cells, both in vitro and in vivo, by 91%[596].

---

[588] Andrade, Milene Aparecida, Clenia dos Santos Azevedo, Flavia Nader Motta, Maria Lucilia dos Santos, Camila Lasse Silva, Jaime Martins de Santana, and Izabela M. D. Bastos. Essential oils: in vitro activity against Leishmania amazonensis, cytotoxicity and chemical composition. BMC Complementary and Alternative Medicine 16 (2016).

[589] Garcia MCF, Soares DC, Santana RC, Saraiva EM, Siani AC, Ramos MFS, Danelli MDGM, Souto-Padron TC, Pinto-da-Silva LH. The in vitro antileishmanial activity of essential oil from Aloysia gratissima and guaiol, its major sesquiterpene against Leishmania amazonensis. Parasitology. 2018 Aug;145(9):1219-1227.

[590] Yang X, Zhu J, Wu J, Huang N, Cui Z, Luo Y, Sun F, Pan Q, Li Y, Yang Q. (-)-Guaiol regulates autophagic cell death depending on mTOR signaling in NSCLC. Cancer Biol Ther. 2018 Aug 3;19(8):706-714.

[591] Yang, Qingyuan, Jianchun Wu, Yingbin Luo, Nan Huang, Ni Zhen, Yun Zhou, Fenyong Sun, Zhi Li, Qiuhui Pan, and Yan Li. (-)-Guaiol regulates RAD51 stability via autophagy to induce cell apoptosis in non-small cell lung cancer. Oncotarget 7.38 (2016): 62585-62597.

[592] Zeng Y, Pan Y, Zhang B, Luo Y, Tian J, Wang Y, Ju X, Wu J, Li Y. Integrating Network Pharmacology, Molecular Docking, and Experimental Validation to Investigate the Mechanism of (-)-Guaiol Against Lung Adenocarcinoma. Med Sci Monit. 2022 Jul 25;28:e937131.

[593] Yang, X., Yang, J., Gu, X. et al. (−)-Guaiol triggers immunogenic cell death and inhibits tumor growth in non-small cell lung cancer. Mol Cell Biochem 478, 1611–1620 (2023).

[594] Ali A, Tabanca N, Demirci B, Blythe EK, Ali Z, Baser KH, Khan IA. Chemical composition and biological activity of four salvia essential oils and individual compounds against two species of mosquitoes. J Agric Food Chem. 2015 Jan 21;63(2):447-56.

[595] Winnett, V., J. Sirdaarta, A. White, F. Clarke, and I. Cock. Inhibition of Klebsiella pneumoniae growth by selected Australian plants: natural approaches for the prevention and management of ankylosing spondylitis. Inflammopharmacology 25.2 (2017): 223-235.

[596] Apel MA, Lima ME, Sobral M, Young MC, Cordeiro I, Schapoval EE, Henriques AT, Moreno PR. Anti- inflammatory activity of essential oil from leaves of Myrciaria tenella and Calycorectes sellowianus. Pharm Biol. 2010 Apr;48(4):433-8.

**Antioxidant:* Guaiol has been shown to exhibit significant antioxidant activity and bacteriostatic effects against staphylococcus aureus, Enterobacter cloacae, bacillus cereus, and klebsiella pneumoniae[597], an activity mentioned previously in this section.

IMPLICATIONS FOR HEALTH & HUMAN NUTRITION

To obtain dietary guaiol, consider cooking or preparing foods with spices like basil, black pepper, cinnamon, ginger, parsley, sage, and cubeb. Eating raw cannabis leaves or juicing cannabis strains that are high in guaiol likely allows absorption of at least a limited amount of this sesquiterpene. Combustion or vaporization of strains with moderate amounts of guaiol like Barbara Bud, Bast 3, Blue Kush, Blueberry Peaches, Bubba Kush, Cannatonic, Chocolope, Deadhead OG, Elvis, Goldmeyer, Goo, Jean Guy, Liberty Haze, Mickey Kush, Pineapple Express, Purple Punch, Shiatsu Kush, Sour Diesel, Super Silver Haze, Whitaker Blues, and White Widow, among others, is likely to permit consumption of trace levels of this compound, however, tinctures and oils are a healthier option.

Other methods of obtaining guaiol include working with isolated guaiol or essential oils made from the plants listed in the "Occurrences in Plants" section above, however, extreme care should be exercised as these compounds can be toxic in some situations.

[597] Mendes, Renata de F., Nicolas de C. C. Pinto, Josiane M. Silva, Jucelia B. Silva, Raquel C. dos S. Hermisdorf, Rodrigo L. Fabri, Luciana M. Chedier, and Elita Scio. The essential oil from the fruits of the Brazilian spice Xylopia sericea A. St.-Hil. presents expressive in-vitro antibacterial and antioxidant activity. Journal of Pharmacy and Pharmacology: An International Journal of Pharmaceutical Science 69.3 (2017): 341-348.

Guaiol Review

Answer the following questions to test your knowledge of this terpene:

Question #1: What type of terpene is guaiol?

a. Acyclic sesquiterpene
b. Monocyclic sesquiterpene
c. Bicyclic sesquiterpenoid
d. Tricyclic sesquiterpene

Question #2: Which of the following best describes the primary flavor & aroma profile of guaiol?

a. Spicy, turpentine, fuel, pine
b. Depends on the isomer
c. Floral, woody, green, citrus
d. Hops, cloves, wood, spicy
e. Rose, floral, woody, tea

Question #3: How common is guaiol in cannabis?

a. Top three
b. Top five
c. Top ten
d. Top fifteen

Question #4: What is the chemical formula for guaiol?

a. $C_{15}H_{26}0$
b. $C_{15}N_{26}0$
c. $C_{15}H_{26}C$
d. $N_{15}H_{26}0$

Question #5: Name two uses of guaiol in plants:

1 ________________________________ 2 ________________________________

Question #6: Name two potential medical uses of guaiol:

1 ________________________________ 2 ________________________________

For the answer key to all individual compound chapters please go here: http://thebigbookofterps.com/bbot-quiz-answer-key/

CARYOPHYLLENE OXIDE

Type: Bicyclic sesquiterpenoid
Chemical Formula: $C_{15}H_{24}0$
Molecular Weight: 220.35 g/mol
Boiling Point: 279.68 °C. @ 760.00 mm Hg (estimated)[598]
Flash Point: >212.00 °F. TCC (>100.00 °C.)
Melting Point: 60.00 to 62.00 °C. @ 760.00 mm Hg
Solubility: Soluble in alcohol ethanol, fats, oils, DMS0, dimethyl formamide
Oral LD50: 5,000 mg/kg (rat)[599]
Biological Role: Antifungal, indirect defense, antifeedant
Therapeutic Role: Anticancer, analgesic, mosquito control
Commercial Use: Food additive
Occurrence in Cannabis: Top twenty
Aroma/Flavor: Woody, dry, cedar, carrot

Occurs in Cannabis Strains: Unfortunately, as of late 2023 many U.S. laboratories do not test for caryophyllene oxide, or they make no distinction between this compound and beta-caryophyllene. It is expected that there will be more occurrences of this compound recorded in cannabis as laboratories and regulatory bodies ensure that this compound is included in testing protocols and standards. Cannabis varieties that are known to contain this sesquiterpenoid include Bast 3, Blue Cheese, Blue Dream, Cheese, Chemdawg, Durban Poison, Ghost Train Haze, Girl Scout Cookies, Goldmeyer, Island Sweet Skunk, Maui Waui, Purple Punch, Purps, Speakeasy, Whitaker Blues.

INTRODUCTION

Identical in structure to beta-caryophyllene except for a single double-bonded oxygen atom, caryophyllene oxide exhibits many properties that diverge sharply from its non-oxygenated counterpart. Most importantly, caryophyllene oxide is not a cannabinoid because it has no affinity for CB1 or CB2 endocannabinoid receptors, unlike beta-caryophyllene.

Caryophyllene oxide is often used in the prepared foods industry, where it imparts woody, spicy, dry, and carrot-like characteristics in baked goods, desserts, condiments, and other prepared foods. These same properties are also apparent in cannabis strains that are high in caryophyllene oxide, which are few in number. After reviewing hundreds of laboratory tests, we determined that the Purple Punch variety had the highest concentration of caryophyllene oxide, featuring this sesquiterpene prominently as the second terpene in the strain sample by concentration.

---

598 The Good Scents Company Data Sheet for Beta-caryophyllene oxide, from: http://www.thegoodscentscompany.com/data/rw1023631.html. Accessed August 31, 2023.

599 Sigma Aldrich Safety Data Sheet for (−)-Caryophyllene oxide, from: https://www.sigmaaldrich.com/US/en/sds/ALDRICH/W509647. Accessed August 31, 2023.

CHEMICAL STRUCTURE

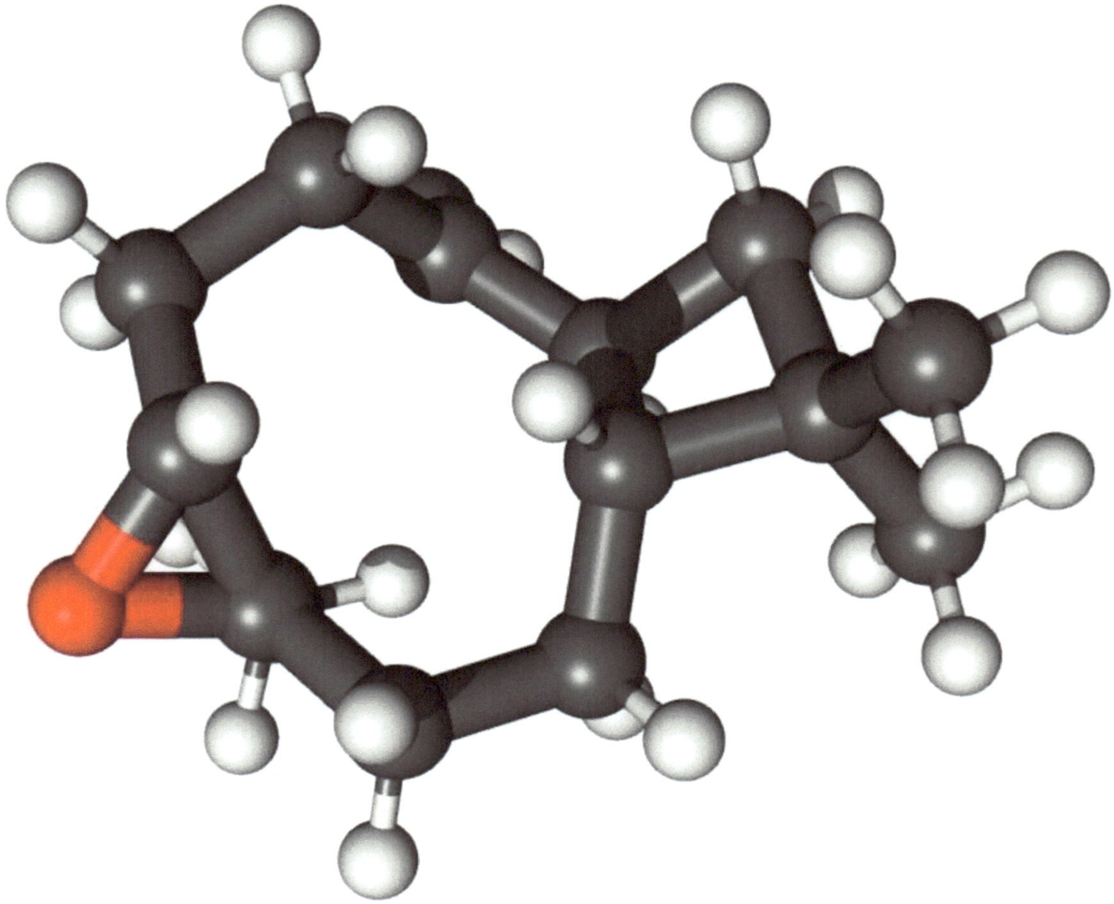

Those who work in the cannabis industry should note that much of the beta-caryophyllene-oxide content in finished cannabis products will likely occur because of oxidation during harvesting, drying, curing, packaging, and storage operations, as opposed to occurring in the plant naturally. As an oxidation product or metabolite of beta-caryophyllene, caryophyllene oxide is a highly reactive epoxide based on a typical carbon skeleton. This compound is comprised of three linked units of isoprene, making this a sesquiterpene, and the presence of two cyclobutene rings further classify this compound as bicyclic. Caryophyllene oxide has fifteen carbon atoms, twenty-four hydrogen atoms, and one oxygen atom, with the scientific notation of $C_{15}H_{24}O$.

Although otherwise indistinguishable from beta-caryophyllene, the addition of an oxygen atom to this bicyclic sesquiterpene changes its properties in such a way that the molecule no longer selectively binds to CB1 endocannabinoid receptors. This also means that, while caryophyllene oxide shares many of the same traits as beta-caryophyllene, it also features individual traits not found in its counterpart, some of which are discussed below.

Interestingly, caryophyllene oxide generally co-occurs in cannabis strains with beta-caryophyllene, however, in a few cases, only caryophyllene oxide appeared in the top ten terpenes by concentration in a strain.

Although often discussed chiefly as caryophyllene oxide, this compound is technically beta-caryophyllene oxide, and in some cases, reference is made to the molecule as such. However, the names are interchangeable, as caryophyllene oxide is identical to beta-caryophyllene with the exception of the oxygen atom. This fact makes this bicyclic sesquiterpene officially beta-caryophyllene oxide.

OCCURRENCE IN PLANTS

Caryophyllene oxide is known to occur in over one hundred plants, and it is likely that researchers will continue to find more plant species that produce this oxygenated terpenoid. The following plants produce caryophyllene oxide, but in varying concentrations:

Allspice	Alpinia galanga	Angelica
Artemisia species	Ayou	Baccharis coridifolia
Basil	Bay	Betel
Black pepper	Bois de rose	Cabocla
Cajuput	Cananga	Cannabis
Caraway	Carrot	Cassis
Cedarleaf	Celery	Chamomile
Champaca	Cinnamon	Clary sage
Clove	Coriander	Cornmint
Cotton	Croton cajucara	Cumin
Curcuma amada	Curry	Cypress
Erigeron graveolens	Eucalyptus species	Featherfew
Ferula	Feverfew	Fig
Fleabane	Galangal	Geranium
Ginger	Gingergrass	Guava
Hop	Horsemint	Hyptis
Jambu	Juniper	Kachur
Labdanum	Laurel	Lavandin
Lavender	Layana	Lemon
Lemongrass	Lime	Mangrove
Manuka	Marjoram	Marsypianthes chamaedrys
Mastic	Melissa	Mikania
Mugwort	Myrtle	Oregano
Origanum	Orthodon dianthera	Palmarosa
Patchouli	Pennyroyal	Peppermint
Phlomis fruticosa	Phoebe	Pimenta
Pine	Polyalthia australis	Prangos
Pteronia	Ravintsara	Rosemary
Safflower	Sage	Salvia species
Savin	Savory	Spearmint
Stachys recta	Sugandha kokila	Tagete
Tansy	Thyme	Thymus richardii
Vassoura	Verbena	Water mint
Wormwood	Yarrow	Ylang ylang
Zanthoxylum species		

BIOLOGICAL ACTIVITY IN PLANTS

Although many different biological uses are likely, the current body of literature on how plants use caryophyllene oxide is small, with a central focus on the compound's potent antifungal properties.

Antifungal Agent: Caryophyllene oxide has been shown to exhibit antimicrobial activity against fusarium culmorum[600], a fungal plant pathogen that often attacks plants like cannabis. As a primary constituent in the

[600] Ellouze, Ines, Manef Abderrabba, Nassereddine Sabaou, Florence Mathieu, Ahmed Lebrihi, and Jalloul Bouajila. Season's Variation Impact on Citrus aurantium Leaves Essential Oil: Chemical Composition and Biological Activities. Journal of Food Science 77.9 (2012): T173-T180.

essential oil of erigeron ramosus, caryophyllene oxide contributed to the strong inhibition of mycelial growth of fusarium oxysporum, fusarium solani, botrytis cinerea, and several other phytopathogenic fungi[601]. Fusarium species often plague cannabis sativa, and botrytis species are responsible for some types of budrot in cannabis. Finally, as a primary constituent in the essential oil of piper chaba hunter, caryophyllene oxide contributed to a potent detrimental effect on spore germination of several phytopathogenic fungi[602].

**Indirect Defense:* Plants may use caryophyllene oxide to attract predators of herbivores. As a primary compound in the essential oil of cotton, this bicyclic sesquiterpene helps attract the parasitoid campoletis sonorensis[603] - a wasp that preys on the fall armyworm, a common pest of cotton. Interestingly, in the study referenced, beta-caryophyllene was not attractive to the parasitoids.

**Antifeedant:* As isolated from the East African plant, uvaria sp., caryophyllene oxide was found to exhibit antifeedant properties[604], similar to the bitter terpenes found in broad leaves of plants like cannabis.

USES IN INDUSTRY

The primary industrial or commercial use of caryophyllene oxide is as a food additive or flavor ingredient. This includes use in baked goods, cereal, cheese, condiments, egg products, fats and oils, fish products, frozen dairy, fruit ices, processed fruits, soft candy, sweeteners, spices, alcoholic beverages, soft drinks, and other prepared foods.

Much to the chagrin of the cannabis industry - including the longstanding black markets for hashish - caryophyllene oxide is the substance that government and police forces have used to train detection dogs for hash-smuggling interdiction work[605].

Interestingly, both beta-caryophyllene and beta-caryophyllene oxide have been proposed as environmental decontaminants. In a 2016 study, scientists showed that these compounds significantly reduced the revertant colonies of several strains of bacteria induced by a methanol extract made from discarded cigarette butts[606] - a common and troubling environmental contaminant.

POTENTIAL USES IN MEDICINE

Caryophyllene oxide offers the potential for many new medicines and treatments, but these should be tempered with the warning that this compound may affect the pharmacokinetics of concurrently administered drugs after it was found that caryophyllene oxide was a strong inhibitor of cytochromes P4503A[607] (a human gene locus involved in drug metabolization) in both rat and human cells.

---

[601] Rahman, Atiqur, M. Hossain, and Sun Kang. Control of phytopathogenic fungi by the essential oil and methanolic extracts of Erigeron ramosus (Walt.) B.S.P. European Journal of Plant Pathology 128.2 (2010): 211-219.

[602] Rahman, Atiqur, Sharif Al-Reza, and Sun Kang. Antifungal Activity of Essential Oil and Extracts of Piper chaba Hunter Against Phytopathogenic Fungi. Journal of the American Oil Chemists' Society 88.4 (2010): 573-579.

[603] Elzen GW, Williams HJ, Vinson SB. Isolation and identification of cotton synomones mediating searching behavior by parasitoid Campoletis sonorensis. J Chem Ecol. 1984 Aug;10(8):1251-64.

[604] F. Bettarini, G. E. Borgonovi, T. Fiorani, I. Gagliardi. Antiparasitic compounds from East African plants: Isolation and biological activity of anonaine, matricarianol, canthin-6-one and caryophyllene oxide. International Journal of Tropical Insect Science Volume 14, Issue 1 February 1993, pp. 93-99.

[605] Stahl, E; Kunde, R (1973). Die Leitsubstanzen der Haschisch-Suchhunde. Kriminalistik: Z Gesamte Kriminal Wiss Prax. 27: 385-389.

[606] Di Giacomo, Silvia, Gabriela Mazzanti, and Antonella Di Sotto. Mutagenicity of cigarette butt waste in the bacterial reverse mutation assay: The protective effects of Beta-caryophyllene and Beta-caryophyllene oxide.

[607] Nguyen, Linh Thuy, Zuzana Mysliveckova, Barbora Szotakova, Alena Spicakova, Katerina Lnenickova, Martin Ambroz, Vladimir Kubicek, Kristyna Krasulova, Pavel Anzenbacher, and Lenka Skalova. The inhibitory ef fects of Beta-caryophyllene, Beta-caryophyllene oxide and a-humulene on the activities of the main drug-metabolizing enzymes in rat and human liver in vitro. Chemico-Biological Interactions 278 (2017): 123-128.

**Analgesic:* Caryophyllene oxide exhibits promising analgesic properties. As a primary constituent in the essential oil from the rhizome (lateral roots) of cyperus rotundus (nut grass), beta-caryophyllene oxide contributed to significant analgesic effects[608], exhibiting central and peripheral analgesic properties comparable to aspirin[609] in rats.

**Antiprotozoal:* As a primary constituent of the essential oil made from aframomum sceptrum (part of the ginger family), caryophyllene oxide contributed to remarkable antiprotozoal activity against trypanosoma brucei - the cause of sleeping sickness, and trichomonas vaginalis[610] - a parasitic protozoan that infects the urogenital tract of humans.

**Antifungal:* In an in vitro experimental model of onychomycosis - a fungus that normally affects toenails and can become pathogenic if not treated - a treatment containing 8% caryophyllene oxide completely inhibited the development of the dermatophytes responsible for this condition[611].

**Anti-cancer:* Caryophyllene oxide exhibits chemoprotective and anti-cancer properties. For instance, both beta-caryophyllene and beta-caryophyllene oxide treatment reduced the expression of pro-cancer genes and proteins, while increasing the levels of those with proapoptotic properties and simultaneously enhancing the efficacy of classic cancer treatment drugs[612]. Beta-caryophyllene oxide was also shown to increase reactive oxygen species generation from mitochondria and appeared to interfere with multiple signaling cascades involved in tumorigenesis[613]. Beta-caryophyllene and caryophyllene oxide have been shown to induce apoptosis in lymphoma and neuroblastoma cells, down-regulating anti-apoptotic genes while up-regulating pro-apoptotic genes[614]. Additionally, as the primary constituent of a guava leaf hexane fraction, caryophyllene oxide contributed to the inhibition of the AKT/mTOR/S6K1 signaling pathway and induced apoptosis in prostate cancer cells[615]. Caryophyllene oxide was also shown to exhibit significant antiproliferative effects against A549 lung cancer cells via cell cycle arrest and apoptosis unrelated to oxidative stress[616], and, as the primary constituent in the bio-guided fractionation of the leaf hexane total extract of Hymenaea courbaril, the compound induced early and late apoptosis and depolarized the mitochondrial membrane in PC-3 androgen-independent prostate cancer cells, with low toxicity

---

[608] Chen Y, Zhao YY, Wang XY, Liu JT, Huang LQ, Peng CS. GC-MS analysis and analgesic activity of essential oil from fresh rhizoma of Cyperus rotundus. Zhong Yao Cai. 2011 Aug;34(8):1225-9.

[609] Chavan, M.J., P.S. Wakte, and D.B. Shinde. Analgesic and anti-inflammatory activity of Caryophyllene oxide from Annona squamosa L. bark. Phytomedicine 17.2 (2010): 149-151.

[610] Cheikh-Ali, Zakaria, Marcelline Adiko, Sylvie Bouttier, Christian Bories, Timothee Okpekon, Erwan Poupon, and Pierre Champy. Composition, and Antimicrobial and Remarkable Antiprotozoal Activities of the Essential Oil of Rhizomes of Aframomum sceptrum K. Schum. (Zingiberaceae). Chemistry & Biodiversity 8.4 (2011): 658-667.

[611] Yang, Depo, Laura Michel, Jean-Pierre Chaumont, and Joelle Millet-Clerc. Use of caryophyllene oxide as an antifungal agent in an in vitro experimental model of onychomycosis. Mycopathologia 148.2 (2004): 79-82.

[612] Fidyt, Klaudyna et al. Beta-caryophyllene and Beta-caryophyllene oxide-natural compounds of anticancer and analgesic properties. Cancer medicine vol. 5,10 (2016): 3007-3017.

[613] Park, Kyung-Ran, Dongwoo Nam, Hyung-Mun Yun, Seok-Geun Lee, Hyeung-Jin Jang, Gautam Sethi, Somi K. Cho, and Kwang Seok Ahn. Beta-caryophyllene oxide inhibits growth and induces apoptosis through the suppression of PI3K/AKT/mTOR/S6K1 pathways and ROS-mediated MAPKs activation. Cancer Letters 312.2 (2011):178-188.

[614] Sain S, Naoghare PK, Devi SS, Daiwile A, Krishnamurthi K, Arrigo P, Chakrabarti T. Beta caryophyllene and caryophyllene oxide, isolated from Aegle marmelos, as the potent anti-inflammatory agents against lymphoma and neuroblastoma cells. Antiinflamm Antiallergy Agents Med Chem. 2014 Mar;13(1):45-55.

[615] Ryu, Nae Hyung et al. A hexane fraction of guava Leaves (Psidium guajava L.) induces anticancer activity by suppressing AKT/mammalian target of rapamycin/ribosomal p70 S6 kinase in human prostate cancer cells. Journal of medicinal food vol. 15,3 (2012): 231-41.

[616] Shabana SM, Gad NS, Othman AI, Mohamed AF, El-Missiry MA. β-caryophyllene oxide induces apoptosis and inhibits proliferation of A549 lung cancer cells. Med Oncol. 2023 May 26;40(7):189.

toward normal cells[617]. Finally, beta-caryophyllene-oxide might be useful in the prevention of metastasis after it was shown to inhibit MMP-2, p-p38, and p-ERK expression levels in HT1080 human fibrosarcoma cells[618].

**Mosquito Control/Antimalarial:* As a primary constituent in the essential oil from the leaves of syzygium lanceolatum, caryophyllene oxide contributed to acute toxicity against six mosquito species studied, while exhibiting "extremely" low environmental and aquatic toxicity[619]. As a major component in the essential oil of Artemisia indica, caryophyllene oxide contributed to significant antimalarial activity, possibly interfering with the ability of the virus to replicate[620].

**Acaricidal:* As isolated from the essential oil of Psidium cattleianum (strawberry guava), beta-caryophyllene oxide was shown to be 7-17 times more effective than the other components studied, including the go-to treatment for house mites, synthetic acaricide benzyl benzoate, and 14-27 times more active than DEET[621] - one of the most-used insect repellant products in the world. Although dust mites might seem like a minor issue, getting rid of these pests lowers allergic disease occurrence.

**Treatment for Herpes:* The complete essential oil of star anise and its primary constituents - including beta-caryophyllene oxide - were shown to exhibit anti-herpes simplex virus activity by direct inactivation of free virus particles[622]. Star anise oil is used anecdotally to relieve symptoms of herpes and to reduce the duration of outbreaks, and this study appears to confirm these actions.

**Trypanocidal - Chagas Disease Treatment:* When combined in a 1:4 mixture with the terpenoid lupenone, caryophyllene oxide contributed to synergistic effects that reduced by more than 80% the number of trypanosoma cruzi amastigote nests in heart tissue and muscles of infected mice[623]. T. cruzi, a flagellate protozoan, is the cause of chagas disease, a serious illness that often leads to death.

**Antibacterial Agent:* As a primary constituent in the essential oil made from Dennettia tripetala (Pepperfruit), caryophyllene oxide showed better antibacterial activity than the reference antibiotic, vancomycin, including against methicillin-resistant Staphylococcus aureus (MRSA) infections[624].

---

[617] Delgado C, Mendez-Callejas G, Celis C. Caryophyllene Oxide, the Active Compound Isolated from Leaves of Hymenaea courbaril L. (Fabaceae) with Antiproliferative and Apoptotic Effects on PC-3 Androgen-Independent Prostate Cancer Cell Line. Molecules. 2021 Oct 12;26(20):6142.

[618] Jo HW, Kim MM. β-Caryophyllene oxide inhibits metastasis by downregulating MMP-2, p-p38 and p-ERK in human fibrosarcoma cells. J Food Biochem. 2022 Dec;46(12):e14468.

[619] Giovanni Benelli, Mohan Rajeswary, Marimuthu Govindarajan. Towards green oviposition deterrents? Effectiveness of Syzygium lanceolatum (Myrtaceae) essential oil against six mosquito vectors and impact on four aquatic biological control agents. Environmental Science and Pollution Research. April 2018, Volume 25, Issue 11, pp 10218-10227.

[620] Tasdemir D, Tierney M, Sen R, Bergonzi MC, Demirci B, Bilia AR, Baser KH, Brun R, Chatterjee M. Antiprotozoal Effect of Artemisia indica Extracts and Essential Oil. Planta Med. 2015 Aug;81(12-13):1029-37.

[621] Oh, Min-Seok, Ji-Yeon Yang, Min-Gi Kim, and Hoi-Seon Lee. Acaricidal activities of Beta-caryophyllene oxide and structural analogues derived from Psidium cattleianum oil against house dust mites. Pest Management Science 70.5 (2014): 757-762.

[622] Astani, Akram et al. Screening for antiviral activities of isolated compounds from essential oils. Evidence-based complementary and alternative medicine: eCAM vol. 2011 (2011): 253643.

[623] Polanco-Hernandez, Glendy, Fabiola Escalante-Erosa, Karlina Garcia-Sosa, Maria E. Rosado, Eugenia Guzman-Marin, Karla Y. Acosta-Viana, Alberto Gimenez-Turba, Efrain Salamanca, and Luis M. Pena-Rodriguez. Synergistic Effect of Lupenone and Caryophyllene Oxide against Trypanosoma cruzi. Evidence-Based Complementary and Alternative Medicine 2013.

[624] Adesida, S.A., Iyebeye, M.I., Aina, O.O. et al. Chemical composition and healing potential of essential oil of Dennettia tripetala on methicillin-resistant Staphylococcus aureus: infected wound model. Beni-Suef Univ J Basic Appl Sci 11, 91 (2022).

IMPLICATIONS FOR HUMAN HEALTH & NUTRITION

The best way to obtain caryophyllene oxide is by cooking, preparing, or serving foods with a wide variety of spices that contain this bicyclic sesquiterpene, including allspice, basil, bay, celery seed, chamomile, cinnamon, clove, coriander, cumin, ginger, horsemint, marjoram, oregano, rosemary, safflower, sage, savory, and thyme. Snacking on carrots, figs, or guava fruits is also a good way to obtain this terpene. Adding black pepper to finished foods, or making teas from lemongrass, pennyroyal, or peppermint can help to supplement daily caryophyllene oxide intake. Responsibly drinking beer may be another way of obtaining dietary caryophyllene oxide, as the hops in these beverages are likely to contain this terpene.

Cannabis strains containing caryophyllene oxide include Bast 3, Blue Cheese, Blue Dream, Cheese, Chemdawg, Durban Poison, Ghost Train Haze, Girl Scout Cookies, Goldmeyer, Island Sweet Skunk, Maui Waui, Purple Punch, Purps, Speakeasy, and Whitaker Blues. Smoking or vaporizing these strains - especially at lower temperatures - can result in the uptake of this compound. However, we recommend against smoking or vaporizing and instead suggest eating, juicing, or blending these strains raw.

Experimentation with the pure compound caryophyllene oxide or with essential oils made from plants listed in the “Occurrences in Plants” section above should be undertaken with care, as these compounds may be toxic in some situations.

Caryophyllene Oxide Review

Answer the following questions to test your knowledge of this terpene:

Question #1: What type of terpene is caryophyllene oxide?

a. Monocyclic sesquiterpene
b. Linear sesquiterpene
c. Bicyclic sesquiterpenoid
d. Acyclic sesquiterpene

Question #2: Which of the following best describes the primary flavor & aroma profile of caryophyllene oxide?

a. Citrus, fruity, dry, terpenic
b. Piney, wood, spicy, sweet
c. Lemon, sharp, fuel, wood
d. Woody, dry, cedar, carrot

Question #3: How common is caryophyllene oxide in cannabis?

a. Top ten
b. Top three
c. Top twenty
d. Less common

Question #4: What is the chemical formula for caryophyllene oxide?

a. $C_{10}H_{16}O$
b. C15H24O2
c. $C_{15}H_{24}O$
d. $C_{20}H_{32}O$

Question #5: Name two uses of caryophyllene oxide in plants:

1 ______________________ 2 ______________________

Question #6: Name two potential medical uses of caryophyllene oxide:

1 ______________________ 2 ______________________

For the answer key to all individual compound chapters please go here: http://thebigbookofterps.com/bbot-quiz-answer-key/

PHYTOL

Type: Acyclic diterpenoid
Chemical Formula: $C_{20}H_{40}O$
Molecular Weight: 296.308 g/mol
Boiling Point: 333.00 to 335.00 °C. @ 760.00 mm Hg[625]
Flash Point: >212.00 °F. TCC (>100.00 °C.)
Solubility: Insoluble in water, soluble in ethanol, DMSO, dimethyl formamide
Oral LD50: >5,000 mg/kg (rat)[626]
Biological Role: Unknown, recycled for production of chlorophyll
Therapeutic Role: Anticancer agent, antioxidant, anti-inflammatory, antiallergen
Commercial Use: Fragrance, food additive, chemical precursor
Occurrence in Cannabis: Top twenty
Aroma/Flavor: Floral, balsamic, waxy, green

Occurs in Cannabis Strains: As a breakdown product of chlorophyll, phytol is likely found in most cannabis plants, although many laboratories do not test for this constituent. Cannabis laboratory tests have confirmed the presence of phytol in Bast 3, Blue Cheese, Blue Dream, Cannatonic, Cheese, Chemdawg, Goo, Island Sweet Skunk, OG Kush, Purps, Sour Diesel, Speakeasy, and Whitaker Blues.

INTRODUCTION

Phytol is a diterpene derived from the metabolism or degradation of chlorophyll; an important pigment in the process of photosynthesis. Most green plants produce chlorophyll, which means that most also contain phytol, including cannabis. However, it is important to understand that it's unlikely that plants deliberately produce phytol; instead, it is a breakdown product and as such likely serves little to no biological purpose. Additionally, discrepancies in the way phytol is reported in the laboratory environment leave many questions about the production and presence of this terpene in plants, and particularly in the presence of essential oils and extractions. Details are discussed in the "Occurrence & Biological Role" section below.

Despite the ambiguity of this compound, phytol exhibits a unique aroma and flavor profile that is primarily floral and balsamic, and to a lesser degree, waxy and green. These attributes have cemented phytol a place in the perfume and fragrance industries, which regularly use the compound in the formulation of products. Phytol is also used as a flavoring agent in the prepared and processed foods industries.

Interestingly, phytol serves many roles in the world of insects. As a primary constituent in scent compounds produced in the wings of cabbage white butterflies, phytol appears to act as an aphrodisiac for conspecific females, increasing the mating success of young males treated with the compound when compared to untreated males[627]. Similar results were found by other researchers studying the pheromones of males of the ephestia elutella species of butterfly[628]. Perhaps even more inventively, male beetles of the species Aesopida malasiaca use the volatile

---

[625] The Good Scents Company Data Sheet for Phytol, from: http://www.thegoodscentscompany.com/data/rw1040391.html. Accessed August 31, 2023.

[626] Biolandes Chemical Safety Data Sheet for Phytol, from: https://www.biolandes.com/wp-content/uploads/fiche-securite-F1529.pdf. Accessed August 31, 2023.

[627] Yildizhan, Selma, Joop van Loon, Anna Sramkova, Manfred Ayasse, Cristian Arsene, Cindy ten Broeke, and Stefan Schulz. Aphrodisiac Pheromones from the Wings of the Small Cabbage White and Large Cabbage White Butterflies, Pieris rapae and Pieris brassicae. ChemBioChem 10.10 (2009): 1666-1677.

[628] P. L. PhelanP. J. SilkC. J. NorthcottS. H. TanT. C. Baker. Chemical identification and behavioral characterization of male wing pheromone ofEphestia elutella (Pyralidae) Journal of Chemical Ecology January 1986, Volume 12, Issue 1, pp 135-146.

emissions produced by wounding of host plants like blueberry bushes - including the primary constituents of beta-caryophyllene and phytol - to locate conspecific females[629]. Finally, larvae of the sumac flea beetle take advantage of several compounds obtained from their host tree, the smooth sumac - notably phytol - which the larvae excrete in feces, retaining the excrement on their backs as a chemical shield against predatory insects[630].

CHEMICAL STRUCTURE

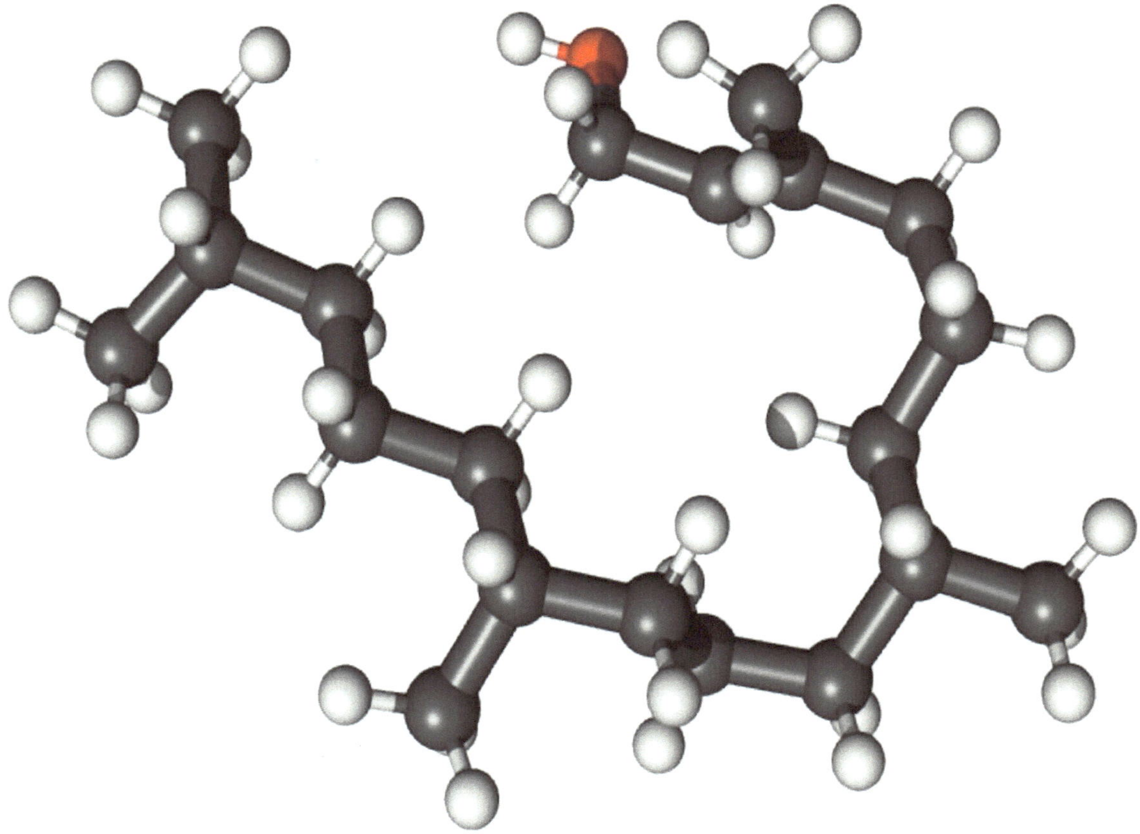

Phytol is an acyclic (containing no rings) diterpenoid alcohol produced as chlorophyll breaks down in the tissue of green leaves. This compound is based on four linked units of isoprene containing twenty carbon atoms, forty hydrogen atoms, and one oxygen atom, notated as $C_{20}H_{40}O$. Phytol features methyl groups at positions 3, 7, 11, and 15 (methyl groups change the functionality of a molecule), and has thirteen rotatable bonds, one double bond, and one covalent bond. There are at least two isomers of phytol, including the cis- and trans- variations.

[629] Fujiwara-Tsujii, Nao, Hiroe Yasui, Sadao Wakamura, Ikuko Hashimoto, and Makoto Minamishima. The white- spotted longicorn beetle, Anoplophora malasiaca (Coleoptera: Cerambycidae), with a blueberry as host plant, utilizes host chemicals for male orientation. Applied Entomology and Zoology 47.2 (2012): 103-110.

[630] Vencl, Fredric V., and Timothy C. Morton. The shield defense of the sumac flea beetle, Blepharida rhois (Chrysomelidae: Alticinae). Chemoecology 8.1 (1998): 25-32.

OCCURRENCE & BIOLOGICAL ROLE

Because phytol is produced from chlorophyll used in photosynthesis, it is reasonable to assume that most green plants produce this compound. However, research indicates that phytol is not only constantly produced[631], but also constantly recycled for the biosynthesis of new chlorophyll molecules in synechocystis[632], and is likely used similarly by other plants.

Ultimately, this probably means that phytol does not actually serve any primary or secondary biological roles in plants. Additionally, this compound generally does not appear in the essential oil of most plants, including cannabis. In fact, in a study that analyzed the terpene profiles of more than 460 samples of cannabis strains, phytol did not appear in any of the tests[633], with the study authors concluding that:

"There is a need for a widely accepted and unified method for terpene analysis in cannabis, to make comparison between different studies more reliable."

Research for this textbook turned up similar discrepancies. For instance, in terpene test results for multiple samples of the same strain, but submitted by different sources in different regions, the author found that if phytol occurred in one sample, it never occurred in the others. More research is clearly needed. Nevertheless, what is certain is that, despite phytol seemingly not having any direct biological roles or uses in plants, the compound has been shown to offer great potential in human medicine, see more on this in the "Potential Uses in Medicine" section below.

USES IN INDUSTRY

Because of its floral aroma, phytol is used in the perfume and fragrance industries, and is added to many different products as a fragrance or masking agent, including detergents, shampoos, toilet cleaners, cosmetics, household cleaners, and more.

In the prepared and processed foods industry, phytol is used as a flavoring agent in baked goods, alcoholic and nonalcoholic beverages, breakfast cereals, cheese, condiments, fish products, frozen dairy items and fruit ices, packaged desserts, gravy, imitation dairy products, jelly, processed fruits, seasonings, soups, snack foods, sauces, sorbet and sherbet, spices, and other prepared foods.

Phytol is also used as a precursor chemical in the manufacture of synthetic versions of vitamins E and K, and subsequently these synthetic vitamins are added to lotions and other skin care products.

Interestingly, phytol has been proposed as a method of detecting refined versus unrefined edible oils, where the absence of cis-phytol can serve as a clear marker of unrefined edible oils[538]. This indicates that something in the refining process alters the oil in such a way that cis-phytol is produced.

Commercial samples of phytol often appear as a colorless or clear liquid, but in some cases as a white solid.

---

[631] Vavilin, Dmitrii, and Wim Vermaas. Continuous chlorophyll degradation accompanied by chlorophyllide and phytol reutilization for chlorophyll synthesis in Synechocystis sp. PCC 6803. Biochimica et Biophysica Acta (BBA) - Bioenergetics 1767.7 (2007): 920-929.

[632] Hazekamp A, Fischedick JT. Cannabis - from cultivar to chemovar. Drug Test Anal. 2012 Jul-Aug; 4(7-8):660-7.

[633] Vetter W, Schroder M, Lehnert K. Differentiation of refined and virgin edible oils by means of the trans- and cis-phytol isomer distribution. J Agric Food Chem. 2012 Jun 20;60(24):6103-7.

POTENTIAL USES IN MEDICINE

Phytol is well-represented in the literature and shows promise as an anti-cancer agent, an anti-inflammatory agent, and as a treatment for tuberculosis, among other therapeutic options as detailed below:

**Treatment for Breast Cancer:* While researching treatments for estrogen-dependent breast cancer, scientists found that trans-phytol potently inhibited estrogen biosynthesis in human ovarian granulosa-like KGN cells[634].

**Treatment for Head & Neck Cancer:* As a primary constituent in the active ethyl acetate fraction of pereskia bleo (rose cactus leaves), phytol exhibited strong cytotoxicity against human nasopharyngeal epidermoid carcinoma cell line (KB)[635], a type of cancer that begins in the upper part of the throat and can quickly spread to the neck, nasal passages, mouth, and other parts of the head.

**Anticancer Agent:* Phytol has been shown to significantly reduce the viability and rate division of sarcoma 180 (S-180) and human leukemia (HL-60) cell lines, inducing apoptosis and necrosis of cells, as well as exhibiting aneugenic and clastogenic effects[636].

**Anti-tumor Agent:* In a study that evaluated the antitumor activity of phytol, researchers found that the compound modulated the pro-carcinogen used to induce cancer, induced genotoxicity and apoptosis in breast cancer cells, and exhibited a DNA-repair capacity in the lymphocytes of mice[637].

**Treatment for Stomach Cancer:* As the primary constituent in the methanolic extract of citrus unshiu leaves, phytol contributed to considerable cytotoxicity in human gastric adenocarcinoma cells, inducing non-apoptotic cell death[638], also known as autophagic cell death.

**Anticonvulsant:* In a study of pilocarpine-induced seizures in mice, phytol decreased the percentage of seizures while increasing the time to onset of seizures, and protected the animals against status epilepticus[639] (a continuous seizure lasting more than thirty minutes.) This has important implications in the treatment of epilepsy.

**Treatment for Allergies:* As a primary constituent in the essential oil of artemisia princeps pamp. and as a singular compound, phytol inhibited the expression of allergic cytokines in a murine model, significantly reducing scratching behavior[640].

---

[634] Guo J, Yuan Y, Lu D, Du B, Xiong L, Shi J, Yang L, Liu W, Yuan X, Zhang G, Wang F. Two natural products, trans-phytol and (22E)-ergosta-6,9,22-triene-3,5a,8a-triol, inhibit the biosynthesis of estrogen in human ovarian granulosa cells by aromatase (CYP19). Toxicol Appl Pharmacol. 2014 Aug 15;279(1):23-32.

[635] Malek, Sri Nurestri Abdul et al. Cytotoxic components of Pereskia bleo (Kunth) DC. (Cactaceae) leaves. Molecules (Basel, Switzerland) vol. 14,5 1713-24. 6 May. 2009.

[636] de Alencar MVOB, Islam MT, da Mata AMOF, Dos Reis AC, de Lima RMT, de Oliveira Ferreira JR, de Castro E Sousa JM, Ferreira PMP, de Carvalho Melo-Cavalcante AA, Rauf A, Hemeg HA, Alsharif KF, Khan H. Anticancer effects of phytol against Sarcoma (S-180) and Human Leukemic (HL-60) cancer cells. Environ Sci Pollut Res Int. 2023 Jul;30(33):80996-81007.

[637] de Alencar, Marcus Vinfcius Oliveira Barros, Muhammad Torequl Islam, Rosalia Maria Torres de Lima, Marcia Fernanda Correia Jardim Paz, Antonielly Campinho Reis, Ana Maria Oliveira Ferreira Mata, Jose Williams Gomes de Oliveira Filho, Gilberto Santos Cerqueira, Paulo Michel Pinheiro Ferreira, Joao Marcelo de Castro e Sousa, Mohammad S. Mubarak, and Ana Amelia de Carvalho Melo-Cavalcante. Phytol as an anticarcinogenic and antitumoral agent: An in vivo study in swiss mice with DMBA-Induced breast cancer. IUBMB Life 71.2 (2019): 200-212.

[638] Song, Yeon, Sabina Shrestha, Rajendra Gyawali, Dong-Sun Lee, and Somi Cho. Citrus unshiu leaf extract containing phytol as a major compound induces autophagic cell death in human gastric adenocarcinoma AGS cells. Journal of the Korean Society for Applied Biological Chemistry 58.2 (2015): 257-265.

[639] Costa, J.P., P.B. Ferreira, D.P. De Sousa, J. Jordan, and R.M. Freitas. Anticonvulsant effect of phytol in a pilocarpine model in mice. Neuroscience Letters 523.2 (2012): 115-118.

[640] Ryu KR, Choi JY, Chung S, Kim DH. Anti-scratching behavioral effect of the essential oil and phytol isolated from Artemisia princeps Pamp. in mice. Planta Med. 2011 Jan;77(1):22-6.

**Antioxidant:* In addition to suppressing melanogenesis in murine melanoma cells, phytol was shown to significantly attenuate melanin production and increase reactive oxygen species production in cells[641], indicating a significant antioxidant effect.

**Immune System Enhancement (adjuvant):* Phytol has been shown to be an excellent adjuvant (a substance that enhances immune response), superior to most commercial adjuvants[642]. However, because the compound is relatively toxic, several compounds derived from phytol have been produced that show similar immune system enhancement effects, but without toxicity[643].

**Antituberculosis:* Although phytol exhibited natural antitubercular activity against the M. tuberculosis H37Rv strain, several logically enhanced semisynthetic analogues derived from this compound were produced that were even more potent than phytol[644].

**Anti-inflammatory:* In a study that investigated the anti-inflammatory properties of phytol, scientists found that the compound significantly reduced swelling, inhibited neutrophil migration, and interfered with the activities of pro-inflammatory mediators[645] in a murine model.

**Anti-schistosomiasis:* Schistosomiasis or snail fever (caused by parasitic flatworms of the genus schistosoma) is a disease that affects millions of people worldwide. With only one drug to treat this illness, phytol has shown great promise after it reduced the motor activity of worms and caused their death in mice, including 100% of all male and female adult parasites at a dose of 100 µg/mL[646].

**Treatment for Migraine:* Researchers seeking to understand the traditional Chinese use of silkworm larvae feces as an analgesic agent discovered that phytol was the primary constituent responsible for the potent effects. Phytol was found to preferentially block the inactivated state of Nav1.7 sodium channels, induced a hyperpolarizing shift of voltage-dependent inactivation and slowed the recovery from inactivation, and displayed state-dependent inhibition of neuronal sodium channels, and decreased neuronal excitability in trigeminal ganglion neurons[647].

**Treatment for Osteoporosis:* As the primary constituent isolated from a fractionation of Stevia rebaudiana leaves, phytol inhibited osteoclast differentiation through TRAP-positive and F-actin formation, suppressed the expression of anti-nuclear factor of activated T cells-c1 (NFATc1) and c-Fos, inhibited oxidative stress by suppressing reactive oxygen species (ROS) accumulation while recovering antioxidant enzymes, and ameliorated osteoclast-specific differentiation, function, and oxidative stress through Nrf2 regulation by siRNA transfection[648].

---

641 Ko, Gyeong-A., and Somi Kim Cho. Phytol suppresses melanogenesis through proteasomal degradation of MITF via the ROS-ERK signaling pathway. Chemico-Biological Interactions 286 (2018): 132-140.

642 Aachoui, Youssef, Roshni Roy Chowdhury, Richard W. Fitch, and Swapan K. Ghosh. Molecular signatures of phytol-derived immunostimulants in the context of chemokine-cytokine microenvironment and enhanced immune response. Cellular Immunology 271.2 (2011): 227-238.

643 Aachoui, Youssef, Michael L. Schulte, Richard W. Fitch, and Swapan K. Ghosh. Synthetic adjuvants for vaccine formulations: Evaluation of new phytol derivatives in induction and persistence of specific immune response. Cellular Immunology 271.2 (2011): 308-318.

644 Saikia, Dharmendra, Swati Parihar, D. Chanda, S. Ojha, J.K. Kumar, C.S. Chanotiya, K. Shanker, and Arvind S Negi. Antitubercular potential of some semisynthetic analogues of phytol. Bioorganic & Medicinal Chemistry Letters 20.2 (2010): 508-512.

645 Silva, Renan O., Francisca Beatriz M. Sousa, Samara R.B. Damasceno, Nathalia S. Carvalho, Valdelania G. Silva, Francisco Rodrigo M.A. Oliveira, Damiao P. Sousa, Karoline S. Aragao, Andre L.R. Barbosa, Rivelilson M. Freitas, and Jand Venes R. Medeiros. Phytol, a diterpene alcohol, inhibits the inflammatory response by reducing cytokine production and oxidative stress. Fundamental & Clinical Pharmacology 28.4 (2014): 455-464.

646 de Moraes, Josue et al. Phytol, a diterpene alcohol from chlorophyll, as a drug against neglected tropical disease Schistosomiasis mansoni. PLoS neglected tropical diseases vol. 8,1 e2617. 2 Jan. 2014.

647 Song J, Jiang M, Jin Y, Li H, Li Y, Liu Y, Yu H, Huang X. Phytol from Faeces Bombycis alleviated migraine pain by inhibiting Nav1.7 sodium channels. J Ethnopharmacol. 2023 Apr 24;306:116161.

648 Kim EN, Trang NM, Kang H, Kim KH, Jeong GS. Phytol Suppresses Osteoclast Differentiation and Oxidative Stress through Nrf2/HO-1 Regulation in RANKL-Induced RAW264.7 Cells. Cells. 2022 Nov 14;11(22):3596.

IMPLICATIONS FOR HUMAN HEALTH & NUTRITION

One of the best ways to obtain dietary phytol is to eat plenty of green leafy plants. Because chlorophyll absorbs light in the blue spectrum and reflects light in the green spectrum, it is safe to assume that most green leafy plants contain chlorophyll. And because chlorophyll metabolizes into phytol, it is likely that common green leafy vegetables like lettuce, spinach, arugula, dandelion greens, chard, kale, and others contain at least some amount of phytol, depending on the species, season, and other variables. Cooking, preparing, and serving foods with herbs where the leaves are the primary constituents - like rosemary, cilantro, parsley, oregano, etcetera, is a good way to consume this terpene.

Cannabis strains that have been shown to contain phytol include Bast 3, Blue Cheese, Blue Dream, Cannatonic, Cheese, Chemdawg, Goo, Island Sweet Skunk, OG Kush, Purps, Sour Diesel, Speakeasy, and Whitaker Blues. The leaves of these strains can be eaten raw, blended, juiced, or full-spectrum extracts can be made from the flowers. Although we do not recommend vaporizing or smoking cannabis, it is possible to obtain trace amounts of phytol in the smoke stream from these products.

Most commercial phytol is available for research purposes only, but some sellers do offer small amounts of this acyclic diterpenoid. Experimentation with pure compounds should be carried out with the utmost caution.

Phytol Review

Answer the following questions to test your knowledge of this terpene:

Question #1: What type of terpene is phytol?

a. Bicyclic monoterpene
b. Linear sesquiterpene
c. Acyclic diterpenoid
d. Acyclic monoterpene

Question #2: Which of the following best describes the primary flavor & aroma profile of phytol?

a. Lemon, sharp, fuel, sweet
b. Floral, balsamic, waxy, green
c. Hops, cloves, wood, spicy
d. Citrus, fruity, dry, turpentine

Question #3: How common is phytol in cannabis?

a. Top three
b. Top ten
c. Top fifteen
d. Top twenty

Question #4: What is the chemical formula for phytol?

a. $C_{24}H_{42}O$
b. $C_{15}H_{40}O$
c. $C_{40}H_{20}O$
d. $C_{20}H_{40}O$

Question #5: From what pigment in plants is phytol derived?

____

Question #6: Name two potential medical uses of phytol:

1 ______________________________ 2 ____________________________________

For the answer key to all individual compound chapters please go here: http://thebigbookofterps.com/bbot-quiz-answer-key/

CAMPHENE

Type: Bicyclic monoterpene
Chemical Formula: $C_{10}H_{16}$
Molecular Weight: 136.238 g/mol
Boiling Point: *Camphene:* 159.00 to 160.00 °C. @ 760.00 mm Hg[649]
D-camphene: 159.00 °C. @ 760.00 mm Hg[650]
L-camphene: 157.50 to 158.50 °C. @ 760.00 mm Hg[651]
Flash Point: *Camphene:* 95.00 °F. TCC (35.00 °C.)
D- and L-camphene: 98.00 °F. TCC (36.67 °C.)
Melting Point: *Camphene:* 51.20 °C. @ 760.00 mm Hg
D-camphene: 52.00 °C. @ 760.00 mm Hg
L-camphene: 35.00 to 38.00 °C. @ 760.00 mm Hg
Solubility: *Camphene:* Soluble in alcohol, dipropylene glycol, fixed oils, slightly soluble in water
D-camphene: Soluble in paraffin oil, water
L-camphene: Soluble in alcohol, slightly soluble in water
Oral LD50: >5,000 mg/kg (rat)[652]
Biological Role: Larvicidal, ovidical, insecticidal
Therapeutic Role: Antibacterial, antioxidant, cholesterol-lowering agent
Commercial Use: Fragrance, food additive
Occurrence in Cannabis: Top twenty
Aroma/Flavor: *Camphene:* Camphoreous, woody, terpenic, citrus, mint
D-camphene: Fresh, woody, pine, camphor, citrus
L-camphene: Fresh, woody, fir, terpenic

Occurs in Cannabis Strains: Because camphene isn't part of most laboratory testing programs, it's likely that this compound occurs more frequently in cannabis varieties than current testing data shows. Cannabis varieties that are known to contain this monoterpene include ACDC, Blue Dream, Chocolate Mint OG, Covelo Sour Diesel, Deathstar, Ghost Cookies, Girl Scout Cookies, Miracle Alien Cookies, Northern Lights, Purple Punch 2.0, and Sour Diesel.

INTRODUCTION

Camphene is a bicyclic monoterpene with an interesting history as a former fuel for oil lamps. Occurring in the top twenty terpenes by concentration in many cannabis strains, this compound exhibits an aroma and flavor profile that is camphoraceous (camphor-like), piney, woody, and green, with lesser notes of mint, spice, and citrus, depending on the isomer. This terpene has often been described as possessing a cooling attribute, although many describe the overall scent and aroma as insipid. Camphene can be found in classic cannabis strains like ACDC and Girl Scout cookies, among others. Because camphene often occurs in the top twenty terpenes by concentration in cannabis, but most laboratories only register the top ten terpenes, occurrences of this compound are likely underreported.

---

649 The Good Scents Company Data Sheet for Camphene, from: http://www.thegoodscentscompany.com/data/rw1006292.html. Accessed September 1, 2023.

650 The Good Scents Company Data Sheet for D-camphene, from: http://www.thegoodscentscompany.com/data/rw1106261.html. Accessed September 1, 2023.

651 The Good Scents Company Data Sheet for L-camphene, from: http://www.thegoodscentscompany.com/data/rw1056891.html. Accessed September 1, 2023.

652 Agilent Technologies Safety Data Sheet for Camphene, from: https://www.agilent.com/cs/library/msds/WRK-105G_NAEnglish.pdf. Accessed September 1, 2023.

Once the most popular type of oil lamp fuel prior to the Civil War, camphene lamps were inexpensive and burned brightly, but they were also explosive, leading one Boston area newspaper to call camphene-burning lamps an "abomination, a swift and ready means of destruction for private families" in 1859[653].

Interestingly, camphene is also used as a host-location cue in some insects. A 2016 study showed that the polyphagous yellow peach moth uses the chemical signature from the volatile emissions of chestnut trees - where camphene is a primary constituent - to locate its host[654]. The adults and larvae of the insect then feed on chestnut fruits.

CHEMICAL STRUCTURE

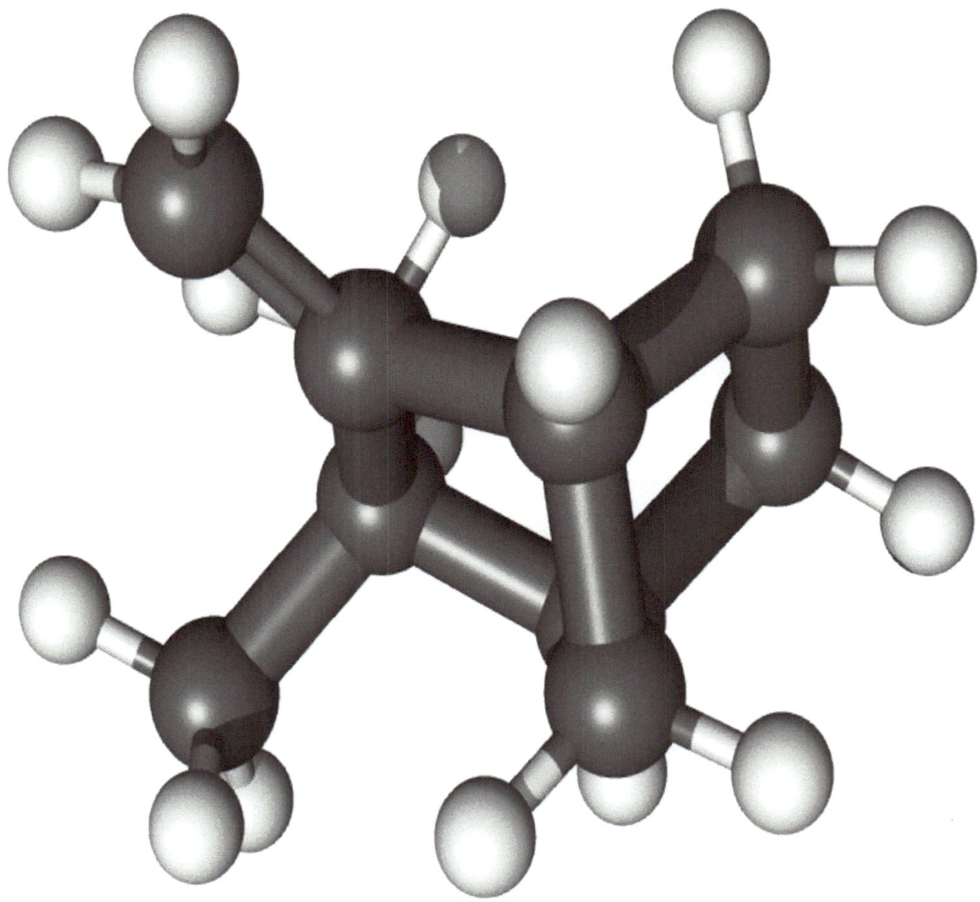

Based on two linked units of isoprene and derived from geranyl pyrophosphate, camphene has ten carbon atoms and sixteen hydrogen atoms, notated like most monoterpenes as $C_{10}H_{16}$. Camphene's molecular skeleton features two rings fused together, which classifies this compound as a bicyclic monoterpene. The functionality of camphene is enhanced by a methyl group at positions 2 and a methylidene group at position 3, and by one double bond. Camphene occurs naturally as a racemic version of its two optical isomers, D-camphene and L-camphene.

---

[653] Camphene to Kerosene Lamps. American Oil & Gas Historical Society. From: https://aoghs.org/products/camphene-to-kerosene-lamps/ Accessed April 22, 2019.

[654] Du, Yanli et al. Host Preference and Performance of the Yellow Peach Moth (Conogethes punctiferalis) on Chestnut Cultivars. PloS one vol. 11,6 e0157609. 21 Jun. 2016.

OCCURRENCE IN PLANTS

Camphene is found in many different plants, with the racemic mixture of the D- and L- version occurring most often:

Camphene:

Achillea tenuifolia	Alpinia galanga	Angelica
Anise	Armoise	Artemisia species
Ayou	Basil	Bay
Bell pepper	Betel	Black currant
Blood orange	Boldo	Boldus
Bombycilaena erecta	Cajuput	Calamus
Cannabis	Caraway	Cardamom
Carqueja	Carrot	Cascarilla
Cassia	Cassis	Cedarleaf
Cedarwood	Celery	Chamomile
Cinnamon	Citronella	Clary sage
Coriander	Cornmint	Costmary
Croton species	Cumin	Curcuma amada
Curry	Cypress	Dill
Elyonurus muticus	Eucalyptus species	Featherfew
Fennel	Ferula	Feverfew
Fir	Fleabane	Galangal
Ginger	Gingergrass	Grapefruit
Guava	Hinoki	Horsemint
Hypericum dogonbadanicum	Hyssop	Juniper
Kachur	Kewda	Labdanum
Laurel	Lavandin	Lavender
Layana	Lemon	Lemongrass
Levisticum officinale	Lime	Litsea
Lovage	Mace	Mandarin
Marjoram	Mastic	Melissa
Microstrobos species	Myrtle	Nepeta
Neroli	Nutmeg	Orange
Oregano	Otanthus maritimus	Parsley
Parsnip	Pepper tree	Peppermint
Pine	Piper cernum	Ravintsara
Rosemary	Sage	Salvia
Savin	Savory	Snake root
Spearmint	Sugandha kokila	Sunflower
Tanacetum annum	Tangerine	Tansy
Tarragon	Taxodium distichum	Tejpat
Thyme	Turmeric	Valerian
Water mint	Wormseed	Wormwood
Yarrow	Yuzu	Zanthoxylum species

D-camphene:

Carrot	Cypress	Dill
Eucalyptus	Fennel	Ginger

Lavender	Lemon	Marjoram
Nutmeg	Orange	Parsley
Pepper	Pine	Sage
Tarragon	Thyme	Turmeric

L-camphene:

Citronella	Pine	Valerian

BIOLOGICAL ACTIVITY IN PLANTS

The primary biological use of camphene in plants appears to be as an ovicidal and larvicidal agent against herbivorous insects. For instance, as the primary constituent in the essential oil of wedelia prostrata (a sunflower-like plant native to China), camphene was more toxic against larvae of the crop pest spodoptera litura (the tobacco cutworm) than any of the other compounds found in the EO[655]. As a primary constituent in the rhizome essential oil of crepe ginger, camphene contributed to significant toxicity against larvae of the cotton bollworm[656], a synthetic-insecticide-resistant pest. In a 2013 study, camphene showed contact and fumigant toxicity against all stages of the red palm weevil, but toxicity was especially potent against eggs and larvae[657]. Finally, as an individual constituent isolated from the essential oil of ajowan, camphene caused 100% mortality against the larvae of aedes aegypti mosquitos and daphnia magna water fleas[658].

USES IN INDUSTRY

Most commercial samples of camphene are produced by isomerization of alpha-pinene, and generally appear as a white or clear crystalline solid. Interestingly, researchers have developed a new type of phase-change building materials based on a mixture of 60% camphene and 40% stearic acid, blended in a substrate that includes marble powder[659]. These materials are designed to store thermal energy and can be used in HVAC and other applications.

Additionally, camphene is used extensively in the perfume industry, where it appears in citrus, lavender, pine, and other types of fragrances. This bicyclic monoterpene is found in some blended essential oils meant for use in aromatherapy applications[660]. In this industry, the maximum recommended level of camphene in a fragrance concentrate is 9%, according to The Good Scents Company.

[655] Benelli, Giovanni, Marimuthu Govindarajan, Mohamad AlSalhi, Sandhanasamy Devanesan, and Filippo Maggi. High toxicity of camphene and y-elemene from Wedelia prostrata essential oil against larvae of Spodoptera litura (Lepidoptera: Noctuidae). Environmental Science and Pollution Research 25.11 (2017): 10383-10391.

[656] Benelli, Giovanni, Marimuthu Govindarajan, Mohan Rajeswary, Baskaralingam Vaseeharan, Sami A. Alyahya, Naiyf S. Alharbi, Shine Kadaikunnan, Jamal M. Khaled, and Filippo Maggi. Insecticidal activity of camphene, zerumbone and a-humulene from Cheilocostus speciosus rhizome essential oil against the 0ld-World bollworm, Helicoverpa armigera. Ecotoxicology and Environmental Safety 148 (2018): 781-786.

[657] Aziza Sharaby, Mona EL-Dosary. Possibility Using Camphene as Biorational Insecticide against the Red Palm Weevil Rhynchophorus ferrugineus (Colioptera: Curculionedae). International Journal of Science and Research ISSN: 2319-7064 (2013): 6.14.

[658] Seo SM, Park HM, Park IK. Larvicidal activity of ajowan (Trachyspermum ammi) and Peru balsam (Myroxylon pereira) oils and blends of their constituents against mosquito, Aedes aegypti, acute toxicity on water flea, Daphnia magna, and aqueous residue. J Agric Food Chem. 2012 Jun 13;60(23):5909-14.

[659] Canbay, Canan, Zuhal Genc, Sureyya Acar, Memet Sekerci, and Murat Genc. Preparation and Thermodynamic Properties of Camphene/Stearic Acid Composites as Phase-Change Materials in Buildings. International Journal of Thermophysics 35.8 (2014): 1526-1537.

[660] Tadtong S, Kamkaen N, Watthanachaiyingcharoen R, Ruangrungsi N. Chemical Components of Four Essential Oils in Aromatherapy Recipe. Nat Prod Commun. 2015 Jun;10(6):1091-2.

When added to fragrances, camphene appears in personal hygiene products, dishwashing gels and liquids, household cleaners, furniture cleaners, and air care products. This compound is also part of adhesives and sealants, masking agents, paints, odorants and deodorants, and mothballs.

The packaged and prepared foods industry uses camphene as a food additive and flavoring ingredient in non-alcoholic beverages, frozen desserts and dairy items, hard candy, and other processed foods.

During the '70s and '80s, a chlorine- and camphene-based insecticide called Toxaphene was widely used on cotton and other crops in the United States, until reports of its severe environmental toxicity led to a complete ban of the substance in the US in 1990[661]. However, the ban came only after hundreds of millions of pounds of Toxaphene had already been used over a period of 20 years.

POTENTIAL USES IN MEDICINE

Although lauded principally for its ability to lower cholesterol and triglycerides, camphene shows medical promise in several other areas, including as a treatment for breast cancer, a treatment for neurodegenerative disorders, and as a potent antibacterial agent, among other properties.

**Antioxidant:* While researching the antinociceptive (pain-blocking) properties of several compounds, scientists found that camphene had the strongest antioxidant and scavenging activities of the group[662]. Coincidentally, the study also concluded that high doses of camphene significantly reduced nociceptive behavior in mice.

**Treatment for Cardiovascular Disease:* As the primary constituent in the essential oil from the leaves of pinus koraiensis (Korean pine), camphene contributed to the activation, reduced oxidation, and up-regulation of LDL (low density lipoprotein) receptors and attenuated the expression of FAS (fatty acid synthase) and other compounds involved in lipid metabolism in human liver carcinoma cells[663]. As isolated from the essential oil of chios mastic gum, camphene was shown to significantly reduce total cholesterol, LDL cholesterol, and triglycerides, exhibiting effects comparable to mevinolin, a well-known cholesterol-lowering agent[664].

**Antibacterial Agent:* As a primary constituent in the essential oil of seseli rigidum (a perennial native to Bulgaria), camphene contributed to significant antibacterial activity against staphylococcus aureus, S. epidermidis, micrococcus luteus, and enterococcus faecalis[665] - all bacteria that commonly infect humans. Another study found that, as a primary constituent in the essential oil of F. crassifolia (kumquat) peel, camphene again contributed to toxic effects against staphylococcus aureus, but also against E. coli, S. typhimurium, and B. cereus[666].

---

[661] Toxaphene (PDF). Report on Carcinogens. National Toxicology Program, Department of Health and Human Services. 13. October 2, 2014. Retrieved from https://ntp.niehs.nih.gov/ntp/roc/content/profiles/toxaphene.pdf April 23, 2019.

[662] Quintans-Junior, Lucindo et al. Antinociceptive Activity and Redox Profile of the Monoterpenes (+)-Camphene, p-Cymene, and Geranyl Acetate in Experimental Models. ISRN toxicology vol. 2013 459530. 14 Jan. 2013.

[663] Kim, Ji-Hyun, Hyo-Jung Lee, Soo-Jin Jeong, Min-Ho Lee, and Sung-Hoon Kim. Essential Oil of Pinus koraiensis Leaves Exerts Antihyperlipidemic Effects via Up-regulation of Low-density Lipoprotein Receptor and Inhibition of Acyl-coenzyme A: Cholesterol Acyltransferase. Phytotherapy Research 26.9 (2012): 1314-1319.

[664] Vallianou, Ioanna et al. Camphene, a plant-derived monoterpene, reduces plasma cholesterol and triglycerides in hyperlipidemic rats independently of HMG-CoA reductase activity. PloS one vol. 6,11 (2011): e20516.

[665] Marcetic M, Bozic D, Milenkovic M, Lakusic B, Kovacevic N. Chemical composition and antimicrobial activity of essential oil of different parts of Seseli rigidum. Nat Prod Commun. 2012 Aug;7(8):1091-4.

[666] Wang, Yong-Wei et al. Chemical composition and antimicrobial activity of the essential oil of kumquat (Fortunella crassifolia Swingle) peel. International journal of molecular sciences vol. 13,3 (2012): 3382-9.

**Bone Repair:* Camphene has been used to produce a porous and strong bioactive glass scaffolding that can repair defects in load-bearing bones[667].

**Hepatoxicity Prevention/Treatment:* Hepatoxicity occurs when toxic substances overwhelm the liver. In an in vivo study with rats, a blend of camphene and geraniol accorded protection against cellular damage and hepatoxicity[668], which could have interesting implications for synergistic use with other substances.

**Treatment for Breast Cancer:* As a primary constituent in the essential oil of melaleuca armillaris (bracelet honey myrtle), camphene contributed to strong cytotoxicity against human breast cancer cells[669].

**Treatment for Alzheimer's:* As a primary constituent in the essential oil of Salvia leriifolia benth, camphene contributed to promising antioxidant activity, and cholinesterase inhibitory activity without toxicity, even at high doses[670].

IMPLICATIONS FOR HUMAN HEALTH & NUTRITION

Eating foods that naturally contain camphene is the best way to obtain this compound. Consider regularly snacking on fruits like black currants, blood oranges, grapefruit, guava, lime, mandarin, oranges, tangerines, and yuzu. Make vegetables like bell peppers, carrots, celery, and parsnip a common feature in meals. When cooking, preparing, or serving food, consider using spices and herbs like basil, bay, black pepper, cinnamon, coriander, cumin, dill, fennel, ginger, marjoram, nutmeg, oregano, rosemary, sage, tarragon, thyme, and turmeric.

Cannabis strains that produce camphene include ACDC, Blue Dream, Chocolate Mint OG, Covelo Sour Diesel, Deathstar, Ghost Cookies, Girl Scout Cookies, Miracle Alien Cookies, Northern Lights, Purple Punch 2.0, and Sour Diesel. While it may be possible to consume camphene in the smoke or vapors of cannabis, we recommend eating cannabis leaves, flowers, or other parts whole, or consider juicing or blending raw cannabis.

Experimentation with pure forms of camphene or with essential oils that are high in this compound should be undertaken with the utmost care, as this compound is extremely flammable and can be explosive.

---

[667] Liu, Xin et al. Porous and strong bioactive glass (13-93) scaffolds prepared by unidirectional freezing of camphene-based suspensions. Acta biomaterialia vol. 8,1 (2011): 415-23.

[668] Brijesh Kumar Singh, Madhulika Tripathi, Bhushan P. Chaudhari, Pramod K. Pandey, and Poonam Kakkar. Natural Terpenes Prevent Mitochondrial Dysfunction, Oxidative Stress and Release of Apoptotic Proteins during Nimesulide-Hepatotoxicity in Rats. PLoS One. 2012; 7(4): e34200.

[669] Chabir, Naziha, Mehrez Romdhane, Alexis Valentin, Beatrice Moukarzel, Hajer Naceur Ben Marzoug, Nadia Ben Brahim, Mohamed Mars, and Jalloul Bouajila. Chemical Study and Antimalarial, Antioxidant, and Anticancer Activities of Melaleuca armillaris (Sol Ex Gateau) Sm Essential Oil. Journal of Medicinal Food 14.11 (2011): 1383- 1388.

[670] Loizzo MR1, Menichini F, Tundis R, Bonesi M, Conforti F, Nadjafi F, Statti GA, Frega NG, Menichini F. In vitro biological activity of Salvia leriifolia benth essential oil relevant to the treatment of Alzheimer's disease. J Oleo Sci. 2009;58(8):443-6.

Camphene Review

Answer the following questions to test your knowledge of this terpene:

Question #l: What type of terpene is camphene?

a. Monocyclic monoterpene
b. Bicyclic sesquiterpene
c. Bicyclic monoterpene
d. Monocyclic sesquiterpene

Question #2: Which of the following best describes the primary flavor & aroma profile of camphene?

a. Depends on the isomer
b. Camphoraceous, piney, woody, green
c. Hops, cloves, wood, spicy
d. Lemon, sharp, fuel, spice
e. Citrus, fruity, dry, terpenic

Question #3: How common is camphene in cannabis?

a. Top three
b. Top ten
c. Top twenty
d. Less common

Question #4: Which is the chemical formula for camphene?

a. $C_{10}H_{18}$
b. $C_{10}H_{16}$
c. $C_{10}H_{16}O$
d. $C_{10}H_{18}O$

Question #5: Name two uses of camphene in plants:

1 ______________________ 2 ______________________

Question #6: Name two potential medical uses of camphene:

1 ______________________ 2 ______________________

For the answer key to all individual compound chapters please go here: http://thebigbookofterps.com/bbot-quiz-answer-key/

BORNEOL

Type: Bicyclic monoterpenoid
Chemical Formula: $C_{10}H_{18}0$
Molecular Weight: 154.253 g/mol
Boiling Point: *Borneol:* 212.00 to 213.00 °C. @ 760.00 mm Hg[671]
D-borneol: 212.00 °C. @ 760.00 mm Hg[672]
L-borneol: 203.00 to 204.00 °C. @ 760.00 mm Hg[673]
Isoborneol: 212.00 °C. @ 760.00 mm Hg (estimated)[674]
Flash Point: *Borneol:* 150.00 °F. TCC (65.56 °C.)
D-borneol: 150.00 °F. TCC (65.56 °C.)
L-borneol: 150.00 °F. TCC (65.56 °C.)
Isoborneol: 150.00 °F. TCC (65.60 °C.) (estimated)
Melting Point: *Borneol:* 202.00 to 208.00 °C. @ 760.00 mm Hg
D-borneol: 207.00 to 208.00 °C. @ 760.00 mm Hg
L-borneol: 206.00 to 208.00 °C. @ 760.00 mm Hg
Solubility: *Borneol:* Soluble in alcohol, weakly soluble in water
D-borneol: Soluble in alcohol, ether, slightly soluble in water
L-borneol: Soluble in alcohol, dipropylene glycol, propylene glycol, water
Isoborneol: Soluble in alcohol, slightly soluble in water
Oral LD50: 5,800 mg/kg (rat)[675]
Biological Role: Insecticidal, antimicrobial
Therapeutic Role: Neurological agent, anti-inflammatory, analgesic
Commercial Use: Fragrance, food additive
Occurrence in Cannabis: Top twenty
Aroma/Flavor: *Borneol:* Balsamic, pine, woody, camphor
D-borneol: Pine, camphor, earthy
L-borneol: Woody, earthy, pine, camphor

Occurs in Cannabis Strains: Borneol and its various isomers are often not included in laboratory testing analyses of cannabis varieties in the United States, however, some tests have reported this compound in several different strains including Angel Food Cake, Cannatonic, Cheese, Girl Scout Cookies, Golden Haze, K-13 Haze, Lemon Skunk, Sour Diesel, and Tosh OG.

INTRODUCTION

Of all the terpenes, only borneol has the esteemed distinction of consistent therapeutic use by humans for many centuries. In fact, this compound is among the first single-organic-ingredient natural medicines ever produced[676]. Used in traditional Chinese medicine for thousands of years, today companies like Wuzhou Huangpu Chemical Pharmaceutical Co, Ltd., which has been producing borneol flakes for more than 70 years[677], are commonplace in China.

---

[671] The Good Scents Company Data Sheet for Borneol, from: http://www.thegoodscentscompany.com/data/rw1039711.html. Accessed September 1, 2023.

[672] The Good Scents Company Data Sheet for D-borneol, from: http://www.thegoodscentscompany.com/data/rw1106071.html. Accessed September 1, 2023.

[673] The Good Scents Company Data Sheet for L-borneol, from: http://www.thegoodscentscompany.com/data/rw1011571.html. Accessed September 1, 2023.

[674] The Good Scents Company Data Sheet for Isoborneol, from: http://www.thegoodscentscompany.com/data/rw1412971.html. Accessed September 1, 2023.

[675] Fisher Scientific Material Safety Data Sheet for L-borneol, from: https://fscimage.fishersci.com/msds/33824.htm. Accessed September 1, 2023.

[676] Sheng LS., Du LD., Qiang GF., Du GH. Borneol. In: Natural Small Molecule Drugs from Plants. pp 187-192 Springer, Singapore (2018).

[677] "About Us" web page of Wuzhou Huangpu Chemical Pharmaceutical Co, Ltd. Published 11/25/2016 at: http://eng.wzhphg.com/engywm/21.htm. Accessed

Medicinally, borneol is known to attenuate the effects of stroke, and exerts a significant influence on the Blood Brain Barrier (BBB), which carries important implications for many types of medicines and treatments. See more about the medical potential of borneol in the "Potential Uses in Medicine" section below.

With a flavor and aroma profile that is primarily balsamic, camphor, spice, mint, and wood, depending on the isomer, borneol also possesses earthy notes of pepper, and lends some of these qualities to classic cannabis strains like Cheese, Girl Scout Cookies, Lemon Skunk, and Sour Diesel, among others.

Interestingly, male Orsilochides leucoptera (the shield bug, a type of stink bug) emit borneol as a pheromone to attract females[678].

CHEMICAL STRUCTURE

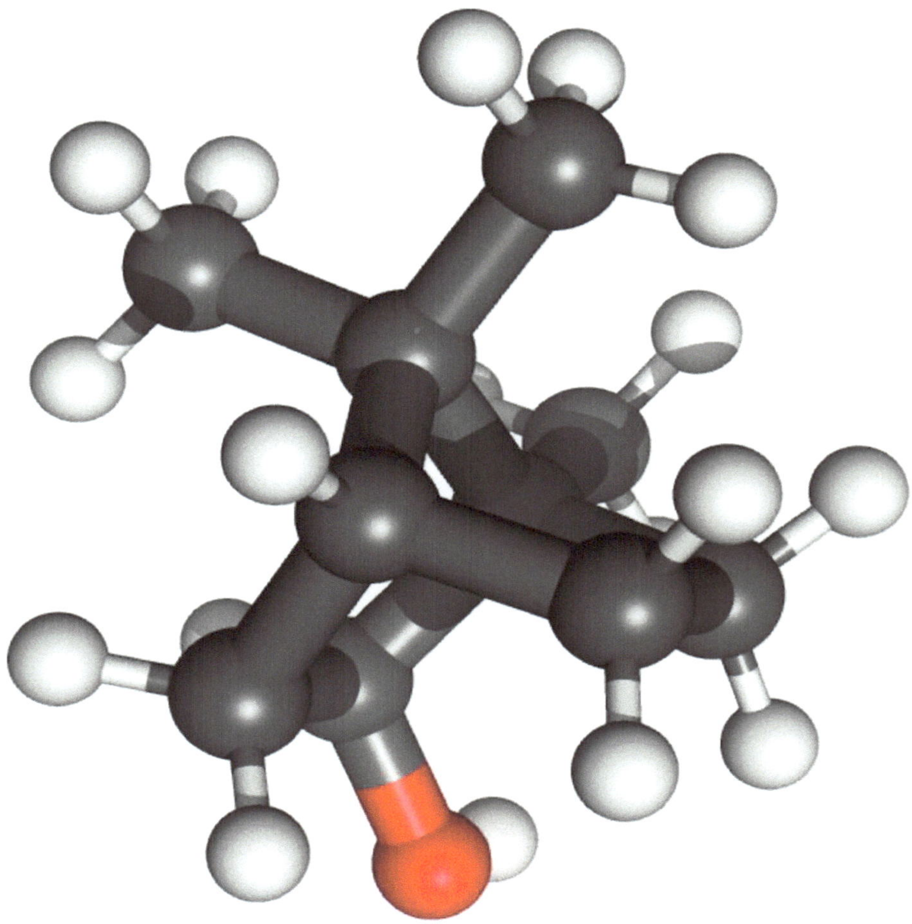

Borneol is an oxygenated terpene formed by two linked units of isoprene containing ten carbon atoms, eighteen hydrogen atoms, and one oxygen atom, scientifically notated as $C_{10}H_{18}O$. This compound is derived from geranyl pyrophosphate catalyzed by bornyl diphosphate. The molecular skeleton of borneol features a hydroxyl group at

May 10, 2019.

[678] de Melo DJ, da Silva MR, Moliterno AAC, Zarbin PHG. Camphor and Borneol as the Male-Produced Sex Pheromone of the Shield Bug, Orsilochides leucoptera (Hemiptera: Scutelleridae). J Chem Ecol. 2020 Jun;46(5-6):490-496.

position 2 (hydroxyl and other groups change the functionality of a molecule), and two rings, which classifies this compound as a bicyclic monoterpene.

There are several variations of borneol; the enantiomers d-borneol and l-borneol, which are common in nature, and in commercial samples of this terpene. Another borneol isomer sometimes found in cannabis and other plants is isoborneol, although this compound has not been well-studied.

OCCURRENCE IN PLANTS

Although it generally only occurs within the top twenty terpenes by concentration in certain cannabis strains, borneol is produced by many different plants, including:

Borneol:

Angelica	Anise	Armoise
Artemisia species	Ayou	Basil
Calamus	Cannabis	Carrot
Chamomile	Cinnamon	Cistus
Citronella	Clary sage	Coriander
Croton species	Curcuma amada	Cypress
Davana	Erigeron graveolens	Eucalyptus species
Featherfew	Feaverfew	Fir
Geranium	Ginger	Guava
Hinoki	Ho	Labdanum
Laurel	Lavandin	Lavender
Layana	Lemon	Lemongrass
Lime	Mace	Mangrove
Marjoram	Mastic	Myrtle
Nutmeg	Oregano	Phoebe
Pine	Ravintsara	Rosemary
Sage	Salvia	Santolina
Savory	Tanacetum annum	Tansy
Thyme	Turmeric	Valerian
Wormwood	Yarrow	Zingiber species

D-borneol:

Ginger	Lavender	Myristica fragrans
Nutmeg	Olibanum	Rosemary

L-borneol:

Cardamom	Carrot	Cinnamon
Citronella	Coriander	Cornmint
Geranium	Ginger	Lime
Mace	Nutmeg	Oakmoss
Olibanum	Pine	Rosemary
Sage	Snake root	Thyme
Valerian		

Isoborneol:

Ginger	Oregano	Rosemary
Sage	Thyme	Turmeric

BIOLOGICAL ACTIVITY IN PLANTS

Literature discussing the biological roles of borneol in plants is scarce, however, deductions can be made based on observation of this compound in the laboratory. For instance, in the study referenced below about the insect repellency effects of borneol, researchers searching for an antileishmanial agent discovered that borneol was repellant against a beetle and a type of louse. More research is needed to learn how this bicyclic monoterpene is used by plants.

***Insecticidal Agent:** As a primary constituent in the essential oil of artemisia subdigitata, borneol contributed to contact and fumigant toxicity against sitophilus zeamais motsch[679] (the maize weevil, a common pest of stored grains).

***Insect Repellant:** In a study of the essential oils of select Chinese herbs, borneol was a primary compound shown to exhibit strong repellency against insects like L. bostrychophila (the booklouse), and the red flour beetle[680] (tribolium castaneum).

***Antimicrobial:** As a primary constituent in the essential oil of rosmarinus officinalis (rosemary), borneol was found to be the most potent antimicrobial agent of all the other primary terpenes, exhibiting significant antimicrobial effects against gram-positive bacteria staphylococcus aureus and bacillus subtilis, gram-negative bacteria escherichia coli and pseudomonas aeruginosa, the yeast candida albicans, and the fungus apergillus niger[681].

***Root Growth Modulator:** Researchers theorize that Arabidopsis thaliana (thale cress) possesses a sensing mechanism that is able to detect different enantiomers of the same molecule after tests showed that d-borneol reduced root length, increased root tip width, induced growth of root tip hair, and reduced auxin signaling in the root tips[682], while l-borneol had no effect on the plants.

USES IN INDUSTRY

Available commercially as colorless or white lumps, crystals, powder, or flakes, borneol smells sharply of camphor, which has made this compound appealing to the perfume and fragrance industries, where it is used in scented products including cosmetics, detergents and soaps, air-care products, furniture cleaners, dish-washing and laundry liquids, etcetera.

[679] Chu, Sha-Sha et al. Chemical composition and insecticidal activity against Sitophilus zeamais of the essential oils derived from Artemisia giraldii and Artemisia subdigitata. Molecules (Basel, Switzerland) vol. 17,6 7255-65.

[680] Evaluation of Repellency of Some Chinese Medicinal Herbs Essential Oils Against Liposcelis bostrychophila (Psocoptera: Liposcelidae) and Tribolium castaneum (Coleoptera: Tenebrionidae). Journal of Economic Entomology 106.1 (2013): 513-519.

[681] Santoyo S, Cavero S, Jaime L, Ibanez E, Senorans FJ, Reglero G. Chemical composition and antimicrobial activity of Rosmarinus officinalis L. essential oil obtained via supercritical fluid extraction. J Food Prot. 2005 Apr;68(4):790-5.

[682] Fukuda K, Uefune M, Fukaki H, Yamauchi Y, Hara-Nishimura I, Ozawa R, Matsui K, Sugimoto K, Okada K, Imai R, Takahashi K, Enami S, Wurst R, Takabayashi J. Aerial (+)-borneol modulates root morphology, auxin signalling and meristematic activity in Arabidopsis roots. Biol Lett. 2022 May;18(5):20210629.

Borneol is also used in the prepared foods industry, where it may be found as a flavoring or perfuming agent in baked goods, non-alcoholic beverages, chewing gum, frozen dessert items, hard candy, soft candy, gelatins and puddings, relishes and condiments, and more.

Formerly produced by tapping dryobalanops aromitaca (a species of tree), or blumea balsamifera, today borneol is generally manufactured synthetically from camphor or pinene. In addition to its use in the fragrance, perfume, and prepared foods industries, this compound is also used in at least one over-the-counter antibacterial product in the form of a spray made by a Chinese company and sold in the United States[683]. The manufacturer claims that the spray kills 99.9% of bacteria on the skin.

POTENTIAL USES IN MEDICINE

Borneol has long been used in traditional Chinese medicine, and today the compound's full medical value is just beginning to be discovered around the world. However, it must be noted that use of this terpene should be carried out with caution, as borneol is quite toxic - just one teaspoon could kill a small adult[684]. Nevertheless, no instances of borneol poisoning, overdoses, or fatalities exist in the literature - a promising fact considering the extremely long duration of use of this compound.

Treatment aid for Stroke, Dementia: Borneol has been shown to directly attenuate the pathological conditions of stroke in rats[685]. Because of its ability to easily cross the blood brain barrier, borneol can also be used to significantly increase the brain concentrations of Nimodipine[686], a calcium channel blocker used as a treatment for ischemic damage. In a Chinese study published in the Journal of Molecular Neuroscience, borneol was shown to exhibit synergistic effects with tetramethylpyrazine phosphate, which together "improved the ultrastructure of neurons, reduced the apoptosis index, and reduced the intracellular calcium content in both the cortex and hippocampus[687]." These effects can provide protection against ischemic damage in the cortex and hippocampus regions of the brain.

Easily Crosses the BBB (Blood Brain Barrier): One of the most significant challenges in treating brain disease is the impermeability of the blood brain barrier. Borneol opens this barrier via several mechanisms, including increasing the expression of intercellular cell adhesion molecule-l[688] (important in inflammation and immune responses), and by reducing oxidative reactions, inhibiting the occurrence of inflammation, and inhibiting apoptosis[689], among other actions. Borneol has also been shown to protect the structural integrity of the blood brain barrier from pathological damage[690]. In efforts to improve the permeability of formononetin – an isoflavonoid with

---

683 DailyMed, US National Library of Medicine. SHUNFA ANTI-BACTERIA SPRA- didecyldimethyl ammonium, triclosan, borneol, benzalkonium bromide spray From: https://dailymed.nlm.nih.gov/dailymed/drugInfo.cfm?setid= 92cf6f49-d10b-49da-9d36-0aaef95ad6c2 Accessed May 3, 2019.

684 Gosselin, R.E., H.C. Hodge, R.P. Smith, and M.N. Gleason. Clinical Toxicology of Commercial Products. 4th ed. Baltimore: Williams and Wilkins, 1976., p. II-169.

685 Pan Xu, Ying Li, Shou-Ying Du, Yang Lu, Jie Bai, Qing-li Guo. Comparative pharmacokinetics of borneol in cerebral ischemia-reperfusion and sham-operated rats. Journal of Zhejiang University SCIENCE B. January 2014, Volume 15, Issue 1, pp 84-91.

686 Wu, Chun, Qiongfeng Liao, Meicun Yao, Xinjun Xu, Yuting Zhou, Xueying Hou, and Zhiyong Xie. Effect of natural borneol on the pharmacokinetics and distribution of nimodipine in mice. European Journal of of Drug Metabolism and Pharmacokinetics 39.1 (2013): 17-24.

687 Yu, Bin, Ming Ruan, Tao Liang, Shi-Wen Huang, Yun Yu, Hai-Bo Cheng, and Xiang-Chun Shen. The Synergic Effect of Tetramethylpyrazine Phosphate and Borneol for Protecting Against Ischemia Injury in Cortex and Hippoc ampus Regions by Modulating Apoptosis and Autophagy. Journal of Molecular Neuroscience 63.1 (2017): 70-83.

688 Tao Wu, MD, Aiqin Zhang, PhD, Hongyang Lu, PhD, and Qiaoyuan Cheng, PhD. The Role and Mechanism of Borneol to Open the Blood-Brain Barrier. Integrative Cancer Therapies 2018, Vol. 17(3) 806-812.

689 Zi-xian Chen, Qing-qing Xu, Chun-shuo Shan, Yi-hua Shi, Yong Wang, Raymond Chuen-Chung Chang, and Guo-qing Zheng. Borneol for Regulating the Permeability of the Blood-Brain Barrier in Experimental Ischemic Stroke: Preclinical Evidence and Possible Mechanism Oxidative Medicine and Cellular Longevity Volume 2019, Article ID 2936737, 15 pages.

690 Qun-Lin Zhang, Bingmei M. Fu & Zhang-Jin Zhang (2017) Borneol, a novel agent that improves central nervous system drug delivery by enhancing blood-brain barrier permeability, Drug Delivery, 24:1, 1037-1044.

potent neuroprotective activity but poor brain delivery, researchers found that borneol increased the permeability of the blood-brain barrier to elevate formononetin concentration in the brain[691], while other work has shown that borneol modulates both the blood brain barrier and the blood tumor barrier to improve the therapeutic efficacy of 5-flourouracil[692] (a chemotherapeutic agent used in the treatment of skin cancer).

Treatment for Epilepsy: Borneol has been shown to counteract neuronal damage caused by pentylenetetrazole-induced kindling in a murine model, significantly suppressing the process of epileptogenesis including the retardation of tonic seizures and other progressive changes linked to kindling[693]. In the study referenced, the authors found that borneol's effects were comparable to diazepam (the most widely known branded version of this drug is Valium), the standard antiepileptic medication. Borneol has also been shown to dose-dependently attenuate acute epileptic seizure in maximal-electroshock seizure and pentylenetetrazol-induced seizure models without obvious side-effects on motor function, while also retarding kindling-induced epileptogenesis[694].

Brain Nanoparticle Enhancer: When combined with borneol, nanoparticles meant to deliver a load of Huperzine A (a cholinesterase inhibitor sometimes used in the treatment of Alzheimer's disease) to the brains of memory-impaired rats were significantly enhanced[695], which indicates that this bicyclic monoterpene may be used to increase the brain targeting efficiency of nanoparticles.

Anti-inflammatory: In an experimental model of acute inflammation, administration of borneol greatly reduced carrageenan-induced paw edema (swelling) in laboratory mice[696]. The study authors noted that borneol also reduced neutrophil migration and cytokine release, important factors in inflammation. As a primary constituent in the essential oil of salvia officinalis L., borneol contributed to anti-inflammatory activity in human gingival fibroblasts[697] (gum tissue).

Treatment for Inflammatory Bowel Disease: As the primary constituents in the essential oil of a novel lavandula cultivar, 1,8-cineole and borneol strongly attenuated the pathological conditions of colitic mice by reducing intestinal tissue damage, decreasing infiltration of neutrophils and macrophages, and promoting the maintenance of microbiota loads while controlling pathogen loads[698]. These findings have implications for other inflammatory bowel diseases in addition to colitis, such as Crohn's disease and irritable bowel syndrome (IBS).

DNA Protective Agent: Borneol has been shown to reduce the level of exogenous oxidative DNA damage in rat hepatocytes, while significantly increasing levels of intracellular glutathione[699] (involved in regulating apoptosis).

---

[691] Li JY, Wang JX, Li QH, Li XF, Xiao JP, Li SS, Shen XP, Zhang WD, Shen YH. Natural borneol enhances the anti-cerebral ischaemia efficacy of formononetin in MCAO/R rats by promoting its delivery in the brain. J Pharm Pharmacol. 2022 Nov 4;74(11):1598-1608.

[692] Lin JF, Liu YS, Huang YC, Chi CW, Tsai CC, Tsai TH, Chen YJ. Borneol and Tetrandrine Modulate the Blood-Brain Barrier and Blood-Tumor Barrier to Improve the Therapeutic Efficacy of 5-Fluorouracil in Brain Metastasis. Integr Cancer Ther. 2022 Jan-Dec;21:15347354221077682.

[693] Tambe, Rufi, Pankaj Jain, Sachin Patil, Priya Ghumatkar, and Sadhana Sathaye. Antiepileptogenic effects of borneol in pentylenetetrazole-induced kindling in mice. Naunyn-Schmiedeberg's Archives of Pharmacology 389.5 (2016): 467-475.

[694] Wang, Y., Qiu, Xy., Liu, Jy. et al. (+)-Borneol enantiomer ameliorates epileptic seizure via decreasing the excitability of glutamatergic transmission. Acta Pharmacol Sin 44, 1600–1611 (2023).

[695] Zhang, Lin, Limei Han, Jing Qin, Weiyue Lu, and Jianxin Wang. The Use of Borneol as an Enhancer for Targeting Aprotinin-Conjugated PEG-PLGA Nanoparticles to the Brain. Pharmaceutical Research 30.10 (2013): 2560-2572.

[696] Vasconcelos RM, Leite FC, Leite JA, Rodrigues Mascarenhas S, Rodrigues LC, Piuvezam MR. Synthesis, acute toxicity and anti-inflammatory effect of bornyl salicylate, a salicylic acid derivative. Immunopharmacol Immunotoxicol. 2012 Dec;34(6):1028-38.

[697] Ehrnhofer-Ressler MM, Fricke K, Pignitter M, Walker JM, Walker J, Rychlik M, Somoza V. Identification of 1,8-cineole, borneol, camphor, and thujone as anti-inflammatory compounds in a Salvia officinalis L. infusion using human gingival fibroblasts. J Agric Food Chem. 2013 Apr 10;61(14):3451-9.

[698] J. Baker, K. Brown E. Rajendiran A. Yip D. DeCoffe C. Dai E. Molcan S. A. Chittick S. Ghosh S. Mahmoudand D., L. Gibson. Medicinal lavender modulates the enteric microbiota to protect against Citrobacter rodentium-induced colitis. AJP - Gastrointestinal and Liver Physiology 303.7 (2012)G825.

[699] Horvthov, Eva, Katarna Kozics, Annamria Srankov, ubica Hunkov, Elika Glov, Andrea evoviov, and Darina Slameov. Borneol administration protects

**Acaricidal Agent:* As a primary constituent in the essential oil of plants including rosemary, spearmint, marjoram, and basil, borneol contributed to up to 100% repellency effects against Ixodes Ricinus ticks[700]. In field trials, cloths were sprayed with the essential oil and dragged across a field in Stockholm, Sweden, yielding a 68% repellency rate.

**Antifungal Synergist:* In addition to easily crossing the blood brain barrier, borneol has been shown to cross the corneal permeation barrier of the eyes[701], which might allow much more effective treatment of diseases of the cornea, particularly fungal infections.

**Treatment for Hypertension:* Borneol has exhibited a protective action on hepatic metabolism by attenuating nitric oxide deficient hypertension via an antioxidant mechanism[702]. The compound also reduced blood pressure, sympathetic vasomotor hyperactivity, and serum oxidative stress, and improved baroreflex sensitivity in rats with renovascular hypertension[703].

**Analgesic Agent:* The hydroxyl group at position 2 of the borneol molecule contributes to inhibition of the human TRPA1 (transient receptor potential channel), leading to significant analgesic and anti-inflammatory effects[704].

**Increase Bioavailability of Turmeric:* In a recent study, borneol was shown to increase the concentration and bioavailability of curcumin (turmeric) in rats by up to twenty times[705]. This is especially important information considering the sudden resurgence of the use of turmeric in the treatment of many medical conditions.

**Chemotherapeutic Agent Enhancer:* Borneol has been shown to obviously promote the cellular uptake of piperlongumine, improving the anti-glioma properties of the chemotherapeutic agent without side effects by triggering ROS-mediated apoptosis and cycle arrest in C6 glioma cells[706]. When combined with anticancer drug cisplatin, d-borneol reduced tumor volume and weight, enhanced tumor-inhibiting effects, and alleviated cisplatin-induced damage to the liver and kidney in vivo[707], while other work by the same researchers has demonstrated that d-borneol and cisplatin co-treatment inhibited tumor growth and reduced cisplatin-induced liver and kidney toxicity in a murine xenograft model of non-small cell lung cancer[708].

---

primary rat hepatocytes against exogenous oxidative DNA damage. Mutagenesis 27.5 (2012): 581-588.

[700] Chemical Composition and Repellency of Essential Oils From Four Medicinal Plants Against Ixodes ricinus Nymphs (Acari: Ixodidae). Journal of Medical Entomology 49.5 (2012): 1067-1075.

[701] Liu, Jingjing, Shaoying Fu, Nan Wei, Yongsheng Hou, Xiaoning Zhang, and Hao Cui. The effects of combined menthol and borneol on fluconazole permeation through the cornea ex vivo. European Journal of Pharmacology 688.1 (2012): 1-5.

[702] Saravanakumar, Murugesan, Jeganathan Manivannan, Jeganathan Sivasubramanian, Thangarasu Silambarasan, Elumalai Balamurugan, and Boobalan Raja. Molecular metabolic fingerprinting approach to investigate the effects of borneol on metabolic alterations in the liver of nitric oxide deficient hypertensive rats. Molecular and Cellular Biochemistry 362.2 (2011): 203-209.

[703] Luz MS, Gadelha DDA, Andrade KJS, Travassos RA, Ribeiro JD, Carvalho-Galvão A, Cruz JC, Balarini CM, Braga VA, França-Falcão MS. Borneol reduces sympathetic vasomotor hyperactivity and restores depressed baroreflex sensitivity in rats with renovascular hypertension. Hypertens Res. 2022 May;45(5):802-813.

[704] Takaishi, Masayuki, Kunitoshi Uchida, Fumitaka Fujita, and Makoto Tominaga. Inhibitory effects of monoterpenes on human TRPA1 and the structural basis of their activity. The Journal of Physiological Sciences 64.1 (2013): 47-57.

[705] Ying Gao, Guo Chen, Xiaojiao Luan, Meijuan Zou, Hongyu Piao, Gang Cheng. Improved Oral Absorption of Poorly Soluble Curcumin via the Concomitant Use of Borneol. AAPS PharmSciTech May 2019, 20:150.

[706] Wang M, Ding L, Zhang C, Yu H, Ma X, Wang X, Zhong F, Zhang Q. Natural borneol serves as an adjuvant agent to promote the cellular uptake of piperlongumine for improving its antiglioma efficacy. Eur J Pharm Sci. 2023 Feb 1;181:106347.

[707] Li J, Yuan J, Li Y, Wang J, Xie Q, Ma R, Wang J, Ren M, Lu D, Xu Z. d-Borneol enhances cisplatin sensitivity via autophagy dependent EMT signaling and NCOA4-mediated ferritinophagy. Phytomedicine. 2022 Nov;106:154411.

[708] Li J, Yuan J, Li Y, Wang J, Gong D, Xie Q, Ma R, Wang J, Ren M, Lu D, Xu Z. d-Borneol enhances cisplatin sensitivity via p21/p27-mediated S-phase arrest and cell apoptosis in non-small cell lung cancer cells and a murine xenograft model. Cell Mol Biol Lett. 2022 Jul 26;27(1):61.

Treatment for Alzheimer's Disease: Recent research has shown that borneol could accelerate the lymphatic clearance of amyloid-Beta peptides from the brain by enhancing meningeal lymphatic drainage[709]. It is the accumulation of these peptides that drives the pathogenesis of Alzheimer's disease.

IMPLICATIONS FOR HUMAN HEALTH & NUTRITION

While it is not as easy to obtain dietary borneol as some of the other more common monoterpenes like myrcene, pinene, limonene, or linalool, there are still plenty of common foods that contain borneol. When cooking, preparing, or serving food, consider working with borneol-rich spices and herbs such as anise, basil, bay, black pepper, cinnamon, coriander, marjoram, nutmeg, oregano, rosemary, saffron, sage, or thyme.

Vegetables that contain borneol include carrots and ginger - both of which can be eaten raw or used in hundreds of different recipes, while limes and persimmon are examples of fruits that contain this monoterpene. Various preparations of valerian root may also be good sources of borneol.

Cannabis strains that feature borneol as a top twenty terpene by concentration include Angel Food Cake, Cannatonic, Cheese, Girl Scout Cookies, Golden Haze, K-13 Haze, Lemon Skunk, Sour Diesel, and Tosh OG. These varieties can be eaten raw, juiced, blended, or processed into oil or other extractions that preserve terpenes and flavonoids.

Essential oils of other plants containing borneol like the ones mentioned previously in this chapter are also potential sources of the compound, depending on the manufacturer and application. Borneol is also available in raw form, and as part of the active ingredients of many over-the-counter remedies and supplements. Extreme care should be exercised when working with purified borneol or essential oils made from the "Occurrences in Plants" section above, as these compounds may be toxic in some situations.

---

[709] Wu Y, Zhang T, Li X, Wei Y, Li X, Wang S, Liu J, Li D, Wang S, Ye T. Borneol-driven meningeal lymphatic drainage clears amyloid-β peptide to attenuate Alzheimer-like phenotype in mice. Theranostics. 2023 Jan 1;13(1):106-124.

Borneol Review

Answer the following questions to test your knowledge of this terpene:

Question #1: What type of terpene is borneol?

a. Acyclic monoterpene
b. Linear monoterpene
c. Bicyclic monoterpenoid
d. Monocyclic monoterpene

Question #2: Which of the following best describes the primary flavor & aroma profile of borneol?

a. Woody, carrot, spice, dry
b. Depends on the isomer
c. Sweet, fir, turpentine, coniferous
d. Camphor, spice, mint, wood
e. Citrus, fruity, dry, terpenic

Question #3: How common is borneol in cannabis?

a. Top ten
b. Top three
c. Top fifteen
d. Top twenty

Question #4: What is the chemical formula for borneol?

a. $C_{10}H_{16}0$
b. $C_{15}H_{24}0_2$
c. $C_{10}H_{18}0$
d. $C_{20}H_{32}0$

Question #5: Name two uses of borneol in plants:

1 ____ 2 ____

Question #6: Name two potential medical uses of borneol:

1 ____ 2 ____

For the answer key to all individual compound chapters please go here: http://thebigbookofterps.com/bbot-quiz-answer-key/

CARENE

Type: Bicyclic monoterpene
Chemical Formula: $C_{10}H_{16}$
Molecular Weight: 136.238 g/mol
Boiling Point: *delta-3-carene:* 170.00 °C. @ 760.00 mm Hg[710]
(+)-3-carene: 170.00 to 172.00 °C. @ 760.00 mm Hg[711]
2-carene: 167.00 to 168.00 °C. @ 760.00 mm Hg[712]
delta-2-carene: 167.00 to 168.00 °C. @ 760.00 mm Hg[713]
Flash Point: *delta-3-carene:* 115.00 °F. TCC (46.11 °C.)
(+)-3-carene: 131.00 °F. TCC (55.00 °C.)
2-carene: 101.00 °F. TCC (38.30 °C.) (estimated by TGSC)
delta-2-carene: 101.00 °F. TCC (38.00 °C.)
Melting Point: *delta-3-carene:* 23.00 to 24.00 °C. @ 760.00 mm Hg
Solubility: *delta-3-carene:* Soluble in alcohol, organic solvents, benzene, ether, slightly soluble in ethanol, water
(+)-3-carene: Soluble in alcohol, weakly soluble in water
2-carene: Soluble in alcohol, weakly soluble in water
delta-2-carene: Soluble in alcohol, weakly soluble in water
Oral LD50: 4,800 mg/kg (rat)[714]
Biological Role: Insecticidal, antimicrobial agent, antifeedant, allelopathic agent
Therapeutic Role: Antioxidant, antifungal, mosquito repellant, anthelmintic
Commercial Use: Fragrance, food additive
Occurrence in Cannabis: Top twenty
Aroma/Flavor: *delta-3-carene:* Citrus, pine, terpenic, herbal
(+)-3-carene: Sweet

Occurs in Cannabis Strains: While more and more laboratories are including standards to test for carene and its isomers, overall this compound is underrepresented in cannabis testing and likely occurs more frequently than currently reported. At this time the most commonly reported isomer of carene in cannabis testing is delta-3-carene. Carene has been quantified in the following varieties; Arjan's Ultra Haze #2, Banana Lemonade, Durban Poison, Gelato, Ghost Cookies, Ghost Train Haze, Jack Herer, Mickey Kush, Skunk #1, Super Lemon Haze, Super Silver Haze

INTRODUCTION

Carene is a bicyclic monoterpene with a wide range of biological roles in many different plants. Exhibiting an aroma and flavor profile characterized as citrus, fir, pine, turpentine, and coniferous, with lesser notes of wood and citrus, depending on the isomer, carene is found in classic cannabis strains like Durban Poison, Jack Herer, Skunk #1, and Super Lemon Haze.

Carene is found in high concentrations in the air surrounding coniferous forests and is a primary constituent inhaled during the practice of Japanese forest bathing. In one study, measurements of the ambient air in and above a Scots

---

710 The Good Scents Company Data Sheet for Delta-3-carene, from: http://www.thegoodscentscompany.com/data/rw1014471.html. Accessed September 2, 2023.

711 The Good Scents Company Data Sheet for (+)-3-carene, from: http://www.thegoodscentscompany.com/data/rw1382041.html. Accessed September 2, 2023.

712 The Good Scents Company Data Sheet for 2-carene, from: http://www.thegoodscentscompany.com/data/rw1056931.html. Accessed September 2, 2023.

713 The Good Scents Company Data Sheet for delta-2-carene, from: http://www.thegoodscentscompany.com/data/rw1483071.html. Accessed September 2, 2023.

714 Thermo Fisher Scientific Safety Data Sheet for 3-carene, from: https://www.fishersci.com/store/msds?partNumber=AC367960250&productDescription=3-CARENE+25ML&vendorId=VN00032119&countryCode=US&language=en. Accessed September 2, 2023.

pine forest in Sweden showed that 3-carene was the primary constituent at 32%, with alpha-pinene at 29%[715]. The study authors also noted that wet trees emitted more 3-carene than dry trees.

Interestingly, the use of this bicyclic monoterpene extends to the insect world. The bark beetle (dendroctonus rhizophagus) is a significant pest of pine trees, with research showing that the insect uses the emissions of 3-carene to locate its preferred host trees[716], and to locate potential mates that have already found host trees. Additionally, the soldiers of several types of termites secrete 3-carene from frontal glands, which acts as an alarm pheromone and exhibits a toxic effect on ants and other termites[717].

CHEMICAL STRUCTURE

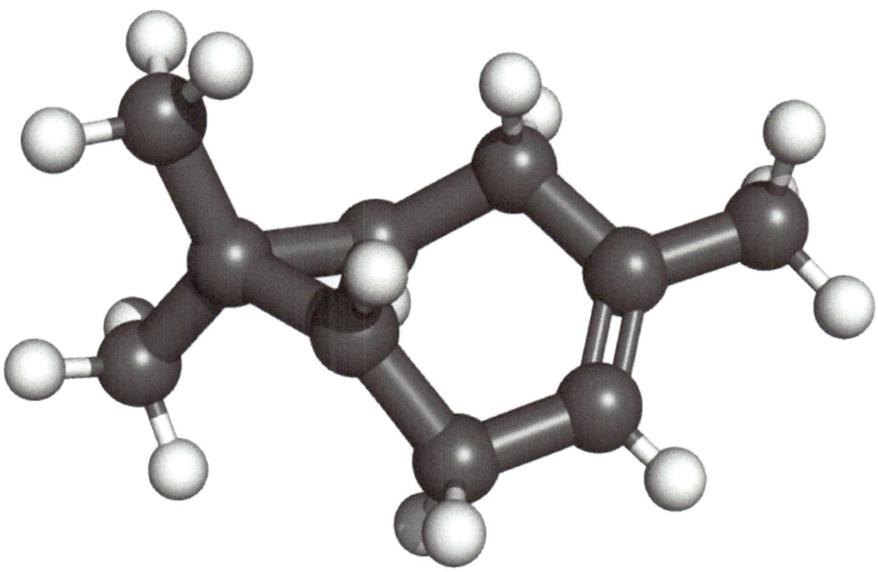

Carene is a classic monoterpene based on two linked units of isoprene derived from acetic acid and geranyl pyrophosphate. With ten carbon atoms and sixteen hydrogen atoms, the scientific notation for this compound is $C_{10}H_{16}$. Containing a cyclohexane ring and a cyclopropane ring fused together, carene is classified as a bicyclic monoterpene. Carene features one endocyclic double bond.

This compound occurs in racemate and enantio-enriched forms, and may be referred to as 3-carene, car-3-ene, and delta-3-carene. The cannabis industry primarily refers to delta-3-carene, however, there are numerous isomers of this compound, including:

delta-3-carene
(+)-3-carene
2-carene
(1R)-(-)-3-carene
acetyl carene

[715] R. Janson. Monoterpene concentrations in and above a forest of Scots pine. Journal of Atmospheric Chemistry. April 1992, Volume 14, Issue 1-4, pp 385-394.

[716] Cano-Ramirez, Claudia, Francisco Armendariz-Toledano, Jorge Macias-Samano, Brian Sullivan, and Gerardo Zuniga. Electrophysiological and Behavioral Responses of the Bark Beetle Dendroctonus rhizophagus to Volatiles from Host Pines and Conspecifics. Journal of Chemical Ecology 38.5 (2012): 512-524.

[717] Irena Valterova, Jan Vrkoc, and Tobjorn Norin. The enantiomeric composition of monoterpene hydrocarbons in the defensive secretions of Nasutitermes termites (Isoptera): Inter- and intraspecific variations. CHEMOECOLOGY. June 1993, Volume 4, Issue 2, pp 120-123.

delta-2-carene
4(5)-carene
(1S,3S)-(E)-4-carene

OCCURRENCE IN PLANTS

Carene is produced by many plants, though not nearly as many occurrences of this compound appear in the literature as other monoterpenes like pinene, myrcene, and limonene. However, with more study, it may become clear that many other plants also produce carene.

Delta-3-carene:

Abies holophylla
Angelica
Anise
Basil
Black pepper
Blood orange
Boldo
Boldus
Cajuput
Cannabis
Cassia
Cassis
Cedarwood
Cinnamon
Citronella
Costmary
Curry
Cypress
Dill
Ferula
Feverfew
Fir
Hyptis
Juniper
Laurel
Lavandin
Lavender
Lemon
Lime
Mace
Mandarin
Marjoram
Mastic
Michelia alba
Myristica fragrans
Nutmeg
Orange
Peucedanum petiolare
Phoebe
Pimento
Pine
Plectranthus glandulosus
Prangos
Ravintsara
Rosemary
Sage
Salvia
Savin
Tagete
Tansy
Thyem
Valerian
Vassoura
Wormseed
Wormwood
Yarrow
Zingiber officinale

(+)-3-carene:

Galangal
Marjoram
Pinus pumillio

2-carene:

Allspice
Anise
Bay laurel
Bell pepper
Black currant
Black pepper
Caraway
Carrot
Cashew
Cayenne
Celery
Chamomile
Cinnamon
Coriander
Cumin
Dill
Fennel
Ginger
Horsemint
Hyssop
Lemon
Lovage
Mandarin
Marjoram
Myristica fragrans
Nutmeg
Orange
Oregano
Parsley
Rosemary
Saffron
Sage
Savory
Spearmint
Tarragon
Thyme

Tomato	Turmeric	

Delta-2-carene:

Angelica	Camel grass	Cascarilla
Ferula	Lime	Mandarin
Nepeta	Protium heptaphyllum	

BIOLOGICAL ACTIVITY IN PLANTS

Carene and its isomers serve several known biological roles in plants, the most notable of which is the compound's function as an insecticidal agent. Carene also serves antibacterial, antifungal, antifeedant, and allelopathic roles in plants.

Insecticidal Agent: Some populations of Sitka spruce trees (Picea sitchensis) significantly increase production of 3-carene and primary flavonoids in response to attack by white pine weevils (Pissodes strobi, which lay their eggs inside trees like the Sitka spruce, and the resulting larvae then eat the tree from the inside until it is dead), which provides the trees with considerable resistance against the pest, compared to populations of the tree that do not increase production of 3-carene and primary flavonoids[718]. In mangoes, monoterpene volatiles - including the primary constituent 3-carene - provide strong resistance to infestations of fruit flies (anastrepha obliqua)[719]. Researchers also found that 3-carene and the other primary monoterpene volatiles in mango - limonene and a-pinene - increased the firmness of the fruit, which makes it harder for fruit flies to penetrate the flesh.

Antibacterial Agent: Growth of symbiotic bacteria from the bark beetle (D. ponderosae) was completely inhibited in the presence of alpha-pinene and delta-3-carene[720]. Monoterpenes are also well-known to affect symbiotic fungi of the same beetle species.

Antifungal Agent: The Norway spruce - an important conifer species in Europe - has been shown to increase production of 3-carene by a massive 1,200 times in response to inoculation with the wood-rotting pathogenic fungus, heterobasidion[721]. While the study referenced didn't specifically examine the antifungal properties of 3-carene against the fungi, such an enormous spike in the tree's production of the compound seems likely to be the result of a strong antifungal defense.

Antifeedant: In a study that examined herbivore browsing of juniper shrubs (Juniperus communis), a chemotype that produced significant delta-3-carene caused sheep to completely avoid the shrub, and instead feed on other juniper shrubs where the primary terpene constituent was sabinene in one variety, and alpha-pinene in another[722]. This is an important consideration for cannabis breeders because it indicates that even very closely related cultivars

[718] Hall, Dawn E., Jeanne A. Robert, Christopher I. Keeling, Dominik Domanski, Alfonso Lara Quesada, Sharon Jancsik, Michael A. Kuzyk, Britta Hamberger, Christoph H. Borchers, and Jorg Bohlmann. An integrated genomic, proteomic and biochemical analysis of (+)-3-carene biosynthesis in Sitka spruce (Picea sitchensis) genotypes that are resistant or susceptible to white pine weevil. The Plant Journal 65.6 (2011): 936-948.

[719] Vazquez-Luna, A., F. Rivera-Cabrera, L. J. Perez-Flores, and R. Diaz-Sobac. Effect of Rootstock on Mango Fruit Susceptibility to Infestation by Anastrepha obliqua. Journal of Economic Entomology 104.6 (2011): 1991-1998.

[720] Adams, Aaron, Celia Boone, Jorg Bohlmann, and Kenneth Raffa. Responses of Bark Beetle-Associated Bacteria to Host Monoterpenes and Their Relationship to Insect Life Histories. Journal of Chemical Ecology 37.8 (2011): 808-817.

[721] Danielsson, Marie et al. Chemical and transcriptional responses of Norway spruce genotypes with different susceptibility to Heterobasidion spp. infection. BMC plant biology vol. 11 154. 8 Nov. 2011, doi:10.1186/1471-2229-11-154.

[722] Marko, Gabor, Ildiko Novak, Jeno Bernath, and Vilmos Altbacker. Both Gas Chromatography and an Electronic Nose Reflect Chemical Polymorphism of Juniper Shrubs Browsed or Avoided by Sheep. Journal of Chemical Ecology 37.7 (2011): 705-713.

or strains might not have similar properties, and, ultimately, those properties are largely determined by variances in phytochemicals like terpenes and flavonoids.

**Allelopathic Agent:* Of forty-seven monoterpenes examined for allelopathic properties (properties or behavior that influences the growth or success of nearby plants), only (+)-3-carene inhibited the growth of Lactuca sativa (lettuce) seedlings, and when combined with p-cymene, inhibition rates of the plants was 87%[723].

USES IN INDUSTRY

Carene is a highly flammable colorless or light-yellow liquid used as an additive in the perfume and fragrance industries. Due to its potent and unique aroma profile, it is recommended at not more than 1% in fragrance concentrates, according to The Good Scents Company.

Carene is also used in the prepared and packaged foods industry, and can be found in baked goods, non-alcoholic beverages, alcoholic beverages, breakfast cereals, cheese, chewing gum, condiments, confectionary, fish products, frozen dairy, fruit ices, gelatins, puddings, gravy, hard candy, instant coffee, tea, jams and jellies, meat products, milk products, nut products, processed fruits, processed vegetables, seasonings and flavorings, soft candy, soups, sugar substitutes, sauces, and other food items.

Interestingly, one study found that the volatile compounds of ripened sausages in storage for several months featured delta-3-carene as the primary constituent, primarily derived from the spices in the prepared meat[724]. Delta-3-carene has also been proposed as a bait for a specialized type of trap designed to capture fruit flies (B. dorsalis)[725], a major pest of mango crops and stored mangoes.

Most commercial samples of carene and its variations appear as a colorless liquid that smells of fir and turpentine.

POTENTIAL USES IN MEDICINE

The medical potential of carene is limited by several compounding factors. For instance, inhalation of carene causes headaches, confusion, and respiratory distress, while ingestion may injure kidneys and lungs, according to the United States Coast Guard (1999). Extreme care should be exercised when handling and formulating with this flammable bicyclic monoterpene. Despite the toxic nature of carene in humans, the compound still offers considerable potential for use in medicine, including as an antifungal, an antioxidant, and as a treatment for Alzheimer's, among other therapeutic uses.

**Treatment for Alzheimer's:* As a primary constituent in the essential oil of salvia fruticose (sage), and as an individual component, 3-carene contributed to and caused direct time-dependent inhibition of the enzyme butyrylcholinesterase[726], a target in the treatment of Alzheimer's disease. It has been theorized that inhibition of this enzyme can improve memory functions.

**Anthelminitic Agent:* The growth of eggs and larvae migration of Strongyloides ratti - a common gastrointestinal parasitic worm in rats - was inhibited by the essential oil of Zanthoxylum zanthoxyloides, where 3-carene was

---

[723] Vokou, Despina, Panagiota Douvli, George Blionis, and John Halley. Effects of Monoterpenoids, Acting Alone or in Pairs, on Seed Germination and Subsequent Seedling Growth. Journal of Chemical Ecology 29.10 (2004): 2281-2301.

[724] Summo, Carmine, Francesco Caponio, Francesco Tricarico, Antonella Pasqualone, and Tommaso Gomes. Evolution of the volatile compounds of ripened sausages as a function of both storage time and composition of packaging atmosphere. Meat Science 86.3 (2010): 839-844.

[725] Kamala Jayanthi, Pagadala, Christine Woodcock, John Caulfield, Michael Birkett, and Toby Bruce. Isolation and Identification of Host Cues from Mango, Mangifera indica, That Attract Gravid Female Oriental Fruit fly, Bactrocera dorsalis. Journal of Chemical Ecology 38.4 (2012): 361-369.

[726] Savelev, Sergey U., Edward J. Okello, and Elaine K. Perry. Butyryl- and acetyl-cholinesterase inhibitory activities in essential oils of Salvia species and their constituents. Phytotherapy Research 18.4 (2004): 315-324.

among the primary constituents[727]. The inhibition rate was 7 times higher than that of thiabendazole, a common parasiticide, and the authors concluded that one possible way of treating helminth infections in populations of animals or people is to include the essential oil or its components as an additive in feedstock and food.

**Antifungal Agent:* In a study that examined the antifungal activities of several monoterpenoids, 3-carene was the most potent, inhibiting the proteolytic (disintegration of proteins) activity of Candida albicans by 92%[728].

**Mosquito Repellant:* As the primary constituent in the essential oil from several parts of echinophora lamondiana, delta-3-carene contributed to mosquito-biting deterrent effects comparable to that of DEET[729] - one of the world's most popular and most effective mosquito repellant products.

**Antioxidant:* As the primary constituent in the essential oil of the mountain pine (Pinus mugo), delta-3-carene contributed to significant antioxidant effects in lipophilic (dissolving in fats or lipids) environments[730].

IMPLICATIONS FOR HUMAN HEALTH & NUTRITION

Carene is obtained easily in diets that are rich in fruits like blood oranges and other orange varieties, mandarins, lemons, and limes. When cooking, preparing, or serving food, consider working with spices and herbs like anise, basil, black pepper, cinnamon, cumin, dill, marjoram, nutmeg, rosemary, sage, and thyme, among others.

Because carene is found in hop plants, beers brewed and containing high amounts of hops may also contain this bicyclic monoterpene. Valerian root - a common natural supplement used for centuries and available at most health food stores - also contains carene. Teas based on wintergreen leaves may also contain this terpene.

Cannabis varieties that are known to produce delta-3-carene, 3-carene, or carene include Arjan's Ultra Haze #2, Durban Poison, Gelato, Ghost Train Haze, Jack Herer, Mickey Kush, Skunk #1, Super Lemon Haze, and Super Silver Haze. It is possible to acquire trace amounts of carene in the stream of smoke or vapor from cannabis, however, we recommend juicing, blending, or eating cannabis directly.

Working with isolated carenes or with essential oils made from the "Occurrences in Plants" section above should be undertaken with extreme caution, as these compounds can be toxic in some situations.

---

[727] Olounlade, P., E. Azando, M. Hounzangbe-Adote, T. Ha, E. Leroy, C. Moulis, N. Fabre, J. Magnaval, H. Hoste, and A. Valentin. In vitro anthelmintic activity of the essential oils of Zanthoxylum zanthoxyloides and New bouldia laevis against Strongyloides ratti. Parasitology Research 110.4 (2011): 1427-1433.

[728] Nikitina, L., V. Startseva, N. Artemova, L. Dorofeeva, I. Kuznetsov, S. Lisovskaya, N. Glushko, and M. Kutyreva. Synthesis and antifungal activity of monoterpenoids of the carane series. Pharmaceutical Chemistry Journal 45.11 (2012): 664-667.

[729] Ali, Abbas, Nurhayat Tabanca, Gulmira Ozek, Temel Ozek, Zeki Aytac, Ulrich R. Bernier, Natasha M. Agramonte, K. Husnu Can Baser, and Ikhlas A. Khan. Essential Oils of Echinophora lamondiana (Apiales: Umbelliferae): A Relationship Between Chemical Profile and Biting Deterrence and Larvicidal Activity Against Mosquitoes (Diptera: Culicidae). Journal of Medical Entomology 52.1 (2015): 93-100.

[730] Grassmann J, Hippeli S, Vollmann R, Elstner EF. Antioxidative properties of the essential oil from Pinus mugo. J Agric Food Chem. 2003 Dec 17;51(26):7576-82.

Carene Review

Answer the following questions to test your knowledge of this terpene:

Question #1: What type of terpene is carene?

a. Monocyclic monoterpene
b. Linear monoterpene
c. Bicyclic monoterpene
d. Acyclic monoterpene

Question #2: Which of the following best describes the primary flavor & aroma profile of carene?

a. Citrus, fruity, dry, terpenic
b. Sweet, fir, turpentine, coniferous
c. Depends on the isomer
d. Lemon, sharp, fuel, wood
e. Woody, carrot, spice, dry

Question #3: How common is carene in cannabis?

a. Top ten
b. Top three
c. Top fifteen
d. Top twenty

Question #4: What is the chemical formula for carene?

a. $C_{10}H_{16}O$
b. C15H24O2
c. $C_{10}H_{16}$
d. $C_{20}H_{32}O$

Question #5: Name two uses of carene in plants:

1 ____________________________ 2 ____________________________

Question #6: Name two potential medical uses of carene:

1 ____________________________ 2 ____________________________

For the answer key to all individual compound chapters please go here: http://thebigbookofterps.com/bbot-quiz-answer-key/

PHELLANDRENE

Type: Monocyclic monoterpene
Chemical Formula: $C_{10}H_{16}$
Molecular Weight: 136.238 g/mol
Boiling Point: *phellandrene:* 170.00 to 173.00 °C. @ 760.00 mm Hg (estimated)[731]
alpha-phellandrene: 175.00 to 176.00 °C. @ 760.00 mm Hg[732]
(S)-(+)-alpha-phellandrene: 171.00 to 172.00 °C. @ 760.00 mm Hg (estimated)[733]
(R)-(-)-alpha-phellandrene: 171.00 to 173.00 °C. @ 760.00 mm Hg[734]
beta-phellandrene: 171.00 to 174.00 °C. @ 760.00 mm Hg[735]
Flash Point: *phellandrene:* 111.00 °F. TCC (43.70 °C.) (estimated by TGSC)
alpha-phellandrene: 117.00 °F. TCC (47.22 °C.)
(S)-(+)-alpha-phellandrene: 111.00 °F. TCC (43.70 °C.) (estimated by TGSC)
(R)-(-)-alpha-phellandrene: 110.00 °F. TCC (43.33 °C.)
beta-phellandrene: 111.00 °F. TCC (43.89 °C.)
Solubility: *phellandrene:* Soluble in alcohol, insoluble in water
alpha-phellandrene: Soluble in alcohol, weakly soluble in water
(S)-(+)-alpha-phellandrene: Soluble in alcohol, weakly soluble in water
(R)-(-)-alpha-phellandrene: Soluble in alcohol, dipropylene glycol, paraffin oil
beta-phellandrene: Soluble in alcohol, weakly soluble in water
Oral LD50: 5,700 mg/kg (rat)[736]
Biological Role: Insecticidal, antifungal, antioxidant
Therapeutic Role: Anticancer agent, antibacterial, analgesic
Commercial Use: Fragrance, flavoring agent
Occurrence in Cannabis: Top twenty
Aroma/Flavor: *phellandrene:* Mint
alpha-phellandrene: Citrus, terpenic, green, woody, pepper
(S)-(+)-alpha-phellandrene: Dill
(R)-(-)-alpha-phellandrene: Terpenic, spicy
beta-phellandrene: Mint, terpenic

Occurs in Cannabis Strains: Phellandrene is a top twenty terpene in cannabis that many laboratories either don't test for, or, if they do, they don't specify an isomer. In the cannabis industry, it's likely that occurrences of phellandrene in laboratory testing indicate the racemic mixture of the alpha- and beta- isomers. These compounds have been tested as occurring in varieties of cannabis including Ace of Spades, Black Widow, Durban Poison, Jack Herer, Sage, Super Silver Haze, and Trainwreck.

---

731 The Good Scents Company Data Sheet for Phellandrene, from: http://www.thegoodscentscompany.com/data/rw1399431.html. Accessed September 3, 2023.

732 The Good Scents Company Data Sheet for Phellandrene, from: http://www.thegoodscentscompany.com/data/rw1009891.html. Accessed September 3, 2023.

733 The Good Scents Company Data Sheet for (S)-(+)-alpha-phellandrene, from: http://www.thegoodscentscompany.com/data/rw1374391.html. Accessed September 3, 2023.

734 The Good Scents Company Data Sheet for (R)-(-)-alpha-phellandrene, from: http://www.thegoodscentscompany.com/data/rw1374381.html. Accessed September 3, 2023.

735 The Good Scents Company Data Sheet for Beta-phellandrene, from: http://www.thegoodscentscompany.com/data/rw1103551.html. Accessed September 3, 2023.

736 Cayman Chemical Safety Data Sheet for Alpha-phellandrene, from: https://cdn.caymanchem.com/cdn/msds/23179m.pdf. Accessed September 3, 2023.

INTRODUCTION

Phellandrene is an acyclic monoterpene (also referred to as a monocyclic monoterpene) that is considered a secondary phytochemical in cannabis, generally occurring in the top twenty terpenes by concentration in strains that contain this compound. With a flavor and aroma profile that is primarily citrus, pepper, and terpenic, with notes of dill, wood and mint, depending on the isomer, phellandrene is found in classic cannabis strains like Durban Poison, Jack Herer, Super Silver Haze, and Trainwreck, among others.

Interestingly, Litopenaeus vannamei (white shrimp) appears to use alpha-phellandrene as an immunomodulator, with research showing the shrimp uses this monoterpene to boost its immunity, in particular against Vibrio alginolyticus[737], a gram-negative marine bacterium that can cause seafood poisoning in humans.

CHEMICAL STRUCTURE

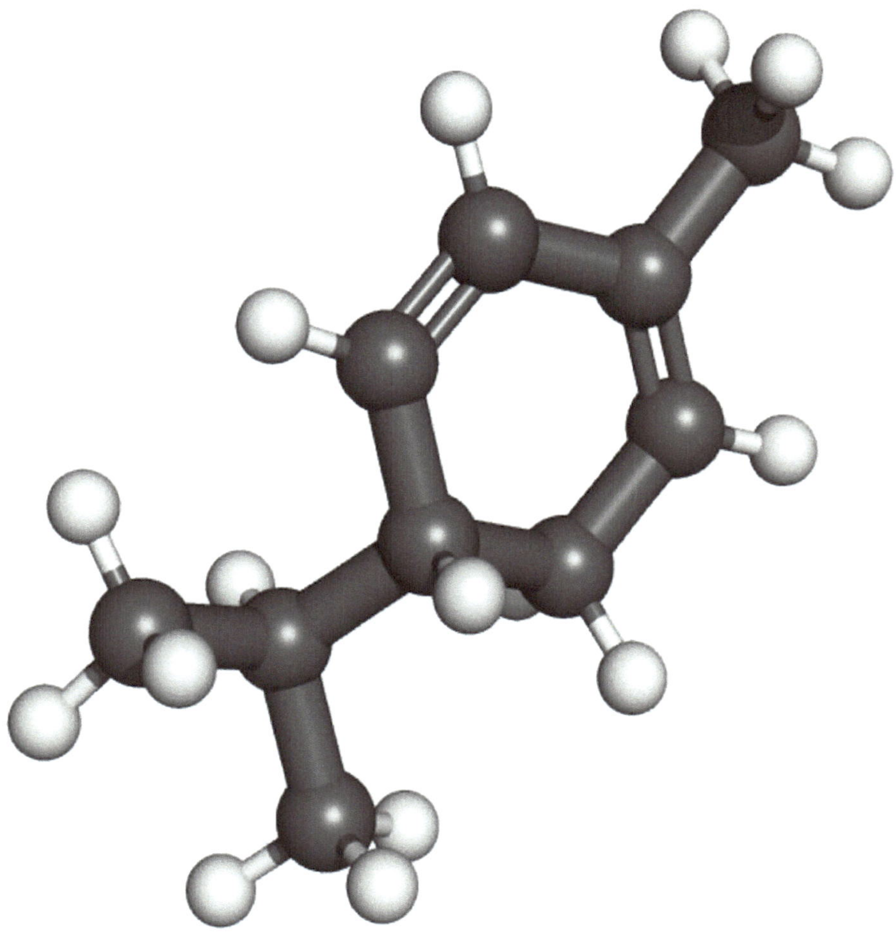

A classic monoterpene, phellandrene consists of ten carbon atoms and sixteen hydrogen atoms, notated as $C_{10}H_{16}$. Like all the monoterpenes, phellandrene is formed from geranyl pyrophosphate.

[737] Wu CC, Lin CL, Huang CY, Hsieh S, Liu CH, Hsieh SL. α-Phellandrene enhances the immune response and resistance against Vibrio alginolyticus in white shrimp (Litopenaeus vannamei). Fish Shellfish Immunol. 2019 Jan;84:1108-1114.

The molecular skeleton of this compound features one ring, with two double bonds present in the ring. The ring occurs in two variations; in one, the double bonds occur entirely within the wall of the ring, making this an endocyclic compound. In the second variation, one of the carbon atoms does not form part of the ring, making it an exocyclic compound. The endocyclic version is alpha-phellandrene, and the exocyclic version of the compound is beta-phellandrene; these are just two of several isomers of phellandrene, including all those listed in this chapter, and (-)-beta-phellandrene, among others that might not yet be reported in the literature. The image of phellandrene above is the alpha- variation.

OCCURRENCE IN PLANTS

Although named after the plant formerly known as Eucalyptus phellandra, phellandrene occurs in and can be extracted from a wide variety of plants. Interestingly, this monocyclic monoterpene is also produced by certain fungi, including the Fomes fomentarius variety of shelf or tree mushroom, which likely uses the release of volatile emissions where phellandrene is the primary constituent to attract beneficial insects[738]. Phellandrene can be found in the following plants and plant parts:

Phellandrene:

Mint
Cinnamon
Peppermint
Anise
Dill
Sage
Basil
Fennel
Tarragon

Alpha-phellandrene:

Allspice
Artemisia species
Black pepper
Cabloca
Cananga
Cardamom
Cassis
Cinamon
Cornmint
Cubeb
Cypress
Elemi
Ferula
Garlic
Grapefruit
Hyssop
Laurel
Layana
Levisticum officinale
Mace
Marjoram
Myristica fragrans
Oregano
Parsnip
Angelica
Basil
Blood orange
Cajuput
Cannabis
Carrot
Cayenne
Citronella
Cotton
Cumin
Dill
Eucalyptus species
Feverfew
Geranium
Ho
Juniper
Lavandin
Lemon
Lime
Mandarin
Mastic
Nutmeg
Papaya
Pennyroyal
Anise
Bay
Boldo
Calamus
Caraway
Cascarilla
Celery
Coriander
Croton flavens
Curry
Echinophora cinerea
Fennel
Fir
Ginger
Horsemint
Kewda
Lavender
Lemongrass
Lovage
Mango
Melaleuca
Orange
Parsley
Pepper tree

---

738 Faldt, Jenny, Mats Jonsell, Goran Nordlander, and Anna-Karin Borg-Karlson. Volatiles of Bracket Fungi Fomitopsis pinicola and Fomes fomentarius and Their Functions as Insect Attractants. Journal of Chemical Ecology 25.3 (2004): 567-590.

Phoebe
Pimenta
Pine
Plumcot
Prangos species
Protium heptaphyllum
Pteronia
Ravintsara
Ridolfia segetum
Rosemary
Safflower
Sage
Salvia
Savory
Sugandha kokila
Tagete
Tanacetum annum
Tangerine
Tansy
Tarragon
Thyme
Turmeric
Wormseed
Wormwood
Yarrow
Yuzu

(S)-(+)-alpha-phellandrene:

Cinnamon
Dill
Elemi
Fennel
Gingergrass
Tea
Turmeric

Beta-phellandrene:

Allspice
Angelica
Anise
Bay
Bergamot
Black currant
Black pepper
Buchu
Cajuput
Carrot
Cassis
Celery
Cinnamon
Coriander
Cornmint
Cotton
Cubeb
Cumin
Curry
Cypress
Dill
Echinophora cinerea
Elemi
Eucalyptus species
Fennel
Ferula
Fir
Garlic
Ginger
Grapefruit
Hyssop
Juniper
Kewda
Labdanum
Laurel
Lemon
Lemongrass
Levisticum officinale
Lime
Lovage
Mace
Mandarin
Mango
Manuka
Marjoram
Mastic
Mayur pankh
Microstrobos species
Mikania
Nectarine
Nutmeg
Orange
Oregano
Papaya
Parsley
Parsnip
Pennyroyal
Pine
Plumcot
Prangos
Rosemary
Sage
Salvia
Savory
Skimmia laureola
Sugandha kokila
Tangerine
Tarragon
Tea
Thyme
Tomato
Valerian
Wormseed
Wormwood
Zanthoxylum species
Zingiber officinale

BIOLOGICAL ACTIVITY IN PLANTS

**Insecticidal Agent:* As the primary constituent in the essential oil of schinus areira (a flowering tree and shrub in the sumac family), alpha-phellandrene contributed to contact toxicity against winged adult rose-grain aphids[739]. As part of the essential oil of cinnamon, alpha-phellandrene was also shown to contribute to contact toxicity in larvae, pupae, and adult yellow mealworm beetles[740], and achieved up to 90% mortality in contact toxicity bioassays against grain weevils[741].

**Antioxidant:* Iranian researchers found that the essential oil of thymus kotschyanus (an herb), which features alpha-phellandrene as a primary constituent, exhibited free radical scavenging activity[742], an important defense against both biotic and abiotic stress in plants.

**Antibacterial Agent:* As the primary constituent in the essential oil of machilus japonica (a flowering plant in the kauraceae family), alpha-phellandrene contributed to the strong suppression of growth of gram-positive bacteria and yeast[743].

**Insect Repellant:* Cannabis growers will be interested to know that alpha-phellandrene (as part of the essential oil of the Brazilian peppertree) has been shown to exhibit strong repellant activity against tetranychus urticae[744], the red spider mite, an opportunistic pest of cannabis.

**Antifungal Agent:* Alpha-phellandrene has been shown to alter the morphology of penicillium cyclopium (a pathogenic plant fungus) by spilling the cytoplasmic material and distorting the mycelia, increasing the permeability of cell membranes, and decreasing the pH of the pathogen, completely inhibiting the growth of the fungus in high doses[745].

USES IN INDUSTRY

The primary industrial use of phellandrene is as a fragrance additive or perfuming agent, where it is recommended at around 8-10% of fragrance composition in products like detergents, fabric softeners, soaps, toiletries, antiperspirants, and cosmetics, among others.

Beta-phellandrene is often used as a flavoring agent to add a citrusy, peppery, or minty taste to foods like baked goods, beverages, frozen dairy items, fruit ices, and hard candy.

---

[739] Sanchez Chopa, Carolina, and Lilian R Descamps. Composition and biological activity of essential oils against Metopolophium dirhodum (Hemiptera: Aphididae) cereal crop pest. Pest Management Science 68.11 (2012): 1492-1500.

[740] Martinez LC, Plata-Rueda A, Colares HC, Campos JM, Dos Santos MH, Fernandes FL, Serrao JE, Zanuncio JC. Toxic effects of two essential oils and their constituents on the mealworm beetle, Tenebrio molitor. Bull Entomol Res. 2018 Dec;108(6):716-725.

[741] Plata-Rueda, Angelica, Juliana Mendona Campos, Gabriela da Silva Rolim, Luis Carlos Martinez, Marcelo Henrique Dos Santos, Flavio Lemes Fernandes, Jose Eduardo Serrao, and Jose Cola Zanuncio. Terpenoid constituents of cinnamon and clove essential oils cause toxic effects and behavior repellency response on granary weevil, Sitophilus granarius. Ecotoxicology and Environmental Safety 156 (2018): 263-270.

[742] Asbaghian S, Shafaghat A, Zarea K, Kasimov F, Salimi F. Comparison of volatile constituents, and antioxidant and antibacterial activities of the essential oils of Thymus caucasicus, T. kotschyanus and T. vulgaris. Nat Prod Commun. 2011 Jan;6(1):137-40.

[743] Ho CL, Su YC. Composition, antioxidant and antimicrobial activities of the leaf essential oil of Machilus japonica from Taiwan. Nat Prod Commun. 2012 Jan;7(1):109-12.

[744] do Nascimento AF, da Camara CA, de Moraes MM, Ramos CS. Essential oil composition and acaricidal activity of Schinus terebinthifolius from Atlantic Forest of Pernambuco, Brazil against Tetranychus urticae. Nat Prod Commun. 2012 Jan;7(1):129-32.

[745] Zhang, Ji-hong, He-long Sun, Shao-yang Chen, Li Zeng, and Tao-tao Wang. Anti-fungal activity, mechanism studies on a-Phellandrene and Nonanal against Penicillium cyclopium. Botanical Studies 58.1 (2017): 1-9.

This acyclic monoterpene appears as a pale yellow to clear liquid, and can be extracted from a number of plant sources, most notably false fennel (ridolfia segetum), dill, cinnamon, and cloves. The alpha-phellandrene isomer has the potential to form explosive gases when it oxidizes at elevated temperatures, so great care should be used when handling isolate alpha-phellandrene.

POTENTIAL USES IN MEDICINE

Phellandrene has been shown to offer significant potential for medical use, especially as a treatment for several different types of cancer including breast, prostate, and liver cancer, as well as leukemia, and as an antibacterial agent.

**Breast & Prostate Cancer Treatment:* As a primary constituent in the essential oil of solanum erianthum (a plant in the nightshade family), alpha-phellandrene (together with alpha-terpinolene) contributed to potent inhibitory effects against Hs 578T human breast cancer cells, and PC-3 prostate tumor cells[746].

**Treatment for Liver Cancer:* In one study, alpha-phellandrene induced J5 human liver tumor cell necrosis but not apoptosis[747], with the same authors later finding that alpha-phellandrene induces cell autophagy (consumption of the body's own tissue as a metabolic process, occurring in starvation and certain diseases) of J5 liver cancer cells[748].

**Treatment for Leukemia:* Alpha-phellandrene was shown to induce morphological changes in leukemia cells resulting in decreased cell viability, while increasing the production of reactive oxygen species (ROS), leading to cell apoptosis[749].

**Analgesic:* Phellandrene offers potential as an analgesic agent after it was found that the compound produces a pronounced antinociceptive (blocking the reception of pain signals) effect in rodents, even at very low oral doses, with no toxicity[750].

**Accelerated MRSA-Infected Wound Healing:* As the primary constituent in the essential oil of dill (anethum graveolens), alpha-phellandrene contributed to the significant inhibition of bacterial growth and reduced the area of wounds in comparison to the control group, reducing the inflammatory phase and accelerating healing in MRSA-infected wounds in mice through several other mechanisms[751].

**Antibacterial Agent:* As a primary constituent in the essential oil of lantana camara (a flowering plant in the verbena family), alpha-phellandrene contributed to the inhibition of growth of Enterococcus faecalis and Staphylococcus aureus bacteria[752], and, as a primary constituent in the fruit EO of alchornea cordifolia, again was shown to exhibit potent antibacterial activity against S. aureus[753].

---

[746] Essien, E. E., I. A. Ogunwande, W. N. Setzer, and O. Ekundayo. Chemical composition, antimicrobial, and cytotoxicity studies on S. erianthum and S. macranthum essential oils. Pharmaceutical Biology 50.4 (2012): 474- 480.

[747] Hsieh SL, Li YC, Chang WC, Chung JG, Hsieh LC, Wu CC. Induction of necrosis in human liver tumor cells by a-phellandrene. Nutr Cancer. 2014;66(6):970-9.

[748] Hsieh LC, Hsieh SL, Chen CT, Chung JG, Wang JJ, Wu CC. Induction of a-phellandrene on autophagy in human liver tumor cells. Am J Chin Med. 2015;43(1):121-36.

[749] Lin, Jen-Jyh, Shu-Chun Hsu, Kung-Wen Lu, Yi-Shih Ma, Chih-Chung Wu, Hsu-Feng Lu, Jaw-Chyun Chen, Jaung-Geng Lin, Ping-Ping Wu, and Jing-Gung Chung. Alpha-phellandrene-induced apoptosis in mice leukemia WEHI-3 cells in vitro. Environmental Toxicology 31.11 (2016): 1640-1651.

[750] Lima, David F., Marcela S. Brandao, Joelma B. Moura, Joseana M. R. S. Leitao, Fernando A. A. Carvalho, Leiz, M. C. V. Miura, Jose R. S. A. Leite, Damiao P. Sousa, and Fernanda R. C. Almeida. Antinociceptive activity of the monoterpene a-phellandrene in rodents: possible mechanisms of action. Journal of Pharmacy and Pharmacology: An International Journal of Pharmaceutical Science 64.2 (2012): 283-292.

[751] Manzuoerh R, Farahpour MR, Oryan A, Sonboli A. Effectiveness of topical administration of Anethum graveolens essential oil on MRSA-infected wounds. Biomed Pharmacother. 2019 Jan; 109:1650-1658.

[752] Tesch NR, Mora F, Rojas L, Diaz T, Velasco J, Yanez C, Rios N, Carmona J, Pasquale S. Chemical composition and antibacterial activity of the essential oil

Hair Loss Prevention: Alpha-phellandrene has been shown to increase cyclic adenosine monophosphate in human papilla cells (hair follicle cells), promoting cell proliferation, and increasing growth factors[754]. This could be useful in hair loss prevention treatments, or in restoration.

IMPLICATIONS FOR HUMAN HEALTH & NUTRITION

Although phellandrene is a lesser or secondary terpene in cannabis, it is easy to acquire this compound as part of a regular, healthy diet. Consider eating fruits like blood oranges, grapefruit, lemons, limes, mandarins, mangoes, oranges, papaya, plumcots, tangerines, and yuzu. Include vegetables like bell peppers, carrots, cayenne peppers, garlic, and parsnip as snacks, as part of meals, or as garnishes and flavor enhancers. When preparing, cooking, or serving food, consider working with spices like allspice, angelica, anise, basil, bay, black pepper, caraway, cardamom, cinnamon, cumin, dill, fennel, ginger, nutmeg, oregano, parsley, rosemary, safflower, sage, and thyme, all of which are good sources of phellandrene.

Cannabis strains that have been shown to contain phellandrene include Ace of Spades, Black Widow, Durban Poison, Jack Herer, Sage, Super Silver Haze, and Trainwreck. Smoking or vaporizing these strains might allow you to consume trace amounts of this acyclic monoterpene, however, we recommend other, healthier methods of consumption. You can eat these strains raw or juice them, use oils, or you can use concentrates or other cannabis preparations that have preserved terpenes, or products with added terpenes.

When working with isolated phellandrene or essential oils made from the plants listed in the "Occurrences in Plants" section above, extreme caution should be used, as these substances can be toxic in some situations.

---

of Lantana camara var. moritziana. Nat Prod Commun. 2011 Jul;6(7):1031-4.

[753] Essien EE, Newby JS, Walker TM, Setzer WN, Ekundayo O. Characterization and Antimicrobial Activity of Volatile Constituents from Fresh Fruits of Alchornea cordifolia and Canthium subcordatum. Medicines (Basel). 2015 Dec 29;3(1). pii: E1.

[754] Kang W, Park S, Choi D, Son B, Park T. Activation of cAMP Signaling in Response to α-Phellandrene Promotes Vascular Endothelial Growth Factor Levels and Proliferation in Human Dermal Papilla Cells. Int J Mol Sci. 2022 Aug 11;23(16):8959.

Phellandrene Review

Answer the following questions to test your knowledge of this terpene:

Question #1: What type of terpene is phellandrene?

a. Monocyclic monoterpene
b. Linear monoterpene
c. Bicyclic monoterpene
d. Acyclic monoterpene

Question #2: Which of the following best describes the primary flavor and aroma profile of phellandrene?

a. Woody, carrot, spice, dry
b. Citrus, pepper, terpenic, wood
c. Lemon, sharp, fuel, wood
d. Woody, cedar, sweet, fresh
e. Depends on the isomer

Question #3: How common is phellandrene in cannabis?

a. Top ten
b. Top three
c. Top fifteen
d. Top twenty

Question #4: What is the chemical formula for phellandrene?

a. $C_{10}H_{16}0$
b. $C_{15}H_{24}0_2$
c. $C_{10}H_{16}$
d. $C_{20}H_{32}0$

Question #5: Name two uses of phellandrene in plants:

1 ________________________________ 2 ________________________________

Question #6: Name two potential medical uses of phellandrene:

1 ________________________________ 2 ________________________________

For the answer key to all individual compound chapters please go here: http://thebigbookofterps.com/bbot-quiz-answer-key/

TERPINENE

Type: Monocyclic monoterpene
Chemical Formula: $C_{10}H_{16}$
Molecular Weight: 136 238 g/mol
Boiling Point: *Alpha-terpinene:* 173.00 to 175.00 °C. @ 760.00 mm Hg[755]
Beta-terpinene: 173.00 to 174.00 °C. @ 760.00 mm Hg[756]
Gamma-terpinene: 181.00 to 183.00 °C. @ 760.00 mm Hg[757]
Flash Point: *Alpha-terpinene:* 115.00 °F. TCC (46.11 °C.)
Beta-terpinene: 115.00 °F. TCC (46.50 °C.)
Gamma-terpinene: 125.00 °F. TCC (51.67 °C.)
Melting Point: *Gamma-terpinene:* -10.00 °C. @ 760.00 mm Hg
Solubility: *Alpha-terpinene:* Soluble in alcohol, fixed oils, slightly soluble in water
Beta-terpinene: Soluble in alcohol, weakly soluble in water
Gamma-terpinene: Soluble in alcohol, fixed oils, water
Oral LD50: 3,652 mg/kg (rat)[758]
Biological Role: Antioxidant, antimicrobial, allelopathic agent, insecticidal
Therapeutic Role: Anti-inflammatory, antioxidant, antifungal, acaricidal
Commercial Use: Food additive, fragrance
Occurrence in Cannabis: Top twenty
Aroma/Flavor: *Alpha-terpinene:* Woody, citrus, terpenic, herbal
Gamma-terpinene: Terpenic, citrus, tropical, oily, lime

Occurs in Cannabis Strains: Because many laboratories and testing services in the cannabis industry do not differentiate between the isomers of terpinene, it's likely that the distribution of isomers may change accordingly as new laboratory standards are implemented. Laboratory testing that largely omits isomerization shows that terpinene occurs in cannabis varieties including Blue Dragon, Chocolate Mint OG, Dark Star, Durban Poison, Ghost Cookies, Gorilla Glue, Grape LA, Mickey Kush, Northern Lights, Purple Nepal, Sentinel, Sour Diesel, Trainwreck, and Warwick #1.

INTRODUCTION

Terpinene is a monocyclic monoterpene consisting of several isomers, with the alpha- version occurring in cannabis within the top twenty terpenes by concentration. This terpene exhibits woody, terpenic, citrus or lemony characteristics with nuances and notes that depend on the isomer. Terpinene is common in the peel of many citrus fruits and is used commercially to add scent to cleaners, solvents, glues, and other industrial products. This terpene is also used in the perfume and fragrance industries.

Terpinene is added to some foods and is known to occur in classic cannabis strains including Durban Poison, Gorilla Glue, Northern Lights, Trainwreck, and Sour Diesel. The delta- isomer of terpinene, terpinolene, occurs in

755 The Good Scents Company Data Sheet for Alpha-terpinene, from; http://www.thegoodscentscompany.com/data/rw1016031.html. Accessed September 4, 2023.

756 The Good Scents Company Data Sheet for Beta-Terpinene, from: http://www.thegoodscentscompany.com/data/rw1399141.html. Accessed September 4, 2023.

757 The Good Scents Company Data Sheet for Gamma-terpinene, from: http://www.thegoodscentscompany.com/data/rw1017031.html. Accessed September 4, 2023.

758 Inventory Multi-tiered Assessment and Prioritisation Report for Gamma-terpinene, from: https://www.industrialchemicals.gov.au/sites/default/files/Terpinene%2C%20terpinolene%20and%20phellandrene_Human%20health%20tier%20II%20assessment.pdf. Accessed September 4, 2023.

many more strains, often within the top ten terpenes by concentration. Accordingly, we have allotted an entire chapter to terpinolene in the preceding sections of this text.

CHEMICAL STRUCTURE

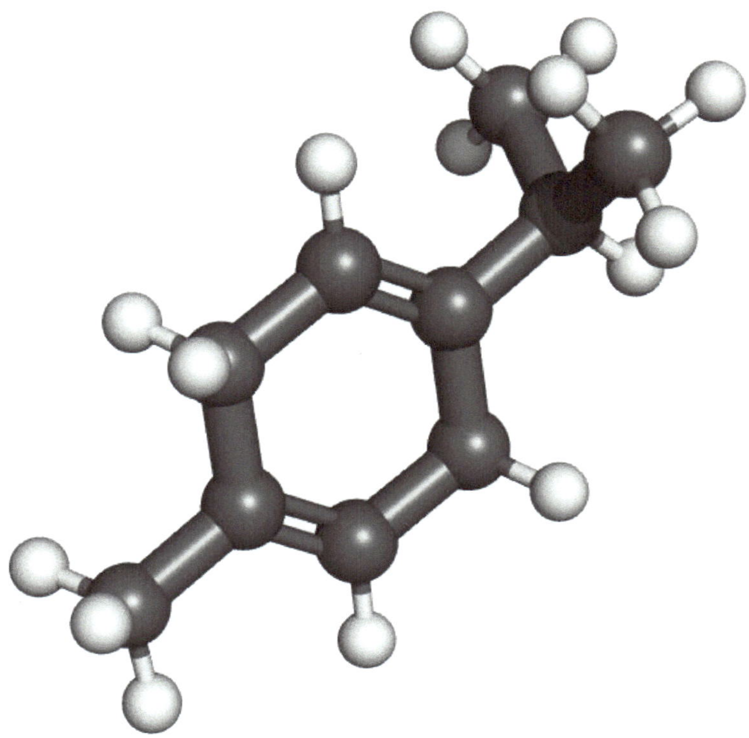

Formed from geranyl pyrophosphate in the mevalonate pathway, terpinene is a monocyclic monoterpene comprised of ten carbon atoms and sixteen hydrogen atoms arranged with one menthane ring containing two endocyclic double bonds. The four known isomers of terpinene are differentiated by the placement of the carbon double bond:

Alpha-terpinene - Double carbon bonds are located at positions 1 and 3 on the molecule's menthane skeleton.

Beta-terpinene - Double carbon bonds are located at positions 1(7) and 3 on the molecule's menthane skeleton, but it should be noted that this isomer has not been found in nature. Instead, it has been manufactured chemically from sabinene (sabinene is included as an independent chapter in this text).

Gamma-terpinene - Double carbon bonds are located at positions 1 and 4 of the molecule's menthane skeleton.

Delta-terpinene - This naturally occurring isomer of terpinene is also called terpinolene, covered in an earlier chapter as the tenth most common terpene in cannabis. The double carbon bonds in this isomer occur at positions 1 and 4(8) of the molecule's menthane skeleton.

OCCURRENCE IN PLANTS

Commonly found in the citrus peels of fruits like limes, lemons, oranges, grapefruit, and mandarins, terpinene also occurs in many other plants:

Alpha-terpinene:

Ajowan
Allspice
Angelica
Anise
Armoise
Artemisia species
Basil
Bay
Bergamot
Black currant
Black pepper
Blood orange
Boldo
Boldus
Buchu
Cabbage
Cabloca
Cajuput
Calamus
Cananga
Cannabis
Caraway
Cardamom
Carrot
Cassia
Cedarwood
Celery
Chamomile
Cinnamon
Citronella
Clary sage
Coriander
Crithmum maritimum
Cumin
Curry
Cypress
Dill
Eucalyptus species
Featherfew
Fennel
Ferula
Fir
Frankincense
Geranium
Ginger
Grapefruit
Hinoki
Horsemint
Hyptis
Hyssop
Juniper
Kewda
Labdanum
Lavandin
Layana
Lemon
Lemongrass
Levisticum
Lime
Lippia aff. gracillis
Lovage
Mace
Mandarin
Manuka
Marjoram
Mastic
Microstrobos
Mikan
Myristica fragrans
Neroli
Nutmeg
Orange
Oregano
Papaya
Parsley
Pepper tree
Peppermint
Pimenta
Pine
Pomelo
Prangos
Pteronia
Ravintsara
Rosemary
Saffron
Sage
Salvia
Santolina
Savin
Savory
Snake root
Spearmint
Sugandha kokila
Sunflower
Tagete
Tanacetum annum
Tangerine
Tansy
Tarragon
Tea tree
Thyme
Turmeric
Vassoura
Wormseed
Wormwood
Yarrow
Yuzu
Zanthoxylum bracchyacanthum

Beta-terpinene:

Artemisia campestris
Basil
Cordia curassavica
Cumin
Cypress
Myrtle
Oregano
Savory

Gamma-terpinene:

Achillea tenuifolia
Ajowan
Allspice
Amomum testaceum
Angelica
Anise

Armoise
Artemisia species
Basil
Bay
Bergamot
Betel
Black pepper
Blood orange
Buchu
Cabbage
Cajuput
Calamus
Cananga
Caraway
Cardamom
Carrot
Cascarilla
Cassia
Cassis
Cedar
Celery
Chamomile
Cinnamon
Clary sage
Coriander
Cornmint
Crithmum maritimum
Croton flavens
Cubeb
Cumin
Curcuma mangga
Curry
Cypress
Dill
Echinophora cinerea
Eucalyptus species
Featherfew
fennel
Ferula
Feverfew
Fir
Foeniculum vukgare
Frankincense
Galangal
Ginger
Glycosmis pentaphylla
Grapefruit
Guava
Heracleum species
Hinoki
Ho
Horsemint
Hypericum dogonbadanicum
Hyptis
Hyssop
Juniper
Kewda
Labdanum
Laurel
Lavandin
Lavender
Layana
Lemon
Lemongrass
Levisticum officinale
Lime
Lippia aff. gracillis
Lovage
Mace
Mandarin
Manuka
Marjoram
Mastic
Mayur pankh
Melissa
Metha longifolia
Microstrobos species
Mikan
Mint
Myristica fragrans
Myrrhinium atropurpureum
Myrtle
Nepeta
Neroli
Nutmeg
Orange
Oregano
Origanum
Papaya
Parsley
Parsnip
Peppermint
Pimenta
Pimento
Pine
Plectranthus glandulosus
Prangos
Ravintsara
Rosemary
Saffron
Sage
Salvia
Santolina
Satureja khuzistanica
Savin
Savory
Snake root
Spearmint
Start fruit
Sugandha kokila
Tagete
Tanacetum annum
Tangerine
Tansy
Tarragon
Tea tree
Thevetia peruviana
Thyme
Trachyspermum roxburghianum
Turmeric
Valerian
Vassoura
Wormseed
Wormwood
Yarrow
Yuzu
Zanthoxylum bracchycanthum

BIOLOGICAL ACTIVITY IN PLANTS

Terpinene alpha- and gamma- isomers serve numerous plant roles, including allelopathic functions not yet reported in the delta- isomer, terpinolene.

Attract Pollinators: In work that could one day lead to the development of an effective repellant against the common housefly (musca domestica), researchers discovered that the flies are attracted to the primary constituents -

delta-terpinene (terpinolene) and alpha-terpinene - of the aroma from a perennial succulent, caralluma europaea[759]. As a known pollinator of the plant, the study suggests that houseflies use the volatile emissions of C. europaea as a chemical location signal.

Nematocidal Agent: Of thirty-four terpenes tested, alpha-terpinene was among the most effective against caenorhabditis elegans[760] (roundworms that feed primarily on bacteria).

Insecticidal Agent: As a primary constituent in the essential oil of marjoram (majorana hortensis moench) leaves, terpinene exhibited significant insecticidal activity against Egyptian cotton leafworms and black bean aphids[761].

Antimicrobial Agent: As the primary constituent in the essential oil of whorled savory (satureja thymbra), gamma-terpinene showed excellent antimicrobial activity against 8 bacteria and 8 fungal species, with the whole essential oil exhibiting stronger antifungal activity than the commercial fungicide bifonazole[762].

Antioxidant: Echoing some of the findings of the research mentioned above, a study of the flavedo peel extract of unripe shiikuwasha (citrus depressa hayata) fruits showed that gamma-terpinene was almost wholly responsible for the significant antioxidant activity of the essential oil[763].

Allelopathic Agent: Wormseed (chenopodium ambrosioides L) has been shown to attack nearby maize plants via a volatile mixture consisting primarily of alpha-terpinene and cymene, which causes significant DNA damage to the target species via the root system, including increased accumulation of ROS, and apoptosis of receptor plant cells[764].

USES IN INDUSTRY

Terpinene is available commercially as a colorless or pale yellow, clear, or oily liquid with an aroma of turpentine. While this monocyclic monoterpene can be extracted from many different plants, it is generally produced industrially in the laboratory using alpha-pinene as a precursor.

Terpinene is often used by the prepared foods industry to add flavor to meat products, baked goods, soft candies, puddings, frozen dairy products, beverages, and more. This monocyclic monoterpene is also used in the fragrance and perfume industries for consumer products such as perfume, lotion, soaps, cosmetics, shampoos, toothpaste, and other personal care products.

Despite the extensive use of terpinene by the prepared foods and fragrance industries, the most common use of this phytochemical is to add scent to industrial and commercial products like paint, solvents, laundry soap, air fresheners, window cleaners, emulsifiers, flooring materials, degreasers, and many other products. As just one testament to its striking versatility, terpinene is added to both foaming agents *and* anti-foaming agents.

---

[759] Zito, Pietro, Salvatore Guarino, Ezio Peri, Maurizio Sajeva, and Stefano Colazza. Electrophysiological and behavioural responses of the housefly to "sweet" volatiles of the flowers of Caralluma europaea (Guss.) N.E. Br. Arthropod-Plant Interactions 7.5 (2013): 485-489.

[760] Abdel-Rahman FH1, Alaniz NM, Saleh MA. Nematicidal activity of terpenoids. J Environ Sci Health B. 2013;48(1):16-22.

[761] Mostafa A. Abbassy, Samir A. M. Abdelgaleil, & Rasha Y. A. Rabie. Insecticidal and synergistic effects of Majoranahortensis essential oil and some of its majorconstituents. Entomologia Experimentalis et Applicata 131(3):225 - 232 • May 2009.

[762] Giweli, Abdulhmid et al. Antimicrobial and antioxidant activities of essential oils of Satureja thymbra growing wild in Libya. Molecules (Basel, Switzerland) vol. 17,5 4836-50. 26 Apr. 2012.

[763] Asikin, Yonathan, Ikuko Taira, Sayuri Inafuku, Hidekazu Sumi, Masayoshi Sawamura, Kensaku Takara, and Koji Wada. Volatile Aroma Components and Antioxidant Activities of the Flavedo Peel Extract of Unripe Shiikuwasha (Citrus depressa Hayata). Journal of Food Science 77.4 (2012): C469-C475.

[764] Jiao Li, Yaqiang He, Danwei Ma, Bing He, Yanan Wang, Bin Chen. Volatile allelochemicals of Chenopodium ambrosioides L. induced mitochondrion-mediated Ca2+-dependent and Caspase-dependent apoptosis signaling pathways in receptor plant cells. Plant and Soil April 2018, Volume 425, Issue 1-2, pp 297-308.

Agricultural and other researchers have taken an interest in using terpinene as a food preservation agent. In one study of the antioxidant and antimicrobial attributes of carum and ferula oils, the essential oil of carum copticum - where gamma-terpinene was the primary constituent - proved to be exceptionally antibacterial when it reduced the growth of both gram-positive and gram-negative bacteria[765]. Similar research into the essential oils of eucalyptus showed that gamma-terpinene comprised an impressive 98% of the EO, which possessed remarkable antibacterial properties[766] against E. coli and S. aureus. In both studies, the authors recommended further investigation into the development of terpinene as an antimicrobial agent in packaged foods.

Interestingly, researchers developing novel methods of gamma-terpinene production for use as an alternative biofuel were successful in engineering E. coli bacteria for this purpose "by introducing a heterologous mevalonate (MVA) pathway combined with the geranyl diphosphate synthase gene and [gamma]-terpinene synthase gene[767]."

POTENTIAL USES IN MEDICINE

Potential medical applications of terpinene are promising, but public consumption of this phytochemical should not occur without addressing two important caveats that warrant further study. Most importantly, alpha-terpinene has been found to be cytotoxic to brain cells of Wistar rats, reducing cell viability, inflicting DNA damage, and altering other brain systems in a way that is likely to lead to memory loss[768]. Additionally, as the primary constituent of tea tree oil, terpinene was found to oxidize into known allergenic volatiles, with the study authors questioning the compound's use in personal care products[769]. These two studies have important implications for the cannabis industries, where hygienic products, toiletries, edibles, extracts, concentrates, and other cannabis products are all being formulated with added terpenes.

**Acaricidal Agent:* With tick-borne illnesses reaching near-epidemic levels in the United States and other parts of the world, scientists are looking at terpenes as potential acaricidal agents. In one study that examined the acaricidal activity of the essential oil of Juniperus communis leaves, alpha-terpinene was found to contribute to significant repellant activity comparable to that of DEET[770], repelling the nymphs of Lone Star and Blacklegged ticks. However, more research is needed, as the EO of J. communis was not active enough to stop adult females of the two previously mentioned species from biting.

****Treatment for Acne:*** As an active component of an extract made from linum usitatissimum, alpha-terpinene was found to be active against both aerobic and anaerobic bacteria including S. aureus, S. epidermidis, and P. acnes[771], which commonly cause facial and other acne.

---

[765] Kavoosi, Gholamreza, Asad Tafsiry, Ali Asghar Ebdam, and Vahid Rowshan. Evaluation of Antioxidant and Antimicrobial Activities of Essential Oils from Carum copticum Seed and Ferula assafoetida Latex. Journal of Food Science 78.2 (2013): T356-T361.

[766] Djenane, D., J. Yangi.ela, T. Amrouche, S. Boubrit, N. Boussad, and P. Roncales. Chemical composition and antimicrobial effects of essential oils of Eucalyptus globulus, Myrtus communis and Satureja hortensis against Escherichia coli O157:H7 and Staphylococcus aureus in minced beef. Food Science and Technology International 17.6 (2011): 505-515.

[767] Qi C, Zhao H, Li W, Li X, Xiang H, Zhang G, Liu H, Wang Q, Wang Y, Xian M, Zhang H. Production of γ-terpinene by metabolically engineered Escherichia coli using glycerol as feedstock. RSC Adv. 2018 Sep 3;8(54):30851-30859.

[768] Baldissera MD, Souza CF, Grando TH, Sagrillo MR, De Brum GF, Nascimento K Peres DS, Maciel MF, Silveira SO, Da Luz SCA, Doleski PH, Leal DBR, da Silva AS, Monteiro SG. Memory deficit, toxic effects and activity of Na(+), K(+)-ATPase and NTPDase in brain of Wistar rats submitted to orally treatment with alpha-terpinene. Environ Toxicol Pharmacol. 2016 Sep; 46:1-8.

[769] Rudback J, Bergstrom MA, Borje A, Nilsson U, Karlberg AT. a-Terpinene, an antioxidant in tea tree oil, autoxidizes rapidly to skin allergens on air exposure. Chem Res Toxicol. 2012 Mar 19;25(3):713-21.

[770] Carroll JF, Tabanca N, Kramer M, Elejalde NM, Wedge DE, Bernier UR, Coy M, Becnel JJ, Demirci B, Baser KH, Zhang J, Zhang S. Essential oils of Cupressus funebris, Juniperus communis, and J. chinensis (Cupressaceae) as repellents against ticks (Acari: Ixodidae) and mosquitoes (Diptera: Culicidae) and as toxicants against mosquitoes. J Vector Ecol. 2011 Dec; 36(2):258-68.

[771] Nand P, Drabu S, Gupta RK. Antimicrobial investigation of Linum usitatissimum for the treatment of acne. Nat Prod Commun. 2011 Nov;6(11) 1701-4.

Anti-inflammatory: In murine inflammation models that examined tumor growth and swelling in the lungs and paws, gamma-terpinene alleviated inflammation parameters such as edema and pro-inflammatory cytokine production[772], showing clear anti- inflammatory properties that have important implications for inflammation-based diseases, including cancers.

Treatment for Vaginitis: In a scientific confirmation of the use of tea tree oil to ameliorate vaginitis, the primary constituents of the essential oil of tea tree - including three isomers of terpinene - proved antibacterial, antifungal, and antiprotozoal against trichomonas vaginalis[773], a common sexually transmitted parasitic infection.

Antioxidant & Anticholinasterase: As a primary constituent in the essential oil of orignanum majorana (marjoram), gamma-terpinene contributed to significant radical scavenging and anticholinesterase activities[774], offering remarkable potential as a natural antioxidant and anticholinesterase.

Antifungal Agent: Because many strains of fungi are becoming increasingly resistant to antifungal medications, novel approaches to the treatment of fungal infections are being heavily researched. Alpha-terpinene has shown promise in this regard; the terpene strongly inhibits genes involved in ergosterol biosynthesis[775] - the same pathways targeted by drugs such as clotrimazole, fluconazole, and miconazole. Gamma-terpinene has also been shown to exhibit antifungal properties after researchers determined that the compound potently controlled infections of Sporothrix brasiliensis in cats and dogs[776].

Treatment for Surra: In livestock, infection with the parasite trypanosoma evansi (a disease called Surra) occurs frequently, and conventional medications are not always ideal. Alpha-terpinene offers a potential alternative to the chemotherapy in use today after it was shown to cure more than 57% of T. evansi infections when used in conjunction with some commercial medications, which are often curative at only 14%[777].

Treatment for Cancer Pain: In combination with beta-cyclodextrin, gamma-terpinene contributed to significant effects against sarcoma 180 cells in animals, reducing the levels of interleukin-1 beta, tumor necrosis factor alpha, and inducible nitric oxide synthase in the tumor and c-Fos protein in the spinal cord, while also reducing $Ca2+$ current (calcium ions) and presenting favorable chemical interactions with different voltage-dependent calcium channels[778].

Antileishmanial Agent: Leishmaniasis is a serious illness caused by the bite of certain sand flies, and, according to the World Health Organization, there are nearly 1 million new cases diagnosed each year, with up to 50,000 deaths. Many of the most common terpenes found in cannabis are potential antileishmanial agents, and terpinene is no exception. Researchers seeking novel treatments for leishmaniasis found that gamma-terpinene exhibited potent

---

[772] Ramalho TR, Oliveira MT, Lima AL, Bezerra-Santos CR, Piuvezam MR. Gamma-Terpinene Modulates Acute Inflammatory Response in Mice. Planta Med. 2015 Sep;81(14) 1248-54.

[773] Azimi H, Fallah-Tafti M, Karimi-Darmiyan M, Abdollahi M. A comprehensive review of vaginitis phytotherapy. Pak J Biol Sci. 2011 Nov 1;14(21) 960-6.

[774] Mossa AT, Nawwar GA. Free radical scavenging and antiacetylcholinesterase activities of Origanum majorana L. essential oil. Hum Exp Toxicol. 2011 Oct; 30(10) 1501-13.

[775] Parveen, Meher, Md. Kamrul Hasan, Junko Takahashi, Yoshinori Murata, Emiko Kitagawa, Osamu Kodama, and Hitoshi Iwahashi. Response of Saccharomyces cerevisiae to a monoterpene: evaluation of antifungal potential by DNA microarray analysis. Journal of Antimicrobial Chemotherapy 54.1 (2004): 46-55.

[776] Waller SB, Ripoll MK, Gonçalves HP, Dalla Lana DF, de Faria RO, Meireles MCA, Fuentefria AM, de Mello JRB, Cleff MB. Are γ-terpinene, 1,8-cineole, p-coumaric acid, and quercetin active against wild-type and non-wild-type Sporothrix brasiliensis to itraconazole? Braz J Microbiol. 2023 Mar;54(1):531-541.

[777] Baldissera MD, Grando TH, Souza CF, Gressler LT, Stefani LM, da Silva AS, Monteiro SG. In vitro and in vivo action of terpinen-4-ol, y-terpinene, and a-terpinene against Trypanosoma evansi. Exp Parasitol. 2016 Mar; 162:43-8.

[778] Pina LTS, Rabelo TK, Trindade GGG, Almeida IKS, Oliveira MA, Dos Santos PL, Souza DS, de Menezes-Filho JER, de Vasconcelos CML, Santos SL, Scotti L, Scotti MT, Araújo AAS, Quintans JSS, Quintans LJ, Guimarães AG. γ-Terpinene complexed with β-cyclodextrin attenuates spinal neuroactivity in animals with cancer pain by Ca2+ channel block. J Pharm Pharmacol. 2022 Nov 4;74(11):1629-1639.

activity against Leishmania major promastigotes and amastigotes, upregulating inducible nitric oxide synthase and JAK-1 (a human tyrosine kinase signaling protein) while downregulating interleukin-10 and transforming growth factor-beta while activating macrophages to kill the parasite[779].

IMPLICATIONS FOR HUMAN HEALTH & NUTRITION

Terpinene can be acquired nutritionally from many different edible plants. Consider snacking on fruits and vegetables like blood oranges, cabbage shoots, carrots, celery, grapefruit, lemons, limes, mandarins, oranges, papaya, parsnip, tangerines, sunflower, or yuzu. When cooking and preparing foods, consider working with spices like allspice, anise, basil, bay, black pepper, cardamom, cinnamon, coriander, cumin, dill, eucalyptus, fennel, ginger, marjoram, nutmeg, oregano, parsley, peppermint, rosemary, thyme, or sage.

Acquisition of terpinene by directly smoking or vaporizing cannabis products is an ineffective method of obtaining this terpene because of its tendency to autoxidize at low temperatures. However, trace amounts of terpinene can be obtained in the side stream or vapors of cannabis strains such as Blue Dragon, Chocolate Mint OG, Dark Star, Durban Poison, Ghost Cookies, Gorilla Glue, Grape LA, Mickey Kush, Northern Lights, Purple Nepal, Sentinel, Sour Diesel, Trainwreck, and Warwick #1, among others. For extracts and concentrates that do not contain terpinene, this monocyclic monoterpene can be purchased and added to these products. Consuming raw, juiced, or blended cannabis or cannabis oils may also be effective methods of obtaining terpinene.

As always, extreme caution should be taken when working with isolated terpinene or essential oils made from plants listed in the "Occurrences in Plants" section above, as these substances can be toxic in some situations.

[779] Nooshadokht M, Mirzaei M, Sharifi I, Sharifi F, Lashkari M, Amirheidari B. In silico and in vitro antileishmanial effects of gamma-terpinene: Multifunctional modes of action. Chem Biol Interact. 2022 Jul 1;361:109957.

Terpinene Review

Answer the following questions to test your knowledge of this terpene:

Question #1: What type of terpene is terpinene?

a. Monocytic monoterpene
b. Linear monoterpene
c. Mononuclear monoterpene
d. Monocyclic monoterpene

Question #2: Which of the following best describes the primary flavor & aroma profile of terpinene?

a. Depends on the isomer
b. Citrus, fruity, dry, terpenic
c. Sweet, fir, turpentine, coniferous
d. Citrus, woody, terpenic, spicy
e. Woody, carrot, spice, dry

Question #3: How common is terpinene in cannabis?

a. Top three
b. Top five
c. Top ten
d. Top twenty

Question #4: What is the chemical formula for terpinene?

a. $C_{10}H_{16}O$
b. $C_{15}H_{24}O_2$
c. $C_{10}H_{16}$
d. $C_{20}H_{32}O$

Question #5: Name two biological roles of terpinene in plants:

1 ____ 2 ____

Question #6: Name two potential medical uses of terpinene:

1 ____ 2 ____

For the answer key to all individual compound chapters please go here: http://thebigbookofterps.com/bbot-quiz-answer-key/

FARNESENE

Type: Acyclic sesquiterpene
Chemical Formula: $C_{15}H_{24}$
Molecular Weight: 204.36 g/mol
Boiling Point: *(E,E)-alpha-Farnesene:* 260.00 to 262.00 °C. @ 760.00 mm Hg[780]
(Z,E)-alpha-Farnesene: 277.00 to 279.00 °C. @ 760.00 mm Hg[781]
(E,Z)-alpha-Farnesene: 279.00 to 280.00 °C. @ 760.00 mm Hg (estimated)[782]
(Z,Z)-alpha-Farnesene: 279.00 to 280.00 °C. @ 760.00 mm Hg (estimated)[783]
Beta-farnesene: 272.00 to 273.00 °C. @ 760.00 mm Hg (estimated)[784]
cis-beta-farnesene: 272.50 °C. @ 760.00 mm Hg (estimated)[785]
trans-beta-farnesene: 206.00 °C. @ 760.00 mm Hg[786]
Flash Point: *(E,E)-alpha-Farnesene:* >212.00 °F. TCC (>100.00 °C.)
(Z,E)-alpha-Farnesene: 236.00 °F. TCC (113.20 °C.) (estimated by TGSC)
(E,Z)-alpha-Farnesene: 236.00 °F. TCC (113.20 °C.) (estimated by TGSC)
(Z,Z)-alpha-Farnesene: 236.00 °F. TCC (113.20 °C.) (estimated by TGSC)
Beta-farnesene: 229.00 °F. TCC (109.70 °C.) (estimated by TGSC)
cis-beta-farnesene: 229.00 °F. TCC (109.70 °C.) (estimated by TGSC)
trans-beta-farnesene: 230.00 °F. TCC (110.00 °C.)
Solubility: *All reported isomers:* Soluble in alcohol, insoluble in water
Oral LD50: >5000 mg/kg (rat)[787]
Biological Role: Indirect defense, insecticidal
Therapeutic Role: Anticancer agent, neuroprotective agent
Commercial Use: Flavoring agent, fragrance
Occurrence in Cannabis: Top thirty
Aroma/Flavor: *(E,E)-alpha-Farnesene:* Woody, green, vegetable, herbal, citrus
cis-beta-farnesene: Citrus, green
trans-beta-farnesene: Woody, citrus, herbal, sweet

Occurs in Cannabis Strains: Blue Dragon, Boy Scout Cookie, Eagle Scout, Hemp Oil, Tangerine Kush, White Rhino

---

[780] The Good Scents Company Data Sheet for Alpha-farnesene, from: http://www.thegoodscentscompany.com/data/rw1044061.html. Accessed September 5, 2023.

[781] The Good Scents Company Data Sheet for (Z,E)-alpha-farnesene, from: http://www.thegoodscentscompany.com/data/rw1560831.html. Accessed September 5, 2023.

[782] The Good Scents Company Data Sheet for (E,Z)-alpha-farnesene, from: http://www.thegoodscentscompany.com/data/rw1560821.html. Accessed September 5, 2023.

[783] The Good Scents Company Data Sheet for (Z,Z)-alpha-farnesene, from: http://www.thegoodscentscompany.com/data/rw1560841.html. Accessed September 5, 2023.

[784] The Good Scents Company Data Sheet for Beta-farnesene, from: http://www.thegoodscentscompany.com/data/rw1428561.html. Accessed September 5, 2023.

[785] The Good Scents Company Data Sheet for cis-beta-farnesene, from: http://www.thegoodscentscompany.com/data/rw1551951.html. Accessed September 5, 2023.

[786] The Good Scents Company Data Sheet for: trans-beta-farnesene, from: http://www.thegoodscentscompany.com/data/rw1053891.html. Accessed September 5, 2023.

[787] Amyris, Inc. Safety Data Sheet for trans-beta-farnesene, from: https://farnesene.net/wp-content/uploads/2014/03/amyris-safety-data-sheet.pdf. Accessed September 5, 2023.

INTRODUCTION

Farnesene is a sesquiterpene that occurs primarily as alpha- and beta- isomers, the latter of which is found more frequently in cannabis. With an aroma and flavor profile that is woody, green, citrus, sweet, and herbal, depending on the isomer, farnesene is unique among the terpenes. With scant cannabis-strain specific information available as of 2023, our research collected few occurrences of farnesene, most of which were the beta-isomer. Beta-farnesene occurred as the first terpene by concentration in a sample of Blue Dragon, and as the second terpene in an unidentified sample of hemp oil.

The scarcity of the reporting of farnesene in testing is most likely because testing facilities are not setting up or calibrating these standards in their laboratory equipment. In fact, many cannabis terpene testing services only offer the top 3-5 terpenes by concentration, though top labs offer around 10.

Interestingly, farnesene is an important semiochemical in the insect world. For example, this sesquiterpene is used by termites as an alarm pheromone[788], by codling moths to locate and properly disperse in habitats[789], and by fire ants as a trail pheromone to orient worker ants along a supply route[790]. As explained in the "Biological Activities in Plants" section below, insects and plants also use farnesene to interact with each other.

CHEMICAL STRUCTURE

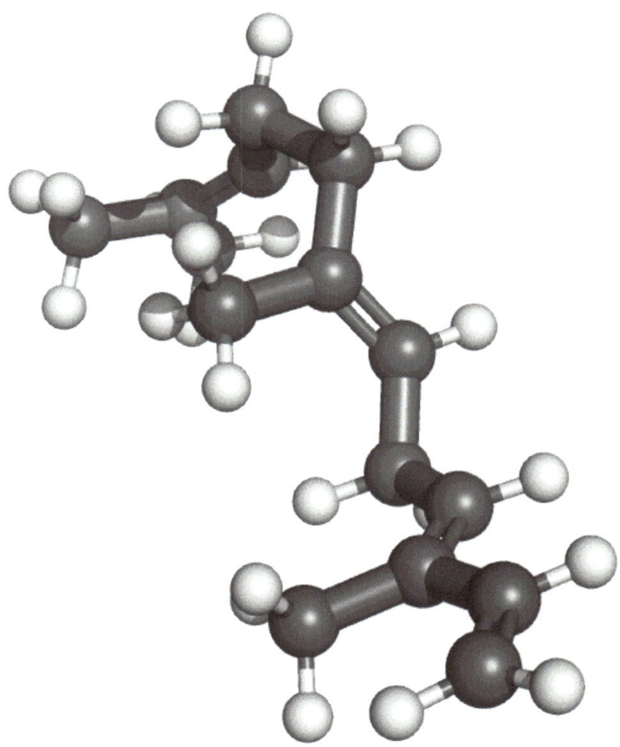

[788] Jan Sobotnik, Robert Hanus, Blanka Kalinova, Rafal Piskorski, Josef Cvacka, Thomas Bourguignon, Yves Roisin. (E,E)-a-Farnesene, an Alarm Pheromone of the Termite Prorhinotermes canalifrons. Journal of Chemical Ecology April 2008, Volume 34, Issue 4, pp 478-486.

[789] Hern, Alan, and Silvia Dorn. Sexual dimorphism in the olfactory orientation of adult Cydia pomonella in response to a-farnesene. Entomologia Experimentalis Et Applicata 92.1 (1999): 63-72.

[790] Suckling, David, Lloyd Stringer, Barry Bunn, Ashraf El-Sayed, and Robert Vander Meer. Trail Pheromone Disruption of Red Imported Fire Ant. Journal of Chemical Ecology 36.7 (2010): 744-750.

As a sesquiterpene, farnesene is comprised of three units of isoprene arranged with fifteen carbon atoms and twenty-four hydrogen atoms, scientifically notated as $C_{15}H_{24}$. Farnesene occurs as alpha- and beta- isomers which each have four double bonds, with one set of these bonds being slightly different in the beta- version. The alpha-form can occur as four stereoisomers differentiated by the arrangement of the internal double bonds, while the beta-form can occur as two stereoisomers differentiated by the arrangement of the central double bonds. Farnesene is classified as an acyclic sesquiterpene because the molecular skeleton of this terpene contains no rings, as detailed in the image above. To be clear, alpha-farnesene is (E,E)-alpha-farnesene.

OCCURRENCE IN PLANTS

(E,E)-alpha-Farnesene:

Apple
Bay
Cajuput
Cannabis
Champaca
Fig
Ginger
Guava
Lavender
Mandarin
Nutmeg
Orthodon dianthera
Persea americana
Skimmia laureola
Thyme
Yuzu
Artemisia campestris
Blood orange
Calamus
Carrot
Clary sage
Fleabane
Grape
Horsemint
Lemon
Manuka
Orange
Pear
Pine
Spearmint
Witch hazel
Basil
Cabocla
Cananga
Chamomile
Cotton
Gardenia
Grapefruit
Jasmin
Lime
Mikan
Oregano
Pepper
Roselle
Tea
Ylang ylang

(Z,E)-alpha-farnesene:

Bay
Carrot
Daphne mezereum
Mikania
Tobacco
Butterfly bush
Chu-lan tree
Ficus
Orthodon dianthera
Valerian
Cananga
Cinnamon
Laurel
Perilla

(E,Z)-alpha-farnesene:

Ambrosia artemisifolia
Eryngium caucasicum
Orthodon dianthera

(Z,Z)-alpha-Farnesene:

Potato

Beta-farnesene:

Basil
Cannabis
Cardamom

Cis-beta-farnesene:

Angelica	Anise	Basil
Bergamot	Black pepper	Carrot
Cedarwood	Chamomile	Cinnamon
Clary sage	Corn	Cumin
Elyonurus muticus	Fig	Fleabane
Geranium	Ginger	Guava
Horsemint	Lavender	Lemon
Lime	Marjoram	Nepeta betonicifolia
Oregano	Parsley	Parsnip
Peppermint	Phlomis fruticosa	Pomelo
Rosemary	Safflower	Salvia
Sideritis species	Tanacetum annum	Tansy
Water mint	Wormwood	

Trans-beta-farnesene:

Alpinia galanga	Ambrette	Anise
Artemisia	Basil	Bergamot
Blood orange	Cajuput	Calamus
Carrot	Cascarilla	Cedarwood
Cedrela	Celery	Chamomile
Costmary	Featherfew	Feverfew
Fleabane	Galangal	Ginger
Grapefruit	Guaiacwood	Herniaria icana
Juniper	Labdanum	Lavandin
Lavandula	Lavender	Lemon
Levisticum	Lime	Mandarin
Marsypianthes chamaedrys	Mentha longifolia	Nepeta racemosa
Orange	Parsley	Peppermint
Persea americana	Phlomis fruticosa	Phoebe
Pine	Plectranthus glandulosus	Prangos
Rose	Rosemary	Snake root
Thyme	Turmeric	Wormwood
Yarrow	Yuzu	

BIOLOGICAL ACTIVITY IN PLANTS

The biological role of farnesene in plants is not well understood, but it is likely that plants use this sesquiterpene for several purposes. The primary functions of farnesene appear to be for indirect defense, and to facilitate interactions with insects and other plants. Remarkably, one group of researchers that set out to determine if farnesene offered protection against abiotic stress instead concluded that if the molecule served that purpose, the activity is secondary to the primary function of the sesquiterpene, which the scientists concluded was limited to "functions in biotic interactions[791]."

[791] Palmer-Young EC, Veit D, Gershenzon J, Schuman MC (2015) The Sesquiterpenes(E)-1-Farnesene and (E)-a- Bergamotene Quench Ozone but Fail to Protect the Wild Tobacco Nicotiana attenuata from Ozone, UVB, and Drought Stresses. PLoS ONE 10(6): e0127296.

Indirect Defense: Plants have been shown to use farnesene for indirect defense. Whether these plants developed this ability as an evolutionary survival mechanism, or whether insects have evolved to detect the volatiles emitted by their prey's favorite host plants is debatable. What is known is that the release of farnesene from plants when under attack by various insects acts as a beacon or call for help, which is often answered by predatory or parasitic insects. For instance, the carnivorous mite neoseiulus womersleyi uses the emission of farnesene from plants to detect its prey, tatranychid mites[792] (farnesene is released along with other terpenes and phytochemicals by some plants when under attack by mites or other insects). Another example is the herbivorous insect Euschistus heros. This pest causes soy plants to release a blend of farnesene-rich volatiles in response to herbivory, which the parasitic wasp telenomus podisi then uses to locate its favorite prey; the eggs of E. heros[793]. Attack by the litchi (a fruit plant native to China) pest Tessaratoma papillosa also causes the release of volatile blends where farnesene is a primary constituent, which the parasitic wasp Anastatus japonicus uses to find the eggs of T. papillosa[794]. Interestingly, merely the act of egg-laying by insects causes plants to change the content and concentration of their volatile emissions. In other words, some plants can detect when an insect lays eggs on its leaves or branches, and respond with what appears to be a chemical call for help.

Insecticidal Agent: Camellia sinensis leaves emit volatiles where alpha-farnesene is a primary constituent (along with linalool, nerolidol, and other terpenes) when attacked by insects like the smaller tea tortrix (a moth), and the kanzawa spider mite[795].

Insect Repellant & Allelopathic Agent: In one of the most fascinating natural uses of terpenes, potato plants have been shown to emit substantial quantities of (E)-beta-farnesene, an isomer of farnesene that aphids emit when injured or dying to warn other aphids away[796]. This activity has appeared in both wild and domestic cultivars of potato plants[797]. This information should be of great interest to cannabis cultivators, whose crops are often besieged by aphids.

Increase Cold Tolerance: Overexpression of alpha-farnesene ((E,E)-alpha-farnesene) in Arabadopsis has been shown to lead to an increased resistance to cold weather[798], information that cannabis cultivators and geneticists might find valuable for operations in northerly climes.

USES IN INDUSTRY

Farnesene is often used in the fragrance and personal care industries, but its most common use is as a flavoring agent in a wide variety of foods including baked goods, beverages, breakfast cereal, condiments, cheese, gum, relish, oils, frozen dairy, fruit ices, gelatins & puddings, gravies, hard candy, soft candy, instant coffee, jam & jelly, milk products, nut products, processed fruits & vegetables, seasonings, spices, snack foods, and sauces.

---

[792] Ishiwari, Hayato, Takahisa Suzuki, and Taro Maeda. Essential Compounds in Herbivore-Induced Plant Volatiles that Attract the Predatory Mite Neoseiulus womersleyi. Journal of Chemical Ecology 33.9 (2007): 1670-1681.

[793] Michereff, Mirian, Raul Laumann, Miguel Borges, Miguel Michereff-Filho, Ivone Diniz, Austeclinio Farias Neto, and Maria Moraes. Volatiles Mediating a Plant-Herbivore-Natural Enemy Interaction in Resistant and Susceptible Soybean Cultivars. Journal of Chemical Ecology 37.3 (2011): 273-285.

[794] Wang Y, Chen Q, Guo J, Li J, Wang J, Wen M, Zhao H, Ren B. Molecular basis of peripheral olfactory sensing during oviposition in the behavior of the parasitic wasp Anastatus japonicus. Insect Biochem Mol Biol. 2017 Oct; 89:58-70.

[795] Dong F, Yang Z, Baldermann S, Sato Y, Asai T, Watanabe N. Herbivore-induced volatiles from tea (Camellia sinensis) plants and their involvement in intraplant communication and changes in endogenous nonvolatile metabolites. J Agric Food Chem. 2011 Dec 28;59(24):13131-5.

[796] R. W. Gibson & J. A. Pickett. Wild potato repels aphids by release of aphid alarm pheromone. Naturevolume 302, pages 608-609 (1983).

[797] Ave, Dirk A., Peter Gregory, and Ward M. Tingey. Aphid repellent sesquiterpenes in glandular trichomes of Solanum berthaultii and S. tuberosum. Entomologia Experimentalis Et Applicata 44.2 (1987).

[798] Zhang, X., Chen, X., Teixeira da Silva, J.A. et al. Characterization of sandalwood (E,E)-α-farnesene synthase whose overexpression enhances cold tolerance through jasmonic acid biosynthesis and signaling in Arabidopsis. Planta 258, 54 (2023).

Researchers have shown great interest in farnesene, in one case engineering E. coli bacteria to produce alpha-farnesene[799] in a microbial factory, and in another producing farnesene photosynthetically from the cyanobacterium anabaena sp. using only CO2, mineralized water, and light[800]. Recent work has also successfully produced beta-farnesene from engineered Yarrowia lipolytica waste lipid oil feedstock[801]. Finally, farnesene is promising agriculturally after it was shown to synergize the effects of the commercial aphid insecticide, imidacloprid[802].

Energy interests in the United States are attracted to farnesene as a potential biofuel. In fact, the California biotech company Amyris received tens of millions of dollars in grants from the US Department of Energy to develop farnesene as a biofuel[803]. Unfortunately, while the company was successful at producing farnesene from proprietary engineered yeasts, the cost to produce the sesquiterpene has been significantly more than what could be recouped in the biodiesel markets.

Commercial farnesene appears as a pale or translucent yellow liquid. This sesquiterpene can be purchased from chemical supply and other specialty companies.

POTENTIAL USES IN MEDICINE

Farnesene has potential in several areas of medicine, although this isoprenoid molecule is not nearly as well-studied medically as many of the other sesquiterpenes covered in this book. Interestingly, farnesene shares a medically valuable trait with several top twenty terpenes; it is antileishmanial. Leishmaniasis (also called Black Fever) is a disease spread by sandflies that affects millions of people. Farnesene, myrcene, bisabolol, nerolidol, ocimene, and guaiol have all been shown to possess significant antileishmanial activity; see more below and in the respective chapters for each terpene.

Antileishmanial Agent: A widespread disease sometimes causing severe or even fatal complications, Leishmaniasis treatments are highly sought after. As a primary constituent in the essential oil from the berries of croton macrostachyus (a tree native to Africa), alpha-farnesene contributed to high efficacy against promastigotes and axenic amastigotes of L. donovani and L. aethiopica[804] - two parasites that cause leishmaniasis.

Treatment for Breast Cancer: As the primary constituent in the essential oil from the leaves of cedrelopsis grevei (katrafay bush), (E)-beta-farnesene showed significant activity against MCF-7 breast cancer cells[805], as well as possessing anti-inflammatory and antioxidant properties - both important in the fight against cancer.

Defense Against Oxidative Stress: As the primary constituent in an essential oil found in millions of kitchens and restaurants worldwide - olive oil - alpha-farnesene was shown to protect against oxidative stress and prevent thyroid and ovary damage in female rats by enhancing the antioxidant defense system and promoting endocrine function[806].

---

799 Wang, Chonglong, Sang-Hwal Yoon, Hui-Jeong Jang, Young-Ryun Chung, Jae-Yean Kim, Eui-Sung Choi, and Seon-Won Kim. Metabolic engineering of Escherichia coli for a-farnesene production. Metabolic Engineering 13.6 (2011): 648-655.

800 Charles Halfmann, Liping Gu, William Gibbons, Ruanbao Zhou. Genetically engineering cyanobacteria to convert CO2, water, and light into the long-chain hydrocarbon farnesene. Applied Microbiology and Biotechnology. December 2014, Volume 98, Issue 23, pp 9869-9877.

801 Liu Y, Zhang J, Li Q, Wang Z, Cui Z, Su T, Lu X, Qi Q, Hou J. Engineering Yarrowia lipolytica for the sustainable production of β-farnesene from waste oil feedstock. Biotechnol Biofuels Bioprod. 2022 Oct 3;15(1):101.

802 Liang-LiangCui, Jie Dong, Frederic Francis, Ying-Jie Liu, Stephanie Heuskin, Georges Lognay, Ju-Lian Chen, Claude Bragard, John F.Tooker, Yong Liua. E-B-farnesene synergizes the influence of an insecticide to improve control of cabbage aphids in China. Crop Protection Volume 35, May 2012, Pages 91-96.

803 Daniel Grushkin. The Rise And Fall Of The Company That Was Going To Have Us All Using Biofuels 08-08-12 Fast Company https://www.fastcompany.com/3000040/rise-and-fall-company-have-us-all-using-biofuels.

804 Tariku Y, Hymete A, Hailu A, Rohloff J. Constituents, antileishmanial activity and toxicity profile of volatile oil from berries of Croton macrostachyus. Nat Prod Commun. 2010 Jun;5(6):975-80.

805 Afoulous S, Ferhout H, Raoelison EG, Valentin A, Moukarzel B, Couderc F, Bouajila J. Chemical composition and anticancer, antiinflammatory, antioxidant and antimalarial activities of leaves essential oil of Cedrelopsis grevei. Food Chem Toxicol. 2013 Jun;56:352-62.

**Neuroprotective Agent:* Farnesene has been shown to suppress the cytotoxicity and genotoxicity of H_2O_2-induced toxicity in rat cells while simultaneously reducing oxidative stress[807].

**Treatment for Alzheimer's Disease:* In a human neuroblastoma cell line model of Alzheimer's disease, farnesene afforded significant neuroprotection against beta-amyloid toxicity, decreasing necrotic cell death up to 3-fold[808].

**Cavity Prevention Agent:* As the primary constituent in the essential oil from the leaves of eucalyptus globules, alpha-farnesene was shown to significantly inhibit the bacteria that commonly cause dental cavities, including lactobacillus acidophilus, lactobacillus casei, staphylococcus aureus, and streptococcus mutans[809].

**Anti-inflammatory:* As a primary constituent in the essential oil of Artemisia dracunculus, farnesene inhibited Ca2+ (calcium ion channel) influx induced in human neutrophils, and human neutrophil chemotaxis, offering potential as an immunomodulatory agent of human neutrophils[810] (a type of white blood cell).

IMPLICATIONS FOR HUMAN HEALTH & NUTRITION

Farnesene is found in many different foods that are easily available to most people. To obtain dietary farnesene eat fruits such as apples, blood oranges, grapes, grapefruit, guava, lemons, limes, mandarins, oranges, pears, or yuzu. Carrots and potatoes are also potential sources of this acyclic sesquiterpene. When preparing, cooking, or serving food, consider working with spices like basil, bay leaf, chamomile, ginger, jasmine, lavender, nutmeg, oregano, rosemary, spearmint, or thyme.

One excellent way to obtain dietary farnesene is to use olive oil, which is high in this terpene. Olive oil can be added to nearly any recipe, used as a dressing, or taken directly. Tea leaves, which are also high in farnesene, offer an easy way to gain this terpene in your diet by consuming something that millions of people drink every day - hot tea.

Farnesene is known to occur in cannabis strains such as Blue Dragon, Boy Scout Cookie, Eagle Scout, Hemp Oil, Tangerine Kush, and White Rhino. Consumption of cannabis or cannabis products based on these strains may allow ingestion of trace amounts of this sesquiterpene. We recommend consuming these chemovars raw in juices or salads, or using oils or tinctures, instead of smoking or vaporizing cannabis directly.

When working with isolated farnesene or essential oils made from the "Occurrences in Plants" section above, extreme care should be used as these substances can be toxic in some situations.

---

[806] Mekircha, Fatiha, Samira Chebab, Rosita Gabbianelli, and Essaid Leghouchi. The possible ameliorative effect of Olea europaea L. oil against deltamethrin-induced oxidative stress and alterations of serum concentrations of thyroid&reproductive hormones in adult female rats. Ecotoxicology and Environmental Safety 161 (2018): 374-382.

[807] Hasan Turkez, Piera Sozio, Fatime Geyikoglu, Abdulgani Tatar, Ahmet Hacimuftuoglu, Antonio Di Stefano. Neuroprotective Effects of Farnesene Against Hydrogen Peroxide-Induced Neurotoxicity In vitro. Cellular and Molecular Neurobiology January 2014, Volume 34, Issue 1, pp 101-111.

[808] Arslan ME, Türkez H, Mardinoğlu A. In vitro neuroprotective effects of farnesene sesquiterpene on alzheimer's disease model of differentiated neuroblastoma cell line. Int J Neurosci. 2021 Aug;131(8):745-754.

[809] Ishnava, Kalpesh B et al. Anticariogenic and phytochemical evaluation of Eucalyptus globules Labill. Saudi journal of biological sciences vol. 20,1 (2013): 69-74.3.

[810] Schepetkin IA, Özek G, Özek T, Kirpotina LN, Khlebnikov AI, Klein RA, Quinn MT. Neutrophil Immunomodulatory Activity of Farnesene, a Component of Artemisia dracunculus Essential Oils. Pharmaceuticals (Basel). 2022 May 23;15(5):642.

Farnesene Review

Answer the following questions to test your knowledge of this terpene:

Question #1: What type of terpene is farnesene?

a. Monocyclic monoterpene
b. Bicyclic sesquiterpene
c. Acyclic sesquiterpene
d. Monocyclic sesquiterpene

Question #2: Which of the following best describes the primary flavor & aroma profile of (E,E)-alpha-farnesene?

a. Spicy, turpentine, dry
b. Woody, green, vegetable, herbal, citrus
c. Lemon, sharp, fuel
d. Citrus, fruity, dry, terpenic

Question #3: How common is farnesene in cannabis?

a. Top thirty
b. Top ten
c. Top twenty
d. Top forty

Question #4: What is the chemical formula for farnesene?

a. $C_{10}H_{16}$
b. $C_{15}H_{24}$
c. $C_{15}H_{32}$
d. $C_{15}H_{26}$

Question #5: Name two uses of farnesene in plants:

1 ____ 2 ____

Question #6: Name two potential medical uses of farnesene:

1 ____ 2 ____

For the answer key to all individual compound chapters please go here: http://thebigbookofterps.com/bbot-quiz-answer-key/

1,8 CINEOLE / EUCALYPTOL

Type: Bicyclic monoterpenoid
Chemical Formula: $C_{10}H_{18}0$
Molecular Weight: 154.25 g/mol
Boiling Point: 176.00 to 177.00 °C. @ 760.00 mm Hg[811]
Flash Point: 122.00 °F. TCC (50.00 °C.)
Melting Point: 1.00 to 2.00 °C. @ 760.00 mm Hg
Solubility: Soluble in ethanol, glycerin, ether, propylene glycol, fixed oils, alcohol, insoluble in water
Oral LD50: 4,300 mg/kg (rat)[812]
Biological Role: Insecticidal, allelopathic agent
Therapeutic Role: Anticancer, antiviral, neuroprotective agent, antifungal, antibacterial
Commercial Use: Flavoring agent, industrial fragrance
Occurrence in Cannabis: Top thirty
Aroma/Flavor: Eucalyptus, mint, cooling, camphoreous

Occurs in Cannabis Strains: ACDC, Ambition, Bubba Kush, Chemdawg, Covelo Sour Diesel, Girl Scout Cookies, Headband, Snow Dome, Super Silver Haze, Tower

INTRODUCTION

Eucalyptol is a bicyclic monoterpene found in a large variety of plants, including cannabis, but it most notably occurs in the eucalyptus plant, which is the primary commercial source of this compound. This terpenoid is also known as 1,8-cineole, which is how the phytochemical is referred to in the scientific community. However, in the cannabis industry this terpene is generally referred to as eucalyptol. The names are interchangeable, but for this text, we will refer to the compound by its scientific name, 1,8-cineole.

With an aroma and flavor profile that is minty, cooling, camphorous, and spicy, 1,8-cineole is found in classic cannabis strains like ACDC, Bubba Kush, Chemdawg, Girl Scout Cookies, Headband, and Super Silver Haze. In our research we also found terpene test results showing 1,8-cineole in a sample of Covelo Sour Diesel, and in two different samples of hemp oil. Aurora Cannabis in Colorado reports several of the company's strains as containing 1,8-cineole; Ambition, Snow Dome, and Tower.

Used in the food and fragrance industries to add flavor and aroma to a multitude of products, 1,8-cineole also offers many potential medical uses, including as a treatment for cancer, and as a powerful weapon in the fight against antibiotic-resistant bacteria.

---

[811] The Good Scents Company Data Sheet for 1,8-cineole, from: http://www.thegoodscentscompany.com/data/rw1056361.html. Accessed September 6, 2023.

[812] Thermo Fisher Scientific Safety Data Sheet for Cineole, from: https://www.fishersci.com/store/msds?partNumber=AC110340050&productDescription=CINEOLE+99%25+5G&vendorId=VN00032119&countryCode=US&language=en. Accessed September 6, 2023.

CHEMICAL STRUCTURE

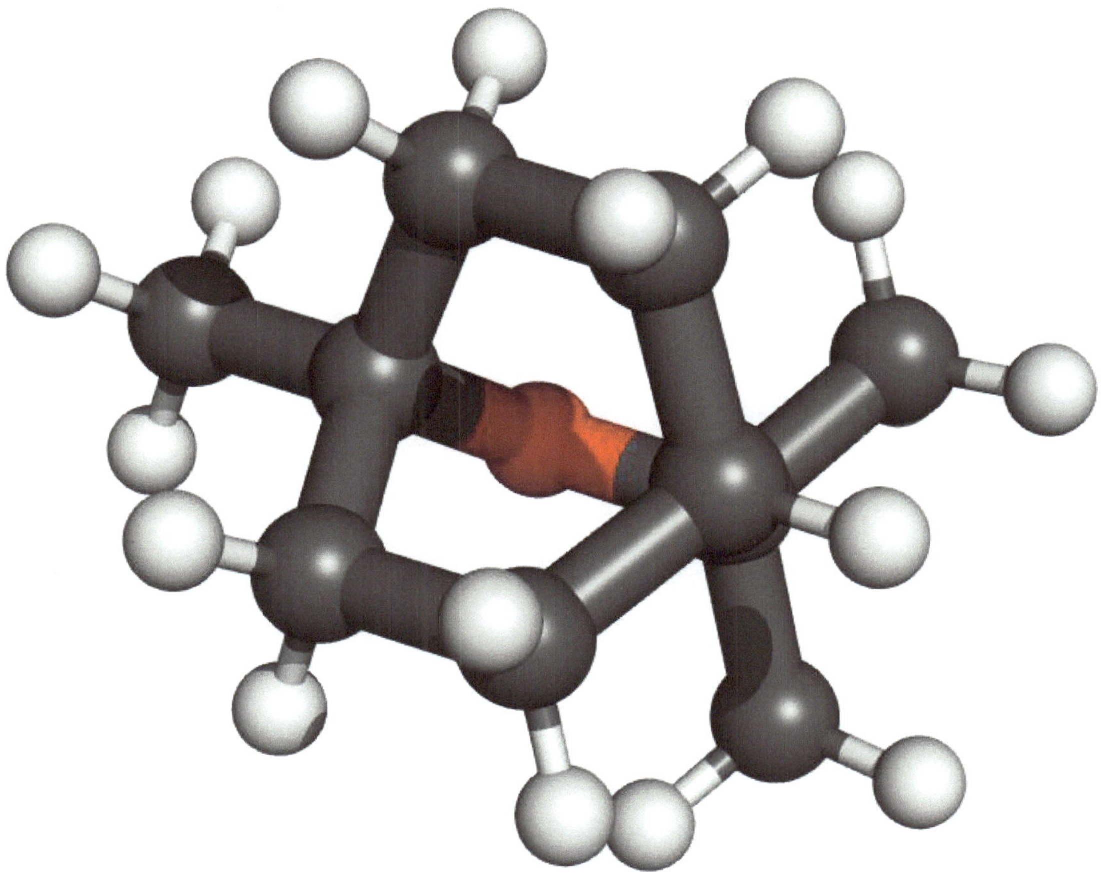

1,8-cineole is comprised of two units of isoprene containing ten carbon atoms, eighteen hydrogen atoms, and an oxygen atom, with the scientific notation of $C_{10}H_{18}O$. As an oxygenated molecule, 1,8-cineole is categorized as a terpenoid. The molecular skeleton of this compound features a central ring of oxane that contains the oxygen atom, and an additional ring which further classifies it as a bicyclic monoterpenoid.

OCCURRENCE IN PLANTS

Although widely known as the primary constituent in the essential oil of eucalyptus trees and shrubs, 1,8-cineole occurs in many different plants:

Ajowan
Anise
Basil
Blackberry
Boldo
Cajuput
Cananga
Cardamom
Cassia
Cinnamon
Corn silk
Dill
Allspice
Armoise
Bay Bilberry
Blueberry
Boldus
Calamus
Cannabis
Carrot
Cayenne
Clary sage
Cornmint
Eucalyptus species
Alpinia
Artemisia
Black currant
Bois de rose
Buchu
Camphor laurel
Caraway
Cascarilla
Chamomile
Coriander
Cumin
Featherfew

Fennel
Galangal
Geranium
Ginger
Ho leaf
Horsemint
Hyssop
Kachur
Kewda
Labdanum
Laurel
Lavandin
Lavender
Layana
Lemon
Lemongrass
Lime
Lovage
Mandarin
Marjoram
Mastic
Melissa
Mikan
Mugwort
Mustard
Myrtle
Nutmeg
Oregano
Pennyroral
Peppermint
Pimento
Pine
Pteronia
Roselle
Rosemary
Rue
Saffron
Sage
Salvia
Savin
Savory
Snake root
Spearmint
Spikenard
Tagete
Tansy
Tarragon
Tea tree
Thyme
Turmeric
Valerian
Verbena
Watermelon
Wormwood
Ylang ylang

BIOLOGICAL ACTIVITY IN PLANTS

The biological roles of 1,8-cineole in plants are not well-studied, however, it is known that this terpenoid has insecticidal and allelopathic properties.

**Insecticidal Agent:* As the primary constituent in the essential oil of Curcuma wenyujin (an herbaceous perennial plant found primarily in three provinces of China), 1,8-cineole showed contact and fumigant toxicity against the booklouse, liposcelis bostrychophila badonnel[813]. This monocyclic monoterpene has also been found to possess strong fumigant toxicity against adult German cockroaches[814] (blattella germanica). Norway Spruce trees (picea abies) engage their enemies, European Spruce bark beetles (ips typographus), in chemical warfare using 1,8-cineole as an effective weapon. When these bark beetles attack en masse, Norway Spruce trees exhibited significant increases in production and emissions of 1,8-cineole; trees that do this quickly enough and in sufficient quantity are often able to repel a mass attack of bark beetles[815].

**Allelopathic Agent:* As the primary constituent of an essential oil made from salvia, 1,8-cineole inhibited the growth of and DNA synthesis in the root system of brassica campestris[816] (field mustard). As an individual compound, 1,8-cineole severely decreased mitosis, moderately decreased root growth, and halved germination in both echinochloa crusgalli (barnyard grass), and cassia obtusifolia[817] (American sicklepod), common weeds. As the

---

[813] Zhi Long Liu, Na Na Zhao, Chun Ming Liu, Ligang Zhou, and Shu Shan Du. Identification of Insecticidal Constituents of the Essential Oil of Curcuma wenyujin Rhizomes Active against Liposcelis bostrychophila Badonnel. Molecules. 2012 Oct; 17(10): 12049-12060.

[814] Yeom HJ, Kang JS, Kim GH, Park IK. Insecticidal and acetylcholine esterase inhibition activity of Apiaceae plant essential oils and their constituents against adults of German cockroach (Blattella germanica). J Agric Food Chem. 2012 Jul 25;60(29):7194-203.

[815] Schiebe C, Hammerbacher A, Birgersson G, Witzell J, Brodelius PE, Gershenzon J, Hansson BS, Krokene P, Schlyter F. Inducibility of chemical defenses in Norway spruce bark is correlated with unsuccessful mass attacks by the spruce bark beetle. Oecologia. 2012 Sep;170(1):183-98.

[816] Koitabashi R, Suzuki T, Kawazu T, Sakai A, Kuroiwa H, Kuroiwa T. 1,8-Cineole inhibits root growth and DNA synthesis in the root apical meristem of Brassica campestris L. J Plant Res. 1997 Mar; 110(1):1-6.

[817] Romagni, Joanne, Stacy Allen, and Franck Dayan. Allelopathic Effects of Volatile Cineoles on Two Weedy Plant Species. Journal of Chemical Ecology 26.1 (2004): 303-313.

primary constituent in the essential oils of eucalyptus dundasii, 1-8-cineole inhibited the germination and seedling growth of lolium rigidum gaudin (annual ryegrass) and hordeum glaucum steud[818] (barley grass).

USES IN INDUSTRY

1,8-cineole has been produced commercially from eucalyptus for more than 150 years. This terpene appears in commercial or industrial preparations as a colorless liquid possessing a camphoreous odor. Because of its minty and somewhat spicy flavor, this terpene is used in packaged foods like candies, baked goods, and chewing gum. Cough drops, cough syrups and other cough remedies often feature this monoterpenoid to mask the flavor of more bitter ingredients in these products.

1,8-cineole is also used heavily in other industries, generally to add scent to a huge variety of products like cosmetics, arts & crafts items, toys, cigarettes, dishwashing liquids, disinfectants, floor cleaners, fragrances, laundry detergent, pharmaceutical drugs, dry cleaning products, food additives, industrial cleaners, air fresheners, personal care products, shampoo, toothpaste, mouthwash, personal lubricants, pesticides, air care products, and much more. In fact, there are over 9,000 patents for products with eucalyptol listed in PubChem (a scientific database) alone. Chances are that you already use or are familiar with products that feature 1,8-cineole, such as Febreze car vent clip air fresheners, Seventh Generation fabric softener sheets, Mrs. Meyer's Clean Day dish soap, Vaseline Intensive Care lotion, Noxzema Plus cleansing cream, Tiger Balm tiger liniment, Aussie Curing muddy revitalizer, Cover Girl classic color believable blush, Vick's Vaporub, and other products from major companies like Gillette and Olay[819].

Agricultural researchers have shown increased interest in 1,8-cineole for its insecticidal properties after it was found that the compound was effective against the stored grain pest, Tribolium confusum (confused flour beetle), increasing the lipid and carbohydrate contents in adults, decreasing the relative growth rate, relative consumption rate and efficiency of conversion of ingested food, digestive enzymes, alpha-amylase, chitinase, protease, and lipase[820].

Interestingly, 1,8-cineole has been successfully produced via photosynthesis in cyanobacteria - without supplementation of a carbon source - by engineering a 1,8-cineole synthase gene in Streptomyces clavuligerus, then introducing and overexpressing this synthase in the cyanobacterium Synechococcus elongatus[821].

POTENTIAL USES IN MEDICINE

Although there are potentially many uses for 1,8-cineole in medicine, it should be noted that just a few milliliters of preparations containing high amounts of this monocyclic monoterpene can be toxic to humans if ingested, especially essential oils of eucalyptus and cajuput. Additionally, it has been shown that when lactating women consume preparations containing 1,8-cineole, the terpene rapidly becomes present in breast milk, changing the aroma of the milk and appearing in the exhaled breath of test subjects just minutes after exposure[822]. This is important information to women who are nursing, although the appearance of this terpene in breast milk does not appear to be dangerous to babies.

---

[818] Wu, Hanwen & Zhang, Jinbiao & Stanton, Rex & An, Min & Lemerle, Deirdre. (2011). Allelopathic effects of Eucalyptus dundasii on germination and growth of ryegrass and barley grass. Allelopathy Journal. 28. 87-94.

[819] From: https://hpd.nlm.nih.gov/cgi-bin/household/brands?tbl=chem&id=560 Accessed September 29, 2019.

[820] Tine, S., Tine-Djebbar, F., Debab, A. et al. Insecticidal efficacy and physiological effects of Eucalyptus globulus essential oil and its constituent, 1,8-Cineole against Tribolium confusum (Jacquelin du Val, 1868) (Coleoptera, Tenebrionidae). J Plant Dis Prot 130, 769–780 (2023).

[821] Sakamaki Y, Ono M, Shigenari N, Chibazakura T, Shimomura K, Watanabe S. Photosynthetic 1,8-cineole production using cyanobacteria. Biosci Biotechnol Biochem. 2023 Apr 24;87(5):563-568.

[822] Kirsch F., Beauchamp J., Buettner A. Time-dependent aroma changes in breast milk after oral intake of a pharmacological preparation containing 1,8-cineole. Clin Nutr. 2012 Oct;31(5):682-92.

Smoking or vaporizing cannabis strains that are high in 1,8-cineole most likely does not pose a health danger in this regard, although using essential oils made from cannabis strains where this terpene is in the top five by concentration may be toxic, depending on several variables.

Treatment for Cancer: 1,8-cineole offers potential as a treatment for various types of cancer. For example, as one of the primary constituents in the essential oil from the leaves of Meriandra benghalensis, 1,8-cineole contributed to the growth inhibition of HT29 (human colorectal adenocarcinoma) tumor cells, inhibiting growth of the tumors by 71%[823]. In fact, patents for the tumor-regressing properties of 1,8-cineole have been filed with the US patent office, such as US Patent 20190030102, which is described as: "A method of preparing a hydrogenated cannabis oil and administering a therapeutically effective amount of the hydrogenated cannabis oil composition to the cancer patient[824]."

Antibacterial Agent (including antibiotic-resistant strains): Researchers are racing to discover or develop alternative antibiotic therapies since the prevalence and virulence of antibiotic-resistant bacteria has been rapidly increasing worldwide. 1,8-cineole is particularly interesting as a potent new antibacterial agent after it was shown to exhibit high antimicrobial activity against methicillin-resistant Staphylococcus aureus (MRSA), and vancomycin-resistant Enterococcus (VRE)[825]. Researchers only discovered this activity after testing the active components of Olbas Topfen, a common over-the-counter product based on the essential oils of eucalyptus, cajuput, and several other plants.

Prevention & Treatment of Influenza: Influenza has long been a killer, largely because of complications of the lungs caused by the virus. In mice, 1,8-cineole has been shown to attenuate pulmonary inflammation responses, alleviate lung lesions, and offer protective effects by inhibiting early recruitment of leukocytes to the lungs and suppressing inflammatory responses[826]. Additional research has echoed these findings, concluding that 1,8-cineole inhibits acute pulmonary inflammation in a murine model[827].

Antifungal Agent: As the primary constituent in the essential oil of lavandula luisieri (Spanish lavender), 1,8-cineole was found to be active against eighteen strains of yeast and filamentous fungi, including inhibiting dimorphic transition in Candida albicans at concentrations 16 times lower than the minimum inhibitory concentration (MIC)[828] , with researchers suggesting that the monoterpene may be useful in the treatment of candidosis, a fungal infection caused by C. albicans.

Antioxidant: 1,8-cineole shows potential to be used as a treatment for high cholesterol after it was demonstrated to possess lipid-lowering and anti-inflammatory properties in zebrafish, with antioxidant properties found both in vitro and in vivo[829].

---

[823] Ali NA, Wursterb M, Denkert A, Arnold N, Fadail I, Al-Didamony G, Lindequist U, Wessjohann L, Setzer WN. Chemical composition, antimicrobial, antioxidant and cytotoxic activity of essential oils of Plectranthus cylindraceus and Meriandra benghalensis from Yemen. Nat Prod Commun. 2012 Aug;7(8):1099-102.

[824] US Patent 20190030102 Hydrogenation of Cannabis Oil. Scialdone; Mark Andrew. Filed: August 30, 2018 https://pubchem.ncbi.nlm.nih.gov/patent/US2019030102.

[825] Hamoud, Razan, Frank Sporer, Jurgen Reichling, and Michael Wink. Antimicrobial activity of a traditionally used complex essential oil distillate (Olbas R Tropfen) in comparison to its individual essential oil ingredients. Phytomedicine 19.11 (2012): 969-976.

[826] Li, Yun, Yanni Lai, Yao Wang, Ni Liu, Fengxue Zhang, and Peiping Xu. 1, 8-Cineol Protect Against Influenza- Virus-Induced Pneumonia in Mice. Inflammation 39.4 (2016): 1582-1593.

[827] Zhao, Chunzhen, Jianbo Sun, Chunyan Fang, and Fadi Tang. 1,8-Cineol Attenuates LPS-Induced Acute Pulmonary Inflammation in Mice. Inflammation 37.2 (2013): 566-572.

[828] Zuzarte, M., M.J. Goncalves, M.T. Cruz, C. Cavaleiro, J. Canhoto, S. Vaz, E. Pinto, and L. Salgueiro. Lavandula luisieri essential oil as a source of antifungal drugs. Food Chemistry 135.3 (2012): 1505-1510.

[829] Cho KH. 1,8-cineole protected human lipoproteins from modification by oxidation and glycation and exhibited serum lipid-lowering and anti-inflammatory activity in zebrafish. BMB Rep. 2012 Oct; 45(10):565-70.

**Treatment for Alzheimer's Disease / Skin Penetration Enhancer:* 1,8-cineole has been used as a penetration enhancer for transdermal delivery of huperzine A and ligustrazine phosphate[830], two drugs used to treat Alzheimer's disease. 1,8-cineole also was shown to successfully reduce the mitochondrial membrane potential, ROS, and NO levels in highly toxic amyloid beta (25-35) protein fragment (implicated in memory loss) treated rat adrenal cells[831].

**Treatment for Orofacial Pain:* In a study that evaluated eucalyptol (the researchers in this case referred to the terpene by this term) as a potential antinociceptive agent, scientists found that eucalyptol significantly reduced formalin-induced nociceptive behaviors in mice by acting as a TRPV1 channel (capsaicin receptor) antagonist[832].

**Acaricidal Agent:* The scabies mite sarcoptes scabiei is a significant cause of disease in domesticated and wild animals. 1,8-cineole has been shown to exhibit toxic effects against S. scabiei by disrupting enzymatic activities that regulate the nervous system of the mites, resulting in toxicity that killed 100% of the mites within 120 seconds[833]. 1,8-cineole proved to be more effective than pyrethrin and ivermectin - two common commercial remedies for scabies.

**Mosquito Repellant:* As the primary constituent in the essential oil of hemizonia fitchii (an annual herb of western north America), 1,8-cineole was shown to be moderately repellant against adult aedes aegypti (yellow fever mosquitos), and highly repellant against ovipositioning (egg-laying) females[834].

**Gastroprotective Agent:* In a study that involved ethanol-induced gastric mucosal damage in rats, 1,8-cineole significantly attenuated the gastric injury, inhibiting gastric juice volume and total acid output[835].

**Neuroprotective Agent:* 1,8-cineole was studied in H_2O_2-induced oxidative stress in rat pheochromocytoma cells, and was shown to attenuate the loss of cell viability and changes in cell morphology by inhibiting the intracellular ROS production and markedly enhancing the expression of antioxidant enzymes, while decreasing apoptosis[836].

**Treatment for COPD:* As a potential treatment for chronic obstructive pulmonary disease, eucalyptol (as referred to by the scientists in the study cited here) shows promise after it was demonstrated to promote lung repair (at high doses) in mice with cigarette smoke induced emphysema[837].

---

830 Shi J, Cong W, Wang Y, Liu Q, Luo G. Microemulsion-based patch for transdermal delivery of huperzine A and ligustrazine phosphate in treatment of Alzheimer's disease. Drug Dev Ind Pharm. 2012 Jun;38(6):752-61.

831 Khan, Andleeb, Kumar Vaibhav, Hayate Javed, Rizwana Tabassum, Md. Ahmed, Mohd. Khan, M. Khan, Pallavi Shrivastava, Farah Islam, M. Siddiqui, M. Safhi, and Fakhrul Islam. 1,8-Cineole (Eucalyptol) Mitigates Inflammation in Amyloid Beta Toxicated PC12 Cells: Relevance to Alzheimer's Disease. Neurochemical Research 39.2 (2013): 344-352.

832 Melo Junior, Jose de, Marina de Damasceno, Sacha Santos, Talita Barbosa, Joao Araujo, Antonio Vieira-Neto, Deysi Wong, Roberto Lima-Junior, and Adriana Campos. Acute and neuropathic orofacial antinociceptive effect of eucalyptol. Inflammopharmacology 25 2 (2017): 247-254.

833 Hu, Zhiqiang, Zhenzhen Chen, Zhongqiong Yin, Renyong Jia, Xu Song, Li Li, Yuanfeng Zou, Xiaoxia Liang, Lixia Li, Changliang He, Lizi Yin, Cheng Lv, Ling Zhao, Gang Su, Gang Ye, and Fei Shi. In vitro acaricidal activity of 1,8-cineole against Sarcoptes scabiei var. cuniculi and regulating effects on enzyme activity. Parasitology Research 114.8 (2015): 2959-2967.

834 James A. KlockeMark V. DarlingtonManuel F. Balandrin. 1 ,8-Cineole (Eucalyptol), a mosquito feeding and ovipositional repellent from volatile oil of Hemizonia fitchii (Asteraceae). Journal of Chemical Ecology. December 1987, Volume 13, Issue 12, pp 2131-2141.

835 Santos, F.A., and V.S.N. Rao. 1,8-Cineol, a Food Flavoring Agent, Prevents Ethanol-Induced Gastric Injury in Rats. Digestive Diseases and Sciences 46.2 (2004): 331-337.

836 Porres-Martinez, Maria, Elena Gonzalez-Burgos, M. Emilia Carretero, and M. Pilar G6mez-Serranillos. In vitro neuroprotective potential of the monoterpenes a-pinene and 1,8-cineole against H 2 O 2 -induced oxidative stress in PC12 cells. Zeitschrift fir Naturforschung C 71 (2016).

837 Kennedy-Feitosa, Emanuel, Isabella Cattani-Cavalieri, Marina Valente Barroso, Bruna Romana-Souza, Lycia Brito-Gitirana, and Samuel Santos Valenca. Eucalyptol promotes lung repair in mice following cigarette smoke- induced emphysema. Phytomedicine 55 (2019): 70-79.

**Anti-inflammatory Agent:* As the primary constituent in the essential oil of Eucalyptus globulus, eucalyptol in combination with flurbiprofen exhibited potent anti-inflammatory, analgesic and anti-pyretic effects, likely by down-regulating expression and actions of pro-inflammatory cytokines with subsequent inhibition of cellular events[838].

**Improve Spatial Memory and Anxiety:* 1,8-cineole has been shown to significantly improve spatial memory, reduce anxiety, and reduce oxidative stress in hyperammonemic rats[839].

**Treatment for Cardiac Hypertrophy:* The combined application of 1,8-cineole and beta-caryophyllene synergistically reversed cardiac hypertrophy (thickening of the walls of the heart's left ventricle) in isoprenaline-induced H9c2 cells and mice via modulation of the P13K-AKT signaling pathway[840].

**Treatment for Inflammatory Bowel Disease:* 1,8-cineole has been shown to be a peroxisome proliferator-activated receptor-gamma (PPARγ) agonist that mediates anti-inflammatory action by stimulating colonic PPARγ protein expression and suppressing the NF-κB signaling pathway, and by activating the Nrf2/Keap1 system to exert potent antioxidant effects in HT-29 cells and in dextran sodium sulfate (DSS)-induced colitic mice[841].

IMPLICATIONS FOR HUMAN HEALTH & NUTRITION

Consumption of 1,8-cineole can be accomplished in a variety of different ways, including via the use of many consumer products, as mentioned earlier in this chapter. Actual dietary intake of 1,8-cineole is best acquired via the use and consumption of spices that are high in this bicyclic monoterpene like allspice, anise, basil, black pepper, caraway, cardamom, cinnamon, coriander, cumin, dill, fennel, ginger, mustard, nutmeg, oregano, peppermint, rosemary, sage, saffron, spearmint, tarragon, thyme, and turmeric, among others. Other options include eating black currants, blueberries, lemons, limes, mandarins, and cayenne peppers.

1,8-cineole is known to occur in cannabis strains such as ACDC, Ambition, Bubba Kush, Chemdawg, Covelo Sour Diesel, Girl Scout Cookies, Headband, Snow Dome, Super Silver Haze, and Tower. While smoking or vaping these strains will likely result in the absorption of trace amounts of this terpene, cannabis oil made from these strains might offer concentrated 1,8-cineole, depending on the method of manufacture. Essential oils of eucalyptus and cajuput contain high amounts of 1,8-cineole, but care should be exercised when consuming or using preparations of these oils or purified compounds, as they can be toxic in some situations.

---

[838] Arooj, B., Asghar, S., Saleem, M. et al. Anti-inflammatory mechanisms of eucalyptol rich Eucalyptus globulus essential oil alone and in combination with flurbiprofen. Inflammopharmacol 31, 1849–1862 (2023).

[839] Bahrami T, Yaghmaei P, Yousofvand N. The effects of Ibuprofen and 1, 8- cineol on anxiety and spatial memory in hyperammonemic rats. Metab Brain Dis. 2023 Feb;38(2):613-620.

[840] Bai C, Ma Q, Li Q, Yu L, Zhen D, Liu M, Wei C. Combination of 1,8-cineole and beta-caryophyllene synergistically reverses cardiac hypertrophy in isoprenaline-induced mice and H9c2 cells. Bioorg Chem. 2022 Jul;124:105823.

[841] Venkataraman B, Almarzooqi S, Raj V, Bhongade BA, Patil RB, Subramanian VS, Attoub S, Rizvi TA, Adrian TE, Subramanya SB. Molecular Docking Identifies 1,8-Cineole (Eucalyptol) as A Novel PPARγ Agonist That Alleviates Colon Inflammation. Int J Mol Sci. 2023 Mar 24;24(7):6160.

1,8-cineole Review

Answer the following questions to test your knowledge of this terpene:

Question #1: What type of terpene is 1,8-cineole?

a. Manocyclic monoterpene
b. Megalithic monoterpene
c. Bicyclic monoterpenoid
d. Monocyclic sesquiterpene

Question #2: Which of the following best describes the primary flavor & aroma profile of 1,8-cineole?

a. Lemon, turpentine, pine
b. Woody, green, floral, sweet
c. Eucalyptus, mint, cooling, camphoreous
d. Citrus, fruity, dry, terpenic

Question #3: How common is 1-8-cineole in cannabis?

a. Top thirty
b. Top ten
c. Top twenty
d. Top forty

Question #4: What is the chemical formula for 1-8, cineole?

a. $C_{10}H_{16}$
b. $C_{10}H_{16}O$
c. $C_{10}H_{24}O$
d. $C_{10}H_{18}O$

Question #5: Name two uses of 1,8-cineole in plants:

1 ______________________________________ 2 ____

Question #6: Name two potential medical uses of 1,8-cineole:

1 ______________________________________ 2 ____

For the answer key to all individual compound chapters, please go here: http://thebigbookofterps.com/bbot-quiz-answer-key/

ISOPULEGOL

Type: Monocyclic monoterpenoid
Chemical Formula: $C_{10}H_{18}0$
Molecular Weight: 154.25 g/mol
Boiling Point: 197.00 to 198.00 °C. @ 760.00 mm Hg (estimated)[842]
Flash Point: 173.00 °F. TCC (78.30 °C.) (estimated by TGSC)
Melting Point: 78 °C (173 °F)
Solubility: Soluble in alcohol, oils, ethanol, weakly soluble in water
Oral LD50: 936 mg/kg (rat)[843]
Biological Role: Insecticidal agent
Therapeutic Role: Analgesic, antifungal, anticonvulsant, antidepressant
Commercial Use: Flavoring agent, fragrance
Occurrence in Cannabis: Top thirty
Aroma/Flavor: Mint, cool, eucalyptus, medicinal, herbal

Occurs in Cannabis Strains: Kosher Tangie, Maui Waui, Trainwreck

INTRODUCTION

Isopulegol is a monocyclic monoterpenoid that often occurs in combination with 1,8-cineole (eucalyptol), or citronellal; at least in part because the primary source for all these terpenes is the Eucalyptus family of plants. Found in some cannabis strains in the top forty terpenes by concentration, isopulegol's primary taste and aroma features are minty, cooling, eucalyptus, medicinal, and herbal, with undercurrents of peppermint.

Terpene test results and information regarding cannabis strains that contain this monoterpenoid are rare as of the publishing of this edition. Several unsubstantiated reports of isopulegol appearing in samples of the Kosher Tangie strain were circulating the web in the fall of 2019, but no actual test results could be found or acquired through secondary sources. During research for this book, 127 randomized terpene test results were reviewed, with isopulegol occurring just three times: once as the #6 terpene by concentration in a sample of Trainwreck, once as the #9 terpene by concentration in a sample of Maui Waui, and once as the #9 terpene by concentration in an unidentified variety of hemp oil.

Like many other minor terpenes, isopulegol may appear in cannabis strains more often than is currently known, however, many laboratories have not setup or calibrated standards to detect this monoterpenoid. As cannabis becomes more widely legalized and laboratory testing services improve and become more affordable, it is likely that we will detect more occurrences of isopulegol, especially in strains where eucalyptol is prominent, such as ACDC, Bubba Kush, Chemdawg, Girl Scout Cookies, Headband, and Super Silver Haze.

Interestingly, the ant species pachycondyla marginata produces and uses isopulegol (in a chemical blend containing several other terpenes and compounds) as a chemical trail marker for other ants to follow, and to elicit emigration behavior in other ants[844] when a nest has been deemed unsafe.

---

[842] The Good Scents Company Data Sheet for Isopulegol, from: http://www.thegoodscentscompany.com/data/rw1449811.html. Accessed September 6, 2023.

[843] Carl Roth Safety Data Sheet for Isopulegol, from: https://www.carlroth.com/medias/SDB-8030-MT-EN.pdf?context=bWFzdGVyfHNlY3VyaXR5RGF0YXNoZWV0c3wyNTA0MzR8YXBwbGljYXRpb24vcGRmfHNlY3VyaXR5RGF0YXNoZWV0cy9oMWIvaDJhLzkwODg5ODM1MzE1NTAucGRmfDcyZmE1NzA1MDZhMjQyZTg0ZmFhNzE5NmMyMGU3ZGI3NzAwNjdiYWE3NzkzZjc0NDgxNWE2YmU4NThjYWJlYzc. Accessed September 6, 2023.

[844] Holldobler, B., E. Janssen, H. Bestmann, F. Kern, I. Leal, P. Oliveira, and W. Konig. Communication in the migratory termite-hunting ant Pachycondyla (= Termitopone) marginata (Formicidae, Ponerinae). Journal of Comparative Physiology A 178.1 (2004): 47-53.

CHEMICAL STRUCTURE

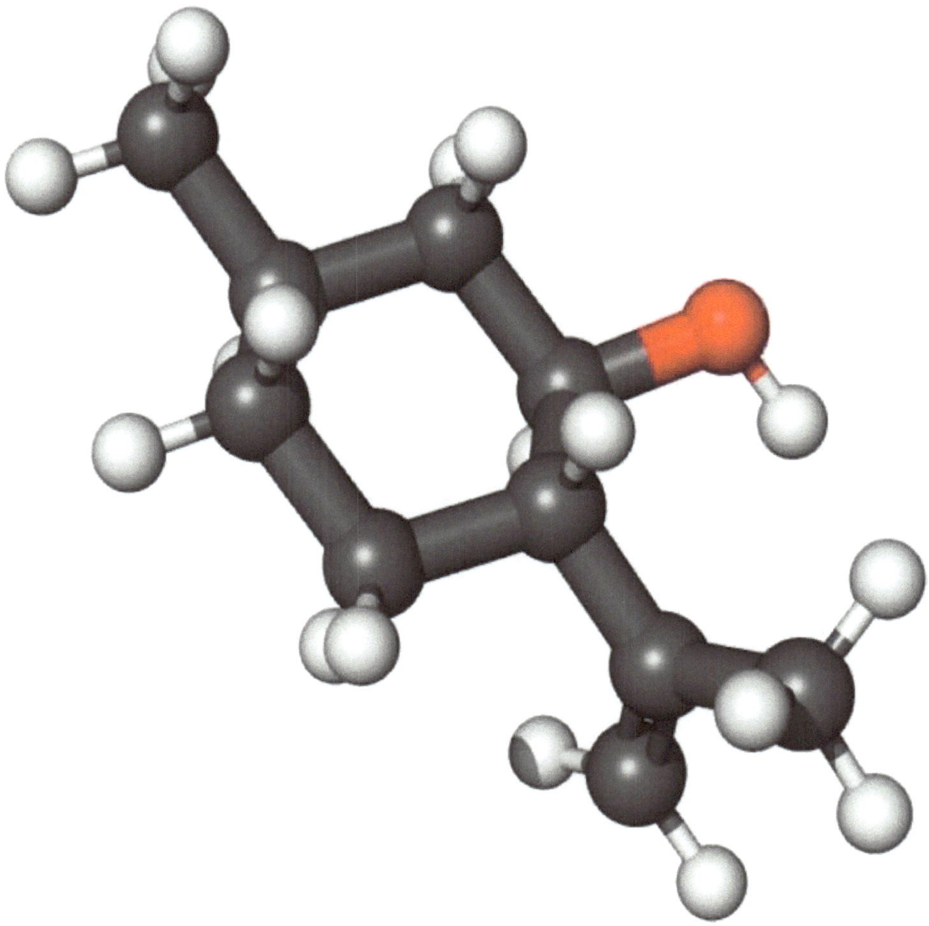

Like most of the monoterpenes, isopulegol is formed from geranyl pyrophosphate and is comprised of two units of isoprene, featuring ten carbon atoms, eighteen hydrogen atoms, and an oxygen atom, scientifically notated as $C_{10}H_{18}O$. Centered around a single ring of oxane, this phytochemical is classified as monocyclic, and because it is oxygenated, the full classification of this terpene is monocyclic monoterpenoid.

OCCURRENCE IN PLANTS

A minor terpene in cannabis, isopulegol is known to occur in a small number of other plants and plant parts, including:

Angelica
Cascarilla
Geranium
Lemongrass
Mint
Pteronia
Wormseed
Bergamot
Cornmint
Grapefruit
Lime
Parsley
Rosemary
Cannabis
Eucalyptus
Laurel
Mandarin
Pennyroyal
Sage

BIOLOGICAL ACTIVITY IN PLANTS

Scant information is available as of publication about the specific biological roles isopulegol plays in plants. However, it seems likely that this terpenoid functions in part to deter insect attack and infestation. For instance, as a primary constituent in the essential oil of Eucalyptus citriodora, isopulegol was shown to contribute to repellency effects against tribolium castaneum[845] - a common pest of stored foods. Other research has shown that the same essential oil exhibits significant insecticidal activity against maize weevils[846] (S. zeamais), and strong fumigant and contact toxicities against srosophila suzukii[847].

USES IN INDUSTRY

Industrial and commercial samples of isopulegol are generally derived from European pennyroyal, or from eucalyptus plants. This terpene is used primarily to add flavor or fragrance to food items like baked goods, beverages, chewing gum, confectionary items, frozen dairy, pudding, and both hard and soft candies. Isopulegol is also added as a fragrance agent to consumer products like disinfectants, perfumes, and cosmetics.

POTENTIAL USES IN MEDICINE

Although little is known about the biological roles of isopulegol in plants, and despite its occurrence in just a handful of cannabis strains and other plants, isopulegol has been found to offer a wide variety of potential applications in medicine; most notably in the treatment of pain, and treatment of drug-resistant tuberculosis.

**Analgesic & Antinociceptive Agent:* Compounds derived from isopulegol have a significant fast-acting and long-lasting analgesic effect in acetic acid-induced writhing in mice[848]. Direct administration of isopulegol has also been shown to exhibit significant antinociceptive behavior in mice, with researchers concluding that the action of this monoterpenoid occurs via the involvement of muscarinic receptors, and opioid and other systems[849].

**Tuberculosis Control:* In the search for novel treatments for multi- and extensively drug resistant tuberculosis, researchers discovered that several monoterpenes - isopulegol among them - exhibited airborne inhibition of tuberculosis at more than 90%[850].

**Treatment for Lead Poisoning:* As a primary constituent in the essential oil of rosemary (rosemarinus officinalis), isopulegol contributed to free radical scavenging and antioxidant effects that significantly suppressed levels of lead acetate poisoning damage in mice, preserving blood cells and hepatic and renal architecture[851].

---

[845] Olivero-Verbel, Jesus, Luz S Nerio, and Elena E Stashenko. Bioactivity against Tribolium castaneum Herbst (Coleoptera: Tenebrionidae) of Cymbopogon citratus and Eucalyptus citriodora essential oils grown in Colombia. Pest Management Science 66.6 (2010): 664-668.

[846] Ngamo Tinkeu LS, Goudoum A, Ngassoum MB, Mapongmetsem PM, Kouninki H, Hance T. Persistance of the insecticidal activity of five essential oils on the maize weevil Sitophilus zeamais (Motsch.) (Coleoptera: Curculionidae). Commun Agric Appl Biol Sci. 2004;69(3):145-7.

[847] Jang, Miyeon, Junheon Kim, Kyungjae Andrew Yoon, Si Hyeock Lee, and Chung Gyoo Park. Biological activity of Myrtaceae plant essential oils and their major components against Drosophila suzukii (Diptera: Drosophilidae). Pest Management Science 73.2 (2017): 404-409.

[848] Nazimova, Ekaterina, Alla Pavlova, Oksana Mikhalchenko, Irina Il'ina, Dina Korchagina, Tat'yana Tolstikova, Konstantin Volcho, and Nariman Salakhutdinov. Discovery of highly potent analgesic activity of isopulegol-derived (2R,4aR,7R,8aR)-4,7-dimethyl-2-(thiophen-2-yl)octahydro-2H-chromen-4-ol. Medicinal Chemistry Research 25.7 (2016): 1369-1383.

[849] Andrade Prospero, Deyna Francelia, Antonio Carlos Reis Filho, Celyane Alves Piauilino, Everton Moraes Lopes, Damiao Pergentino de Sousa, and Fernanda Regina de Castro Almeida. Effects of isopulegol in acute nociception in mice: Possible involvement of muscarinic receptors, opioid system and l-arginine/NO/cGMP pathway. Chemico- Biological Interactions 293 (2018): 55-60.

[850] Ramos Alvarenga RF, Wan B, Inui T, Franzblau SG, Pauli GF, Jaki BU. Airborne antituberculosis activity of Eucalyptus citriodora essential oil. J Nat Prod. 2014 Mar 28;77(3):603-10.

[851] Mohamed WA, Abd-Elhakim YM, Farouk SM. Protective effects of ethanolic extract of rosemary against lead- induced hepato-renal damage in rabbits. Exp Toxicol Pathol. 2016 Sep;68(8):451-61.

Transdermal Drug Delivery Enhancer: Researchers in China have synthesized saturated long-chain esters of isopulegol, which were subsequently shown to significantly increase the permeation of the high blood pressure medication, amlodipine, and the pain/arthritis medication, flurbiprofen, through transdermal patches, with researchers noting the isopulegol-derived compounds offer high efficiency, low volatility, and low skin irritation and toxicity[852].

Antidepressant & Antianxiety Treatment: Administration of isopulegol was shown to reduce anxiety and depression in mice subjected to several different stress-inducing tests, with researchers concluding that isopulegol possibly presents depressor activity on the central nervous system[853].

Treatment for Ulcers: Isopulegol has exhibited gastroprotective effects in ethanol- and indomethacin-induced gastric ulcers in mice[854] in work that may one day have human applications.

Anticonvulsant Agent: Researchers investigated the effects of isopulegol on pentylenetetrazol-induced convulsions in mice, finding that the monocyclic monoterpene prevented pentylenetetrazol-induced increase in lipid peroxidation and preserved catalase activity at normal levels, among other actions, with scientists theorizing that this activity was related to isopulegol's antioxidant properties and modulation of benzodiazepine-sensitive GABA(A) receptors[855].

Antifungal Agent: As a primary constituent in the essential oil of Eucalyptus citriodora growing in Algeria, isopulegol contributed to significant antifungal activity against microsporum canis, microsporum gypseum, trichophyton mentagrophytes, and trichophyton rubrum[856], fungi which typically infect the skin and fur of mammals.

IMPLICATIONS FOR HUMAN HEALTH & NUTRITION

Unlike many other monoterpenes, dietary isopulegol is not easy to obtain. Using essential oils of eucalyptus or pennyroyal in foods, beverages, or other consumables is the easiest method of obtaining isopulegol, however, it should be noted that these oils can be toxic at high doses.

Another method of ingestion is to drink teas made with lemongrass or mint. When cooking, preparing, or serving foods, consider working with spices like mint, parsley, rosemary, and sage. Snacking on grapefruit, mandarins, or limes may also help with isopulegol intake.

Cannabis strains that are reasonably known to contain isopulegol include Kosher Tangie, Maui Waui, and Trainwreck. Strains that are high in eucalyptol or 1,8-cineole may also be good targets to obtain isopulegol, although no studies at this time have been conducted to measure isopulegol content in the smokestream or vapor of cannabis products.

---

[852] Chen, Yang, Dongmei Cun, Peng Quan, Xiaochang Liu, Wenjia Guo, Lihua Peng, and Liang Fang. Saturated Long-Chain Esters of Isopulegol as Novel Permeation Enhancers for Transdermal Drug Delivery. Pharmaceutical Research 31.8 (2014): 1907-1918.

[853] Silva, Maria Izabel Gomes, Manuel Rufino de Aquino Neto, Paulo Florentino Teixeira Neto, Brinell Arcanjo Moura, Jeferson Falcao do Amaral, Damiao Pergentino de Sousa, Silvania Maria Mendes Vasconcelos, and Francisca Clea Florenzo de Sousa. Central nervous system activity of acute administration of isopulegol in mice. Pharmacology Biochemistry and Behavior 88.2 (2007): 141-147.

[854] Silva, Maria, Brinell Moura, Manuel Aquino Neto, Adriana Rocha Tome, Nayrton Rocha, Alyne Carvalho, Danielle Macedo, Silvania Vasconcelos, Damiao Sousa, Glauce Barros Viana, and Francisca Sousa.

[855] Silva MI, Silva MA, de Aquino Neto MR, Moura BA, de Sousa HL, de Lavor EP, de Vasconcelos PF, Macedo DS, de Sousa DP, Vasconcelos SM, de Sousa FC. Effects of isopulegol on pentylenetetrazol-induced convulsions in mice: possible involvement of GABAergic system and antioxidant activity. Fitoterapia. 2009 Dec;80(8):506-13.

[856] Tolba H, Moghrani H, Benelmouffok A, Kellou D, Maachi R. Essential oil of Algerian Eucalyptus citriodora: Chemical composition, antifungal activity. J Mycol Med. 2015 Dec;25(4):e128-33.

Isopulegol Review

Answer the following questions to test your knowledge of this terpene:

Question #1: What type of terpene is isopulegol?

a. Linear monoterpene
b. Monocyclic monoterpene
c. Monocyclic monoterpenoid
d. Monoterpenic bicycloid

Question #2: Which of the following best describes the primary flavor & aroma profile of isopulegol?

a. Citrus, sweet, tropical, spicy
b. Clove, nutmeg, wood, terpenic
c. Pine, earth, dry, woody
d. Mint, cool, eucalyptus, medicinal, herbal

Question #3: How common is isopulegol in cannabis?

a. Top thirty
b. Top five
c. Top ten
d. Top forty

Question #4: What is the chemical formula for isopulegol?

a. $C_{15}H_{26}0$
b. $C_{10}H_{18}0$
c. $C_{10}H_{16}0$
d. $C_{10}H_{22}0$

Question #5: Name two uses of isopulegol in plants:

1 ____ 2 ____

Question #6: Name two potential medical uses of isopulegol:

1 ____ 2 ____

For the answer key to all individual compound chapters please go here: http://thebigbookofterps.com/bbot-quiz-answer-key/

GERANIOL

Type: Acyclic monoterpenoid
Chemical Formula: $C_{10}H_{18}O$
Molecular Weight: 154.25 g/mol
Boiling Point: 229.00 to 230.00 °C. @ 760.00 mm Hg[857]
Flash Point: 226.00 °F. TCC (108.00 °C.)
Melting Point: -16.00 to -14.00 °C. @ 760.00 mm Hg
Solubility: Soluble in alcohol, fixed oils, kerosene, acetone, ether, paraffin oil, propylene glycol, weakly soluble in water, insoluble in glycerin
Oral LD50: 3,600 mg/kg (rat)[858]
Biological Role: Antifungal, nematocidal
Therapeutic Role: Anticancer agent, antifungal, antiulcerogenic
Commercial Use: Fragrance, flavor agent, insecticidal
Occurrence in Cannabis: Top thirty
Aroma/Flavor: Floral, rose, sweet, waxy, citrus

Occurs in Cannabis Strains: Agent Orange, Amnesia Haze, Black Cherry Soda, Dutch Hawaiian, Ghost Cookies, God's Green Crack, Grape LA, Great White Shark, Harlequin, Headband, Island Sweet Skunk, Kimbo Kush, Lavender, Lemon G, Master Kush, Northern Lights, OG Shark, Purple Punch, Strawberry Diesel, Tahoe OG, Trainwreck.

INTRODUCTION

One of the most well-studied monoterpenoids, geraniol is a known anticancer and antifungal agent. An acyclic monoterpenoid, geraniol is highly prized in the perfume and fragrance industries, where its floral, rose, sweet, citrusy, and waxy profile is used to enhance many products, although this isoprenoid compound is typically described overall as rose-like. Geraniol does not often occur in the top ten terpenes by concentration in most cannabis strains, however, it has been reported in God's Green Crack, Northern Lights, and Trainwreck within the top ten. For many cannabis strains, this volatile molecule appears within the top thirty to forty terpenes.

Interestingly, geraniol is the principal active component of honeybee scent gland emissions[859], directing the activity of workers and foragers[860], and helping to determine caste by the chemical induction of queen bees via a larval diet consisting of geraniol[861]. In fact, the attractiveness of geraniol to honeybees is so reliable that French scientists have devised a method of bee census-taking that uses geraniol in bait stations[862] to understand bee population trends.

---

[857] The Good Scents Company Data Sheet for Geraniol, from: http://www.thegoodscentscompany.com/data/rw1006992.html. Accessed September 6, 2023.

[858] European Chemicals Agency Safety Data Sheet for Geraniol, from: https://echa.europa.eu/registration-dossier/-/registered-dossier/14184/7/3/1#:~:text=Lysenko%2C%201962).-,Justification%20for%20classification%20or%20non%2Dclassification,EC)%20No%201272%2F2008. Accessed September 6, 2023.

[859] R. BOCH & D. A. SHEARER. Identification of Geraniol as the Active Component in the Nassanoff Pheromone of the Honey Bee. Nature volume 194, pages704-706 (1962).

[860] J. B. Free. The Attractiveness of Geraniol to Foraging Honeybees. Journal of Apicultural Research Volume 1, 1962 - Issue 1 Pages 52-54 13 Nov 1962.

[861] Jarau S, van Veen JW, Twele R, Reichle C, Gonzales EH, Aguilar I, Francke W, Ayasse M. Workers make the queens in melipona bees: identification of geraniol as a caste determining compound from labial glands of nurse bees. J Chem Ecol. 2010 Jun;36(6):565-9.

[862] Rg Danka, Jl Williams, Te Rinderer. A bait station for survey and detection of honey bees. Apidologie, Springer Verlag, 1990, 21 (4), pp.287-292.

Another insect that uses geraniol is the Sycamore lace bug. This insect emits geraniol as an alarm pheromone, with as little as one nanogram of the compound needed to elicit panicked or evasive behavior in nearby nymphs[863]. Other interesting uses of geraniol include the compound's inclusion in a study that showed that motion sickness causes heightened perception of unpleasant smells[864], and research that programmed mice to developmentally respond to geraniol at a neurological level[865].

CHEMICAL STRUCTURE

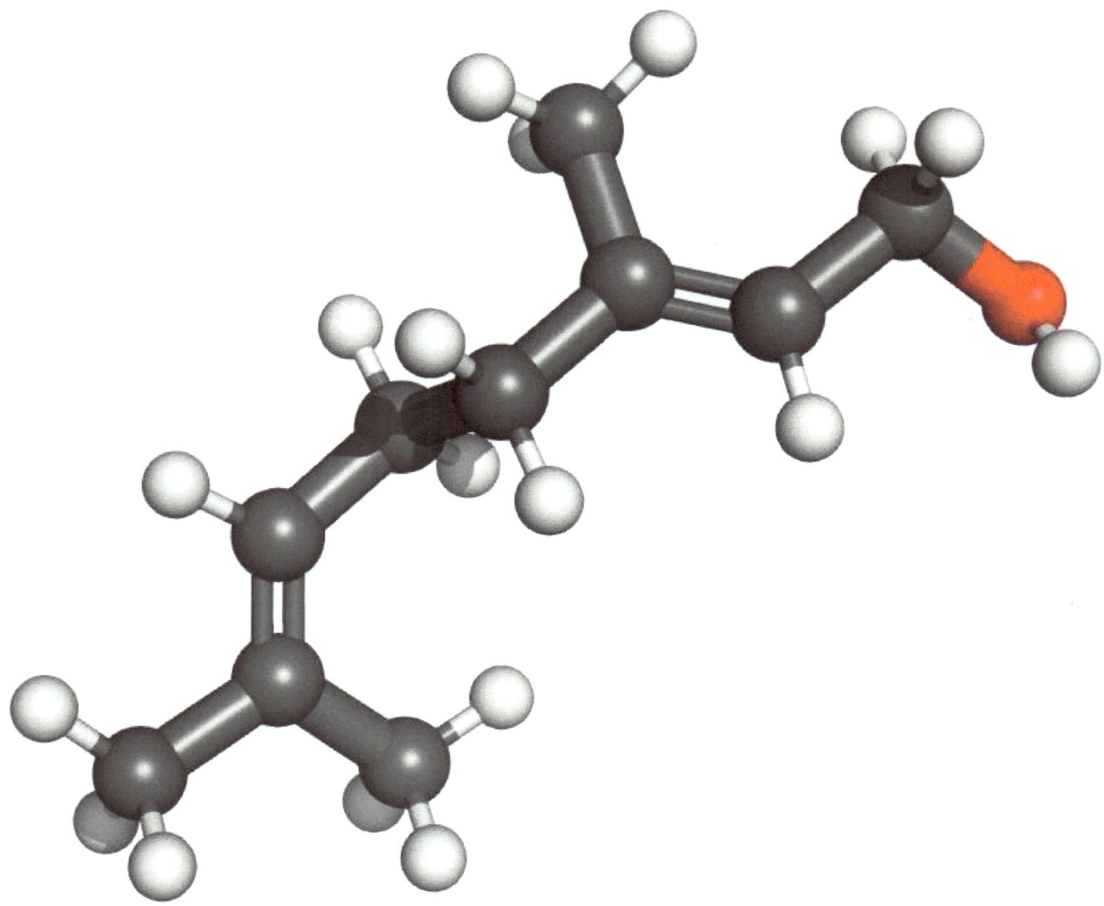

Geraniol is an isoprenoid compound formed from ten carbon atoms, eighteen hydrogen atoms, and one oxygen atom, with a scientific notation of $C_{10}H_{18}O$. Because it has two connected units of isoprene, this volatile molecule is classified as a monoterpene, with a more specific designation as a monoterpenoid because of its oxygenated state.

Geraniol has two pairs of double bonds, and is closely related to geranyl acetate. This phytochemical is referred to as a linear monoterpenoid or an acyclic monoterpenoid to denote the fact that the molecular skeleton of the compound does not contain rings.

---

863 Kuwahara Y, Kawai A, Shimizu N, Tokumaru S, Ueyama H. Geraniol, E-3,7-dimethyl-2,6-octadien-1-ol, as the alarm pheromone of the sycamore lace bug Corythucha ciliata (Say). J Chem Ecol. 2011 Nov; 37(11):1211-5.

864 Paillard AC, Lamore M, Etard O, Millot JL, Jacquot L, Denise P, Quarck G. Is there a relationship between odor and motion sickness? Neurosci Lett. 2014 Apr 30;566:326-30.

865 Eckert M, Schmidt U. The influence of permanent odor stimuli on the postnatal development of neural activity in the olfactory bulbs of laboratory mice. Brain Res. 1985 Jun;352(2):185-90.

Geraniol occurs as several isomers that differ based on the arrangement of atoms and the position of the double bonds. Geraniol and nerol are geometric isomers, with geraniol being the trans- version and nerol the cis- version. The two sets of double bonds means that an additional 4 isomers are possible – the (E,E)-, (E,Z)-, (Z,E)-, and (Z,Z)-versions. Despite the occurrence of these isomers, very little data is reported on them except for the primary geraniol and nerol molecules. As nerol is not known to occur in cannabis, herein we discuss geraniol exclusively.

OCCURRENCE IN PLANTS

Though geraniol is found in many different plants and plant parts, it is primarily sourced from the oils of citronella and palmarosa plants, and is widely known for its presence in rose oil. Other plants that contain geraniol include:

Acacia
Almond
Apple
Apricot
Artemisia
Basil
Basswood
Bay laurel
Bay leaf
Bellary
Bergamot
Betel
Bilberry
Blackberries
Black currant
Black walnut
Blueberries
Bois de rose
Boldo
Boysenberry
Cajuput
Cannabis
Cardamom
Carrot
Catnip
Chamomile
Cinnamon
Citronella
Clary sage
Coriander
Corn
Curry
Dill
Eucalyptus
Fennel
Garlic
Geranium
Ginger
Grapes
Grapefruit
Horsemint
Hyssop
Kewda
Lavandin
Lavender
Lemon
Lemongrass
Lime
Lovage
Mace
Mandarin
Mango
Marjoram
Neroli
Nutmeg
Orange
Oregano
Palmarosa
Parsley
Peppermint
Phoebe
Pine
Plum
Plumcot
Potato
Rose
Rosemary
Sage
Salvia
Savory
Snake root
Spearmint
Strawberry
Tamarind
Tarragon
Tea
Thyme
Watermelon
Witch hazel
Wormseed
Wormwood
Ylang ylang
Zingiber

BIOLOGICAL ACTIVITY IN PLANTS

Although the scientific literature about geraniol is extensive, it mostly focuses on treatment for cancer, so there is little published regarding the biological roles this monoterpenoid serves in plants. However, as a potent and well-known antifungal agent, we can assume other roles that cross-reference with the medical literature and other research that currently exists on geraniol. For example, the antibacterial properties geraniol has shown in some medical research may well mean that the terpene also serves an antibacterial role in plants.

**Antifungal Agent:* As the primary constituent in the essential oil of Pelargonium graveolens (a type of geranium), geraniol significantly inhibited the growth of the fungi aspergillus niger, while synergizing the effects of common antifungal drugs[866]. Geraniol has also been shown to inhibit the growth of both aspergillus flavus and aspergillus ochraceus by more than 98%[867]. Modes of action in both studies referenced here included effects on fungi cell membrane permeability, and by inducing intracellular ROS accumulation.

**Nematocidal Agent:* Geraniol has been shown to exhibit contact toxicity mortality at 100% in the root-knot nematode meloidogyne javanica, while inhibiting egg hatching of the same species by about 70%[868].

**Oviposition Deterrent:* Researchers working to control a devastating pest of eggplant, Leucinodes orbonalis, discovered that one variety – Solanum melongena (Himalayan eggplant) – emitted geraniol as an oviposition deterrent (deters female insects from depositing eggs), which, when applied to and expressed in other varieties, reduced oviposition by >90%[869].

USES IN INDUSTRY

Geraniol is a colorless or clear liquid generally used to add fragrance or flavor to many foods, household products, and industrial products. This acyclic monoterpene is derived by fractional distillation of the essential oils of citronella, palmarosa, and other plants that are rich in geraniol. High-quality geraniol can be obtained synthetically using myrcene or pinene as primary chemical precursors.

As one of the most commonly used substances for the addition of fragrance, geraniol is used in hundreds of products, including absorbents, adsorbents, air fresheners, air treatments, auto soaps, polishes, waxes, grease, cigarettes & tobacco products, laundry soap, degreasers, spot removers, carpet cleaners & detergents, dishwashing soap, softeners, disinfectants, dry-cleaning products, drain cleaners, floor polish, window cleaners, toilet sanitizers, anti-foaming agents, emulsifiers, foaming agents, coagulating agents, industrial and commercial lubricants, toothpaste, shampoo, perfume, paint, deodorant, antiperspirants, hair styling and coloring products, hand & body creams & lotions, makeup & cosmetics, nail polish, along with many other products.

Consumer products containing geraniol are in virtually every grocery, pharmacy, and convenience store in the United States, and include instantly recognizable brands like Febreze, Clorox, Pine-Sol, Lysol, Pledge, Spray N Wash, Air Wick, Glade, Renuzit, Simple Green, Seventh Generation, Garnier, Revlon, Aveda, Nexxus, Noxzema, St. Ives, Loreal, Secret, Boss, TRESemme, Old Spice, and Cutter, among hundreds of other major and lesser-known brands[870].

Geraniol is also used to flavor some foods, including chewing gum, beverages, extracts, colorings, seasonings & spices, condiments, preservatives, gelatins, puddings, gravies, hard candy, soft candy, and some baked goods. Interestingly, researchers have engineered E. coli bacteria to produce geraniol via the biotransformation of glucose[871]. The researchers hoped to manufacture enough geraniol to produce a viable gasoline alternative. Other

---

[866] Shin, Seungwon. Anti- Aspergillus activities of plant essential oils and their combination effects with ketoconazole or amphotericin b. Archives of Pharmacal Research 26.5 (2003): 389-393.

[867] Xi Tang, Ye-Lin Shao, Ya-Jie Tang, and Wen-Wen Zhou. Antifungal Activity of Essential Oil Compounds (Geraniol and Citral) and Inhibitory Mechanisms on Grain Pathogens (Aspergillus flavus and Aspergillus ochraceus) Molecules. 2018 Sep; 23(9): 2108.

[868] Eleni Nasiou, Ioannis O. Giannakou. Effect of geraniol, a plant-based alcohol monoterpene oil, against Meloidogyne javanica European Journal of Plant Pathology November 2018, Volume 152, Issue 3, pp 701-710.

[869] Ghosh R, Metze D, Sant S, Shaikh M, Deshpande A, Firake DM, Pandit S. Chemical ecology of Himalayan eggplant variety's antixenosis: identification of geraniol as an oviposition deterrent against the eggplant shoot and fruit borer. New Phytol. 2023 Mar 14.

[870] Household Products Database: Geraniol: US Department of Health & Human Services: https://hpd.nlm.nih.gov/cgi-bin/household/brands?tbl=chem&id=1926.

[871] Wei Liu, Xin Xu, Rubing Zhang, Tao Cheng, Yujin Cao, Xiaoxiao Li, Jiantao Guo, Huizhou Liu & Mo Xian Engineering Escherichia coli for high-yield

researchers are studying the biotransformation of geraniol by fungi instead of bacteria, focusing their efforts on strains of botrytis cinerea[872].

In addition to all the above uses and products, geraniol also makes an excellent pesticide. In a study that evaluated the efficacy of Terminix AllClear Mosquito Mister, researchers found that the product delivered on its marketing claims, reducing mosquito biting pressure by 90% within 18 feet of the product, and by 67% within 300 feet[873].

Finally, because of geraniol's antibacterial properties, it has been studied as a potential antiperspirant after it was shown to exhibit significant antimicrobial activity against several human axillary odor-causing bacteria, but the molecule was too volatile and evaporated too quickly to be useful in a commercial product[874].

POTENTIAL USES IN MEDICINE

Despite the dozens of developing and potential medical uses for geraniol, thousands of people in the Northern and Southern Hemispheres develop contact allergies (with clinical symptoms) to this monoterpene each year[875], with legislative measures being considered in the US and elsewhere that would reduce the allowable amount of geraniol in consumer products. Apart from this concern, geraniol offers exciting promise in cancer treatment, and several other areas:

**Prevention & Treatment of Pancreatic Cancer:* A novel new treatment for pancreatic cancer - one of the most deadly and difficult to treat cancers - could be possible after it was shown that geraniol completely inhibited PC-1 pancreatic tumor growth in hamsters[876], with no observed collateral toxicity in the animals.

**Treatment & Prevention of Oral Cancers:* Oral administration of geraniol completely prevented the formation of DMBA (7,12-Dimethylbenz[a]anthracene)-induced oral tumors in hamsters, indicating this terpene's potent anti-inflammatory, anti-angiogenic, anti-cell proliferative, and apoptosis-inducing properties[877]. Geraniol also prevented the formation of 4NQO (4-Nitroquinoline 1-oxide)-induced tongue cancer in rats, with researchers noting the molecule's powerful anti-inflammatory effects as its main cancer-fighting property[878].

**Treatment for Liver Cancer:* As one of the primary constituents in the essential oil of senecio chinensis, geraniol contributed to significant antiproliferative (stops it from spreading) and proapoptotic (allows cells to die) activity

---

geraniol production with biotransformation of geranyl acetate to geraniol under fed-batch culture Biotechnology for Biofuels volume 9, Article number: 58 (2016).

872 G. Bock, I. Benda, P. Schreier. Microbial transformation of geraniol and nerol by Botrytis cinerea Applied Microbiology and Biotechnology January 1988, Volume 27, Issue 4, pp 351-357.

873 Revay EE, Junnila A, Kline DL, Xue RD, Bernier UR, Kravchenko VD, Yefremova ZA, Miiller GC. Reduction of mosquito biting pressure by timed-release 0.3% aerosolized geraniol. Acta Trop. 2012 Oct; 124(1):102-5.

874 Tan Wen Nee, Leong Chean Ring, Venoth Arumugam, Judy Loo Ching Yee, Lee Wing Hin, Fahmi Asyadi Md Yusof, Mohd Azizan Mohd Noor,Tong Woei Yenn. Sustained Release Geraniol Nanoparticles Inhibit Human Axillary Odor-Causing Bacteria Arabian Journal for Science and Engineering January 2019, Volume 44, Issue 1, pp 103-109.

875 W.P.Jongeneel, J.E.Delmaar, B.G.H.Bokkers Health impact assessment of a skin sensitizer: Analysis of potential policy measures aimed at reducing geraniol concentrations in personal care products and household cleaning products Environment International Volume 118, September 2018, Pages 235-244.

876 Burke, Yvette, M. Stark, Steven Roach, Stephanie Sen, and Pamela Crowell. Inhibition of pancreatic cancer growth by the dietary isoprenoids farnesol and geraniol. Lipids 32.2 (1997): 151-156.

877 Veerasamy Vinothkumar, Shanmugam Manoharan, Ganapathy Sindhu, Madhavan Ramados Nirmal, Venkatesan Vetrichelvi. Geraniol modulates cell proliferation, apoptosis, inflammation, and angiogenesis during 7,12-dimethylbenz[a]anthracene-induced hamster buccal pouch carcinogenesis Molecular and Cellular Biochemistry October 2012, Volume 369, Issue 1-2, pp 17-25.

878 Arumugam Madankumar, Sasivarnam Tamilarasi, Thandavamoorthy Premkumar, Mani Gopikrishnan, Natesh Nagabhishek, Thiruvengadam Devaki. Geraniol attenuates 4NQO-induced tongue carcinogenesis through downregulating the activation of NF-KB in rats Molecular and Cellular Biochemistry October 2017, Volume 434, Issue 1-2, pp 7-15.

against human hepatocarcinoma cells by weakening the proliferative capacity of the cells, increasing ROS, and regulating the expression of tumor suppressor genes[879].

**Chemoprotective Agent Against Renal Cancer:* Geraniol has been shown to suppress induced renal oxidative stress and carcinogenesis in rats by up regulation of xenobiotic metabolizing enzyme activities, down regulation of serum toxicity markers, and induction of apoptosis[880].

**Treatment for Skin Cancer:* In mice subjected to DMBA-induced skin carcinogenesis, geraniol significantly prevented tumor formation and attenuated a reduction in phase II detoxification agents and antioxidants, thereby inhibiting abnormal cell proliferation[881]. Other researchers have also found geraniol to be an excellent anti-skin cancer agent, writing that the acyclic monoterpenoid works "by attenuating the Ras proliferation pathway and inducing pro-apoptotic state via inhibition of oxidative stress response and inflammation[882]."

**Treatment for Prostate Cancer:* Geraniol potently induced apoptosis and autophagy in PC-3 prostate cancer cells, and enhanced docetaxel (a cancer treatment drug) chemosensitivity in cultured cell or xenograft tumors[883].

**Treatment for Chemotherapeutic Agent Toxicity:* Doxorubicin – a common chemotherapeutic agent – often causes cardiotoxicity in cancer patients taking the drug. However, treatment with geraniol significantly ameliorated cardiac damage and restored serum cardiac injury marker levels in doxorubicin-treated animals[884] via multiple mechanisms, offering a potential new adjuvant for cancer therapy.

**Antifungal Agent:* Geraniol has exhibited potent antifungal activity against several varieties of candida by causing mitochondrial dysfunction (impaired iron homeostasis and genotoxicity) via inhibition of both virulence attributes of hyphal morphogenesis and biofilm formation[885]. Candida varieties similarly affected by geraniol include C. albicans, C. krusei, C. tropicalis, C. parapsilisos, and C. glabrata. Research that examined candida infections in zebrafish echoed these findings, adding that geraniol significantly increases chromosome damage[886], while other researchers found that geraniol enhances the rate of potassium leakage out of whole cells and increases membrane fluidity of not only C. albicans, but also saccharomyces cerevisiae strains[887]. In the fungus trichophyton rubrum, geraniol was shown to cause leakage of intracellular material and inhibit ergosterol biosynthesis[888], a critical

---

[879] Shen, Xionghu, Xian Cui, Hai Cui, Yongmin Jin, Wenbiao Jin, and Honghua Sun. Geraniol and lupeol inhibit growth and promote apoptosis in human hepatocarcinoma cells through the MAPK signaling pathway. Journal of Cellular Biochemistry 120.4 (2019): 5033-5041.

[880] Ahmad ST, Arjumand W, Seth A, Nafees S, Rashid S, Ali N, Sultana S. Preclinical renal cancer chemopreventive efficacy of geraniol by modulation of multiple molecular pathways. Toxicology. 2011 Nov 28; 290(1):69-81.

[881] Manoharan S, Selvan MV. Chemopreventive potential of geraniol in 7,12-dimethylbenz(a) anthracene (DMBA) induced skin carcinogenesis in Swiss albino mice. J Environ Biol. 2012 Mar; 33(2):255-60.

[882] Chaudhary SC, Siddiqui MS, Athar M, Alam MS. Geraniol inhibits murine skin tumorigenesis by modulating COX-2 expression, Ras-ERK1/2 signaling pathway and apoptosis. J Appl Toxicol. 2013 Aug;33(8):828-37.

[883] Kim SH, Park EJ, Lee CR, Chun JN, Cho NH, Kim IG, Lee S, Kim TW, Park HH, So I, Jeon JH. Geraniol induces cooperative interaction of apoptosis and autophagy to elicit cell death in PC-3 prostate cancer cells. Int Oncol. 2012 May;40(5):1683-90.

[884] Younis NS, Elsewedy HS, Soliman WE, Shehata TM, Mohamed ME. Geraniol isolated from lemon grass to mitigate doxorubicin-induced cardiotoxicity through Nrf2 and NF-κB signaling. Chem Biol Interact. 2021 Sep 25;347:109599.

[885] Singh, Shweta, Zeeshan Fatima, and Saif Hameed. Insights into the mode of action of anticandidal herbal monoterpenoid geraniol reveal disruption of multiple MDR mechanisms and virulence attributes in Candida albicans. Archives of Microbiology 198.5 (2016): 459-472.

[886] Singulani JL, Pedroso RS, Ribeiro AB, Nicolella HD, Freitas KS, Damasceno JL, Vieira TM, Crotti AE, Tavares DC, Martins CH, Mendes-Giannini MJ, Pires RH. Geraniol and linalool anticandidal activity, genotoxic potential and embryotoxic effect on zebrafish. Future Microbiol. 2018 Nov; 13:1637-1646.

[887] Martin Bard, Mark R. Albrecht, Navyash Gupta, Cynthia J. Guynn, William Stillwell. Geraniol interferes with membrane functions in strains of Candida and Saccharomyces Lipids Volume 23, Issue 6 June 1988 Pages 534-538.

[888] Pereira Fde O, Mendes JM, Lima IO, Mota KS, Oliveira WA, Lima Ede O. Antifungal activity of geraniol and citronellol, two monoterpenes alcohols, against Trichophyton rubrum involves inhibition of ergosterol biosynthesis. Pharm Biol. 2015 Feb; 53(2):228-34.

function for the proliferation of T. rubrum. Finally, geraniol has been shown to deplete ergosterol content and inhibit biofilm formation in Candida auris[889] – a dangerous multi drug resistant bacteria known as a "superbug."

**Antiulcerogenic:* In one study, ethanol-induced ulcers in rats were inhibited by up to 99% after oral administration of geraniol[890], while another found that - as the primary constituent in the essential oil of lemongrass - geraniol accelerated the gastric healing process by more than 80%[891].

Treatment for Irritable Bowel Syndrome: In a pilot study where irritable bowel syndrome patients were administered geraniol orally, the monoterpene was shown to modulate the intestinal microbial ecosystem, increasing levels of beneficial bacteria like collinsella and faecalibacterium while reducing a chemokine associated with long running IBS[892].

Control of Mosquito-borne Illnesses: As a primary constituent in the essential oil of Pelargonium roseum, geraniol contributed to a significant toxic effect on larvae and eggs raft of culex pipiens, a mosquito disease vector of West Nile and filariasis viruses, while showing a moderate knock-down effect in adults of the same species[893].

Anthelmintic Agent: As the primary constituent in the essential oil of Cymbopogon schoenanthus (camel grass), geraniol was shown to exhibit significant toxicity against trichostrogylus[894], an intestinal worm that typically infects sheep and other livestock.

Insecticidal Agent - Treatment for Head Lice: As a major constituent in the essential oil of geranium maculatum, geraniol contributed to significant contact toxicity against adult female Pediculus humanus capitis[895] - the common head lice. Topical application of this terpene or a blend of geraniol and other terpenes such as citronellol may provide a novel treatment for this extremely common, worldwide condition.

Treatment for Parkinson's Disease: Geraniol has ameliorated MPTP (1-methyl-4-phenyl-1,2,3,6-tetrahydropyridine)-induced Parkinson's Disease in mice, including behavioral, biochemical, immunohistochemical, and immunoblotting manifestations of the condition, with researchers concluding that "[geraniol] confers a potent neuroprotective agent against MPTP-induced dopaminergic denervation and may become a potential therapeutic agent for PD and its progression.[896]"

---

[889] Fatima, T., Fatima, Z. & Hameed, S. Abrogation of efflux pump activity, biofilm formation, and immune escape by candidacidal geraniol in emerging superbug, Candida auris. Int Microbiol (2023).

[890] Carvalho, Katharinne, Flavia Bonamin, Raquel Santos, Larissa Perico, Fernando Beserra, Damiao Sousa, Jose Filho, Lucia Rocha, and Clelia Hiruma-Lima. Geraniol-a flavoring agent with multifunctional effects in protecting the gastric and duodenal mucosa. Naunyn-Schmiedeberg's Archives of Pharmacology 387.4 (2013): 355-365.

[891] Venzon L, Mariano LNB, Somensi LB, Boeing T, de Souza P, Wagner TM, Andrade SF, Nesello LAN, da Silva LM. Essential oil of Cymbopogon citratus (lemongrass) and geraniol, but not citral, promote gastric healing activity in mice. Biomed Pharmacother. 2018 Feb; 98:118-124.

[892] Fernando Rizzello, Chiara Ricci, Michela Scandella, Elena Cavazza, Elisabetta Giovanardi, Maria Chiara Valerii, Massimo Campieri, Antonietta Comparone, Luigia De Fazio, Marco Candela, Silvia Turroni & Enzo Spisni. Dietary geraniol ameliorates intestinal dysbiosis and relieves symptoms in irritable bowel syndrome patients: a pilot study BMC Complementary and Alternative Medicine volume 18, Article number: 338 (2018).

[893] Tabari MA, Youssefi MR, Esfandiari A, Benelli G. Toxicity of B-citronellol, geraniol and linalool from Pelargonium roseum essential oil against the West Nile and filariasis vector Culex pipiens (Diptera: Culicidae). Res Vet Sci. 2017 Oct; 114:36-40.

[894] Katiki LM1, Chagas AC, Bizzo HR, Ferreira JF, Amarante AF. Anthelmintic activity of Cymbopogon martinii, Cymbopogon schoenanthus and Mentha piperita essential oils evaluated in four different in vitro tests. Vet Parasitol. 2011 Dec 29;183(1-2):103-8.

[895] Gallardo A, Picollo MI, Gonzalez-Audino P, Mougabure-Cueto G. Insecticidal activity of individual and mixed monoterpenoids of geranium essential oil against Pediculus humanus capitis (Phthiraptera: Pediculidae). J Med Entomol. 2012 Mar;49(2):332-5.

[896] Karamkolly R. Rekha, Govindasamy P. Selvakumar, Subramaniam Sethupathy, Karunanidhi Santha, Ramu Inmozhi Sivakamasundari. Geraniol Ameliorates the Motor Behavior and Neurotrophic Factors Inadequacy in MPTP-Induced Mice Model of Parkinson's Disease Journal of Molecular Neuroscience November 2013, Volume 51, Issue 3, pp 851-862.

Antibacterial Agent: In a study published in 2019, geraniol was shown to reduce the biofilm biomass of Staphylococcus aureus by up to 100%[897], offering a promising potential new tool against this common but potentially dangerous bacterium. Other work has shown that geraniol affected the growth, prolonged the lag phase, and delayed growth periods in a concentration-dependent manner by suppressing genes used in quorum sensing, thereby reducing virulence in Pseudomonas aeruginosa[898].

Treatment for Heart Attack: In an experimental model of myocardial infarction in rats subjected to isoproterenol-induced oxidative stress and inflammation, geraniol prevented heart rate enhancement, cardiac and lipid peroxidative genes, and reverted the blood pressure and antioxidant status of the animals while also inhibiting the mitogen-activated protein kinase, inflammatory responder, and cardiac fibrotic proteins[899].

IMPLICATIONS FOR HUMAN HEALTH & NUTRITION

Dietary geraniol should be easy for most people to acquire, as this acyclic monoterpenoid occurs in foods and beverages we commonly consume. For instance, when preparing, cooking, or serving food, consider working with spices like basil, bay leaf, cardamom, chamomile, cinnamon, coriander, dill, fennel, garlic, ginger, lemongrass, mace, marjoram, nutmeg, oregano, parsley, peppermint, rosemary, sage, savory, spearmint, tarragon, or thyme.

There is a wide variety of fruits that contain geraniol, including apples, apricots, blackberries, black currants, blueberries, blood oranges, grapes, grapefruit, lemons, limes, mandarins, mangoes, oranges, plums, plumcots, red raspberries, wild strawberries, and watermelons. In addition to all these fruits, some vegetables like carrots and potatoes contain geraniol, as well as black walnuts, and lemongrass.

Most green teas and black teas contain geraniol (and any teas made with the spices mentioned earlier this chapter), and both red and white wines are also known to feature this phytonutrient.

Cannabis strains that contain geraniol include Agent Orange, Amnesia Haze, Black Cherry Soda, Dutch Hawaiian, Ghost Cookies, God's Green Crack, Grape LA, Great White Shark, Harlequin, Headband, Island Sweet Skunk, Kimbo Kush, Lavender, Lemon G, Master Kush, Northern Lights, OG Shark, Purple Punch, Strawberry Diesel, Tahoe OG, and Trainwreck. While we do not recommend smoking or vaporizing cannabis for this purpose, using oils, tinctures, or value-added cannabis products may allow some ingestion of geraniol.

Experimentation with essential oils that are high in geraniol can include using small amounts of this EO in beverages or food, or in topical applications. Extreme care should be used when working with isolated compounds or essential oils, as these substances can be toxic in some situations.

---

[897] Eveline Kelle Ursulino Pontes, Hider Machado Melo, Jose Walter Araujo Nogueira, Nairley Cardoso Sa Firmino, Mario Geraldo de Carvalho, Francisco Eduardo Aragao Catunda Junior, Theodora Thays Arruda Cavalcante. Antibiofilm activity of the essential oil of citronella (Cymbopogon nardus) and its major component, geraniol, on the bacterial biofilms of Staphylococcus aureus Food Science and Biotechnology June 2019, Volume 2 8, Issue 3, pp 633-639.

[898] Li WR, Zeng TH, Zhang ZQ, Shi QS, Xie XB. Geraniol attenuates virulence factors by inhibiting quorum sensing of Pseudomonas aeruginosa. Front Microbiol. 2023 Apr 27;14:1190619.

[899]Zou G, Wan J, Balupillai A, David E, Ranganathan B, Saravanan K. Geraniol enhances peroxiredoxin-1, and prevents isoproterenol-induced oxidative stress and inflammation associated with myocardial infarction in experimental animal models. J Biochem Mol Toxicol. 2022 Aug;36(8):e23098.

Geraniol Review

Answer the following questions to test your knowledge of this terpene:

Question #1: What type of terpene is geraniol?

a. Linear monoterpene
b. Monocyclic monoterpene
c. Monocyclic monoterpenoid
d. Acyclic monoterpenoid

Question #2: Which of the following best describes the primary flavor & aroma profile of geraniol?

a. Citrus, sweet, tropical, spicy
b. Clove, nutmeg, wood, terpenic
c. Floral, rose, sweet, waxy, citrus
d. Mint, cool, herbal, green

Question #3: How common is geraniol in cannabis?

a. Top three
b. Top five
c. Top thirty
d. Top forty

Question #4: What is the chemical formula for geraniol?

a. $C_{15}H_{26}O$
b. $C_{10}H_{18}O$
c. $C_{10}H_{16}O$
d. $C_{10}H_{22}O$

Question #5: Name two uses of geraniol in plants:

1 ____ 2 ____

Question #6: Name two potential medical uses of geraniol:

1 ____ 2 ____

For the answer key to all individual compound chapters please go here: http://thebigbookofterps.com/bbot-quiz-answer-key/

VALENCENE

Type: Bicyclic sesquiterpene
Chemical Formula: $C_{15}H_{24}$
Molecular Weight: 204.35 g/mol
Boiling Point: 274.00 °C. @ 760.00 mm Hg[900]
Flash Point: 212.00 °F. TCC (100.00 °C.)
Melting Point: 274 °C (525.2 °F)
Solubility: Soluble in oils, alcohol, insoluble in water, ethanol
Oral LD50: >5,000 mg/kg (rat)[901]
Biological Role: Insect repellant, insecticidal agent
Therapeutic Role: Anticancer agent, antibacterial, anti-inflammatory
Commercial Use: Flavoring agent, fragrance, chemical precursor
Occurrence in Cannabis: Top thirty
Aroma/Flavor: Citrus, sweet, fresh, grapefruit, woody

Occurs in Cannabis Strains: ACDC, Agent Orange, Cannatonic, Clementine, Covelo Sour Diesel, Girl Scout Cookies, Jack Herer, Jillybean, OG Kush, Sour Diesel, Tangie

INTRODUCTION

Valencene is well-known in the orange juice industry as the primary terpene responsible for the sweet, citrus smell and flavor of citrus fruits, most notably Valencia oranges. This terpene is found almost exclusively in citrus fruits, although it does occur in a few other herbs and essential oils.

Valencene offers potential new treatments for cancer, and is a promising antibacterial agent. Strains like ACDC, Cannatonic, and Covelo Sour Diesel feature this terpene within the top ten terpenes by concentration; though valencene occurs more often in the top twenty to forty terpenes per chemovar.

[900] The Good Scents Company Data Sheet for Valencene, from: http://www.thegoodscentscompany.com/data/rw1006341.html. Accessed September 6, 2023.

[901] Bedoukian Research Safety Data Sheet for Valencene, from:https://search.bedoukian.com/product_images/mxts/806_English_SDS_US.pdf. Accessed September 6, 2023.

CHEMICAL STRUCTURE

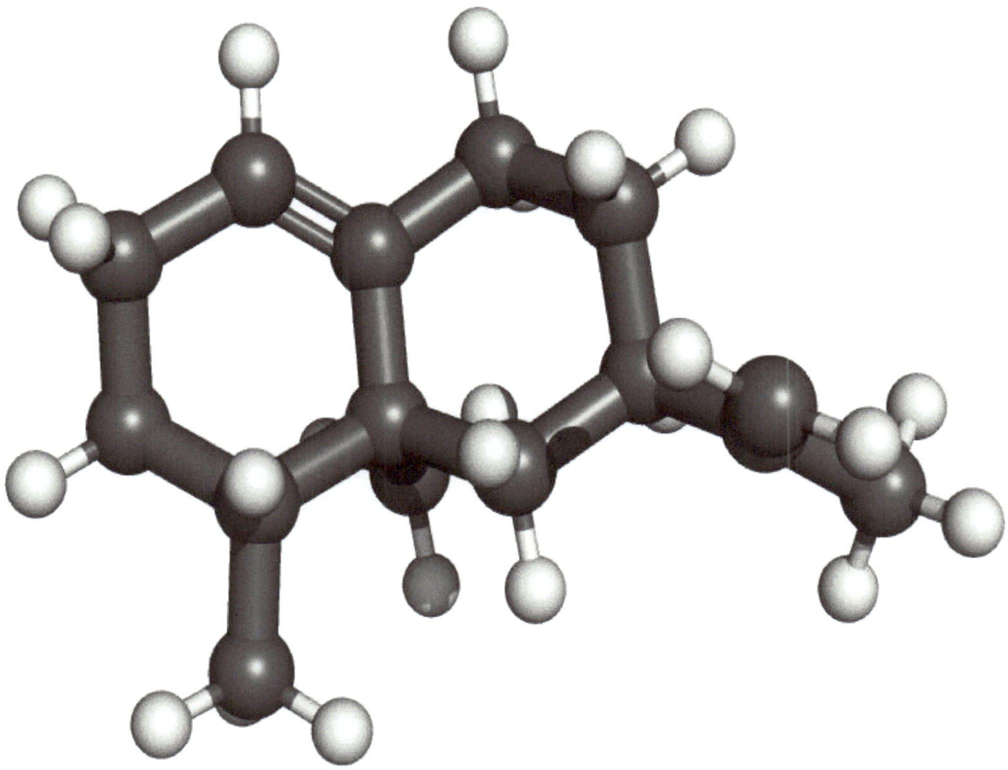

Derived from farnesyl diphosphate, valencene is comprised of three linked units of isoprene, which classifies this compound as a sesquiterpene. The molecular skeleton of this phytonutrient is built from fifteen carbon atoms and twenty-four hydrogen atoms, with a scientific notation of $C_{15}H_{24}$. Valencene features two rings, which further classifies the compound as a bicyclic sesquiterpene, with methyl groups at positions 4a and 5, and two sets of double bonds; one of which is endocyclic, and the other exocyclic.

OCCURRENCE IN PLANTS

Unlike most other terpenes covered in this book, valencene is known to occur in just a few plants, mostly citrus fruits. However, the occurrence of this sesquiterpene varies even from one type of fruit to another; for instance, valencene is rich in the juice of some mandarins, but does not appear at all in others[902].

- Black pepper
- Cannabis
- Cocoa
- Labdanum
- Mandarin
- Rosemary
- Tansy
- Blood orange
- Cedarwood
- Grapefruit
- Lemon
- Orange
- Sage
- Thyme
- Cascarilla
- Clary sage
- Hyptis
- Lime
- Oregano
- Tagete

[902] Yu, Qibin et al. Deficiency of valencene in mandarin hybrids is associated with a deletion in the promoter region of the valencene synthase gene. BMC plant biology vol. 19,1 101. 13 Mar. 2019.

BIOLOGICAL ACTIVITY IN PLANTS

Little is known about the biological roles that valencene plays in the plant kingdom, mostly because little to no research has yet been conducted in this regard. Instead, research into this bicyclic sesquiterpene focuses on its potential medical applications in the treatment of cancer, and as a novel antibacterial agent.

Insect Repellant/Insecticidal Agent: One potential role of valencene in plants is as an insect repellant. As a primary constituent in the essential oil of atractylodes lancea (an herb used in traditional Chinese medicine), valencene showed potent repellant activity against Tribolium castaneum (the red flour beetle); more so than the other terpenes in the EO, with effects comparable to that of DEET[903], one of the most common commercial insect repellants in the world. Valencene has also been shown to exhibit insecticidal (note repellant versus insecticidal effects) activity against several insects, most notably mosquitos[904].

USES IN INDUSTRY

Commercial preparations of valencene appear as pale yellow to clear liquids, and are generally obtained from Valencia oranges. There are hundreds of patents in the United States alone for products containing valencene, and nearly all of these use this bicyclic sesquiterpene as a fragrance or flavoring agent. The most common products that contain valencene include laundry detergents, soaps, de-greasers, spot removers, dry-cleaning chemicals, laundry detergents, soaps, and window/floor cleaners, while the prepared foods industry primarily uses this volatile molecule in spices, extracts, colorings, flavors, condiments, and seasonings.

Interestingly, researchers have produced the highly valuable aromatic of grapefruit, nootkatone, via the biotransformation of valencene by chlorella algae and two mucor fungi species[905]. Other researchers produced nootkatone from the biotransformation of valencene by Escherichia coli[906] bacteria.

POTENTIAL USES IN MEDICINE

Valencene is not as well-studied medically as some other terpenes; nevertheless, it is clear from the scientific research available that this bicyclic sesquiterpene offers several potential important uses in medicine. However, a note of caution; liquid valencene could be severely injurious or fatal if swallowed and aspirated into the lungs. Valencene offers the following potential medical uses:

**Cancer Treatment Synergist & Antiproliferative*: Scientists looking for ways to improve the efficacy of the common anticancer drug Doxorubicin (an antitumor chemotherapy drug derived from Streptomyces bacterium) discovered that, of four major sesquiterpenes, only valencene was effective across all the ovarian and lymphoblast cancer cell lines studied, in some cases increasing the efficacy of Doxorubicin by 6 times[907], while also showing an individual antiproliferative effect. In a similar research project, the same study author, Martin Ambroz, led a team

---

[903] Guo SS, Wang Y, Pang X, Geng ZF, Cao JQ, Du SS. Seven herbs against the stored product insect: Toxicity evidence and the active sesquiterpenes from Atractylodes lancea. Ecotoxicol Environ Saf. 2019 Mar;169:807-813.

[904] Use of Novel Compounds for Pest Control: Insecticidal and Acaricidal Activity of Essential Oil Components from Heartwood of Alaska Yellow Cedar. Journal of Medical Entomology 42.3 (2005): 352-358.

[905] Furusawa M, Hashimoto T, Noma Y, Asakawa Y. Highly efficient production of nootkatone, the grapefruit aroma from valencene, by biotransformation. Chem Pharm Bull (Tokyo). 2005 Nov; 53(11):1513-4.

[906] Zelena K, Krings U, Berger RG. Functional expression of a valencene dioxygenase from Pleurotus sapidus in E. coli. Bioresour Technol. 2012 Mar; 108:231-9.

[907] Ambroz, Martin et al. The Effects of Selected Sesquiterpenes from Myrica rubra Essential Oil on the Efficacy of Doxorubicin in Sensitive and Resistant Cancer Cell Lines. Molecules (Basel, Switzerland) vol. 22,6 1021. 20 Jun. 2017.

that showed that valencene inhibited proliferation of CaCo-2 (human colorectal) cancer cells in rat hepatocytes, noting that valencene was also able to increase the pro- oxidative effect of Doxorubicin in these cells[908].

Antibacterial Agent: As a primary constituent in the essential oil of vitex negundo - a plant often used in Ayurvedic and Unani systems of medicine - valencene contributed to potent antibacterial effects against bacillus subtilis and Escherichia coli bacterial strains[909].

Treatment for Tuberculosis: The antibacterial properties of valencene have also been shown in the treatment of tuberculosis (caused by a type of mycobacteria), where, as a primary constituent in the essential oil of two lantana species, valencene contributed to in vitro antimycobacterial activity[910].

Treatment for Sepsis: As the primary constituent in the essential oil from the rhizomes of cyperus rotundus (nut grass), valencene was shown to exhibit anti-inflammatory properties, increasing the expression of heme oxygenase (an enzyme), which significantly inhibited nitric oxide production in murine tumor cells, while increasing survival rates of mice subjected to induced sepsis[911] (sepsis is a complication of serious infections).

Treatment for Atopic Dermatitis: As isolated from nutgrass, valencene increased both the expression of the skin barrier protein involucrin, and the filament-associated protein filaggrin, reducing atopic dermatitis-like symptoms and modulating the inflammatory response in a murine model[912].

Treatment for Allergies: Once again as the primary constituent of an extract made from Cyperus rotundus (nutgrass), valencene was shown to strongly inhibit several key enzymes responsible for inflammatory responses in rat basophilic leukemia (RBL)-1 cells, while inhibiting delayed-type hypersensitivity reactions in mice[913].

Anthelmintic Agent: As a primary constituent in the essential oil from zanthoxylum zanthoxyloides (Senegal prickly ash), valencene contributed to potent anthelmintic activities against strongyloides ratti (rat threadworms), inhibiting egg hatching at a rate 7 times higher than that of thiabendazole[914] (a commercial anthelmintic), while also inhibiting larval migration.

Anti-inflammatory Agent: Valencene has been shown to significantly inhibit paw edema formation, reduce albumin extravasation, leukocyte recruitment, and the production of myeloperoxidase, interleukin-1-beta, and tumor necrosis factor-alpha in both pleural and peritoneal lavages, while also reducing both granuloma weight and concentration of total proteins in a murine model of chronic inflammation[915].

---

[908] Ambroz, Martin et al. The Influence of Sesquiterpenes from Myrica rubra on the Antiproliferative and Pro- Oxidative Effects of Doxorubicin and Its Accumulation in Cancer Cells. Molecules (Basel, Switzerland) vol. 20,8 15343-58. 21 Aug. 2015.

[909] Khokra, S. L., Prakash, O., Jain, S., Aneja, K. R., & Dhingra, Y. (2008). Essential Oil Composition and Antibacterial Studies of Vitex negundo Linn. Extracts. Indian journal of pharmaceutical sciences, 70(4), 522-526.

[910] Juliao, Lisieux de Santana, Humberto R Bizzo, Andressa M Souza, Maria Cristina S Louren;;o, Pedro Eduardo A Silva, Eliana S Tavares, Luca Rastrelli, and Suzana Guimaraes Leitao. Essential oils from two Lantana species with antimycobacterial activity. Natural product communications 4.12 (2010): 1733-6.

[911] Tsoyi K, Jang HJ, Lee YS, Kim YM, Kim HJ, Seo HG, Lee JH, Kwak JH, Lee DU, Chang KC. (+)-Nootkatone and (+)-valencene from rhizomes of Cyperus rotundus increase survival rates in septic mice due to heme oxygenase-1 induction. J Ethnopharmacol. 2011 Oct 11; 137(3):1311-7.

[912] Yang, In Jun et al. Inhibitory Effect of Valencene on the Development of Atopic Dermatitis-Like Skin Lesions in NC/Nga Mice. Evidence-based complementary and alternative medicine: eCAM vol. 2016 (2016): 9370893.

[913] Jin JH, Lee DU, Kim YS, Kim HP. Anti-allergic activity of sesquiterpenes from the rhizomes of Cyperus rotundus. Arch Pharm Res. 2011 Feb; 34(2):223-8.

[914] Olounlade PA, Azando EV, Hounzangbe-Adote MS, Ha TB, Leroy E, Moulis C, Fabre N, Magnaval JF, Hoste H, Valentin A. In vitro anthelmintic activity of the essential oils of Zanthoxylum zanthoxyloides and Newbouldia laevis against Strongyloides ratti. Parasitol Res. 2012 Apr;110(4):1427-33.

[915] Dantas LBR, Alcântara IS, Júnior CPS, de Oliveira MRC, Martins AOBPB, Dantas TM, Ribeiro-Filho J, Coutinho HDM, Passos FRS, Quintans-Júnior LJ, Almeida JRGS, Cruz-Martins N, Kim B, de Menezes IRA. In vivo and in silico anti-inflammatory properties of the sesquiterpene valencene. Biomed Pharmacother. 2022 Sep;153:113478.

Treatment for Heart Attack: In rats induced to myocardial infarction via the administration of isoproterenol, valencene significantly decreased serum cardiac diagnostic markers, cardiac hypertrophy, conjugated dienes, markers of inflammation, pro-inflammatory cytokines, and myocardial infarct size, while also increasing antioxidant enzymes and anti-inflammatory cytokine genes[916], among other beneficial effects targeting myocardial infarction pathogenesis.

IMPLICATIONS FOR HUMAN HEALTH & NUTRITION

The most obvious way to obtain dietary valencene is to consume citrus fruits like blood oranges, grapefruit, lemons, limes, mandarins, and oranges. When preparing, cooking, or serving foods, consider working with spices like oregano, black pepper, rosemary, sage, or thyme.

Cannabis strains that generally contain at least trace amounts of valencene include ACDC, Agent Orange, Cannatonic, Clementine, Covelo Sour Diesel, Girl Scout Cookies, Jack Herer, Jillybean, OG Kush, Sour Diesel, and Tangie, among others. Juicing or consuming these strains raw is the healthiest option, but low-temperature vaporization may also be an option for some people. Healthier options that do not involve the inhalation of cannabis combustibles or vapor include the use of oils, tinctures, or other value-added products.

Because of the dangers related to aspiration of valencene in human lungs, experimentation with pure valencene or essential oils where this sesquiterpene is a primary constituent should be undertaken with extreme caution.

---

[916] Shervin Prince S, Stanely Mainzen Prince P, Berlin Grace VM. Valencene post-treatment exhibits cardioprotection via inhibiting cardiac hypertrophy, oxidative stress, nuclear factor- κB inflammatory pathway, and myocardial infarct size in isoproterenol-induced myocardial infarcted rats; A molecular study. Eur J Pharmacol. 2022 Jul 15;927:174975.

Valencene Review

Answer the following questions to test your knowledge of this terpene:

Question #1: What type of terpene is valencene?

a. Acyclic sesquiterpene
b. Linear sesquiterpene
c. Monocyclic sesquiterpene
d. Bicyclic sesquiterpene

Question #2: Which of the following best describes the primary flavor & aroma profile of valencene?

a. Hops, cloves, wood, spicy
b. Citrus, woody, sweet, grapefruit
c. Spicy, turpentine, fuel, pine
d. Citrus, fruity, dry, terpenic

Question #3: How common is valencene in cannabis?

a. Top three
b. Top thirty
c. Top ten
d. Top forty

Question #4: What is the chemical formula for valencene?

a. $C_{15}H_{32}0$
b. $H_{24}C0_{20}$
c. $C_{15}H_{24}$
d. $C_{15}H_{26}0$

Question #5: What is the most common commercial source for valencene?

____

Question #6: Name two potential medical uses of valencene:

1 ______________________ 2 ______________________

For the answer key to all individual compound chapters please go here: http://thebigbookofterps.com/bbot-quiz-answer-key/

SABINENE

Type: Bicyclic monoterpene
Chemical Formula: $C_{10}H_{16}$
Molecular Weight: 136.23 g/mol
Boiling Point: *Sabinene:* 163.00 to 165.00 °C. @ 760.00 mm Hg[917]
(+)-sabinene: 163.00 to 164.00 °C. @ 760.00 mm Hg[918]
(-)-sabinene: 164.00 °C. @ 760.00 mm Hg (estimated)[919]
Flash Point: *Sabinene:* 98.00 °F. TCC (36.67 °C.)
(+)-sabinene: 97.00 °F. TCC (36.11 °C.)
(-)-sabinene: 98.00 °F. TCC (36.70 °C.) (estimated by TGSC)
Solubility: *All isomers:* Soluble in alcohol, weakly soluble in water
Oral LD50: 301-2,000 mg/kg (rat)[920]
Biological Role: Allelopathic agent, antifungal, insect repellant
Therapeutic Role: Larvicidal, antibacterial, anti-inflammatory, sedative
Commercial Use: Food additive, fragrance
Occurrence in Cannabis: Top thirty
Aroma/Flavor: *Sabinene:* Woody, spicy, terpenic, citrus

Occurs in Cannabis Strains: Arjan's Haze, Super Silver Haze, Trainwreck

INTRODUCTION

Sabinene is a bicyclic monoterpene often used to add flavor to foods, and sometimes to add fragrance to household and commercial products. When sabinene occurs in cannabis, it is likely only within the top twenty to forty terpenes by concentration in some strains, however, little evidence exists as of fall 2023 that this terpene occurs regularly in cannabis. In fact, of the hundreds of cannabis terpene test results reviewed during research and preparation for this book, only one occurrence of sabinene could be found, as the #10 terpene by concentration in a sample of Trainwreck. Unsubstantiated reports on the internet state that Super Silver Haze and Arjan's Haze also contain this monoterpene.

Herbivores appear to be attracted to the volatile emissions of sabinene, including the African trioza erytreae (a sap-sucking insect), which shows a strong attraction to the monoterpene[921]; and sheep showed a preference for browsing on juniper bushes that were high in sabinene, as opposed to those that were high in delta-3-carene[922]. However, some plants may have developed defenses to cope with these types of herbivory. For instance, research has found

---

[917] The Good Scents Company Data Sheet for Sabinene, from: http://www.thegoodscentscompany.com/data/rw1056021.html. Accessed September 7, 2023.

[918] The Good Scents Company Data Sheet for (R)-sabinene, from: http://www.thegoodscentscompany.com/data/rw1374251.html. Accessed September 7, 2023.

[919] The Good Scents Company Data Sheet for (1S)-(Z)-sabinene, from: http://www.thegoodscentscompany.com/data/rw1552791.html. Accessed September 7, 2023.

[920] Carl Roth Chemical Safety Data Sheet for Sabinene, from: https://www.carlroth.com/medias/SDB-7188-AU-EN.pdf?context=bWFzdGVyfHNlY3VyaXR5RGF0YXNoZWV0c3wyNTQ3MjN8YXBwbGljYXRpb24vcGRmfHNlY3VyaXR5RGF0YXNoZWV0cy9oNWEvaDA3LzkwNDA0MTA5MzUzMjYucGRmfDZkN2VhZGM4NTVhODIzY2U2MmMyZmEyYzkzOWZlMjMzZjUxYmU0OTg2OTYwYWJhODUyNTQwOGFlNDMxMWFlZjQ. Accessed September 7, 2023.

[921] Antwi-Agyakwa, Akua, Ayuka Fombong, Emilie Deletre, Sunday Ekesi, Abdullahi Yusuf, Christian Pirk, and Baldwyn Torto. Lemon Terpenes Tnfluence Behavior of the African Citrus Triozid Trioza erytreae (Hemiptera: Triozidae). Journal of Chemical Ecology OnlineFirst (2019): 1-12.

[922] Marko, Gabor, Tldiko Novak, Jeno Bernath, and Vilmos Altbacker. Both Gas Chromatography and an Electronic Nose Reflect Chemical Polymorphism of Juniper Shrubs Browsed or Avoided by Sheep. Journal of Chemical Ecology 37.7 (2011): 705-713.

that brassica oleracea var. sabauda (Savoy cabbage) lowers its emission rate of sabinene (among several other terpenes) in response to feeding or oviposition by murgantia histrionica[923] (harlequin cabbage bug).

CHEMICAL STRUCTURE

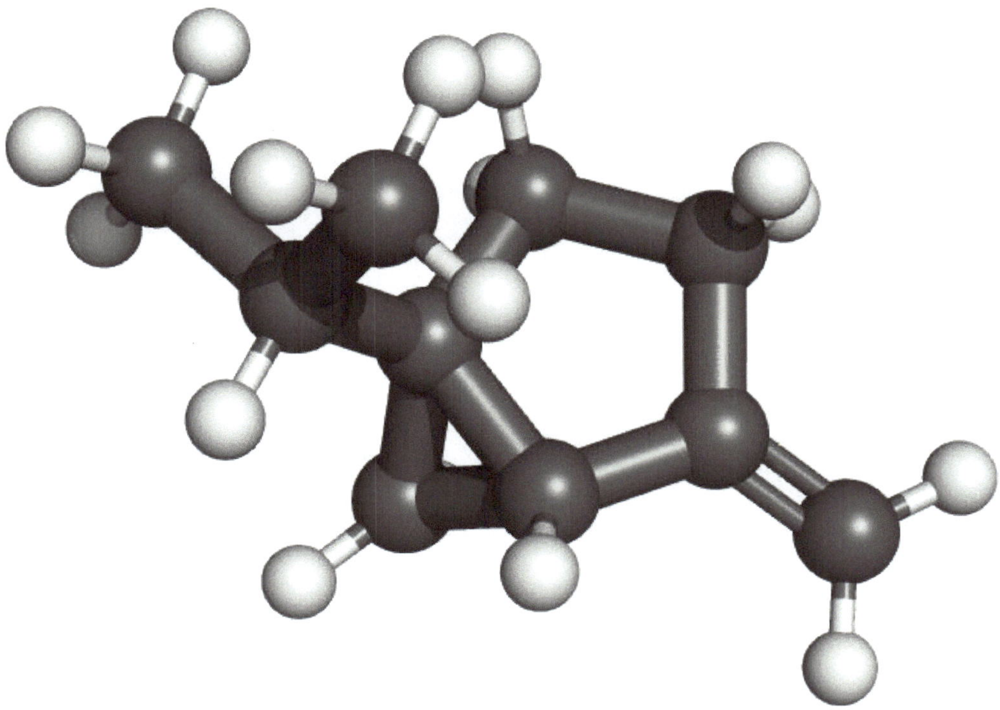

Sabinene is formed from geranyl pyrophosphate and is comprised of two linked units of isoprene with ten carbon atoms and sixteen hydrogen atoms, notated as $C_{10}H_{16}$ and classified as a monoterpene. The carbon skeleton of this isoprenoid compound consists of a strained ring (the triangular, strained ring is visible in the molecule above, in the bottom-center of the carbon skeleton) made from cyclopentane, which is fused to a larger, unstrained ring made of cyclopropane. The presence of these two rings further delineates the molecule as a bicyclic monoterpene. Sabinene contains one exocyclic double bond fused to the cyclopropane ring.

Sabinene occurs as several isomers, including the dextrorotatory ((+)-sabinene) and laevorotatory ((-)-sabinene) versions, and a double bond isomer known as thujene, which to date has not been found in cannabis so we have not included it here.

OCCURRENCE IN PLANTS

Sabinene occurs in a large number of plants, mostly in herbs, spices, and some common essential oils

Sabinene:

Achillea tenuifolia
Ajowan
Allspice
Alpinia galanga
Angelica
Anise
Anise star
Artemisia
Ayou wood

[923] Eric Conti, Claudia Zadra, Gianandrea Salerno, Benedetta Leombruni, Daniela Volpe, Francesca Frati, Cesare Marucchini, Ferdinando Bin. Changes in the volatile profile of Brassica oleracea due to feeding and oviposition by Murgantia histrionica (Heteroptera: Pentatomidae) Eur. J. Entomol. 105: 839-847, 2008.

Basil
Bergamot
Black pepper
Buchu
Cananga
Cardamom
Cassia
Celery
Citronella
Cornmint
Cumin
Dill
Fennel
Fig
Frankincense
Grapefruit
Holm oak
Hyssop
Kewda
Lavandin
Levisticum
Mace
Mastic
Myristica fragrans
Nutmeg
Parsley
Phoebe
Pomelo
Ravintsara
Sage
Savory
Sunflower
Tansy
Thyme
Vassoura
Wormwood
Zanthooxylum

Bay laurel
Betel
Blood orange
Cajuput
Cannabis
Carrot
Cassis
Chamomile
Clary sage
Croton
Curry
Eucalyptus
Ferula
Fir needle
Geranium
Hinoki
Horsemint
Jambu
Labdanum
Layana
Lime
Mandarin
Microstrobos
Neroli
Orange
Parsnip
Pimento
Pteronia
Rose
Salvia
Snakeroot
Tagete
Tarragon
Turmeric
Verbena
Yarrow

Bell pepper
Black currant
Boldus
Calamus
Caraway
Cascarilla
Cedar
Cinnamon
Coriander
Cubeb
Cypress
Featherfew
Feverfew
Fleabane
Ginger
Ho leaf
Hyptis
Juniper
Laurel
Lemon
Lovage
Marjoram
Mikan
Norway spruce
Oregano
Peppermint
Pine
Raspberry
Rosemary
Savin
Spearmint
Tangerine
Tea tree
Valerian
Water mint
Yuzu

(+)-sabinene:

Nutmeg
Sage
Turmeric

(-)-sabinene:

Bay laurel

BIOGICAL ACTIVITY IN PLANTS

The biological roles of sabinene in plants are not well-studied, however, research indicates that this bicyclic monoterpene serves antifeedant, allelopathic, antifungal, and insect repellant functions:

**Antifeedant:* Despite the appeal of sabinene to some herbivores, others are deterred by this bicyclic monoterpene. For instance, the Sitka black-tailed deer and the European roe deer are both repelled by sabinene[924], indicating its likely role as an antifeedant.

**Allelopathic Agent:* As a primary constituent in the essential oil of ravensara aromatica sonn. (clove nutmeg), sabinene contributed to germination inhibition and seedling toxicity in rice (oryza sativa), garden cress (lepidium sativum), and greengram (vigna radiata)[925], with sabinene chemotypes displaying the highest level of toxicity among the several chemotypes tested. Additionally, as the primary constituent in the essential oil of the perennial ground cover nepeta x faassenii (catmint), sabinene exhibited significant inhibitory effects on the seedling growth of lepidium sativum[926] (curly cress).

**Antifungal Agent:* Cypress trees infected with seiridium cardinale (the main fungal pathogen that causes cypress bark cankers) have been shown to increase emission rates of sabinene within the first day of infection, increasing rates of this bicyclic monoterpene until reaching a peak around ten days later[927]. This increase of sabinene did not occur in healthy control trees, or in trees subjected to mechanical wounding.

**Insect Repellant:* As the primary constituent in the essential oil extracted from the rhizomes of zingiber purpureum roscoe, sabinene exhibited strong repellency against tribolium castaneum[928] (red flour beetle), a common pest of stored grains and other foods.

USES IN INDUSTRY

Commercial samples of sabinene appear as a pale yellow to colorless liquid. It derives from essential oils of certain oak and spruce trees. Production of this compound occurs in small batches, as there is currently no efficient high-volume manufacturing process[929], despite previous successful attempts to produce the terpene by feeding carbon to an engineered strain of E. coli bacteria[930], in an effort to produce an alternative type of jet fuel.

The most common use of sabinene is to add flavor to processed food items including dairy products, fats, oils, sherbets, sorbets, processed fruit, processed vegetables, nuts, seeds, cereal, bakery items, meat products, sweeteners, spices, soups, sauces, and beverages. Sabinene is also used to add fragrance to some perfumes, fine chemicals, and advanced biofuels.

---

[924] Gwenael Vourc'h, Michel De Garine-Wichatitsky, Alice Labbe, Dimitri Rosolowski, Jean-Louis Martin, Herve Fritz. Monoterpene Effect on Feeding Choice by Deer Journal of Chemical Ecology December 2002, Volume 28, Issue 12, pp 2411-2427.

[925] Andrianjafinandrasana, S.N. & Andrianoelisoa, H.S. & Jeanson, M.L. & Ramonta, I. & Danthu, Pascal. (2013). Allelopathic effects of volatile compounds of essential oil from Ravensara aromatica Sonnerat chemotypes. Allelopathy Journal. 31. 333-344.

[926] Eom SH, Yang HS, Weston LA. An evaluation of the allelopathic potential of selected perennial groundcovers: foliar volatiles of catmint (Nepeta x faassenii) inhibit seedling growth. J Chem Ecol. 2006 Aug;32(8):1835-48.

[927] Achotegui-Castells, Ander, Roberto Danti, Joan Llusia, Gianni Rocca, Sara Barberini, and Josep Peñuelas. Strong Induction of Minor Terpenes in Italian Cypress, Cupressus sempervirens, in Response to Infection by the Fungus Seiridium cardinale. Journal of Chemical Ecology 41.3 (2015): 224-243.

[928] Wang Y, You CX, Yang K, Wu Y, Chen R, Zhang WJ, Liu ZL, Du SS, Deng ZW, Geng ZF, Han J. Bioactivity of Essential Oil of Zingiber purpureum Rhizomes and Its Main Compounds against Two Stored Product Insects. J Econ Entomol. 2015 Jun;108(3):925-32.

[929] Yujin Cao, Haibo Zhang, Hui Liu, Wei Liu, Rubing Zhang, Mo Xian, Huizhou Liu. Biosynthesis and production of sabinene: current state and perspectives Applied Microbiology and Biotechnology February 2018, Volume 102, Issue 4, pp 1535-1544.

[930] Haibo Zhang, Qiang Liu, Yujin Cao, Xinjun Feng, Yanning Zheng, Huibin Zou, Hui Liu, Jianming Yang, Mo Xian. Microbial production of sabinene-a new terpene-based precursor of advanced biofuel Microbial Cell Factories volume 13, Article number: 20 (2014).

POTENTIAL USES IN MEDICINE

Sabinene exhibits many properties that could be valuable in medicine, including the potential use in sedative, antibacterial, acaricidal, larvicidal, antifungal, and anti-inflammatory applications. However, it is critical to note that liquid sabinene may be fatal if swallowed and aspirated into the lungs; caution must be used when handling preparations of sabinene or essential oils that are high in this monoterpene.

**Sedative:* As the primary constituent in the essential oil made from the leaves of lantana camara, sabinene was shown to exhibit significant sedative effects in mice via inhalation administration, potently decreasing locomotor activity[931].

**Treatment for Salmonella Infections:* In research aimed at finding treatments for multidrug-resistant bacteria, scientists demonstrated that sabinene was a more effective antibacterial agent against salmonella typhimurium (the bacterium that causes typhoid fever and other potentially life-threatening conditions) than Ciprofloxacin[932], the standard drug treatment for salmonella infections.

**Acaracidal Agent:* As the primary constituent in the essential oil made from cunila angustifolia, sabinene exhibited potent acaricidal action against larvae of the cattle tick riphicephalus boophilus microplus, killing nearly 100% of the larvae[933].

**Larvicidal Agent:* As a major component in the essential oil of blumea densiflora, an edible and traditional medicinal plant, sabinene contributed to 100% larval mortality against the mosquito anopheles anthropophagus[934], the primary vector of malaria in China and other East Asian countries. As the primary constituent in the essential oil of severinia monophyla, sabinene exhibited remarkable larvicidal activity against the mosquitos, Ae. aegypti and Ae. albopictus[935] (vectors of dengue, chikungunya, yellow fever, and Zika viruses).

**Antifungal Agent:* As the primary constituent in the essential oil made from the needles of juniperus communis, sabinene exhibited significant antifungal activity against the dermatophytes, microsporum canis, and trichophyton rubrum[936].

**Prevention of Schistosomiasis:* As the primary constituent in the essential oil of piper cubeba, sabinene was shown to kill cercariae (larval stage), schistosomula (advanced larval stage), and adult schistosoma mansoni flatworms[937] - the parasite responsible for causing Schistosomiasis.

---

[931] Godfried Dougnon, Michiho Ito Sedative effects of the essential oil from the leaves of Lantana camara occurring in the Republ ic of Benin via inhalation in mice Journal of Natural Medicines pp 1-11 24 August 2019.

[932] Vimal, Archana, Dharm Pal, Timir Tripathi, and Awanish Kumar. Eucalyptol, sabinene and cinnamaldehyde: potent inhibitors of salmonella target protein l-asparaginase. 3 Biotech 7.4 (2017): 1-4.

[933] Apel, Miriam, Vera Sarda Ribeiro, Sergio Bordignon, Amelia Henriques, and Gilsane Poser. Chemical composition and toxicity of the essential oils from Cunila species (Lamiaceae) on the cattle tick Rhipicephalus (Boophilus) microplus. Parasitology Research 105.3 (2009): 863-868.

[934] Liang Zhu, Yingjuan Tian. Chemical composition and larvicidal activity of Blumea densiflora essential oils against Anopheles anthropophagus: a malarial vector mosquito Parasitology Research November 2011, Volume 109, Issue 5, pp 1417-1422.

[935] Satyal, Prabodh, Ho Hieu, Nguyen Chuong, Nguyen Hung, Le Sinh, Pham The, Thieu Tai, Vu Hien, and William Setzer. Chemical composition, Aedes mosquito larvicidal activity, and repellent activity against Triatoma rubrofasciata of Severinia monophylla leaf essential oil. Parasitology Research 118.3 (2019): 733-742.

[936] Cabral, C., V. Francisco, C. Cavaleiro, M. J. Goncalves, M. T. Cruz, F. Sales, M. T. Batista, and L. Salgueiro. Essential Oil of Juniperus communis subsp. alpina (Suter) Celak Needles: Chemical Composition, Antifungal Activity and Cytotoxicity. Phytotherapy Research 26.9 (2012): 1352-1357.

[937] Magalhaes, Lizandra, Julia Souza, Kamila Wakabayashi, Rosangela Laurentiz, Adriana Vinh6lis, Karen Rezende, Guilherme Simaro, Jairo Bastos, Vanderlei Rodrigues, Viviane Esperandim, Daniele Ferreira, Antonio Crotti, Wilson Cunha, and Marcio Silva. In vitro efficacy of the essential oil of Piper cubeba L. (Piperaceae) against Schistosoma mansoni. Parasitology Research 110.5 (2011): 1747-1754.

**Treatment for Colon Cancer:* As a primary constituent in the essential oil of nutmeg, sabinene contributed to significant in vitro inhibitory effect on the growth of a colon cancer cell line[938] (undifferentiated Caco-2 cells).

**Antibacterial Agent:* As a primary constituent in the essential oil from the leaves, stems, and flowers of astrodaucus orientalis, sabinene contributed to significant antibacterial activity against both gram-positive (bacillus subtilis) and gram-negative (escherichia coli) bacteria[939]. Additionally, as a primary constituent in the essential oil of bay laurel (laurus nobilis linnaeus), sabinene contributed to strong antibacterial activity against all tested foodborne spoilage and pathogenic bacteria[940].

**Prevention of Muscular Atrophy:* Among the 8 terpenes tested from the essential oil of chrysanthemum boreale makino, sabinene was the most effective at preventing and reducing muscular atrophy in rats through several mechanisms, including by decreasing ROS, and inducing predominant recovery of reductions of myotube (muscle fiber) diameters[941].

**Anti-inflammatory Agent:* Because inflammation can cause and complicate a variety of diseases, scientists are constantly researching new potential anti-inflammatory medications. Sabinene showed promise in this regard when, as the primary constituent in the essential oil of a hallabong flower, it was found to inhibit nitric oxide and prostaglandin E2 production, and to suppress the LPS-induced (lipopolysaccharide) expression of cyclooxygenase-2 (COX-2) proteins while downregulating the production of inflammatory cytokines[942], among other anti-inflammatory functions.

IMPLICATIONS FOR HUMAN HEALTH & NUTRITION

Because sabinene occurs in so many different plants, it is reasonably easy to obtain this bicyclic monoterpene dietarily. When preparing, cooking, or serving food, consider working with spices and seasonings that are high in sabinene, such as allspice, anise, basil, black pepper, caraway, cardamom, cinnamon, clary sage, coriander, cumin, dill, fennel, ginger, lovage, mace, marjoram, nutmeg, oregano, parsley, peppermint, rosemary, sage, savory, spearmint, tarragon, thyme, and turmeric. Fruits that may contain this terpene include blood oranges, black currants, grapefruit, lemons, limes, mandarins, oranges, pomelos, red raspberries, tangerines, and yuzu. Vegetables that also likely contain sabinene include bell peppers, carrots, celery, and parsnip.

As mentioned previously, there is scant information at this time concerning which cannabis strains feature sabinene prominently, but it is thought that Arjan's Haze, Super Silver Haze, and Trainwreck contain this molecule. Vaporizing these strains or strains with similar genetics should occur at low temperatures only. Other options include working with essential oils that are high in sabinene, including certain cannabis oils.
Finally, consider cooking with sunflower oil, or use aromatherapy applications with rose oil, tea tree oil, or valerian root oil, among others. As always, extreme care should be exercised when working with purified compounds or complex essential oils, as these substances can be toxic in some situations.

---

938 Piras A, Rosa A, Marongiu B, Atzeri A, Dessi MA, Falconieri D, Porcedda S. Extraction and separation of volatile and fixed oils from seeds of Myristica fragrans by supercritical CO2: chemical composition and cytotoxic activity on Caco-2 cancer cells. J Food Sci. 2012 Apr;77(4):C448-53.

939 Torabbeigi, Marzieh, Parviz Abroomand Azar, Anousheh Sharifan, and Zahra Aghaei Meibodi. Antibacterial activity and comparison of the volatile constituents obtained by several extraction methods from the flowers, stems and leaves of Astrodaucus orientalis.. Natural product communications 7.2 (2012): 249-252.

940 Ramos, C, B Teixeira, I Batista, O Matos, C Serrano, N R Neng, J M F Nogueira, M L Nunes, and A Marques. Antioxidant and antibacterial activity of essential oil and extracts of bay laurel Laurus nobilis Linnaeus (Lauraceae) from Portugal. Natural product research 26.6 (2012): 518-529.

941 Yunkyoung Ryu, Donghyen Lee, Seung Hyo Jung, Kyung-Jin Lee, Hengzhe Jin, Su Jung Kim, Hwan Myung Lee, Bokyung Kim, and Kyung-Jong Won. Sabinene Prevents Skeletal Muscle Atrophy by Inhibiting the MAPK- MuRF-1 Pathway in Rats Int J Mol Sci. 2019 Oct; 20(19): 4955.

942 MJ, Yang KW, Kim SS, Park SM, Park KJ, Kim KS, Choi YH, Cho KK, Lee NH, Hyun CG. Chemical composition and anti-inflammatory effects of essential oil from Hallabong flower. EXCLI J. 2013 Nov 7;12:933-42. eCollection 2013.

Sabinene Review

Answer the following questions to test your knowledge of this terpene:

Question #1: What type of terpene is sabinene?

a. Monocytic monoterpene
b. Linear monoterpene
c. Mononuclear monoterpene
d. Bicyclic monoterpene

Question #2: Which of the following best describes the primary flavor & aroma profile of sabinene?

a. Citrus, fruity, dry, terpenic
b. Woody, spicy, terpenic, citrus
c. Citrus, woody, terpenic, spicy
d. Woody, carrot, spice, dry

Question #3: How common is sabinene in cannabis?

a. Top three
b. Top ten
c. Top twenty
d. Top thirty

Question #4: What is the chemical formula for sabinene?

a. $C_{10}H_{16}O$
b. C15H24O2
c. $C_{10}H_{16}$
d. $C_{20}H_{32}O$

Question #5: Name two biological roles of sabinene in plants:

1 ______________________________ 2 ______________________________

Question #6: Name two potential medical uses of sabinene:

1 ______________________________ 2 ______________________________

For the answer key to all individual compound chapters please go here: http://thebigbookofterps.com/bbot-quiz-answer-key/

CAMPHOR

Type: Monocyclic monoterpenoid
Chemical Formula: $C_{10}H_{16}0$
Molecular Weight: 152.237 g/mol
Boiling Point: *D-camphor:* 207.40 °C. @ 760.00 mm Hg[943]
L-camphor: 207.42 °C. @ 760.00 mm Hg (estimated)[944]
Flash Point: *D-camphor:* 148.00 °F. TCC (64.44 °C.)
L-camphor: 149.00 °F. TCC (65.00 °C.)
Melting Point: *D-camphor:* 178.80 °C. @ 760.00 mm Hg
L-camphor: 178.00 to 180.00 °C. @ 760.00 mm Hg
Solubility: *D-camphor:* Soluble in alcohol, insoluble in water, glycerin
L-camphor: Soluble in alcohol, weakly soluble in water
Oral LD50: 1,310 mg/kg (mouse)[945]
Biological Role: Insecticidal, allelopathic agent, antifeedant, antifungal
Therapeutic Role: DNA repair agent, antifungal, anticonvulsant, circulatory aid
Commercial Use: Fragrance, flavoring agent
Occurrence in Cannabis: Top thirty
Aroma/Flavor: *D-camphor:* Camphoreous, mint, cooling, green, woody
L-camphor: Camphoreous

Occurs in Cannabis Strains: Girl Scout Cookies

INTRODUCTION

Perhaps the most well-known terpene in the world, camphor has been used by humans consistently and prolifically for thousands of years. This cyclic (can also be referred to as monocyclic) monoterpenoid is so unique that it isn't generally labelled with individual flavor or aroma descriptors; the scent and taste of this compound is best described simply as camphoreous. However, while this is the primary feature of camphor, the volatile molecule also has cool, medicinal, and woody or herbaceous nuances, with some reviewers also describing coniferous or piney notes. Interestingly, camphor has been shown to inhibit the bioluminescence of Plasmodium fischeri bacteria[946].

Camphor is theorized to be a secondary terpene in cannabis, likely occurring within the top forty terpenes by concentration in some cannabis strains. Information about camphor in cannabis is scarce; as of the fall of 2023, research turned up not a single mention of camphor in cannabis on the internet, and our search of hundreds of cannabis terpene test results showed just one instance of this terpene within the top ten; Girl Scout Cookies at position #9. Because most laboratories today only report the top five to ten terpenes, it will likely continue to be exceedingly rare that we see this terpene appear in test results.

---

943 The Good Scents Company Data Sheet for D-camphor, from: http://www.thegoodscentscompany.com/data/rw1030211.html. Accessed September 7, 2023.

944 The Good Scents Company Data Sheet for L-camphor, from: http://www.thegoodscentscompany.com/data/rw1056911.html. ACcessed September 7, 2023.

945 Thermo FIsher Scientific Safety Data Sheet for Camphor, from: https://www.fishersci.com/store/msds?partNumber=AC227841000&productDescription=D+-CAMPHOR+100GRD+-C&vendorId=VN00032119&countryCode=US&language=en. Accessed September 7, 2023.

946 Danilov, Vadim, Nina Baranova, Anvar Ismailov, and Nicolai Egorov. The effect of camphor on bacterial bioluminescence. Applied Microbiology and Biotechnology 14.2 (2004): 125-129.

Among our research, camphor often occurs with 1,8-cineole. When one of these terpenes is the primary constituent of a particular essential oil, the other is often the next terpene by concentration.

Camphor has a long history with humans and is well-known to have been in use by ancient Sumerians, Egyptians, and particularly by ancient Indians[947]. Early uses primarily focused on camphor as a fragrance, although many traditional medicinal uses of camphor have been reported in cultures worldwide, with hundreds of camphor oils and other products on the market by the 1700's[948]. Earlier, during the Black Death plague that wiped out nearly two-thirds of humans in Europe in the Middle Ages, camphor was used as a fumigant[949], and to mask the smell of rotting corpses.

CHEMICAL STRUCTURE

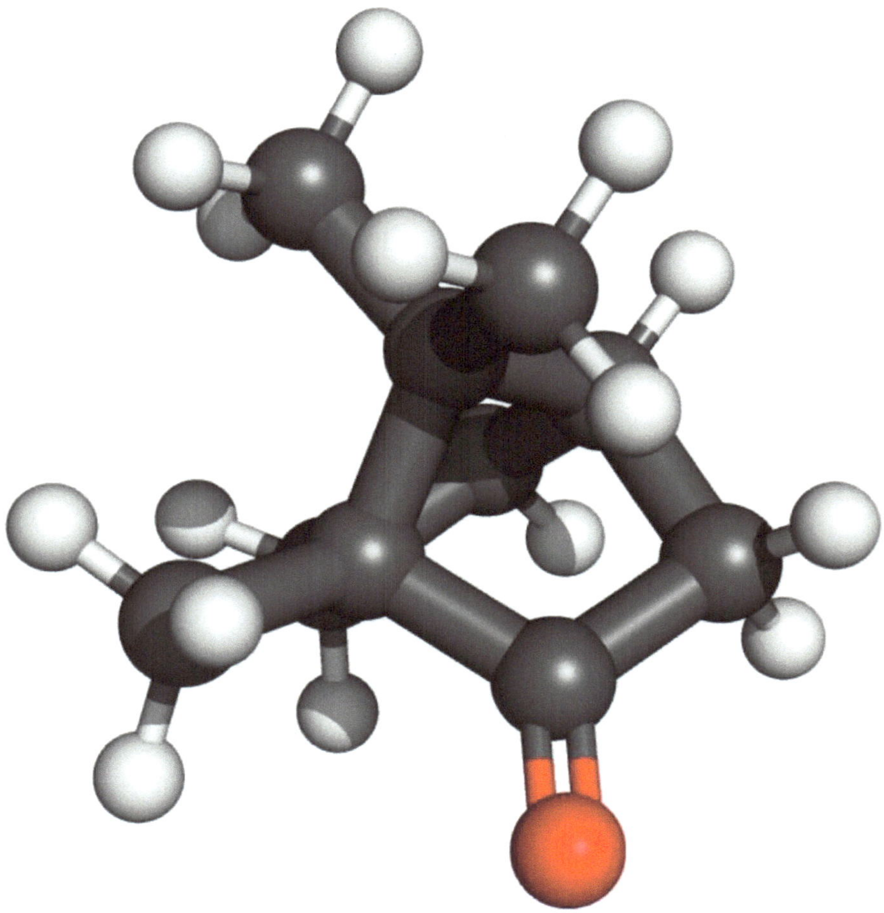

Camphor is derived from geranyl pyrophosphate and oxidation of borneol. This compound is comprised of two linked units of isoprene, with a molecular skeleton containing ten carbon atoms, sixteen hydrogen atoms, and one oxygen atom that is double-bonded, giving camphor a chemical formula of $C_{10}H_{16}O$. This volatile molecule is formed around one central cyclopentane ring, making it monocyclic, and its oxygenated state further classifies camphor as a monoterpenoid.

[947] Donkin, R.A. (1999). Dragon's Brain Perfume: An Historical Geography of Camphor. Brill. ISBN 978-9-004- 10983-4.

[948] Camphor - National Museum of American History - Smithsonian Institute. From: https://americanhistory.si.edu/collections/search/object/nmah_993816 Accessed December 10, 2019.

[949] Chen, Weiyang et al. Camphor--a fumigant during the Black Death and a coveted fragrant wood in ancient Egypt and Babylon--a review. Molecules (Basel, Switzerland) vol. 18,5 5434-54. 10 May. 2013.

OCCURRENCE IN PLANTS

Camphor is found in many plants and trees, although it is primarily sourced from the essential oil of camphor laurel, camphor basil, and other essential oils where camphor is a primary constituent:

D-camphor:

Anise	Apple	Basil
Bay	Black pepper	Cardamom
Carrot	Cayenne	Cinnamon
Coriander	Dill	Fennel
Ginger	Hyssop	Lavandula
Lovage	Mugwort	Nutmeg
Oregano	Rosemary	Saffron
Sage	Sassafras	Savory
Sunflower	Tarragon	Thyme
Turmeric	Wormwood	

L-camphor:

Armoise	Matricaria parthenium	Salvia

BIOLOGICAL ACTIVITY IN PLANTS

Camphor is a potent insecticidal agent in plants, and likely serves many more biological functions than the current scientific literature reports.

Insecticidal Agent: Numerous studies have confirmed the insecticidal activity of camphor, including effects against some of the most important insect pests in the world. For instance, as the primary constituent in the essential oil of cinnamomum camphora, camphor inflicted a nearly 85% mortality rate against red imported fire ants (solenopsis invicta), in addition to exhibiting fumigant toxicity against minor and major worker ants, strongly repelling both groups[950]. As a primary constituent in the essential oil made from the rhizomes of curcuma wenyujin, camphor contributed to both contact toxicity and fumigant toxicity against liposcelis bostrychophila (the common booklouse), and individually was 5 times more potent than the primary constituent in the same oil, 1,8-cineole[951]. Camphor has also been shown to exhibit contact toxicity against callosobruchus maculatus (cowpea weevil), fumigant toxicity against tribolium castaneum[952] (the red flour beetle, in which camphor has been found to cause 100% mortality via fumigant toxicity against larvae, pupae, and adults), and both contact and fumigant toxicity against T. putrescentiae[953] (cheese mites).

[950] Fu, J T et al. Fumigant Toxicity and Repellence Activity of Camphor Essential Oil from Cinnamonum camphora Siebold Against Solenopsis invicta Workers (Hymenoptera:Formicidae). Journal of insect science (Online) vol. 15, 1129. 21 Sep. 2015.

[951] Liu, Zhi Long et al. Identification of insecticidal constituents of the essential oil of Curcuma wenyujin rhizomes active against Liposcelis bostrychophila Badonnel. Molecules (Basel, Switzerland) vol. 17,10 12049-60. 15 Oct. 2012.

[952] Tripathi, Aran, Anil Singh, and Shikha Upadhyay. Contact and fumigant toxicity of some common spices against the storage insects Callosobruchus maculatus (Coleoptera: Bruchidae) and Tribolium castaneum (Coleoptera: Tenebrionidae). International Journal of Tropical Insect Science 29.3 (2009): 151-157.

[953] Lee, Myung-Ji, Sung-Eun Lee, Min-Seung Kang, Bueyong Park, Sang-Guei Lee, and Hoi-Seon Lee. Acaricidal and insecticidal properties of Coriandrum sativum oils and their major constituents extracted by three different methods against stored product pests. Journal of the Korean Society for Applied Biological Chemistry 61.5 (2018): 481-488.

Allelopathic Agent: As the primary constituent in the essential oil from salvia leucophylla (a shrub common to California), camphor inhibited the germination of brassica campestris (field mustard) seeds at high concentrations, while also inhibiting root growth and interfering with the growth of nearby plants via the inhibition of cell proliferation in the root apical meristem[954].

Antifeedant: Scientists in Canada set out to understand why snowshoe hares (lepus americanus), an abundant herbivore in the region, rarely browsed juvenile white spruce trees, but regularly ate leaves and branches from seedling and mature trees. When researchers found that juvenile white spruce trees emitted 4 times as much camphor than mature trees, tests were conducted with rabbits which proved that the hares preferred rabbit chow without high concentrations of camphor[955].

Antifungal Agent: Camphor oil was shown to inhibit the growth of Aspergillus niger (black mold on vegetables and fruits) by 100%, Aspergillus flavus (rot in grains, nuts, and legumes post-harvest) by more than 96%, and penicillium spp. by nearly 85%[956] when applied topically to certain handicrafts.

USES IN INDUSTRY

Commercial samples of camphor appear as waxy, white, colorless, translucent or transparent crystalline powders. Camphor is generally derived from steam distillation of mature cinnamomum camphora trees that have been finely ground, or synthetically from alpha-pinene.

Today the main use of camphor is as a cough suppressant and as a decongestant. This volatile molecule is also used to add fragrance to commercial and household products like air fresheners, makeup, perfume, moth-ball alternatives, foot care products, laundry soaps, dishwashing liquids, etcetera, and as a flavor enhancer in baked goods, condiments, frozen dairy, hard candy, soft candy, chewing gum, and alcoholic beverages, among other grocery items.

There are thousands of patents for products containing camphor. However, after a spate of deaths and injuries involving camphor, the US Food and Drug Administration enacted policy in 1980 that set the allowable limit of camphor in consumer products at not more than 11%, with stipulations that labeling and marketing materials refrain from mentioning camphor, except in ingredient lists. This is because camphor can be extremely toxic to people, more on this is covered in the "Potential Medical Uses" section below.

Interestingly, cinnamomum camphora trees have been shown to decontaminate soil of heavy metals, in one case sequestering 68.8 tons of biomasss per acre, with the trees flourishing despite the heavily contaminated soil conditions[957]. Other research has led to the development of nanospheres constructed from magnetized camphor[958], which could one day be used in several different medical and industrial applications.

---

[954] Nishida, Nami, Satoshi Tamotsu, Noriko Nagata, Chieko Saito, and Atsushi Sakai. Allelopathic Effects of Volatile Monoterpenoids Produced by Salvia leucophylla: Inhibition of Cell Proliferation and DNA Synthesis in the Root Apical Meristem of Brassica campestris Seedlings. Journal of Chemical Ecology 31.5 (2005): 1187-1203.

[955] Sinclair, A R, M K Jogia, and R J Andersen. Camphor from juvenile white spruce as an antifeedant for snowshoe hares. Journal of chemical ecology 14.6 (2013): 1505-14.

[956] Mahilrajan, S., Nandakumar, J., Kailayalingam, R. et al. Screening the antifungal activity of essential oils against decay fungi from palmyrah leaf handicrafts. biol res 47, 35 (2014).

[957] Ho, Jian-Ren, Hwong-Wen Ma, Yi-Chung Wang, Chun-Han Ko, Fang-Chih Chang, Fong-Long Feng, and Ya- Nang Wang. Extraction of heavy metals from contaminated soil by Cinnamomum camphora. Ecotoxicology 23.10 (2014): 1987-1995.

[958] Ali Maleki, Pedram Zand & Zahra Mohseni Design and development of a novel magnetic camphor nanospheres core/shell nanostructure Journal of Nanostructure in Chemistry volume 7, pages149-157(2017).

POTENTIAL USES IN MEDICINE

Camphor can be toxic to humans via several administration routes, including dermally and through inhalation, which makes this substance somewhat dangerous to work with. Just ½ gram to 2 grams of camphor can be fatal to children, and dangerously toxic to adults[959] at slightly higher doses. Camphor is a central nervous system depressant that can lead to seizures and convulsions, delirium, hallucinations, and death. In fact, during research for this book, we uncovered more case studies and toxicology reports about camphor as a toxin to humans than we did studies mentioning the monoterpenoid in some beneficial way. The lead scientist who reviewed a case study of camphor-induced myocarditis (heart tissue inflammation) has succinctly summed-up what we've found in our research:

"The medicinal uses of camphor are unsupported by evidence and safer, more effective alternatives exist. Its use in household products and edibles should be discouraged[960]."

These sentiments are mostly echoed by the US Food & Drug Administration. However, it appears that camphor may still have some limited medical benefit that could be developed with further research.

Decrease Rate of Mitochondria Oxygen Consumption: Camphor has been shown to halve the rate of oxygen consumption of mitochondria in rat livers[961] and kidneys[962], which could be used to oxygenate tumors prior to treatment.

Antifungal Agent: In work that further indicates camphor's effect on mitochondria, this monocyclic terpene was found to markedly decrease cell viability and proliferation and induce apoptotic cell death via the reduction of mitochondrial transmembrane potential and induction of oxidative stress in the fission yeast schizosaccharomyces pombe[963]. As the primary constituent in the essential oil made from the leaves of cedar trees, camphor significantly reduced Candida albicans biofilm formation and inhibited hyphal formation[964].

DNA Repair Agent: Additional evidence that camphor affects DNA includes research that shows that: "by making a small amount of DNA lesions, camphor can stimulate error-free DNA repair processes and act as bioantimutagens[965]." This is important work considering that mutagens can induce permanent changes in the DNA sequence of an organism, with scientists showing that camphor can counteract these effects.

Anticonvulsant Agent: Although camphor has been shown to cause convulsions in humans, it has been shown to reduce the number and severity of convulsions in mice and guinea pigs[966].

---

[959] Sarah McCrea - Monograph for Camphor - National Poisons Information Service (London Centre) Medical Toxicology Unit Guy's & St Thomas' Hospital Trust Avonley Road London SE14 5ER UK. http://www.inchem.org/documents/ukpids/ukpids/ukpid19.htm Accessed December 10, 2019.

[960] Bhaya, Maneesha, and Rajesh Beniwal. Camphor Induced Myocarditis: A Case Report. Cardiovascular Toxicology 7.3 (2007): 212-214.

[961] Guilland-Cumming, D., and G. Smith. The effect of camphor on mitochondrial respiration. Cellular and Molecular Life Sciences 38.2 (2005): 236-237.

[962] Guilland-Cumming, D., and G. Smith. Mitochondrial respiration depressed by camphor: A possible aid in radiotherapy. Cellular and Molecular Life Sciences 35.5 (2005): 659-659.

[963] Agus, Hizlan Hincal et al. Oxidative stress-mediated apoptotic cell death induced by camphor in sod1-deficient Schizosaccharomyces pombe. Toxicology research vol. 8,2 216-226. 27 Nov. 2018.

[964] Manoharan, Ranjith Kumar et al. Antibiofilm and Antihyphal Activities of Cedar Leaf Essential Oil, Camphor, and Fenchone Derivatives against Candida albicans. Frontiers in microbiology vol. 8 1476. 3 Aug. 2017.

[965] Nikolic, Biljana, Dragana Mitic-Culafic, Branka Vukovic-Gacic, and Jelena Knezevic-Vukcevic. Modulation of genotoxicity and DNA repair by plant monoterpenes camphor, eucalyptol and thujone in Escherichia coli and mammalian cells. Food and chemical toxicology : an international journal published for the British Industrial Biological Research Association 49.9 (2011): 2035-45.

[966] A. P. Vereshchagin & S. M. Dionesov The influence of nociceptive stimuli on the convulsant action of camphor Bulletin of Experimental Biology and Medicine volume 49, pages268-270(1960).

**Circulatory Aid:* Camphor has been shown to induce a cold sensation that fades to a warm sensation, while increasing local blood flow to the skin and muscles[967] surrounding the site of application.

**Larvicidal Agent:* Prepared in a solution at 32% concentration, camphor has been shown to kill more than 93% of sheep blowfly larvae via contact toxicity[968].

**Respiratory Stimulant:* Camphor is currently used in veterinary medicine to aid in the treatment of breathing difficulties in horses. In general, this treatment consists of an injection containing 200 mg of camphor suspended in a sesame oil base, with benzyl alcohol as a preservative[969].

IMPLICATIONS FOR HUMAN HEALTH & NUTRITION

For most people, increasing intake of dietary camphor is not necessary, nor medically advisable. Likewise, for cannabis breeders, producing strains with higher amounts of camphor by concentration is not medically desirable. Trace or dietary levels of camphor can be obtained by preparing, cooking, or serving foods with spices and herbs like anise, basil, bay, cinnamon, coriander, dill, fennel, ginger, marjoram, rosemary, sage, savory, thyme, and turmeric. Lemongrass teas might also contain some camphor. For vegetables, consider eating carrots and ginger, while guava and lemon are two fruits that contain at least trace amounts of camphor.

At the time of the publishing of this text, we were only able to find one instance of camphor occurring in a cannabis strain within the top ten terpenes by concentration; that was for Girl Scout Cookies at position #9. This could mean that other strains in the Girl Scout Cookies family - such as Platinum Cookies, or Thin Mint GSC - contain camphor. Always remember to vaporize these or other strains at the appropriate temperature if terpene consumption is your goal. Cannabis oils, tinctures, or other products may also contain camphor in some cases.

---

[967] Kotaka T, Kimura S, Kashiwayanagi M, Iwamoto J. Camphor induces cold and warm sensations with increases in skin and muscle blood flow in human. Biol Pharm Bull. 2014;37(12):1913-8.

[968] Shalaby, Hatem, Rabab El Khateeb, Amira El Namaky, Heba Ashry, Omnia Kandil, and Sanaa Abou El Dobal. Larvicidal activity of camphor and lavender oils against sheep blowfly, Lucilia sericata (Diptera: Calliphoridae).

[969] Univet. Camphor Injections. https://www.drugs.com/vet/camphor-injection-can.html. Accessed December 1, 2019.

Camphor Review

Answer the following questions to test your knowledge of this terpene:

Question #1: What type of terpene is camphor?

a. Manocyclic monoterpene
b. Megalithic monoterpene
c. Monocyclic monoterpenoid
d. Monocyclic sesquiterpene

Question #2: Which of the following best describes the primary flavor & aroma profile of D-camphor?

a. Lemon, turpentine, pine
b. Woody, green, floral, sweet
c. Camphoreous, mint, cooling, green, woody
d. Citrus, fruity, dry, terpenic

Question #3: How common is camphor in cannabis?

a. Top thirty
b. Top ten
c. Top twenty
d. Top forty

Question #4: What is the chemical formula for camphor?

a. $C_{10}H_{16}$
b. $C_{10}H_{16}O$
c. $C_{10}H_{24}O$
d. $C_{10}H_{18}O$

Question #5: Name two uses of camphor in plants:

1 ________________________________ 2 ________________________________

Question #6: Name two potential medical uses of camphor:

1 ________________________________ 2 ________________________________

For the answer key to all individual compound chapters please go here: http://thebigbookofterps.com/bbot-quiz-answer-key/

MENTHOL

Type: Monocyclic monoterpenoid
Chemical Formula: $C_{10}H_{20}O$
Molecular Weight: 156.26 g/mol
Boiling Point: 212.00 to 216.00 °C. @ 760.00 mm Hg[970]
Flash Point: 190.00 °F. TCC (87.78 °C.)
Melting Point: 38 °C (100.4 °F)
Solubility: Slightly soluble in water, ethanol, soluble in alcohol, ether, hexane, various oils, chloroform
Oral LD50: 3,400 mg/kg (mouse)[971]
Biological Role: Insecticidal
Therapeutic Role: Permeability enhancer, analgesic, anticancer agent
Commercial Use: Flavoring agent, fragrance
Occurrence in Cannabis: Top thirty
Aroma/Flavor: Peppermint, cooling, camphoreous, fresh

Occurs in Cannabis Strains: Unconfirmed reports in Green Monster, Wonder Woman OG, Himalayan Gold, Cabbage Patch, Space Needle[972]

INTRODUCTION

Like camphor, menthol is one of the most well-known terpenes in the world, occurring in hundreds of consumer products over more than a century. Menthol is a monocyclic monoterpenoid that is almost always derived from plants in the mint family. With an associated aroma and flavor profile that is cooling, minty, aromatic, and fresh, menthol is used for a wide variety of purposes, including as a topical analgesic agent, as a flavor or fragrance additive, a permeability enhancer, and a cooling agent.

Often a top forty terpene by concentration in some cannabis strains, menthol is a secondary terpene, rarely occurring within the top ten terpenes in a given strain. In fact, in our review of hundreds of cannabis terpene tests, there were no occurrences of menthol. We primarily reviewed the top ten or fewer strains in each test because most laboratories only report five to ten terpenes. This means it is likely that menthol occurs in many strains, but until standardized terpene testing includes the top twenty to forty terpenes in cannabis, we will not see this terpene in most laboratory test results.

---

[970] The Good Scents Company Data Sheet for L-menthol, from: http://www.thegoodscentscompany.com/data/rw1030471.html. Accessed September 7, 2023.

[971] Cayman Chemical Company Safety Data Sheet for Menthol, from: https://cdn.caymanchem.com/cdn/msds/30384m.pdf. Accessed September 7, 2023.

[972] Ionization Labs - The Cannabis Terpene Menthol, from: https://ionizationlabs.com/the-cannabis-terpene- menthol/ Accessed December 12, 2019.

CHEMICAL STRUCTURE

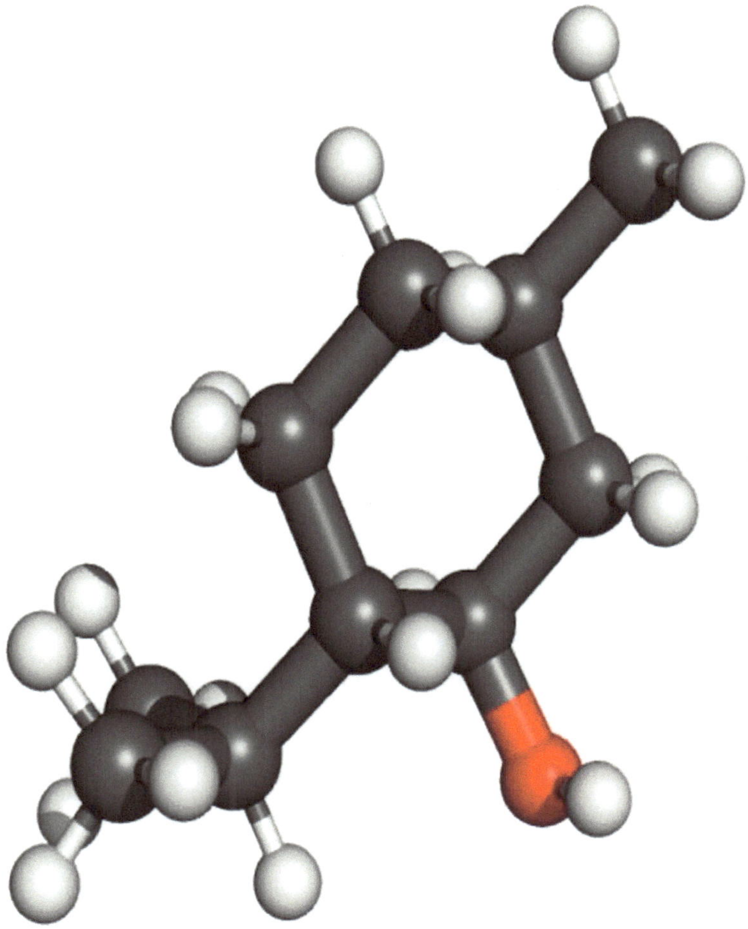

Like most monoterpenes, menthol is derived from geranyl diphosphate and the standard linked isoprene units with ten carbon atoms, twenty hydrogen atoms, and an oxygen atom notated as $C_{10}H_{20}O$. Menthol has one cyclohexane ring, which makes it a monocyclic compound, and its oxygenated state further classifies it as a monoterpenoid. Menthol occurs as 8 isomers, including the following:

(+)-menthol (D-menthol)
(+)-isomenthol
(+)-neomenthol
(+)-neoisomenthol
(-)-menthol (L-menthol)
(-)-isomenthol
(-)-neomenthol
(-)-neisomenthol

L-menthol is the most common variation and the one most likely found in cannabis, however, laboratory testing for the compound is rare in cannabis products and if it does happen it likely doesn't differentiate between isomers.

OCCURRENCE IN PLANTS

Menthol is generally found in plants in the mint family, but it has also been reported in the oils of these plants:

Basil	Black elder	Bitter melon
Cabbage	Coconut	Coriander
Corn	Cornmint	Guava
Horsemint	Lovage	Marjoram
Mint	Pennyroyal	Peppermint
Sage	Spearmint	Sunflower
Tarragon	Tea	Water mint

BIOLOGICAL ACTIVITY IN PLANTS

Scant information exists regarding the biological roles of menthol in plants, as most studies have been focused in other areas, particularly the compound's cooling effect, and other studies as noted in the "Potential Uses in Medicine" section below. One study confirmed the strong possibility that the monocyclic monoterpenoid serves an insecticidal role:

Insecticidal Agent: As isolated from the essential oil of mentha arvensis, menthol was shown to exhibit high toxicity in contact and vapor toxicity assays against several insects, including callosobruchus maculatus F., rhyzopertha dominica F., sitophilus oryzae L. and tribolium castaneum herbst, while also repelling insects in a dose-dependent manner with up to 100% effectiveness[973].

USES IN INDUSTRY

Commercial samples of menthol appear as waxy, clear, or white crystalline solids which are generally derived from the essential oil of plants in the mint family, especially mentha arvensis (wild mint). Researchers have also produced menthol from engineered Saccharomyces cerevisiae[974] (Brewer's yeast).

Menthol is used to add a minty taste to some foods like baked goods, beverages, chewing gum, frozen dairy, fruit ices, hard candy, and more. Hundreds of patents for products containing menthol exist, including lip balms, skin ointments, cooling gels, hair care products, toothpaste and dental floss, and many other products.

Interestingly, menthol has been used experimentally in both Europe and Japan to combat mite infestations in honeybees. Menthol has been shown to increase colony survival rates and reduce parasitic loads after infection with tracheal mites (acarapis woodi)[975], a common problem that can easily destroy bee colonies.

POTENTIAL USES IN MEDICINE

Although menthol has several therapeutic properties that could be beneficial medically, the majority of studies in recent years investigated menthol's links to smoking, tobacco products, addiction, withdrawal, relapse, and other serious public health concerns. For instance, one study used rats to show that self-administration of nicotine was enhanced with menthol, with the study authors concluding that: "The findings suggest that menthol directly facilitates nicotine consumption, thereby contributing to tobacco smoking[976]." Other studies have focused on the

[973] Aggarwal, K., A. Tripathi, Ateeque Ahmad, Veena Prajapati, Neetu Verma, and Sushil Kumar. Toxicity of l- Menthol and Its Derivatives Against Four Storage Insects. International Journal of Tropical Insect Science 21.3 (2001): 229-235.

[974] Lv X, Zhou X, Ma J, Tao M, Liu Y, Li J, Du G, Liu L. Engineered Saccharomyces cerevisiae for the De Novo Biosynthesis of (-)-Menthol. J Fungi (Basel). 2022 Sep 19;8(9):982.

[975] Maeda, Taro, and Yoshiko Sakamoto. Field application of menthol for Japanese honey bees, Apis cerana japonica (Hymenoptera: Apidae), to control tracheal mites, Acarapis woodi (Acari: Tarsonemidae). Experimental and Applied Acarology 70.3 (2016): 299-308.

[976] Biswas, Lisa, Erin Harrison, Yongzhen Gong, Ramachandram Avusula, Jonathan Lee, Meiyu Zhang, Thomas Rousselle, Janice Lage, and Xiu Liu.

social impact of menthol cigarette advertising schemes, which have historically insinuated that menthol cigarettes have some medical value, with 7% of African Americans and 5% of White smokers stating that they smoke menthol cigarettes because they 'are better for you[977].' Still more studies are far graver, with one reporting that the link between menthol tobacco use and mental health problems is clear; "Menthol is disproportionately used among young adult tobacco users with mental health problems, above and beyond the impact of a variety of other mental health and tobacco use risk factors[978]." Finally, researchers in 2022 studying the allergiuc reactions of menthol concluded that: "Menthol is capable of provoking allergic hypersensitivity reactions and disorders, including asthma, allergic rhinitis, and urticaria. It may also trigger an anaphylactic response[979]."

Despite the serious implications menthol and tobacco have, there is significant medical interest in this monoterpenoid. Many clinical trials involving menthol are ongoing, including trials in 2019 that examined menthol to treat painful diabetic peripheral neuropathy, another for general pain relief, and yet another to discover if menthol made nicotine more desirable in tobacco users, among dozens of other clinical trials[980].

While the most well-established medical use of menthol is as a permeability enhancer, the compound also offers medical potential in pain management, cancer treatment, multi-drug resistant antibacterials, and more.

**Permeability & Absorption Enhancer:* Menthol has been shown to be an excellent skin permeability enhancer by fluidizing the stratum corneum lipids, perturbing the ordered lipids, and making the lipids floppier[981], which allows easier permeation of other molecules, particularly medical drugs. Menthol has also been shown to significantly enhance the penetration of dexamethasone in the corneas of rabbits, without toxicity[982]. Menthol has been demonstrated to enhance the intestinal absorption of daidzein (a phytonutrient that occurs in legumes), increasing the relative bioavailability of the compound by 3.65 times[983], while yet another study found that menthol increases the permeability of quercetin[984] - a flavonoid found in cannabis and many other plants - by decreasing the dermal central energy barrier.

**Treatment for Depression:* Menthol was shown to produce anti-depressant like effects in mice, while also exhibiting synergistic effects with other drugs used to treat depression[985].

**Prevention of Colonic Spasms:* Menthol administered intraluminally (in the gastrointestinal tract) was effective in preventing colonic spasms in 70.8% of 65 patients tested during a colonoscopic examination[986].

Enhancing effect of menthol on nicotine self-administration in rats. Psychopharmacology 233.18 (2016): 3417-3427.

[977] Rising, Joshua, and Lori Alexander. Marketing of menthol cigarettes and consumer perceptions. Tobacco Induced Diseases 9.1 (2011): 1-8.

[978] Cohn, Amy, Amanda Johnson, Elizabeth Hair, Jessica Rath, and Andrea Villanti. Menthol tobacco use is correlated with mental health symptoms in a national sample of young adults: implications for future health risks and policy recommendations. Tobacco Induced Diseases 14.1 (2016): 1-10.

[979] Azizli E, Dilber M. Do products containing menthol exacerbate allergic rhinitis? A narrative review. Eur Rev Med Pharmacol Sci. 2022 Dec;26(2 Suppl):61-64.

[980] US National Library of Medicine. ClinicalTrials.gov. Accessed April 4, 2020.

[981] Wang, Huanjie, and Fancui Meng. The permeability enhancing mechanism of menthol on skin lipids: a molecular dynamics simulation study. Journal of Molecular Modeling 23.10 (2017): 1-10.

[982] Xu, Xiaolin, Nannan Yu, Zhengzhong Bai, Yanbin Xun, Di Jin, Zhijian Li, and Hao Cui. Effect of menthol on ocular drug delivery. Graefe's Archive for Clinical and Experimental Ophthalmology 249.10 (2011): 1503-1510.

[983] Shen, Qi, Xi Li, Wenji Li, and Xinyi Zhao. Enhanced intestinal absorption of daidzein by borneol/menthol eutectic mixture and microemulsion.. AAPS PharmSciTech 12.4 (2012): 1044-9.

[984] Huang, Changjiang, Huangjie Wang, Lida Tang, and Fancui Meng. Penetration enhancement of menthol on quercetin through skin: insights from atomistic simulation. Journal of Molecular Modeling 25.8 (2019): 1-9.

[985] Wang, Weidong, Yuanyuan Jiang, Enbo Cai, Bingchen Li, Yan Zhao, Hongyan Zhu, Lianxue Zhang, and Yugang Gao. L-menthol exhibits antidepressive-like effects mediated by the modification of 5-HTergic, GABAergic and DAergic systems. Cognitive Neurodynamics 13.2 (2018): 191-200.

[986] Yoshida, Naohisa, Yuji Naito, Ryouhei Hirose, Kiyoshi Ogiso, Yutaka Inada, Nilesh Fernandopulle, Kazuhiro Kamada, Kazuhiro Katada, Kazuhiko

**Analgesic Agent:* As a topical agent, menthol has been shown to reduce pain by initially stimulating and then desensitizing nociceptors, and by activating the central analgesic pathways[987]. When applied topically to mice, it has also been demonstrated that menthol inhibits phototoxicity-evoked action potentials and reduces pain behavior[988].

**Treatment for Prostate & Other Cancers:* Menthol has been shown to inhibit proliferation and motility of androgen-independent prostate cancer, inhibiting cell growth, inducing cell cycle arrest, and inhibiting the migration of DU145 (human prostate cancer) cells by downregulating the focal-adhesion kinase[989]. Menthol has also been shown to reduce the invasion potential of oral squamous carcinoma cell lines[990].

**Anti-peristalsis Agent:* Useful in the endoscopic treatment of gastric cancers, a menthol spray has been shown to significantly suppress peristalsis[991] (involuntary constriction and relaxation of the muscles of the intestine).

**Prevention & Treatment of Ulcers:* As the primary constituent in the essential oil of wild mint, menthol was shown to provide up to 88.62% protection against induced ulcers in rats[992], without toxicity.

**Multi-Drug Resistant Antibacterial Agent:* Researchers looking for a new agent in the fight against multi-drug resistant bacteria showed that low concentrations of menthol can markedly inhibit the expression of virulence-related extracellular proteins in staphylococcus aureus, while also acting synergistically with beta-lactam antibiotics[993].

**Antifungal Agent:* Menthol has been shown to offer exceptional protection against the T-2 mycotoxin, reducing LDH leakage, halting cell cycle arrest at several phases, promoting cell cycle progression, and down regulating i-NOS in HaCaT cells (immortalized human keratinocytes) exposed to T-2 mycotoxin[994].

IMPLICATIONS FOR HUMAN HEALTH & NUTRITION

It may not be necessary or sensible to seek an increase in dietary menthol. However, some intake of this monocyclic monoterpenoid can be accomplished by preparing, cooking, or serving food using herbs like basil, coriander, cornmint, fennel, peppermint, or spearmint. Making teas from any of the mint family of plants is also an option, as well as pennyroyal tea.

Uchiyama, Osamu Handa, Tomohisa Takagi, Hideyuki Konishi, Nobuaki Yagi, Naoki Wakabayashi, Akio Yanagisawa, and Yoshito Itoh. Prevention of colonic spasm using l-menthol in colonoscopic examination. International Journal of Colorectal Disease 29.5 (2014): 579-583.

[987] Pergolizzi, J. V., R. Taylor, J.-A. LeQuang, R. B. Raffa, The role and mechanism of action of menthol in topical analgesic products. Journal of Clinical Pharmacy & Therapeutics 43.3 (2018): 313-319.

[988] Wright, Lisa, Daniel Baptista-Hon, Fiona Bull, Faith Dalgaty, Michael Gallacher, Sally H Ibbotson, and Tim G Hales. Menthol reduces phototoxicity pain in a mouse model of photodynamic therapy. Pain 159.2 (2019): 284-297.

[989] Wang, Yongzhi, Xinghuan Wang, Zhonghua Yang, Guangbin Zhu, Dong Chen, and Zhe Meng. Menthol Inhibits the Proliferation and Motility of Prostate Cancer DU145 Cells. Pathology & Oncology Research 18.4 (201 2): 903-910.

[990] Okamoto, Yoshihiko, Tsuyako Ohkubo, Tetsuro Ikebe, and Jun Yamazaki. Blockade of TRPM8 activity reduces the invasion potential of oral squamous carcinoma cell lines. International Journal of Oncology 40.5 (2012): 1431-1440.

[991] Fujishiro, Mitsuhiro, Michio Kaminishi, Naoki Hiki, Ichiro Oda, Junko Fujisaki, Noriya Uedo, Mitsuru Kaise, Satoshi Tanabe, Mikitaka Iguchi, Nobuyuki Matsuhashi, Sachiyo Nomura, Hisao Tajiri, Naohisa Yahagi, and Hiroaki Suzuki. Efficacy of spraying l-menthol solution during endoscopic treatment of early gastric cancer: a phase III, multicenter, randomized, double-blind, placebo-controlled study. Journal of Gastroenterology 49.3 (2013): 446-454.

[992] Rozza, A.L., C.A. Hiruma-Lima, R.K. Takahira, C.R. Padovani, and C.H. Pellizzon. Effect of menthol in experimentally induced ulcers: Pathways of gastroprotection. Chemico-Biological Interactions 206.2 (2013): 272- 278.

[993] Qiu, Jiazhang, Mingjing Luo, Jing Dong, Jianfeng Wang, Hongen Li, Xiaoliang Wang, Yanhong Deng, Haihua Feng, and Xuming Deng. Menthol diminishes Staphylococcus aureus virulence-associated extracellular proteins expression. Applied Microbiology and Biotechnology 90.2 (2011): 705-712.

[994] Rachitha P, Krupashree K, Brindhadevi K, Pal A, Chinnathambi A, Alahmadi TA, Shanmuganathan R, Karuppusamy I, Raghavendra VB. Convalescent action of menthol against T-2 mycotoxin-induced toxicity: An in vitro study with HaCaT cells. Environ Res. 2023 Jun 15;227:115690.

At this time, it is unknown precisely which stabilized cannabis strains contain menthol, however, future editions of this book will make updates in this regard as new information or data becomes available. Adding menthol to cannabis products - in similar ways as the tobacco industry has done - is inadvisable.

Menthol Review

Answer the following questions to test your knowledge of this terpene:

Question #1: What type of terpene is menthol?

a. Linear monoterpene
b. Monocyclic sesquiterpene
c. Bicyclic monoterpene
d. Monocyclic monoterpenoid

Question #2: Which of the following best describes the primary flavor & aroma profile of menthol?

a. Hops, cloves, wood, spicy
b. Balsam, pepper, wood, spice
c. Peppermint, cooling, camphoreous, fresh
d. White, smoke, menthol, camphor

Question #3: How common is menthol in cannabis?

a. Top three
b. Top thirty
c. Top ten
d. Top forty

Question #4: What is the chemical formula for menthol?

a. $C_{15}H_{26}0$
b. $C_{10}H_{20}0$
c. $C_{10}H_{16}$
d. $N_{15}H_{26}$

Question #5: Describe the appearance of commercial samples of menthol:

Question #6: Name two potential medical uses of menthol:

1 ______________________________ 2 ______________________________

For the answer key to all individual compound chapters please go here: http://thebigbookofterps.com/bbot-quiz-answer-key/

CEDRENE

Type: Tricyclic sesquiterpene
Chemical Formula: $C_{15}H_{24}$
Molecular Weight: 204.35 g/mol
Boiling Point: *Alpha-cedrene:* 261.00 to 262.00 °C. @ 760.00 mm Hg[995]
Beta-cedrene: 263.00 to 264.00 °C. @ 760.00 mm Hg[996]
Flash Point: *Alpha-cedrene:* 219.00 °F. TCC (103.89 °C.)
Beta-cedrene: 230.00 °F. TCC (110.00 °C.)
Solubility: *Both isomers:* Soluble in alcohol, insoluble in water
Oral LD50: >5g/kg (rat)[997]
Biological Role: Insect repellant
Therapeutic Role: Anticancer, antibacterial, anti-obesity agent
Commercial Use: Flavoring agent, solvent, chemical precursor
Occurrence in Cannabis: Top thirty
Aroma/Flavor: *Alpha-cedrene:* Woody, cedar, sweet, fresh
Beta-cedrene: No aroma or flavor descriptors found

Occurs in Cannabis Strains: Spaceberry (unconfirmed)

INTRODUCTION

Cedrene is a tricyclic sesquiterpene with an aroma and flavor profile that is best described as woody, cedar, sweet, and fresh. This terpene is primarily derived from cedar wood oils, and has numerous applications as flavor enhancers, solvents, fragrance additives and derivatives, and some medical products.

It is currently unknown how prevalent cedrene is in cannabis, but an aggregate of laboratory test standards indicates that this compound falls within the top forty terpenes by concentration in some strains. During research for this book, we reviewed hundreds of cannabis terpene tests, and cedrene was not reported in the top ten terpenes by concentration in any of the samples. This makes it likely that cedrene is a secondary and perhaps minor terpene in cannabis. However, The Farm Genetics in Boulder, Colorado, reports that one of their proprietary strains, Spaceberry, is dominated by cedrene as the primary terpene constituent[998]. Because the listed genetics for this strain consists of Starfighter F2 (F2 denotes a strain that is likely still in development) and Sour Strawberry, cannabis varieties with similar genetics may also contain high amounts of cedrene.

Interestingly, cedrene has been found in submerged wood from ancient Istanbul[999], as well as in Japanese conifer wood from the Miocene period[1000] (between 23 million and 5 million years ago); in both cases cedrene was

[995] The Good Scents Company Data Sheet for (-)-alpha-cedrene, from: http://www.thegoodscentscompany.com/data/rw1032351.html. Accessed September 8, 2023.

[996] The Good Scents Company Data Sheet for (+)-beta-cedrene, from: http://www.thegoodscentscompany.com/data/rw1376091.html. Accessed September 8, 2023.

[997] Foreverest Resources Ltd. Data Sheet for Alpha-cedrene, from: https://foreverest.net/products/extractives-synthetic/alpha-cedrene-90.html. Accessed September 8, 2023.

[998] The Farm Genetics, Boulder, CO, USA. https://thefarmco.com/cannabis-genetics-best-new-strains-2019.

[999] Balaban-U, M., and O. Gonultas. Volatile compounds of archaeological wood from the ancient harbor Thedosius in Istanbul. European Journal of Wood and Wood Products 77.3 (2019): 475-481.

extracted in pristine, perfectly preserved condition, demonstrating the resistance and stability of this complex sesquiterpene.

CHEMICAL STRUCTURE

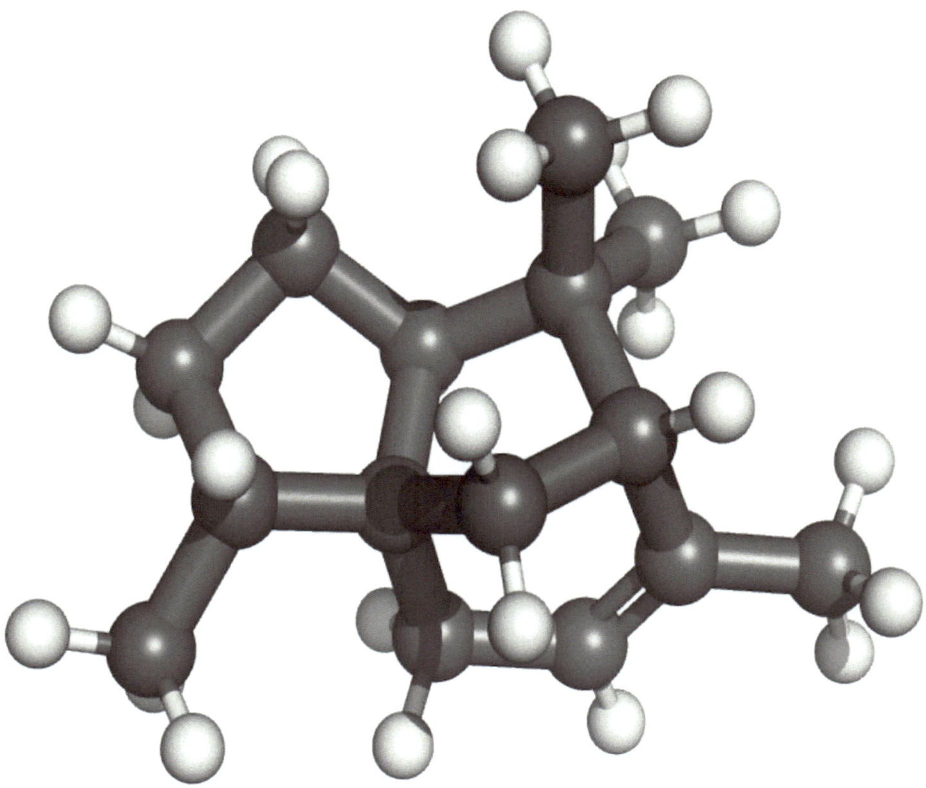

Like most sesquiterpenes, cedrene is formed from farnesyl pyrophosphate via the mevalonate pathway in plants. The molecular skeleton of this terpene consists of three linked units of isoprene totaling fifteen carbon atoms and twenty-four hydrogen atoms, notated as $C_{15}H_{24}$. Cedrene is unique among most sesquiterpenes in cannabis, as it contains three rings, making it a tricyclic sesquiterpene. This isoprenoid compound exists in two isomers which differ only in the position of a double-bond; these are the alpha-cedrene and beta-cedrene isomers. The image above is of alpha-cedrene, which shows the carbon double-bond inside one of the rings - this is the endocyclic isomer.

For the purposes of this book, we'll focus primarily on the alpha- version of cedrene, as it is the most common.

OCCURRENCE IN PLANTS

Although cedrene is closely associated with cedarwood oils, it does occur in other plants, including:

Alpha-cedrene:

Artichoke	Ayou	Basil
Blueberry	Calamus	Cannabis

[1000] Ludwiczuk, Agnieszka, and Yoshinori Asakawa. Terpenoids Preserved in Fossils from Miocene-aged Japanese Conifer Wood. Natural product communications 10.6 (2015): 1051-3.

Cedarwood	Celery	Clary sage
Costus	Cypress	Fir
Ginger	Guava	Lavender
Lemon	Peppermint	Phoebe
Pine	Safflower	Sage
Tarragon	Verbena	Ylang ylang
Zingiber		

Beta-cedrene:

Angelica	Basil	Cedarwood
Guava	Marsypianthes chamaedrys	Prangos
Sideritis species	Ylang ylang	

BIOLOGICAL ACTIVITY IN PLANTS

Scant research has been published on the potential or actual biological roles cedrene plays in plants. One study carried out in China found that cedrene is emitted by the damaged or adulterated branches of platycladus orientalis (Chinese thuja), which attracts the juniper bark borer (semanotus bifasciatus)[1001]. Healthy branches contained no cedrene.

Insect Repellant: As the primary constituent in the essential oil made from several varieties of Rhododendron, cedrene exhibited higher repellency than DEET against three insects; tribolium castaneum, lasioderma serricorne, and liposcelis bostrychophila[1002].

USES IN INDUSTRY

Commercial samples of cedrene appear as clear, colorless, or translucent oily liquids. This tricyclic sesquiterpene is used as a flavor enhancer or masking agent, and can be found in dairy products, fats and oils, edible ices, sherbet, sorbet, processed fruit, processed vegetables, nuts, mushrooms, confectionary items, cereals, bakery items, meats, egg products, sweeteners, spices and seasonings, and other processed foods.

Cedrene is also used industrially as a solvent, and as a precursor to other ingredients in perfumes and fragrances.

POTENTIAL USES IN MEDICINE

Cedrene shows promise as an anticancer, antibacterial, and anti-obesity agent, however, it should be noted that this compound could be fatal if ingested and aspirated into the lungs. Care should be exercised when handling raw cedrene or essential oils that are high in this sesquiterpene.

Anticancer Agent: As a primary constituent in the ethanolic extract of fenugreek (an annual plant with leaves and seeds that are often used in Indian cooking) seeds, cedrene contributed to the induction of autophagy and autophagy-associated death of jurkat (human T lymphocyte) cells by creating vacuoles (small cavities) that led to

[1001] Yue-qin Sun, You-qing Luo, Xiong-fei Yan Comparison of host volatiles under different physiological states and electroantennogram responses of Semanotus bifasciatus (Motschulsky) to these volatiles Forestry Studies in China volume 10, pages64-66(2008).

[1002] Bai, Li, Mei-Ling Jiao, Hong-Yuan Zang, Shan-Shan Guo, Yang Wang, Yu-Li Sang, and Shu-Shan Du. Chemical composition of essential oils from four Rhododendron species and their repellent activity against three stored-product insects. Environmental Science and Pollution Research 26.22 (2019): 23198-23205.

the complete disintegration of plasma membrane[1003]. Surprisingly low concentrations of the fenugreek extract induced death of the jurkat cells. In another study that reviewed the anticancer effects of fenugreek extract, strong selective cytotoxicity was found, but with other cancer cell lines including T-cell lymphoma (TCP), B-cell lymphomas, thyroid papillary carcinoma (FRO) and breast cancer (MCF7), without toxicity to surrounding healthy cells[1004]. Additionally, as the primary component in the essential oil from the heartwood of cunninghamia lanceolate, cedrene was shown to exhibit cytotoxic activity against human lung, liver, and oral cancer cells[1005].

**Antibacterial Agent:* As a primary constituent in the essential oil from the leaves and stems of platycladus orientalis, cedrene contributed to a strong inhibitory response to the bacterium Clostridium difficile and Clostridium perfringens, and moderate inhibitory activity against Escherichia coli[1006]. As the primary constituent in the essential oil made from the wood burl parts of Tetraclinis articulate (thuja), cedrene exhibited considerable inhibitory effects against enterococcus faecalis, Escherichia coli, klebsiella pneumoniae, and streptococcus D.[1007].

**Antiobesity Agent:* Oral administration of cedrene to rats with established induced obesity reversed the weight gain without affecting food consumption, in part by increasing energy expenditure and reducing triglyceride accumulation[1008], among other mechanisms.

IMPLICATIONS FOR HUMAN HEALTH & NUTRITION

Because cedrene is not known to occur in many plants, it can be difficult to increase the dietary amount of this tricyclic sesquiterpene. Consider snacking on foods like artichokes, blueberries, celery, or guava. When preparing, cooking, or serving foods, use herbs and spices like basil, celery seed, ginger, peppermint, safflower, sage, or tarragon. Experimentation with various applications of cedarwood oil, ayou wood oil, and phoebe oil may prove beneficial as well. This text will be updated in future editions with emerging data or information about cannabis strains or chemovars that contain cedrene.

[1003] Al-Daghri, Nasser, Majed Alokail, Khalid Alkharfy, Abdul Mohammed, Sherif Abd-Alrahman, Sobhy Yakout, Osama Amer, and Soundararajan Krishnaswamy. Fenugreek extract as an inducer of cellular death via autophagy in human T lymphoma Jurkat cells. BMC Complementary and Alternative Medicine 12.1 (2012): 1-8.

[1004] Alsemari, Abdulaziz, Fahad Alkhodairy, Ahmad Aldakan, Mai Al-Mohanna, Eman Bahoush, Zakia Shinwari, and Ayodele Alaiya. The selective cytotoxic anti-cancer properties and proteomic analysis of Trigonella Foenum- Graecum. BMC Complementary and Alternative Medicine 14.1 (2014): 1-9.

[1005] Su, Yu-Chang, Kuan-Ping Hsu, Eugene I-Chen Wang, and Chen-Lung Ho. Composition, Anticancer, and Antimicrobial Activities in vitro of the Heartwood Essential Oil of Cunninghamia lanceolata var. konishii from Taiwan. Natural Product Communications 7.9 (2012)1.

[1006] Min-Gi Kim & Hoi-Seon Lee Growth-inhibiting effects and chemical composition of essential oils extracted from Platycladus orientalis leaves and stems toward human intestinal bacteria Food Science and Biotechnology volume 24, pages427-431(2015).

[1007] Mohammed El Moussaouiti, Ahmed Talbaoui, Said Gmouh, Mohammed Aberchane, Abdelaziz Benjouad, Youssef Bakri & D. Pascal Kamdem. Chemical composition and bactericidal evaluation of essential oil of Tetraclinis articulata burl wood from Morocco Journal of the Indian Academy of Wood Science volume 7, pages14-18(2010).

[1008] Tong, Tao, Rina Yu, and Taesun Park. a-Cedrene protects rodents from high-fat diet-induced adiposity via adenylyl cyclase 3. International Journal of Obesity 43.1 (2018): 202-216.

Cedrene Review

Answer the following questions to test your knowledge of this terpene:

Question #1: What type of terpene is cedrene?

a. Acyclic sesquiterpene
b. Bicyclic sesquiterpene
c. Tricyclic sesquiterpene
d. Trescyclic sesquiterpene

Question #2: Which of the following best describes the primary flavor & aroma profile of cedrene?

a. Woody, cedar, sweet, fresh
b. Citrus, pepper, terpenic, wood
c. Lemon, sharp, fuel, wood
d. Woody, carrot, spice, dry

Question #3: How common is cedrene in cannabis?

a. Top ten
b. Top thirty
c. Top fifteen
d. Top forty

Question #4: What is the chemical formula for cedrene?

a. $C_{10}H_{16}O$
b. $C_{15}H_{24}O_2$
c. $C_{15}H_{24}$
d. $C_{20}H_{32}O$

Question #5: What feature makes cedrene a tricyclic sesquiterpene?

____

Question #6: Name two potential medical uses of cedrene:

1 ________________________ 2 ________________________

For the answer key to all individual compound chapters please go here: http://thebigbookofterps.com/bbot-quiz-answer-key/

GERANYL ACETATE

Type: Linear monoterpenoid
Chemical Formula: $C_{12}H_{20}O_2$
Molecular Weight: 196.29 g/mol
Boiling Point: 240.00 to 245.00 °C. @ 760.00 mm Hg[1009]
Flash Point: 219.00 °F. TCC (104.00 °C.)
Solubility: Soluble in alcohol, fixed oils, paraffin oil, slightly soluble in water, propylene glycol
Oral LD50: 6,330 mg/kg (rat)[1010]
Biological Role: Insecticidal, acaricidal
Therapeutic Role: Antifungal, antibacterial, antinociceptive
Commercial Use: Food additive, fragrance
Occurrence in Cannabis: Top forty
Aroma/Flavor: Floral, rose, waxy, green, herbal

Occurs in Cannabis Strains: Afgoo, Gelato, Miami White Kush, Zkittlez

INTRODUCTION

Geranyl acetate is a monoterpenoid that is highly prized in the perfume and fragrance industries for its floral, rose, waxy, green, herbal, and slightly lavender aroma and flavor profile. This isoprenoid compound is also a common flavor additive, and has some applications in commercial soaps, creams, and other products.

Likely occurring in cannabis within the top forty terpenes by concentration, we found just one instance of this terpene in our review of hundreds of cannabis terpene tests. This instance was as the #3 terpene by concentration in a sample of Afgoo. A 2017 study that sought to identify terpenoid chemotypes among high-THC strains found geranyl acetate in Gelato and Miami White Kush strains[1011], while one report published on Leafly indicates that geranyl acetate is the #1 terpene in Zkittlez[1012].

Interestingly, geranyl acetate is emitted by some insects when disturbed or when seeking to attract a mate. For instance, Indian bees (pithitis smaragdula) have been shown to emit a secretion where geranyl acetate is one of the primary constituents, which it appears to use defensively when disturbed or handled[1013]. The sceliphron caementarium species of solitary wasp also emits geranyl acetate defensively[1014], and possibly as a roosting aggregation pheromone. On the opposite end of the biological uses spectrum, male fruit flies have been shown to emit geranyl acetate, which attracts females of the same species[1015].

[1009] The Good Scents Company Data Sheet for Geranyl Acetate, from: http://www.thegoodscentscompany.com/data/rw1030092.html. Accessed September 8, 2023.

[1010] Thermo Fisher Scientific Safety Data Sheet for Geranyl Acetate, from: https://www.fishersci.com/store/msds?partNumber=AAA1986422&productDescription=GERANYL+ACETATE%2C+98%25+100G&vendorId=VN00024248&countryCode=US&language=en. Accessed September 8, 2023.

[1011] Justin T. Fischedick Identification of Terpenoid Chemotypes Among High (-)-trans-delta-9- Tetrahydrocannabinol- Producing Cannabis sativa L. Cultivars Cannabis and Cannabinoid Research Volume 2, No. 1 December 2017.

[1012] Leafly. https://www.leafly.com/news/strains-products/oregon-craft-cannabis-strain-terpenes.

[1013] Hefetz, A., Batra, S.W.T. & Blum, M.S. Chemistry of the mandibular gland secretion of the Indian bee Pithitis smaragdula . J Chem Ecol 5, 753-758 (1979).

[1014] Hefetz, A., Batra, S.W.T. Geranyl acetate and 2-decen-1-ol in the cephalic secretion of the solitary wasp Sceliphron caementarium (Sphecidae; Hymenoptera). Experientia 35, 1138-1139 (1979).

[1015] Heath, R.R., Landolt, P.J., Tumlinson, J.H. et al. Analysis, synthesis, formulation, and field testing of three major components of male mediterranean fruit

CHEMICAL STRUCTURE

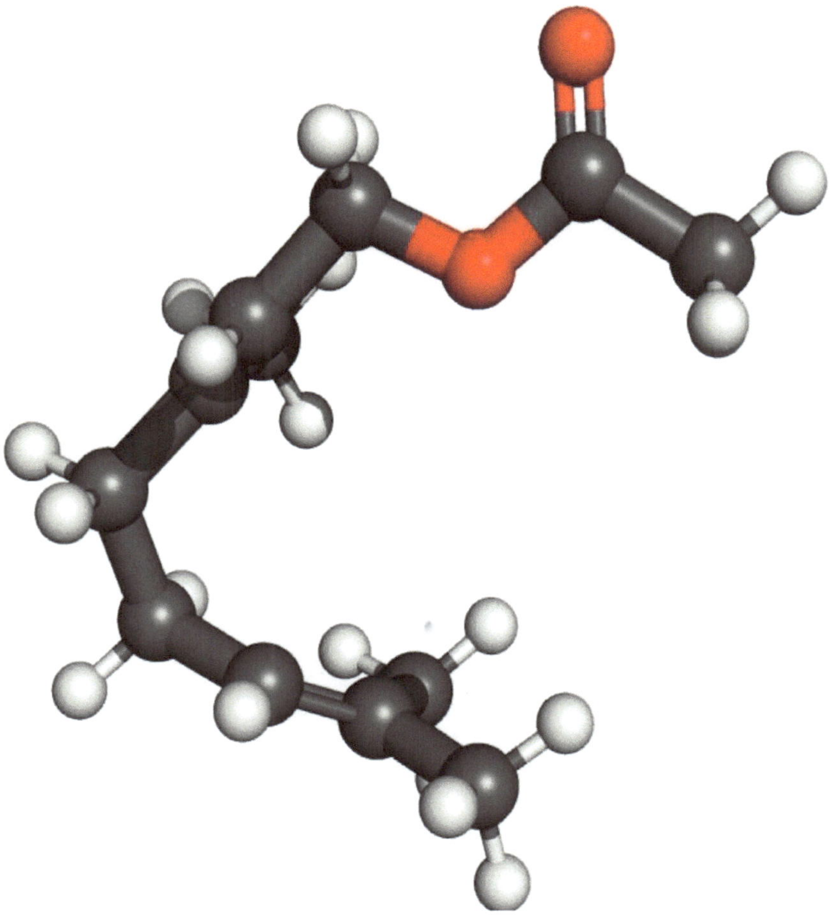

Like many monoterpenes, geranyl acetate is derived from geranyl pyrophosphate with geraniol as a precursor. Comprised of two linked units of isoprene containing twelve carbon atoms, twenty hydrogen atoms, and two oxygen atoms, the geranyl acetate molecule is notated as $C_{12}H_{20}O_2$. This molecular structure classifies this phytonutrient as a non-cyclic monoterpenoid. Derived from the biosynthesis of geraniol, geranyl acetate often co-occurs with geraniol in plants.

OCCURRENCE IN PLANTS

Geranyl acetate is found in a large and diverse number of plants and plant parts, with citronella species being one of the most common sources for essential oils that are high in this monoterpenoid.

Allspice	Almond	Alpinia galanga
Artemisia species	Basil	Basswood
Bay	Bergamot	Black pepper
Blood orange	Cananga	Cannabis
Cardamom	Carrot	Cedarleaf
Celery	Champaca	Chervil
Cinnamon	Citronella	Clary sage

fly pheromone. J Chem Ecol 17, 1925-1940 (1991).

Coriander
Fig
Fir
Frankincense
Geranium
Ginger
Gingergrass
Grape
Grapefruit
Ho leaf
Horsemint
Jasmin
Labdanum
Lavandin
Lavandula
Lavender
Lemon
Lemongrass
Levisticum
Lime
Lovage
Mace
Mandarin
Marjoram
Melissa
Mustard
Myrtle
Neroli
Nutmeg
Orange
Oregano
Palmarosa
Peppermint
Plum
Plumcot
Rose
Rosemary
Sage
Salvia
Savory
Skimmia laureola
Snake root
Stachys recta
Tagete
Tangerine
Thyme
Tomato
Vassoura
Verbena
Wormwood
Ylang ylang
Yuzu
Zingiber officinale

BIOLOGICAL ACTIVITY IN PLANTS

Few studies on the biological roles of geranyl acetate in plants have been published, but it is clear from the body of medical literature that this compound is a potent antimicrobial agent, with these properties and activities likely being similar in other plants.

Insecticidal Agent: As a primary constituent in the essential oil isolated from Mexican mint (plectranthus zeylanicus), geranyl acetate contributed to fumigant and contact toxicity against adult Cowpea weevils (callosobruchus maculatus), while also significantly inhibiting ovipositioning (egg-laying) and adult emergence[1016].

Acaricidal Agent: As a primary constituent in the essential oil of lemon eucalyptus (eucalyptus citriodora), geranyl acetate contributed to high contact and fumigant toxicity against red spider mites[1017] (tetranychus urticae). This information should be especially interesting to cannabis farmers, who often battle various species of mites.

USES IN INDUSTRY

Commercial samples of geranyl acetate appear as colorless or clear liquids with a strong floral aroma of roses or lavender. This monoterpenoid is produced via fractional distillation of essential oils that are high in geranyl acetate, like citronella, eucalyptus, and clary sage, but it can also be produced synthetically with geraniol as a precursor.

Geranyl acetate is used to add flavor or aroma to a wide variety of processed foods, including alcoholic and nonalcoholic beverages, baked goods, chewing gum, frozen dairy, gelatin (pudding), gravies, hard and soft candy, meat products, and syrups, among others.

Geranyl acetate is also heavily used in the perfume and fragrance industries for use in both fine perfumes and in commercial and consumer products like creams, soaps, air care products, cleaners, dishwashing liquids, furniture cleaners, and more.

[1016] Balachandra, B A H E, P U Pathirathna, and P A Paranagama. Control of stored grain pest, Callosobruchus maculatus (F.) (Coleoptera: Bruchidae) using the essential oil isolated from Plectranthus zeylanicus. Natural product research 26.23 (2013): 2219-22.

[1017] Han, Jun, Soon-Il Kim, Byeoung-Ryeol Choi, Sang-Guei Lee, and Young-Joon Ahn. Fumigant toxicity of lemon eucalyptus oil constituents to acaricide-susceptible and acaricide-resistant Tetranychus urticae. Pest Manage ment Science 67.12 (2011): 1583-1588.

Interestingly, geranyl acetate is difficult to produce affordably. Some scientists have engineered brewer's yeast (S. cerevisiae) to produce this and other monoterpenoids[1018] via the biotransformation of various media, while others have engineered E. coli bacteria to inversely produce geraniol from geranyl acetate, and geranyl acetate from geraniol[1019].

POTENTIAL USES IN MEDICINE

Geranyl acetate has been shown to exhibit considerable antimicrobial effects against a variety of important pathogens. This monoterpenoid also shows promise in other areas of medicine, including treatment for colon cancer.

**Antimicrobial Agent:* As a primary constituent in the essential oil of the curry plant (helichrysum italicum), geranyl acetate contributed to antimicrobial activities against the common pathogens staphylococcus aureus, and candida albicans[1020]. These findings were echoed when, as a primary constituent in the essential oil of coriander, geranyl acetate was shown to contribute to antimicrobial activity against S. aureus and C. albicans, but also against Escherichia coli and pseudomonas aeruginosa[1021].

**Antifungal Agent and Synergist:* As a primary constituent in the essential oil of geranium, geranyl acetate was found to be an effective fungicidal agent against candida albicans, and exhibited significant synergism with Fluconazole[1022], a common anti-fungal medication. As a primary constituent in the essential oil of palmarosa (cymbopogon martinii), geranyl acetate was found to be an active ingredient that contributed to the complete inhibition of Brewer's yeast (saccharomyces cerevisiae) cells[1023]. As a primary constituent in the essential oil made from the aerial parts of the curry plant (helichrysum italicum), geranyl acetate was found to be the active agent responsible for fungicidal effects against C. albicans[1024].

**Antibacterial Agent:* Continuing with more proof of geranyl acetate's antimicrobial activities, as a primary constituent in the essential oils made from the leaves of the curry tree (murraya koenigii), geranyl acetate contributed to the inhibition of various pathogens, including proteus mirabilis, staphylococcus aureus, corynebacterium pseudotuberculosis, klebsiella pneumoniae, pseudomonas aeruginosa, and enterobacter aerogenes[1025]. Against the mycobacterium responsible for causing tuberculosis, geranyl acetate was found to exhibit potent and selective inhibition of mycobacterium tuberculosis[1026].

[1018] Wu, T., Li, S., Zhang, B. et al. Engineering Saccharomyces cerevisiae for the production of the valuable monoterpene ester geranyl acetate. Microb Cell Fact 17, 85 (2018).

[1019] Liu, W., Xu, X., Zhang, R. et al. Engineering Escherichia coli for high-yield geraniol production with biotransformation of geranyl acetate to geraniol under fed-batch culture. Biotechnol Biofuels 9, 58 (2016) doi: 10.1186/s13068-016-0466-5.

[1020] Mastelic, J., Politeo, O., Jerkovic, I. et al. Composition and Antimicrobial Activity of Helichrysum italicum Essential Oil and Its Terpene and Terpenoid Fractions. Chem Nat Compd 41, 35-40 (2005).

[1021] Sourmaghi, Mohammad Hossein Salehi et al. Comparison of essential oil composition and antimicrobial activity of Coriandrum sativum L. extracted by hydrodistillation and microwave-assisted hydrodistillation. Journal of food science and technology vol. 52,4 (2015): 2452-7. doi: 10.1007/s13197-014-1286-x.

[1022] Zore GB, Thakre AD, Rathod V, Karuppayil SM. Evaluation of anti-Candida potential of geranium oil constituents against clinical isolates of Candida albicans differentially sensitive to fluconazole: inhibition of growth, dimorphism and sensitization. Mycoses. 2011 Jul;54(4): e99-109.

[1023] Prashar, Anjali, Pauline Hili, Robert G Veness, and Christine S Evans. Antimicrobial action of palmarosa oil (Cymbopogon martinii) on Saccharomyces cerevisiae. Phytochemistry 63.5 (2003): 569-575.

[1024] Djihane, Bouzid et al. Chemical constituents of Helichrysum italicum (Roth) G. Don essential oil and their antimicrobial activity against Gram-positive and Gram-negative bacteria, filamentous fungi and Candida albicans. Saudi pharmaceutical journal: SPJ : the official publication of the Saudi Pharmaceutical Society vol. 25,5 (2017): 780-787.

[1025] Rajendran, Mini Priya et al. Chemical composition, antibacterial and antioxidant profile of essential oil from Murraya koenigii (L.) leaves. Avicenna journal of phytomedicine vol. 4,3 (2014): 200-14.

[1026] Vik, Anders, Anthony James, and Lise-Lotte Gundersen. Screening of terpenes and derivatives for antimycobacterial activity; identification of geranyl geraniol and geranyl acetate as potent inhibitors of Mycobacterium tuberculosis in vitro. Planta medica 73.13 (2007): 1410-2.

***Treatment for Colon Cancer:** Geranyl acetate was shown to induce apoptosis, DNA damage, and cell cycle arrest in human colorectal cancer cells[1027], indicating its potential as an anticancer agent.

***Antinociceptive Agent:** Geranyl acetate has been shown to exhibit significant antinociceptive (blocks the feeling of pain) activity against induced writhing and formalin-induced nociception in mice[1028].

IMPLICATIONS FOR HUMAN HEALTH & NUTRITION

Dietary consumption of geranyl acetate can be as simple as preparing, cooking, or serving foods with spices like allspice, basil, bay laurel, black pepper, cardamom, cinnamon, clary sage, coriander, ginger, mace, marjoram, nutmeg, oregano, peppermint, rosemary, sage, savory, or thyme. This can also be accomplished by eating fruits like blood oranges, figs, grapefruit, lemons, limes, mandarins, oranges, plums, plumcots, tangerines, or vegetables like carrots, celery, ginger, or tomatoes.

Cannabis strains that contain geranyl acetate include Afgoo, Gelato, Miami White Kush, and Zkittlez, among others. While we do not recommend smoking or vaporization of these strains for health reasons, it is possible that some geranyl acetate may be obtained in the smoke, vapor, or side stream. Experimentation with isolated compounds or essential oils that are high in this terpene should be undertaken with caution, as these substances can be toxic in some situations.

[1027] Qi, Fei, Qiang Yan, Zhaozheng Zheng, Jian Liu, Yan Chen, and Guiyang Zhang. Geraniol and geranyl acetate induce potent anticancer effects in colon cancer Colo-205 cells by inducing apoptosis, DNA damage and cell cycle arrest. Journal of B.U.ON. : official journal of the Balkan Union of Oncology 23.2 (2019): 346-352.

[1028] Quintans-J(nior, Lucindo et al. Antinociceptive Activity and Redox Profile of the Monoterpenes (+)-Camphene, p-Cymene, and Geranyl Acetate in Experimental Models. TSRN toxicology vol. 2013 459530. 14 Jan. 2013.

Geranyl Acetate Review

Answer the following questions to test your knowledge of this terpene:

Question #1: What type of terpene is geranyl acetate?

a. Acyclic sesquiterpene
b. Bicyclic sesquiterpene
c. Tricyclic monoterpene
d. Non-cyclic monoterpenoid

Question #2: Which of the following best describes the primary flavor & aroma profile of geranyl acetate?

a. Woody, cedar, sweet, fresh
b. Citrus, pepper, terpenic, wood
c. Floral, rose, waxy, green, herbal
d. Woody, carrot, spice, dry

Question #3: How common is geranyl acetate in cannabis?

a. Top ten
b. Top three
c. Top thirty
d. Top forty

Question #4: What is the chemical formula for geranyl acetate?

a. $C_{10}H_{16}O$
b. $C_{12}H_{20}O_2$
c. $C_{15}H_{24}$
d. $C_{20}H_{32}O$

Question #5: What feature makes geranyl acetate a monoterpenoid?

____

Question #6: Name two potential medical uses of geranyl acetate:

1 ______________________________ 2 ______________________________

For the answer key to all individual compound chapters please go here: http://thebigbookofterps.com/bbot-quiz-answer-key/

CYMENE

Type: Monocyclic monoterpene
Chemical Formula: $C_{10}H_{14}$
Molecular Weight: 134.22 g/mol
Boiling Point: *P-cymene:* 176.00 to 178.00 °C. @ 760.00 mm Hg[1029]
O-cymene: 177.00 to 179.00 °C. @ 760.00 mm Hg[1030]
M-cymene: 174.00 to 176.00 °C. @ 760.00 mm Hg[1031]
Flash Point: *P-cymene:* 117.00 °F. TCC (47.22 °C.)
O-cymene: 123.00 °F. TCC (50.60 °C.) (estimated by TGSC)
M-cymene: 118.00 °F. TCC (47.80 °C.) (estimated by TGSC)
Melting Point: *P-cymene:* -68.90 °C. @ 760.00 mm Hg
Solubility: *P-cymene:* Soluble in water, alcohol, acetone, ethanol, benzene, ether, carbon tetrachloride, oils
O-cymene: Soluble in alcohol, slightly soluble in water
M-cymene: Soluble in alcohol, insoluble in water
Oral LD50: 4,750 mg/kg (rat)[1032]
Biological Role: Insecticidal, antioxidant, allelopathic agent, antifungal
Therapeutic Role: Anticancer, antimicrobial, acaricidal, treatment for lung injury
Commercial Use: Fragrance
Occurrence in Cannabis: Top forty
Aroma/Flavor: *P-cymene:* Woody, terpenic, chemical, citrus

Occurs in Cannabis Strains: Ghost Cookies, Grape LA, Northern Lights

INTRODUCTION

Cymene is a monocyclic monoterpene found in a large variety of plants, including cannabis, where it is generally thought to occur within the top forty terpenes by concentration. While there are no reports on the web as of the fall of 2023 regarding the specific strains that cymene occurs in, our review of hundreds of cannabis terpene laboratory tests showed that this monoterpene occurred as the second terpene in a sample of Northern Lights, the fourth terpene in Ghost Cookies, and tenth in Grape LA.

Cymene has several properties that make it a valuable additive in commercial and personal consumer products like detergents, cleaners, solvents, shampoo, makeup, and many more items. This isoprenoid compound also has multiple potential medical uses, particularly as an antimicrobial agent, and in the treatment of cancer.

1029 The Good Scents Company Data Sheet for para-cymene, from: http://www.thegoodscentscompany.com/data/rw1032712.html. Accessed September 8, 2023.

1030 The Good Scents Company Data Sheet for Ortho-cymene, from: http://www.thegoodscentscompany.com/data/rw1057011.html. Accessed September 8, 2023.

1031 The Good Scents Company Data Sheet for Meta-cymene, from: http://www.thegoodscentscompany.com/data/rw1057021.html. Accessed September 8, 2023.

1032 Carl Roth International Safety Data Sheet for P-Cymene, from: https://www.carlroth.com/medias/SDB-3465-GB-EN.pdf?context=bWFzdGVyfHNlY3VyaXR5RGF0YXNoZWV0c3wzMDg1Mzh8YXBwbGljYXRpb24vcGRmfHNlY3VyaXR5RGF0YXNoZWV0cy9oZmQvaDgzLzkwNDU5NTA3MjYxNzQucGRmfGU0ZWExNzg2Mzc2NjUxMjhhZWE4NWM5N2Q3NjRiMDdlOGJiMzhlYTgyNmQzYzdkNzY4M2I0ZjdmZWYzMzMyZGI. Accessed September 8, 2023.

Cymene exhibits an aroma and flavor profile that is described as woody, terpenic, chemical, and citrus, with spicy nuances. These characteristics are often imparted to the hundreds of various products that feature this monocyclic monoterpene.

CHEMICAL STRUCTURE

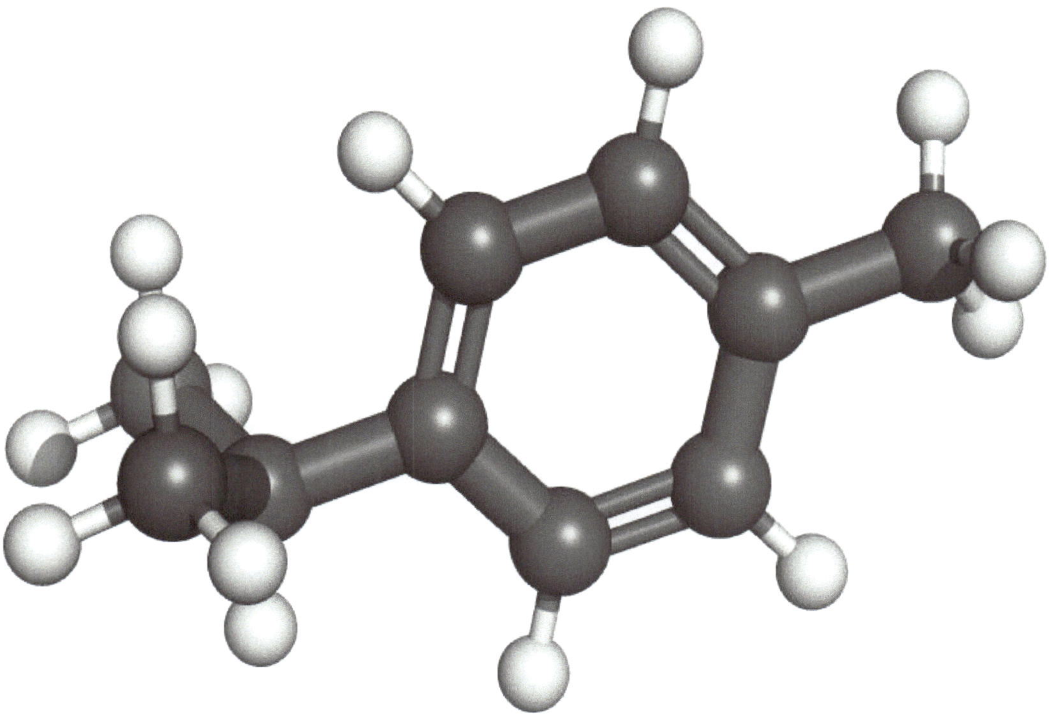

As a monoterpene, cymene is formed from geranyl pyrophosphate in the methyl erythritol phosphate pathway of plants. This terpene consists of two linked units of isoprene containing ten carbon atoms and fourteen hydrogen atoms, chemically notated as $C_{10}H_{14}$. Cymene is further classified as a monocyclic monoterpene because its carbon skeleton contains one ring made of benzene, and, interestingly, three separate sets of endocyclic double-bonds (see the image above).

Cymene contains both methyl and isopropyl groups, and has three isomers; p-cymene, o-cymene, and m-cymene, the latter two of which feature alkyl groups that are ortho- and meta- substituted, respectively. For the purposes of this book, we'll refer primarily to p-cymene, which is the most common, naturally occurring version. Interestingly, when the essential oils of a plant contain cymene as a primary constituent, thymol and carvacrol also occur in close relative concentration proximity.

OCCURRENCE IN PLANTS

Cymene is found in nearly as many plants as some of the top terpenes in cannabis like pinene, myrcene, and beta-caryophyllene, although the primary source of this terpene is the essential oils of cumin and thyme.

Para-cymene:

Achillea tenuifolia
Allspice
Apricot
Agathosma crenulata
Angelica
Artemisia
Ajowan
Anise star
Ayou wood

Cannabis
Cardamom
Cassia
Chamomile
Close
Costmary
Curry
Eucalyptus
Feverfew
Frankincense
Ginko
Hinoki
Bay leaf
Blackberry
Hyssop
Kewda
Lavandin
Lemongrass
Mace
Marjoram
Melon
Neroli
Orange
Parsley
Pepper tree
Phoebe
Boldo
Calamus
Pine
Rose
Sage
Savory
Star fruit
Tangerine
Tea tree
Tomato
Vassora
Yarrow

Camel grass
Carrot
Cedarwood
Cinnamon
Coriander
Cubeb
Cypress
Featherfew
Fig
Geranium
Grapefruit
Ho leaf
Bell pepper
Black currant
Juniper
Labdanum
Layana
Lime
Mandarin
Mastic
Mikan
Nutmeg
Oregano
Parsnip
Peppermint
Blueberry
Buchu
Pimenta
Prangos
Rosemary
Salvia
Snake root
Sunflower
Tansy
Tejpat
Turmeric
Water mint
Yuzu

Caraway
Cascarilla
Celery
Clary sage
Cornmint
Cumin
Dill
Fennel
Fir
Ginger
Guava
Bay laurel
Bergamot
Horsemint
Kachur
Laurel
Lemon
Lovage
Mangrove
Melissa
Myrtle
Onion
Papaya
Pennyroyal
Petitgrain
Bois de rose
Cajuput
Pineapple
Pteronia
Safflower
Savin
Spearmint
Tagete
Tarragon
Thyme
Valerian root
Wormseed
Zinnie elegans

Ortho-cymene:

Arum maculatum
Ferula
Ravintsara
Sesuvium portulacastrum

Citrus aurantium
Labdanum
Rhynchosia minima
Xylopia aethiopica

Cypress
Lavender
Senecio articulatus

Meta-cymene:

Black currant
Ferula

Basil
Gynandropsis gynandra

Cassis
Orange

Plumcot
Radix angelicae pubescentis
Senecio vernalis
Ulva rigida

BIOLOGICAL ACTIVITY IN PLANTS

Cymene appears to be a potent insecticidal agent and antioxidant in plants, and likely serves numerous other purposes, depending on the plant and situation:

**Insecticidal Agent:* As the primary constituent in the essential oil of oregano (origanum vulgare), p-cymene was shown to repel both nymph and adult southern green stink bugs (nezara viridula), while also exhibiting strong fumigant and contact toxicity against adults and nymphs, and inhibiting the egg hatching of this insect at just under 100%[1033]. As a primary constituent in the steam distillate made from the rhizomes of nut grass (cyperus rotundus), p-cymene was shown to exhibit high fumigant and contact toxicity against female German cockroaches (blattella germanica)[1034], which are notoriously resistant to pesticides and other methods of eradication.

**Antioxidant:* As a primary constituent in the essential oil of zataria multiflora (a thyme-like plant), p-cymene showed significant antioxidant capacities, NO (nitric oxide) scavenging, and MDA scavenging activity[1035], and also exhibited antioxidant effects comparable to ascorbic acid, as a primary constituent of the essential oil of the lime berry plant (triphasia trifolia)[1036].

**Allelopathic Agent:* As a main component in the essential oil of Mexican tea (chenopodium ambrosioides), cymene was shown to exhibit significant allelopathic effects on the root tips of nearby maize (corn, zea mays) plants, causing membrane lipid peroxidation, increasing mitochondrial H_2O_2 and MDA contents, decreasing mitochondrial membrane potential, inducing accumulation of ROS, and triggering the mediation of important apoptosis signaling pathways[1037] - a severe battery of effects.

**Antifungal Agent:* As an individual isolated compound, cymene was shown to completely inhibit the growth of candida lusitaniae, a common pathogenic fungi[1038].

USES IN INDUSTRY

Commercial samples of cymene appear as colorless, pale yellow, or clear liquids. This terpene is commonly used in laundry detergent & liquids, personal care products, cosmetics, toiletries, shampoos, furniture cleaners, paint, disinfectants, thinners, lacquers, and varnishes, among other products.

[1033] Werdin Gonzalez, Jorge O, Maria M Gutierrez, Ana P Murray, and Adriana A Ferrero. Composition and biological activity of essential oils from Labiatae against Nezara viridula (Hemiptera: Pentatomidae) soybean pest. Pest Management Science 67.8 (2011): 948-955.

[1034] Chang KS, Shin EH, Park C, Ahn YJ. Contact and fumigant toxicity of Cyperus rotundus steam distillate constituents and related compounds to insecticide-susceptible and -resistant Blattella germanica. J Med Entomol. 2012 May;49(3):631-9.

[1035] Karimian P, Kavoosi G, Saharkhiz MJ. Antioxidant, nitric oxide scavenging and malondialdehyde scavenging activities of essential oils from different chemotypes of Zataria multiflora. Nat Prod Res. 2012;26(22):2144-7.

[1036] Jaramillo Colorado BE, Martelo IP, Duarte E. Antioxidant and repellent activities of the essential oil from Colombian Triphasia trifolia (Burm. f.) P. Wilson. J Agric Food Chem. 2012 Jun 27;60(25):6364-8.

[1037] Jiao Li Yaqiang He, Danwei Ma, Bing He, Yanan Wang, Bin Chen. Volatile allelochemicals of Chenopodium ambrosioides L. induced mitochondrion-mediated Ca2+-dependent and Caspase-dependent apoptosis signaling pathways in receptor plant cells. Plant and Soil April 2018, Volume 425, Issue 1-2, pp 297-308.

[1038] Aznar, Arantxa, Pablo S Fernandez, Paula M Periago, and Alfredo Palop. Antimicrobial activity of nisin, thymol, carvacrol and cymene against growth of Candida lusitaniae. Food Science and Technology International 21.1 (2015): 72-79.

Cymene reacts with oxidants and is known to attack rubber. In fact, researchers have conducted experiments that dissolved polystyrene into p-cymene, and showed that both substances can then be recovered for recycling[1039]. Other research has shown that p-cymene can eradicate populations of Escherichia coli in unpasteurized apple juice[1040], offering the potential for a novel natural food preservative.

POTENTIAL USES IN MEDICINE

Although cymene offers humans and animals several promising medical uses, it should be noted that purified versions of this monocyclic monoterpene can be extremely harmful or fatal if swallowed and aspirated into the lungs. Care should be exercised when working with raw forms of this terpene. Apart from this warning, cymene shows potential for medical use in cancer treatment, treatment of lung conditions, and as a natural but potent antimicrobial agent - including against MRSA strains.

**Treatment of Cancer:* As the primary constituent in a blend of three compounds, p-cymene exhibited powerful anti-tumor effects in breast, ovarian, brain, lung, and colon cancer cells[1041]. In a separate study, a similar cymene-focused compound was shown to induce DNA damage, cell cycle arrest, and cell death in non-small cell lung cancer (NSCLC) cells[1042]. Cymene has also been shown to inhibit cancer pain in mice by more than 60%[1043], a promising figure considering the typical difficulties presented in treating oncological pain.

**Antimicrobial Agent:* As a primary constituent in the essential oil of satureja kitaibelii, p-cymene contributed to significant antimicrobial activity against foodborne microbes, multi-resistant bacteria, and dermatophytes[1044]. P-cymene appeared as the main component of the essential oil of clove basil (ocimum gratissimum), which showed antimicrobial effects against candida albicans and staphylococcus aureus[1045] - two of the most common causes of pathogenic disease in humans. As a main component in the essential oil of the flowering centaury (centaurium erythraea) plant, p-cymene contributed to antimicrobial activity against escherichia coli, salmonella enteritidis, and bacillus cereus[1046]. As a primary constituent in the essential oil made from the aerial parts of plectranthus marrubatus, p-cymene was shown to exhibit antimicrobial activity against a variety of both gram-positive and gram-negative bacteria, and also showed free-radical scavenging activities[1047]. P-cymene has also been shown to exhibit

---

[1039] Kazuyuki Hattori, Sota Shikata, Ryo Maekawa & Masakazu Aoyama. Dissolution of polystyrene into p-cymene and related substances in tree leaf oils Journal of Wood Science volume 56, pages169-171(2010).

[1040] Gabriella Kisk & Sibel Roller Carvacrol and p-cymene inactivate Escherichia coli O157:H7 in apple juice BMC Microbiology volume 5, Article number: 36 (2005).

[1041] Caruso, Francesco, Riccardo Pettinari, Miriam Rossi, Elena Monti, Marzia Bruna Gariboldi, Fabio Marchetti, Claudio Pettinari, Alessio Caruso, Modukuri V. Ramani, and Gottumukkala V. Subbaraju. The in vitro antitumor activity of arene-ruthenium(II) curcuminoid complexes improves when decreasing curcumin polarity. Journal of Inorganic Biochemistry 162 (2016): 44-51.

[1042] Bhatti, Muhammad Zeeshan, Amjad Ali, Hong-Quan Duong, Jiwu Chen, and Faiz-Ur Rahman. Anticancer activity and mechanism of bis-pyrimidine based dimetallic Ru(II)(16-p-cymene) complex in human non-small cell lung cancer via p53-dependent pathway. Journal of Inorganic Biochemistry 194 (2019): 52-64.

[1043] Santos, Wagner B.R., Marlange A.O. Melo, Rafael S. Alves, Renan G. de Brito, Thallita K. Rabelo, Lindaura da S. Prado, Virginia K. dos S. Silva, Daniel P. Bezerra, Jose E.R. de Menezes-Filho, Diego S. Souza, Carla M.L. de Vasconcelos, Luciana Scotti, Marcus Tullius Scotti, Waldecy de Lucca Junior, Lucindo J. Quintans-Junior, and Adriana G. Guimaraes. p-Cymene attenuates cancer pain via inhibitory pathways and modulation of calcium currents. Phytomedicine 61 (2019).

[1044] Mihajilov-Krstev, Tatjana, Dusanka Kitic, Dragan Radnovic, Mihajlo Ristic, Mira Mihajlovic-Ukropina, and Bojan Zlatkovic. Chemical Composition and Antimicrobial activity of Satureja kitaibelii Essential Oil against Pathogenic Microbial Strains. Natural Product Communications 6.8 (2011)1.

[1045] Kpadonou Kpoviessi BG, Ladekan EY, Kpoviessi DS, Gbaguidi F, Yehouenou B, Quetin-Leclercq J, Figueredo G, Moudachirou M, Accrombessi GC. Chemical variation of essential oil constituents of Ocimum gratissimum L. from Benin, and impact on antimicrobial properties and toxicity against Artemia salina leach. Chem Biodivers. 201 2 Jan;9(1):139-50.

[1046] Igor Jerkovic, Dajana Gaso-Sokac, Hrvoje Pavlovic, Zvonimir Marijanovic, Mirko Gugic, Ivana Petrovic, and Spomenka Kovac. Volatile Organic Compounds from Centaurium erythraea Rafn (Croatia) and the Antimicrobial Potential of Its Essential Oil Molecules. 2012 Feb; 17(2): 2058-2072.

[1047] Asres K, Tadesse S, Mazumder A, Bucar F. Composition, antimicrobial and free-radical scavenging activities of the essential oil of Plectranthus marrubatus. Nat Prod Commun. 2012 May;7(5):667-70.

selective antimicrobial activity against oral bacteria, and significantly inhibit bacterial biofilm formation[1048]. Finally, in engineered p-cymene-loaded gellan/PVA nanofibers wound dressings, p-cymene was been shown to exhibit strong antibiofilm activity against Candida albicans and Candida glabrata, without toxicity to NIH3T3 cell lines[1049] (murine).

Treatment for MRSA-infected Wounds: Continuing with the lengthy showcase of the antimicrobial activity of cymene, research also shows that this monocyclic monoterpene is capable of treating wounds infected with MRSA. As a primary constituent in a topical ointment prepared from the essential oil of dill (anethum graveolens) and administered to MRSA-infected wounds induced in mice, p-cymene contributed to the prevention of bacterial growth, the reduction of wound area, reduction of inflammatory phase, and acceleration of several key healing processes[1050].

Treatment for Lung Disease & Injury: P-cymene, together with related compounds carvacrol and thymol, were shown to reduce emphysema and lung inflammation in mice by reducing oxidative stress, alleviating alveolar enlargement, and by reducing macrophages and protein levels in bronchoalveolar lavage fluid[1051], among other modes of action. P-cymene has also been shown to significantly reduce pro-inflammatory cytokines, lung water gain, inflammatory cell infiltration, and lung tissue myeloperoxidase activity in mice subjected to induced acute lung injury[1052], attenuating the development and progression of pulmonary edema. As the primary constituent in the essential oil made from the leaves of pepper-rosmarin (lippia sidoides), p-cymene was shown to reduce the inflammatory changes and the alveolar enlargement observed in a murine emphysema model by reducing NF-KB expression and cytokine release, as well as the control of oxidative stress[1053].

Acaricidal Agent: As a primary constituent in the essential oil of oregano, p-cymene (together with carvacrol and thymol) contributed to significant acaricidal effects against ear mites (psoroptes cuniculi) in rabbits, killing 100% of the mites within one hour. As a main component in the essential oil of clove basil (ocimum gratissimum), p-cymene contributed to high acaricidal activity against several strains of ticks, including rhipicephalus microplus, rhipicephalus sanguineus, ixodes ricinus, and amblyomma cajennense[1054].

Antinociceptive Agent – Treatment for Pain: P-cymene has been shown to significantly reduce behavioral responses to pain in mice[1055], and exhibits anti-inflammatory activities that aid in the reduction of pain perception. Other research has also confirmed the antinociceptive properties of p-cymene in both in vivo and in vitro

---

[1048] Miladi, Hanene, Tarek Zmantar, Bochra Kouidhi, Yasir Mohammed A Al Qurashi, Amina Bakhrouf, Yassine Chaabouni, Kacem Mahdouani, and Kamel Chaieb. Synergistic effect of eugenol, carvacrol, thymol, p-cymene and y-terpinene on inhibition of drug resistance and biofilm formation of oral bacteria. Microbial pathogenesis 112 (2018): 156-163.

[1049] Mishra P, Gupta P, Srivastava R, Srivastava AK, Poluri KM, Prasad R. Exploration of Antibiofilm and In Vivo Wound Healing Activity of p-Cymene-Loaded Gellan/PVA Nanofibers. ACS Appl Bio Mater. 2023 May 15;6(5):1816-1831.

[1050] Manzuoerh, Reza, Mohammad Reza Farahpour, Ahmad Oryan, and Ali Sonboli. Effectiveness of topical administration of Anethum graveolens essential oil on MRSA-infected wounds. Biomedicine & pharmacotherapy = Biomedecine & pharmacotherapie 109 (2019): 1650-1658.

[1051] Games, Ellen et al. Structurally Related Monoterpenes p-Cymene, Carvacrol and Thymol Isolated from Essential Oil from Leaves of Lippia sidoides Cham. (Verbenaceae) Protect Mice against Elastase-Induced Emphysema. Molecules (Basel, Switzerland) vol. 21,10 1390. 20 Oct. 2016.

[1052] Guanghong Xie, Na Chen, Lanan Wassy Soromou, Fang Liu, Ying Xiong, Qianchao Wu, Hongyu Li, Haihua Feng, and Guowen Liu. p-Cymene Protects Mice Against Lipopolysaccharide-Induced Acute Lung Injury by Inhibiting Inflammatory Cell Activation Molecules. 2012 Jul; 17(7): 8159-8173.

[1053] Shang, Xiaofei, Yu Wang, Jiyu Zhang, and Xiaolou Miao. Acaricidal activity of oregano oil and its major component, carvacrol, thymol, and p-cymene, against Psoroptes cuniculi in vitro and in vivo. Veterinary parasitology 226 (2017): 93-96.

[1054] Lima, Aldilene, Marlise Milhomem, Odair Monteiro, Alana Arruda, Joaquim Castro, Yan Fernandes, Jose Maia, and Livio Costa-Junior. Seasonal analysis and acaricidal activity of the thymol-type essential oil of Ocimum gratissimum and its major constituents against Rhipicephalus microplus (Acari: Ixodidae). Parasitology Research 117.1 (2017): 59-65.

[1055] Bonjardim LR, Cunha ES, Guimaraes AG, Santana MF, Oliveira MG, Serafini MR, Araujo AA, Antoniolli AR Cavalcanti SC, Santos MR, Quintans-Junior LJ. Evaluation of the anti-inflammatory and antinociceptive properties of p-cymene in mice. Z Naturforsch C J Biosci. 2012 Jan-Feb;67(1-2):15-21.

experimental models[1056]. When combined with beta-cyclodextrin, p-cymene has been shown to contribute to the significant reduction of cancer pain in an S180 murine cancer pain model[1057].

Scolicidal Agent – Treatment for Tapeworms: As a primary constituent in the essential oil of ajowan (trachyspermum ammi), p-cymene contributed to potent effects against the larval stage of the dog tape worm (hydatid cyst protoscolices), killing 100% of the larvae within 60 minutes[1058].

IMPLICATIONS FOR HUMAN HEALTH & NUTRITION

Because cymene occurs in many different plants, it is easy to obtain this monocyclic monoterpene dietarily. When preparing, cooking, or serving foods, consider working with spices like allspice, basil, black pepper, caraway seed, cinnamon, clary sage, coriander, cornmint, cumin, curry, dill, fennel, ginger, mace, nutmeg, oregano, parsley, peppermint, rosemary, sage, savory, spearmint, tarragon, thyme, or turmeric. Fruits that may contain cymene include apricots, blackberries, black currants, blueberries, figs, grapefruit, guava, lemons, limes, mandarins, melons, oranges, papaya, pineapples, star fruits, tangerines, and yuzu. Vegetables that likely contain this terpene include bell peppers, carrots, celery, ginger, onions, parsnip, and tomatoes. Cannabis strains known to contain cymene include Ghost Cookies, Grape LA, and Northern Lights. Other options include working with essential oils that are high in cymene, particularly EO of cumin, thyme, oregano, or dill. As always, care should be used when working with isolated compounds or complex essential oils, as these substances can be toxic in some situations.

[1056] Lucindo Quintans-Junior, Jose C. F. Moreira, Matheus A. B. Pasquali, Soheyla M. S. Rabie, Andre S. Pires, Rafael Schroder, Thallita K. Rabelo, Joao P. A. Santos, Pollyana S. S. Lima, Socrates C. H. Cavalcanti, Adriano, A. S. Araujo, Jullyana S. S. Quintans, and Daniel P. Gelain. Antinociceptive Activity and Redox Profile of the Monoterpenes (+)-Camphene, p-Cymene, and Geranyl Acetate in Experimental Models ISRN Toxicol. 2013:459530.

[1057] Santos WBR, Pina LTS, de Oliveira MA, Santos LABO, Batista MVA, Trindade GGG, Duarte MC, Almeida JRGS, Quintans-Júnior LJ, Quintans JSS, Serafini MR, Coutinho HDM, Kowalska G, Baj T, Kowalski R, Guimarães AG. Antinociceptive Effect of a p-Cymene/β-Cyclodextrin Inclusion Complex in a Murine Cancer Pain Model: Characterization Aided through a Docking Study. Molecules. 2023 May 31;28(11):4465.

[1058] Moazeni M, Saharkhiz MJ, Hosseini AA. In vitro lethal effect of ajowan (Trachyspermum ammi L.) essential oil on hydatid cyst protoscoleces. Vet Parasitol. 2012 Jun 8;187(1-2):203-8.

Cymene Review

Answer the following questions to test your knowledge of this terpene:

Question #1: What type of terpene is cymene?

a. Monocyclic monoterpene
b. Bicyclic monoterpene
c. Monocyclic sesquiterpene
d. Bicyclic sesquiterpene

Question #2: Which of the following best describes the primary flavor & aroma profile of cymene?

a. Citrus, fruity, dry, terpenic
b. Sweet, floral, tangy
c. Woody, terpenic, chemical, citrus
d. Hops, cloves, wood, spicy

Question #3: How common is cymene in cannabis?

a. Top three
b. Top ten
c. Top twenty
d. Top forty

Question #4: What is the chemical formula for cymene?

a. $C_{10}H_{16}C$
b. $H_{24}CO_{20}$
c. $C_{20}H_{32}$
d. $C_{10}H_{14}$

Question #5: Name two uses of cymene in plants:

1 ______________________________ 2 ______________________________

Question #6: Name two potential medical uses of cymene:

1 ______________________________ 2 ______________________________

For the answer key to all individual compound chapters please go here: http://thebigbookofterps.com/bbot-quiz-answer-key/

PULEGONE

Type: Monocyclic monoterpenoid
Chemical Formula: $C_{10}H_{16}0$
Molecular Weight: 152.23 g/mol
Boiling Point: 220.00 to 222.00 °C. @ 760.00 mm Hg[1059]
Flash Point: 180.00 °F. TCC (82.22 °C.)
Solubility: Soluble in alcohol, oils, ethanol, ether, chloroform, weakly soluble in water
Oral LD50: 470 mg/kg (rat)[1060]
Biological Role: Insecticidal, allelopathic agent, antifeedant
Therapeutic Role: Antimicrobial, anticancer agent
Commercial Use: Fragrance, flavoring agent
Occurrence in Cannabis: Top forty
Aroma/Flavor: Mint, Camphor, sweet, fresh

Occurs in Cannabis Strains: Unsubstantiated reports in OG Kush, Pink Kush

INTRODUCTION

Pulegone is a monocyclic monoterpenoid that occurs as a secondary or minor terpene in cannabis. With a flavor and aroma profile that is minty, camphor, sweet, and fresh with herbaceous notes, pulegone is used to add flavor to some foods, and to add fragrance to perfumes or commercial products. Pulegone also has significant medical value as an antioxidant, anticancer agent, and antibacterial agent[1061], among other uses.

Pulegone is well-known to have been used for centuries by women seeking to abort unwanted pregnancies. For this purpose, preparations of pennyroyal tea or essential oils are consumed, which are high in pulegone.

Pulegone is thought to occur within the top forty terpenes by concentration in cannabis, although reviews of hundreds of cannabis terpene laboratory tests carried out as part of the research for this book uncovered not one occurrence of pulegone. However, unsubstantiated reports on the internet indicate that this terpene may be found in high concentration in OG Kush and Pink Kush.

[1059] The Good Scents Company Data Sheet for (R)-(+)-pulegone, from: http://www.thegoodscentscompany.com/data/rw1006501.html. Accessed September 8, 2023.

[1060] Carl Roth International Safety Data Sheet for Pulegone, from: https://www.carlroth.com/medias/SDB-5168-MT-EN.pdf?context=bWFzdGVyfHNlY3VyaXR5RGF0YXNoZWV0c3wyMzk1NTB8YXBwbGljYXRpb24vcGRmfHNlY3VyaXR5RGF0YXNoZWV0cy9oMjAvaDM1LzkwMjgyNDEzNTg4NzgucGRmfDhmNWM4MzExZjc5ZWIyMGE1MGRmODA3MjFlY2Q3MDc0YjE2ZTNhOGNjNmY0MTM5NmNiMTgxMjhkZDc0NTc1OGU. Accessed September 8, 2023.

[1061] Elansary HO, Abdelgaleil SAM, Mahmoud EA, Yessoufou K, Elhindi K, El-Hendawy S. Effective antioxidant, antimicrobial and anticancer activities of essential oils of horticultural aromatic crops in northern Egypt. BMC Complement Altern Med. 2018 Jul 13;18(1):214.

CHEMICAL STRUCTURE

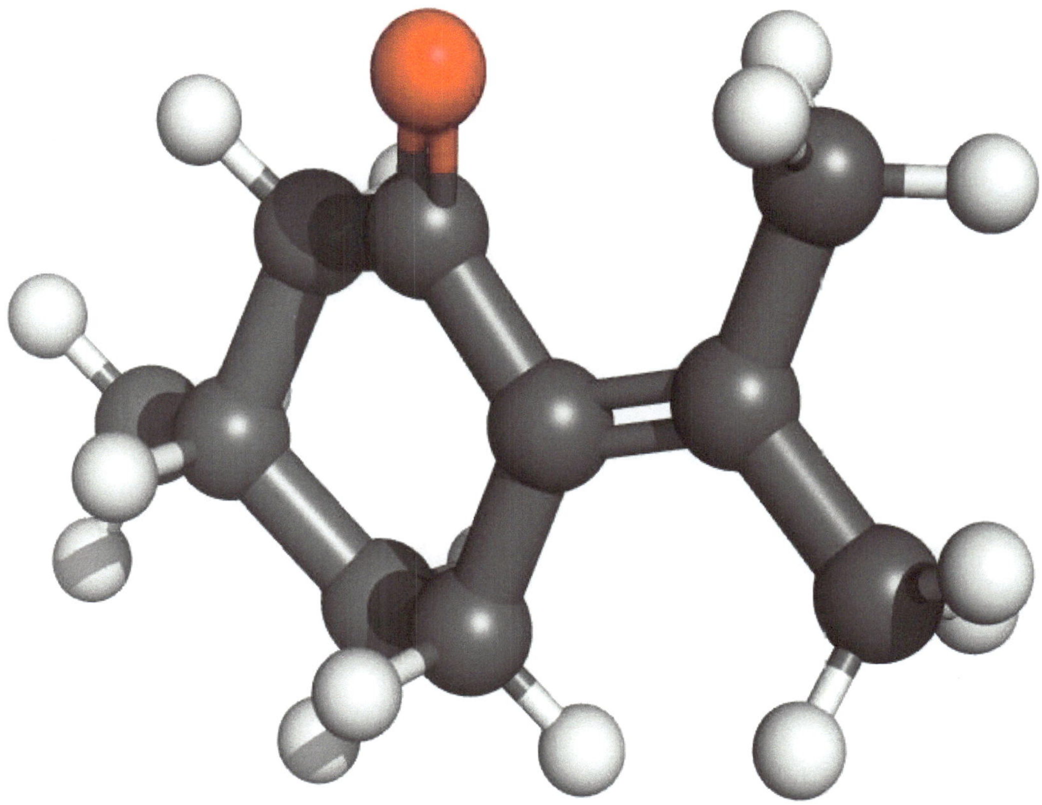

Pulegone is derived from geranyl diphosphate and consists of two linked units of isoprene with a molecular skeleton containing ten carbon atoms, sixteen hydrogen atoms, and one oxygen atom notated as $C_{10}H_{16}O$. Because pulegone has one ring, it is classified as monocyclic, and its oxygenated state further classifies this compound as a monoterpenoid. Pulegone features one exocyclic double bond.

Interestingly, pulegone is initially formed from limonene, while later in the biosynthesis process, this monoterpenoid is crucial in the formation of menthol.

OCCURRENCE IN PLANTS

Pulegone is mostly found in plants in the mint family, including mentha longifolia, muticous mountain mint, California mountain mint, cornmint, horsemint, peppermint, spearmint, Pilose mountain mint, biblical mint, applemint, Turkish calamint, catmint, orange mint, water mint, and other local varieties. Pulegone is also found in these plants:

Agathosma crenulata
Black currant
Buchu
Chamomile
Lemon balm
Origanum
Rosemary
Salvia

Bergamot
Bell pepper
Calamintha nepeta
Cornmint
Lemongrass
Pennyroyal
Rue
Satureja viminea

Blackberry
Blueberry
Cayenne
Horsemint
Mint
Peppermint
Sage
Spearmint

Tea	Thyme	Water mint

BIOLOGICAL ACTIVITY IN PLANTS

Pulegone appears to serve two primary functions in plants: as an insecticidal agent, and as an allelopathic agent. This terpene also has significant antifeedant activity, as described below.

**Insecticidal Agent:* Of forty-two pure monoterpenes, pulegone proved to be the strongest insecticide tested, exhibiting potent fumigation toxicity against the common housefly (musca domestica)[1062], while as the primary constituent in the essential oil of peperina (minthostachys verticillate), similar effects were again observed against the housefly[1063], with other research confirming that the fly's detoxification system actually serves to intensify the effects of the essential oil[1064]. As a primary constituent in the essential oil of Japanese catnip (schizonpeta multifida), pulegone exhibited pronounced contact and fumigant toxicity against the maize weevil (sitophilus zeamais), and the red flour beetle (tribolium castaneum)[1065].

**Allelopathic Agent:* Pulegone has been shown to inhibit the germination of field mustard (brassica rapa) seeds, and radish (R. sativus) seeds[1066], while other research has shown that the compound strongly inhibits the seed germination[1067] and root growth of lettuce (lactuca sativa)[1068]. As a primary constituent in the essential oil of wormwood (artemisia judaica), pulegone also contributed to significant inhibition of the root growth of milk thistle (silybum marianum)[1069].

**Antifeedant:* Pulegone has been shown to prevent aphids from settling on leaves and deterred probing and feeding, while decreasing the duration and effectiveness of feeding[1070]. This is particularly important information for cannabis growers considering that aphids are one of the most common pests encountered in the industry. Pulegone has also been shown to deter house sparrows from feeding[1071].

USES IN INDUSTRY

[1062] Zhilin Zhang, Yongjian Xie, Yong Wang, Zhufeng Lin, Lihua Wang, Guoyuan Li Toxicities of monoterpenes against housefly, Musca domestica L. (Diptera: Muscidae) Environmental Science and Pollution Research November 2017, Volume 24, Issue 31, pp 24708-24713.

[1063] Sara M. Palacios, Alberto Bertoni, Yanina Rossi, Rocio Santander, Alejandro Urzua Insecticidal activity of essential oils from native medicinal plants of Central Argentina against the house fly, Musca domestica (L.) Parasitology Research December 2009, Volume 106, Issue 1, pp 207-212.

[1064] Rossi YE, Canavoso L, Palacios SM. Molecular response of Musca domestica L. to Mintostachys verticillata essential oil, (4R)+-pulegone and menthone. Fitoterapia. 2012 Mar;83(2):336-42.

[1065] Liu ZL, Chu SS, Jiang GH. Toxicity of Schizonpeta multifida essential oil and its constituent compounds towards two grain storage insects. J Sci Food Agric. 2011 Mar 30;91(5):905-9.

[1066] Nobuhiro Sekine & Sakae Shibutani Chemical structures of p-menthane monoterpenes with special reference to their effect on seed germination and termite mortality Journal of Wood Science volume 59, pages229-237(2013).

[1067] Poser, G.L. von, Menut, C., Toffoli, M.E., Verin, P., Sobral, M., Bessiere, J.M., Lamaty, G., Henriques, A.T. Essential oil composition and allelopathic effect of the Brazilian Lamiaceae Hesperozygis ringens (Benth.) Epling and Hesperozygis rhododon Epling Journal of agricultural and food chemistry 1996 v.44 no.7 pp. 1829-1832.

[1068] Fabrizio Araniti, Et. al Allelopathic potential of Artemisia arborescens: Isolation, identification and quantification of phytotoxic compounds through fractionation-guided bioassays Natural Product Research Volume 27, 2013 Issue 10 Pages 880-887 4/4/12.

[1069] Samir A M Abdelgaleil Allelopathic Potential of Essential Oils Isolated from Aromatic Plants on Silybum marianum L. Global Advanced Research Journal of Agricultural Science 3(9):289-297. September 2014.

[1070] Katarzyna Dancewicz, Beata Gabrys, Iwona Dams, Czeslaw Wawrzenczyk Enantiospecific Effect of Pulegone and Pulegone-Derived Lactones on Myzus persicae (Sulz.) Settling and Feeding Journal of Chemical Ecology Volume 34, Article number: 530 (2008).

[1071] Clapperton BK, Porter RE, Day TD, Waas JR, Matthews LR. Designer repellents: combining olfactory, visual or taste cues with a secondary repellent to deter free-ranging house sparrows from feeding. Pest Manag Sci. 2012 Jun;68(6):870-7.

Commercial samples of pulegone appear as colorless oils that are typically obtained from the essential oils of catnip, pennyroyal, and mint species of plants. Used to add flavor to foods such as baked goods, beverages, frozen dairy items, fruit ices, and hard candy, pulegone is also used to lend fragrance to perfumes or consumer products like cleaning agents, perfumes, and aroma therapy items.

POTENTIAL USES IN MEDICINE

Although pulegone offers several potential medical uses, toxicity and other potentially harmful features of this monocyclic monoterpenoid should be evaluated carefully. First, pulegone can cause acute toxicity if swallowed - care should be exercised when handling this terpene raw or in high concentrations in essential oils. Second, pulegone has been shown to be a nervous system depressant with psychoactive properties, with researchers concluding that "pulegone is a psychoactive compound and has the profile of an analgesic drug[1072]." Third, pulegone has been shown to slightly increase incidences of bladder cancer in rats, likely caused by the cytotoxicity of the compound and by increased cell proliferation[1073]. Fourth, pulegone has been shown to reduce the heart contractile force and contraction time in guinea pigs[1074], indicating that this terpenoid could have a negative effect on heart health in certain cases. Finally, a 2022 study found that pulegone was a concerning chemical in tobacco vaporization devices, with researchers noting that "While the FDA flavour ban has reduced sales of JUUL to minors, young users appear to have rapidly adopted other brands, such as Puff,22 which has high concentrations of WS-23 and concerning levels of pulegone. Ironically, the flavour ban may have caused youth to migrate to a potentially more harmful e-cigarette. Switching from flavoured JUUL to Puff e-cigarettes may expose users to increased harm due to the higher levels of WS-23 and pulegone in Puff products[1075]."

Despite these warnings, pulegone still offers potential for the following medical uses:

Treatment for Colon & Breast Cancer: As the primary constituent in the essential oil of the medicinal herb micromeria fruticose, pulegone was shown to exhibit marked antitumor activities against human colon tumor cells and mammary carcinoma[1076] (breast cancer).

Antimicrobial Agent: As a major constituent in the essential oil of micromeria congesta, pulegone contributed to significant antimicrobial effects against bacillus subtilis, bacillus cereus, staphylococcus aureus, escherichia coli, micrococcus luteus, enterococcus faecalis, proteus vulgaris, and yersinia enterecolitica[1077]. As the primary constituent in the essential oil of ziziphora clinopodioides, pulegone contributed to the inhibition of aflatoxin production in aspergillus parasiticus as well as inhibition of the growth of the pathogen itself[1078]. As a major component in the essential oil of hesperozygis myrtoides, pulegone contributed to antimicrobial activity against C.

1072 de Sousa DP, Nobrega FF, de Lima MR, de Almeida RN. Pharmacological activity of (R)-(+)-pulegone, a chemical constituent of essential oils. Z Naturforsch C J Biosci. 2011 Jul-Aug;66(7-8):353-9.

1073 Da Rocha MS, Dodmane PR, Arnold LL, Pennington KL, Anwar MM, Adams BR, Taylor SV, Wermes C, Adams TB, Cohen SM. Mode of action of pulegone on the urinary bladder of F344 rats. Toxicol Sci. 2012 Jul;128(1):1-8.

1074 de Cerqueira SV, Gondim AN, Roman-Campos D, Cruz JS, Passos AG, Lauton-Santos S, Lara A, Guatimosim S, Conde-Garcia EA, de Oliveira ED, de Vasconcelos CM. R(+)-pulegone impairs Ca2+ homeostasis and causes negative inotropism in mammalian myocardium. Eur J Pharmacol. 2011 Dec 15;672(1-3):135-42.

1075 Omaiye EE, Luo W, McWhirter KJ, Pankow JF, Talbot P. Flavour chemicals, synthetic coolants and pulegone in popular mint-flavoured and menthol-flavoured e-cigarettes. Tob Control. 2022 Aug;31(e1):e3-e9.

1076 Shehab NG, Abu-Gharbieh E. Constituents and biological activity of the essential oil and the aqueous extract of Micromeria fruticosa (L.) Druce subsp. serpyllifolia. Pak J Pharm Sci. 2012 Jul;25(3):687-92.

1077 Herken EN, Celik A, Aslan M, Ayd nl k N. The constituents of essential oil: antimicrobial and antioxidant activity of Micromeria congesta Boiss. & Hausskn. ex Boiss. from East Anatolia. J Med Food. 2012 Sep; 15(9):835-9.

1078 Khosravi AR, Shokri H, Minooeianhaghighi M. Inhibition of aflatoxin production and growth of Aspergillus parasiticus by Cuminum cyminum, Ziziphora clinopodioides, and Nigella sativa essential oils. Foodborne Pathogen Dis. 2011 Dec;8(12):1275-80.

albicans, C. glabrata, C. krusei, C. parapsilosis and C. tropicalis[1079], while a separate study that examined EO from the same plant found that pulegone completely inhibited the growth of apergillus flavus[1080].

**Treatment for Vaginosis:* Continuing with the antimicrobial activity of pulegone, this compound was shown to contribute to significant antifungal activity against vaginal candidiasis as the primary compound in the essential oil of applemint (mentha suaveolens)[1081], offering the potential for a novel new treatment of this common pathogen.

**Nematocidal Agent:* As part of the essential oils of several plants and as an individual compound, pulegone has been shown to exhibit high nematocidal activity against the root-knot nematode (meloidogyne incognita)[1082], one of the most common parasites of plants in the world.

**Acaricidal Agent:* As the primary constituent in the essential oil of hesperozygis ringens, pulegone was shown to inhibit egg hatching of the cattle tick (riphicephalus boophilus) by as much as 95%[1083].

**Aquatic & Aquaculture Bactericide*: Once again, as the primary constituent in the essential oil of hesperozygis ringens, pulegone was shown to exhibit antibacterial effects against clinical isolates of the fish pathogenic bacteria A. hydrophila, raoultella ornithinolytica, and citrobacter freundii, significantly increasing the survival rate of infected catfish[1084].

**Treatment for Hypertension:* Pulegone has been shown to reduce systolic blood pressure and heart rate, reverse reduced levels of serum nitric oxide (NO), ameliorate lipid profile and oxidative stress markers, and improve the vascular response to acetylcholine, overall improving liver and kidney functions and lipid profile, among other positive effects in L-NAME (NG-Nitro-L-arginine-methyl ester)-Induced Hypertensive Rats[1085].

IMPLICATIONS FOR HUMAN HEALTH & NUTRITION

As a less common monoterpene, pulegone is not as easy to obtain dietarily than major compounds like pinene, myrcene, or limonene. One of the simplest ways to consume pulegone is to drink black, lemongrass, or mint teas, all of which likely contain significant amounts of this terpenoid. When preparing, cooking, or serving foods, consider using spices and herbs like oregano, cayenne, sage, rosemary, thyme, or any type of mint. Vegetables like bell peppers and cayenne peppers contain pulegone, as well as fruits such as blackberries, black currants, and blueberries.

Unsubstantiated reports that the cannabis strains OG Kush and Pink Kush contain pulegone could mean that strains with these genetics may also contain pulegone. More research and testing is needed.

[1079] Martini MG, Bizzo HR, Moreira Dde L, Neufeld PM, Miranda SN, Alviano CS, Alviano DS, Leitao SG. Chemical composition and antimicrobial activities of the essential oils from Ocimum selloi and Hesperozygis myrtoides. Nat Prod Commun. 2011 Jul;6(7):1027-30.

[1080] Gonzalez-Chavez MM, Cardenas-Ortega NC, Mendez-Ramos CA, Perez-Gutierrez S. Fungicidal properties of the essential oil of Hesperozygis marifolia on Aspergillus flavus link. Molecules. 2011 Mar 15;16(3):2501-6.

[1081] Pietrella D, Angiolella L, Vavala E, Rachini A, Mondello F, Ragno R, Bistoni F, Vecchiarelli A. Beneficial effect of Mentha suaveolens essential oil in the treatment of vaginal candidiasis assessed by real-time monitoring of infection. BMC Complement Altern Med. 2011 Feb 28;11:18.

[1082] Ntalli NG, Ferrari F, Giannakou I, Menkissoglu-Spiroudi U. Phytochemistry and nematicidal activity of the essential oils from 8 Greek Lamiaceae aromatic plants and 13 terpene components. J Agric Food Chem.

[1083] Ribeiro VL, Dos Santos JC, Bordignon SA, Apel MA, Henriques AT, von Poser GL. Acaricidal properties of the essential oil from Hesperozygis ringens (Lamiaceae) on the cattle tick Riphicephalus (Boophilus) microplus. Bioresour Technol. 2010 Apr;101(7):2506-9.

[1084] Rosa IA, Rodrigues P, Bianchini AE, Silveira BP, Ferrari FT, Bandeira Junior G, Vargas APC, Baldisserotto B, Heinzmann BM. Extracts of Hesperozygis ringens (Benth.) Epling: in vitro and in vivo antibacterial activity against fish pathogenic bacteria. J Appl Microbiol. 2019 May;126(5):1353-1361.

[1085] Razzaq MA, Younis W, Malik MNH, Alsahli TG, Alamgeer, Jahan S, Ehsan R, Gasparotto Junior A, Bashir A. Pulegone Prevents Hypertension through Activation of Muscarinic Receptors and Cyclooxygenase Pathway in L-NAME-Induced Hypertensive Rats. Cardiovasc Ther. 2023 May 9;2023:8166840.

Because of the potential dangers of pure pulegone or essential oils that are high in this monoterpene, care should be exercised when working with this compound.

Pulegone Review

Answer the following questions to test your knowledge of this terpene:

Question #1: What type of terpene is pulegone?

a. Monocyclic monoterpenoid
b. Linear sesquiterpene
c. Bicyclic sesquiterpene
d. Acyclic diterpene

Question #2: Which of the following best describes the primary flavor & aroma profile of pulegone?

a. Citrus, fruity, dry, terpenic
b. Piney, wood, spicy
c. Mint, Camphor, sweet, fresh
d. Spicy, cloves, dry

Question #3: How common is pulegone in cannabis?

a. Top ten
b. Top three
c. Top fifteen
d. Less common

Question #4: What is the chemical formula for pulegone?

a. $C_{10}H_{16}O$
b. $C_{12}H_{20}O_2$
c. $C_{15}H_{24}$
d. $C_{20}H_{32}O$

Question #5: Name two uses of pulegone in plants:

1 ______________________________ 2 ______________________________

Question #6: Name two potential medical uses of pulegone:

1 ______________________________ 2 ______________________________

For the answer key to all individual compound chapters please go here: http://thebigbookofterps.com/bbot-quiz-answer-key/

BERGAMOTENE

Type: Bicyclic sesquiterpene
Chemical Formula: $C_{15}H_{24}$
Molecular Weight: 204.35 g/mol
Boiling Point: *Alpha-bergamotene:* 259.00 to 260.00 °C. @ 760.00 mm Hg[1086]
Trans-alpha-bergamotene: 259.00 to 260.00 °C. @ 760.00 mm Hg[1087]
Beta-bergamotene: 260.00 to 261.00 °C. @ 760.00 mm Hg[1088]
Trans-beta-bergamotene: 258.00 to 260.00 °C. @ 760.00 mm Hg[1089]
Flash Point: *All isomers:* 213.00 °F. TCC (100.56 °C.)
Solubility: *All isomers:* Soluble in alcohol, insoluble in water
Oral LD50: 2.848 g/kg (rat)[1090]
Biological Role: Attract pollinators, indirect defense
Therapeutic Role: Anti-inflammatory, antimicrobial, mosquito control agent
Commercial Use: Research chemical
Occurrence in Cannabis: Top forty
Aroma/Flavor: *Trans-alpha-bergamotene:* Woody, warm, green, tea, citrus

Occurs in Cannabis Strains: Unknown

INTRODUCTION

Bergamotene is a bicyclic sesquiterpene that is theorized to be a minor or secondary terpene in cannabis, likely appearing within the top forty terpenes by concentration in varieties where it occurs. In our review of hundreds of cannabis terpene tests, and after an exhaustive search of the web, no occurrences of bergamotene were found. However, most laboratories only test for ten or fewer terpenes in cannabis, and it seems that very few have standards for bergamotene as of 2023. However, we theorize that, should terpene tests include the top forty most common terpenes in cannabis, bergamotene will appear frequently.

Unlike nearly all the other terpenes in this book, bergamotene is not used in the fragrance, perfume, foods, or consumer products industries, and in fact The Good Scents Company pages for all variations of bergamotene report that the compound is not for fragrance use, and not for flavor use[1091]. Nevertheless, this sesquiterpene is a volatile molecule that emits a flavor and aroma profile that is woody, warm, and green, with slight citrus nuances.

Interestingly, Caribbean and Mexican fruit flies emit bergamotene[1092], probably as a mating or aggregation pheromone, and soldier termites of the species prorhinotermes canalifrons have also been shown to emit

---

[1086] The Good Scents Company Data Sheet for Alpha-bergamotene, from: http://www.thegoodscentscompany.com/data/rw1384471.html. Accessed September 9, 2023.

[1087] The Good Scents Company Data Sheet for Trans-alpha-bergamotene, from: http://www.thegoodscentscompany.com/data/rw1384461.html. Accessed September 9, 2023.

[1088] The Good Scents Company Data Sheet for Beta-bergamotene, from: http://www.thegoodscentscompany.com/data/rw1384441.html. Accessed September 9, 2023.

[1089] The Good Scents Company Data Sheet for Trans-beta-bergamotene, from: http://www.thegoodscentscompany.com/data/rw1563121.html. Accessed September 9, 2023.

[1090] AromaDb Database Specification Sheet for Cis-alpha-bergamotene, from: http://bioinfo.cimap.res.in/aromadb/web_compound_detail.php?id=CRMOL465. Accessed September 9, 2023.

[1091] The Good Scents Company, Alpha-Bergamotene: http://www.thegoodscentscompany.com/data/rw1384471.html Accessed December 28 2019.

[1092] Rocca, J R, J L Nation, L Strekowski, and M A Battiste. Comparison of volatiles emitted by male caribbean and mexican fruit flies. Journal of chemical ecology 18.2 (2013): 223-244.

bergamotene via frontal glands[1093]. Other insects are either attracted to or repelled by emissions of bergamotene; read more on this in the "Biological Activity in Plants" section below.

CHEMICAL STRUCTURE

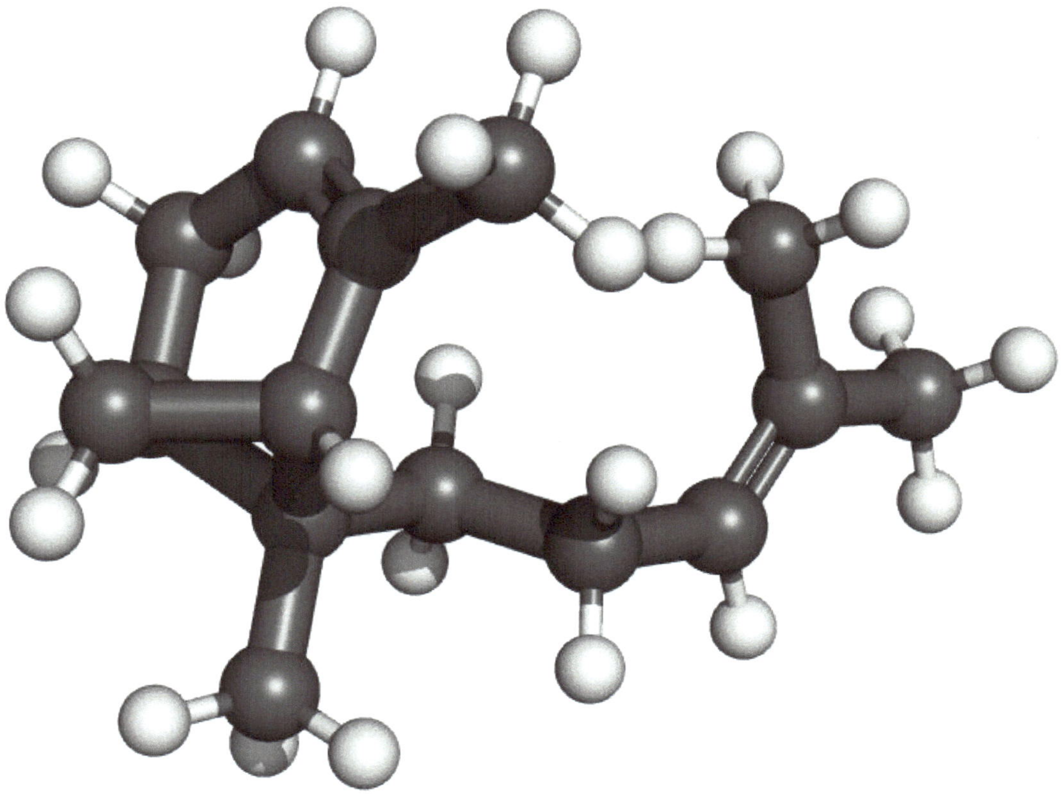

Like most sesquiterpenes, bergamotene is derived from farnesyl pyrophosphate, based on the interlinking of three units of isoprene arranged with fifteen carbon atoms and twenty-four hydrogen atoms notated as $C_{15}H_{24}$. Bergamotene's carbon skeleton contains one strained ring, and one ring with an endocyclic double-bond. These features classify this compound as a bicyclic sesquiterpene, which occurs as two structural isomers: alpha-bergamotene and beta-bergamotene, which are differentiated by the placement of the double-bond; in the image above, alpha-bergamotene and its double bonds are shown. These structural isomers each have stereoisomers, the cis- and trans- versions, with the majority of known bergamotene variations listed below:

Alpha-bergamotene
Cis-alpha-bergamotene
Trans-alpha-bergamotene
Beta-bergamotene
Cis-beta-bergamotene
Trans-beta-bergamotene

[1093] Piskorski, Rafal, Robert Hanus, Sona Vasickova, Josef Cvacka, Jan Sobotnik, Ales Svatos, and Irena Valterova. Nitroalkenes and Sesquiterpene Hydrocarbons from the Frontal Gland of Three Prorhinotermes Termite Species. Journal of Chemical Ecology 33.9 (2007): 1787-1794.

Limited data is available for most of these isomers, which explains the gaps in the data segments at the beginning of this chapter. Future editions will feature updates to these gaps where applicable.

OCCURRENCE IN PLANTS

Alpha-bergamotene:

Alpinia galanga	Basil	Bergamot
Calamus	Carrot	Curcuma amada
Daucus carota	Grapefruit	Guaiacwood
Lemon	Lime	Mace
Nutmeg	Ocimum basilicum	Orange
Parsley	Sugandha kokila	Turmeric
Zingiber officinale		

Trans-alpha-bergamotene:

Anise	Baccharis cordifolia	Basil
Bergamot	Black pepper	Carrot
Celery	Cinnamon	Cumin
Fleabane	Galangal	Helichrysum italicum
Jambu	Lavender	Lemon
Lemongrass	Lime	Lovage
Marsypianthes chamaedrys	Myristica fragrans	Nutmeg
Orthodon dianthera	Persea americana	Phoebe
Sandalwood		

Beta-bergamotene:

Lavender

Trans-beta-bergamotene:

Anise	Lavandin	Parsley
Rose	Snake root	

BIOLOGICAL ACTIVITY IN PLANTS

The primary function of bergamotene in plants appears to be to interact with insects; both as a repellant, and as an attractant.

Dual-Purpose – Attract Pollinators & Indirect Defense: Coyote tobacco (nicotiana attenuate) flowers have been shown to emit bergamotene at night to attract its primary pollinator, the tobacco hawkmoth (manduca sexta), while using similar emissions of bergamotene during the day to attract predatory insects that prey on the larvae of the tobacco hawkmoth, which feed on their host[1094].

[1094] Zhou, W., Kilgler A., McGale, E., Haverkamp, A., Knaden, M., Guo, H., Beran, F., Yon, F., Li, R., Lackus, N., K llner, T. G., Bing, J., Schuman, M. C., Hansson, B. S., Kessler, D., Baldwin, I. T., Xu, S. Tissue-specific emission of (E)-a-bergamotene helps resolve the dilemma when pollinators are also

**Indirect Defense:* After being fed upon by moths or butterflies, maize plants substantially increase emissions of bergamotene, which attract parasitic wasps that feed on the larvae of the herbivores[1095]. Corn seedlings were also shown to release bergamotene in response to herbivory by beet armyworm larvae - these emissions attract the parasitic wasp cotesia marginiventris, which feeds on the larvae[1096]. Other studies have shown that maize can be induced to increase emissions of bergamotene via pretreatment with cis-jasmone, which is thought to activate plant defenses against herbivores[1097].

USES IN INDUSTRY

Bergamotene is generally not used in commercial, medical, or food product industries; its primary use is as a research chemical, where it appears as a light green to colorless or clear liquid. This compound can be extracted from Turkish citrus leaf and schisandra chinensis seeds[1098], and other plants that feature bergamotene in moderate to high amounts.

Interestingly, research by the prepared foods industry has shown that green bell peppers begin to emit bergamotene in increasing amounts as the peppers age in packaging[1099].

POTENTIAL USES IN MEDICINE

Bergamotene offers a unique set of potential medical uses, most notably as anti-inflammatory and antimicrobial agents, among several other uses:

**Antimicrobial Agent:* As a primary constituent in the essential oil made from the inner bark of kielmeyera coriacea, bergamotene contributed to remarkable antimicrobial activity against the anaerobic microorganism prevotella nigrescens[1100]. As a primary constituent in the essential oil made from the flowers of eugenia klotzschiana, bergamotene contributed to moderate and high antioxidant and antimicrobial activity against streptococcus salivarius, S. mutans, S. mitis, and prevotella nigrescens[1101].

**Anti-inflammatory Agent:* The efficacy of the traditional topical use of ointments made from alchornea cordifolia was confirmed in a study that showed bergamotene - as a major component in the essential oil of the plant - contributed to significant anti-inflammatory effects in mice with induced ear edema[1102] (swelling). As a primary

---

herbivores. Current Biology; April 20, 2017.

[1095] Schnee, Christiane et al. The products of a single maize sesquiterpene synthase form a volatile defense signal that attracts natural enemies of maize herbivores. Proceedings of the National Academy of Sciences of the United States of America vol. 103,4 (2006): 1129-34.

[1096] Turlings, T C, J H Tumlinson, R R Heath, A T Proveaux, and R E Doolittle. Isolation and identification of allelochemicals that attract the larval parasitoid,Cotesia marginiventris (Cresson), to the microhabitat of one of its hosts. Journal of chemical ecology 17.11 (2013): 2235-51.

[1097] Oluwafemi, Sunday et al. Priming of Production in Maize of Volatile Organic Defence Compounds by the Natural Plant Activator cis-Jasmone. PloS one vol. 8,6 e62299. 26 Jun. 2013.

[1098] Toronto Research Chemicals. a-cis-Bergamotene, From: https://www.trc-canada.com/product-detail/?B318450 Accessed December 28, 2019.

[1099] Ranjitha, K et al. Effect of pretreatments and modified atmosphere packaging on the shelf life and quality of fresh- cut green bell pepper. Journal of food science and technology vol. 52,12 (2015): 7872-82.

[1100] Martins, Carla de M et al. Chemical Constituents and Evaluation of Antimicrobial and Cytotoxic Activities of Kielmeyera coriacea Mart. & Zucc. Essential Oils. Evidence-based complementary and alternative medicine : eCAM vol. 2015 (2015): 842047.

[1101] Carneiro, Nargella S, Cassia C F Alves, Jose M Alves, Mariana B Egea, Carlos H G Martins, Thayna S Silva, Lizandra C Bretanha, Maira P Balleste, Gustavo A Micke, Eduardo V Silveira, and Mayker L D Miranda. Chemical composition, antioxidant and antibacterial activities of essential oils from leaves and flowers of Eugenia klotzschian a Berg (Myrtaceae). Anais da Academia Brasileira de Ciencias 89.3 (2018): 1907-1915.

[1102] Okoye, F B C, P O Osadebe, C S Nworu, N N Okoye, E O Omeje, and C O Esimone. Topical anti-inflammatory constituents of lipophilic leaf fractions of Alchornea floribunda and Alchornea cordifolia. Natural product research 25.20 (2012): 1941-9.

constituent in the essential oil made from the tree copaifera multijuga, bergamotene contributed to in vivo and in vitro anti-inflammatory effects in rats[1103].

**Detection of Fungal Lung Infections:* Bergamotene can be used with around 93% accuracy to detect the presence of invasive fungal infections in the breath of patients suffering from suspected fungal pneumonia caused by aspergillus fungal species[1104, 1105].

**Mosquito Control Agent:* As a primary constituent in extracts made from the leaves of hyptis suaveolens from West Africa, bergamotene contributed to the significant reduction of probing activity and repellency against malaria and other diseases of the vector mosquito, A. aegypti[1106]. As a main component in the essential oil made from the leaves of guarea scabra, bergamotene contributed to 100% mortality against larvae of A. aegypti[1107] mosquitoes.

**Treatment for Psoriasis:* As the primary constituent in the essential oil of the tree, copaifera langsdorffii, bergamotene attenuated erythema, skin thickness, and scaliness caused by psoriasis by reducing the release of pro-inflammatory cytokines[1108], among other mechanisms.

IMPLICATIONS FOR HUMAN HEALTH & NUTRITION

To obtain bergamotene dietarily, consider using herbs and spices like basil, curcuma amada (used like ginger), mace, nutmeg, parsley, or turmeric when preparing or serving foods. Meals should include mila potatoes and carrots when possible, and bergamotene-rich snacks to consider include grapefruit, kumquat, lemons, limes, and oranges. You can also make lemongrass tea, or work with essential oils that are high in this bicyclic sesquiterpene. Future editions of this book will be updated as laboratory cannabis testing includes more standards for bergamotene.

[1103] Kobayashi, Cristine, Tiago Oselame Fontanive, Barbara Grade Enzweiler, Laura Renata de Bona, Thalita Massoni, Miriam Anders Apel, Amelia Teresinha Henriques, Marc Francois Richter, Patricia Ardenghi, and Edna Sayuri Suyenaga. Pharmacological evaluation of Copaifera multijuga oil in rats. Pharmaceutical biology 49.3 (2011) 306-313.

[1104] Koo, Sophia, Horatio R. Thomas, S. David Daniels, Robert C. Lynch, Sean M. Fortier, Margaret M. Shea, Preshious Rearden, James C. Comolli, Lindsey R. Baden, and Francisco M. Marty. A Breath Fungal Secondary Metabolite Signature to Diagnose Invasive Aspergillosis. Clinical Infectious Diseases 59.12 (2014): 1733-1740.

[1105] Heddergott, C et al. The volatome of Aspergillus fumigatus. Eukaryotic cell vol. 13,8 (2014): 1014-25.

[1106] Evaluation of Extracts and Oils of Mosquito (Diptera: Culicidae) Repellent Plants from Sweden and Guinea- Bissau. Journal of Medical Entomology 43.1 (2006): 113-119.

[1107] Magalhaes, Lyege Amazonas Maciel et al. Chemical composition and larvicidal activity against Aedes aegypti larvae of essential oils from four Guarea species. Molecules (Basel, Switzerland) vol. 15,8 5734-41. 19 Aug. 2010.

[1108] Gelmini, Fabrizio, Giangiacomo Beretta, Cecilia Anselmi, Marisanna Centini, Paolo Magni, Massimiliano Ruscica, Alberto Cavalchini, and Roberto Maffei Facino. GC-MS profiling of the phytochemical constituents of the oleoresin from Copaifera langsdorffii Desf. and a preliminary in vivo evaluation of its antipsoriatic effect. International Journal of Pharmaceutics 440.2 (2013): 170-178.

Bergamotene Review

Answer the following questions to test your knowledge of this terpene:

Question #1: What type of terpene is bergamotene?

a. Acyclic sesquiterpene
b. Monocyclic sesquiterpene
c. Bicyclic sesquiterpene
d. Linear monoterpene

Question #2: Which of the following best describes the primary flavor & aroma profile of bergamotene?

a. Hops, cloves, wood, spicy
b. Woody, warm, green, tea, citrus
c. Spicy, turpentine, fuel, pine
d. Turpentine, pine, woody, balsam

Question #3: How common is bergamotene in cannabis?

a. Top three
b. Top five
c. Top twenty
d. Top forty

Question #4: What is the chemical formula for bergamotene?

a. $C_{15}H_{26}$
b. $C_{15}H_{24}$
c. $C_{10}H_{16}$
d. $N_{15}H_{26}$

Question #5: Name two uses of bergamotene in plants:

1 ______________________________ 2 ______________________________

Question #6: Name two potential medical uses of bergamotene:

1 ______________________________ 2 ______________________________

For the answer key to all individual compound chapters please go here: http://thebigbookofterps.com/bbot-quiz-answer-key/

ELEMENE

Type: Monocyclic sesquiterpene
Chemical Formula: $C_{15}H_{24}$
Molecular Weight: 204.35 g/mol
Boiling Point: *Alpha-elemene:* 254.00 to 255.00 °C. @ 760.00 mm Hg (estimated)[1109]
Beta-elemene: 251.00 to 253.00 °C. @ 760.00 mm Hg (estimated)[1110]
Gamma-elemene: 257.00 to 259.00 °C. @ 760.00 mm Hg (estimated)[1111]
Delta-elemene: 254.00 to 255.00 °C. @ 760.00 mm Hg (estimated)[1112]
Flash Point: *Alpha-elemene:* 211.00 °F. TCC (99.60 °C.) (estimated by TGSC)
Beta-elemene: 209.00 °F. TCC (98.30 °C.) (estimated by TGSC)
Gamma-elemene: 214.00 °F. TCC (101.30 °C.) (estimated by TGSC)
Delta-elemene: 211.00 °F. TCC (99.50 °C.) (estimated by TGSC)
Solubility: *All isomers:* Soluble in alcohol, insoluble in water
Oral LD50: 1.14 ml/kg (man)[1113]
Biological Role: Indirect defense, allelopathic agent
Therapeutic Role: Anticancer agent, cancer drug/treatment synergist
Commercial Use: Research chemical
Occurrence in Cannabis: Top forty
Aroma/Flavor: *Beta-elemene:* Herbal, waxy, fresh
Delta-elemene: Sweet, herbal, lavender, woody

Occurs in Cannabis Strains: Unknown

INTRODUCTION

Elemene is a monocyclic sesquiterpene that is widely known in traditional Chinese medicine and in research circles worldwide, thanks to its considerable anti-cancer properties. In fact, most scientific literature regarding elemene focuses on cancer research, and generally centers on the beta- isomer of this compound.

The occurrence of elemene in cannabis is theoretical as of early 2021, with few to no laboratories testing for this sesquiterpene in relation to cannabis. In our reviews of hundreds of cannabis terpene test results, we did not find any strains that contained elemene, nor did we find any tests published on the internet (particularly in the cannabis space) that showed this terpene, and an extensive review of research articles was unsuccessful in pinpointing any occurrences of elemene in cannabis. Nevertheless, we theorize that elemene is a secondary or trace compound in cannabis, most likely occurring within the top forty terpenes by concentration. As laboratories add this molecule to standards for gas chromatography, mass spectrometry, or other testing, it is likely that we will see regular occurrences of elemene in some or many cannabis strains.

According to The Good Scents Company, beta-elemene features an aroma and flavor profile that is herbal, waxy, and fresh[1114], with presumed "green" notes, while the delta- isomer is sweet, herbal, lavender, and woody.

[1109] The Good Scents Company Data Sheet for Alpha-elemene, from: http://www.thegoodscentscompany.com/data/rw1523941.html. Accessed September 9, 2023.

[1110] The Good Scents Company Data Sheet for Beta-elemene, from: http://www.thegoodscentscompany.com/data/rw1053861.html. Accessed September 9, 2023.

[1111] The Good Scents Company Data Sheet for Gamma-elemene, from: http://www.thegoodscentscompany.com/data/rw1522521.html. Accessed September 9, 2023.

[1112] The Good Scents Company Data Sheet for Delta-elemene, from: http://www.thegoodscentscompany.com/data/rw1511301.html. Accessed September 9, 2023.

[1113] Cayman Chemical Safety Data Sheet for Beta-elemene, from: https://cdn.caymanchem.com/cdn/msds/19641m.pdf. Accessed September 9, 2023.

CHEMICAL STRUCTURE

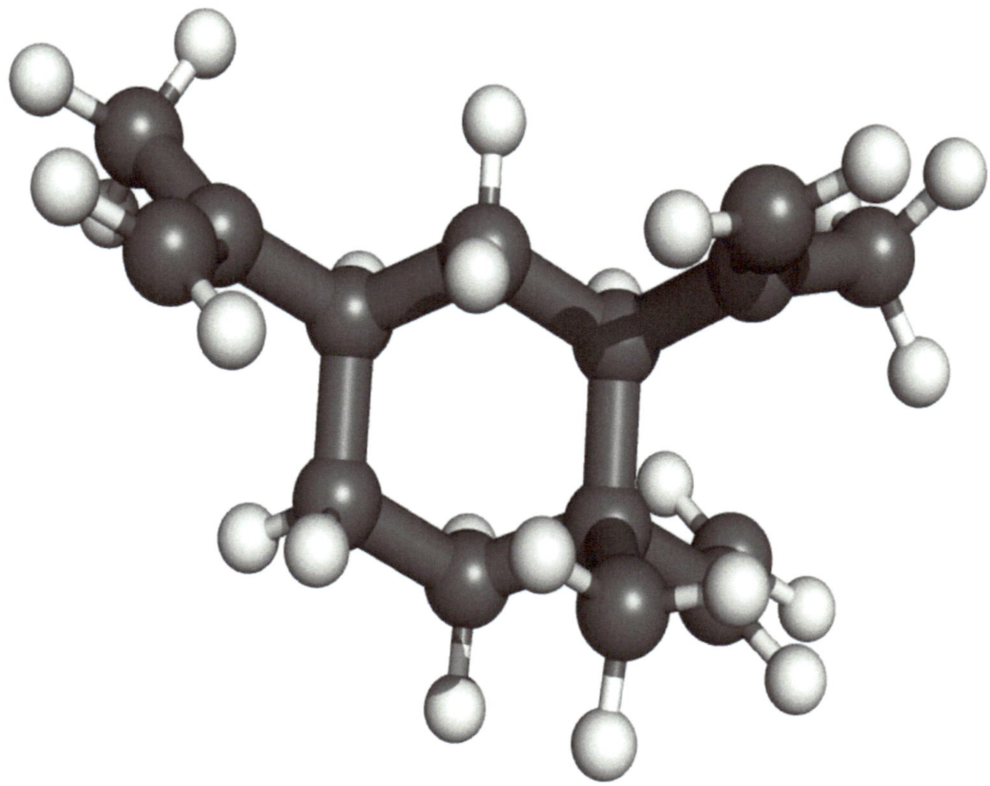

Elemene is based on the typical carbon skeleton of sesquiterpenes; it is derived from farnesyl pyrophosphate via the mevalonate pathway, and consists of three linked units of isoprene arranged with fifteen carbon atoms and twenty-four hydrogen atoms, notated as $C_{15}H_{24}$. Elemene has one set of double bonds, and a cyclohexane ring, which further classifies this sesquiterpene as monocyclic. Elemene has four isomers - alpha, beta, delta, and gamma, but usually occurs as the beta- version, which is the compound most scientific research focuses on.

OCCURRENCE IN PLANTS

Although elemene is generally associated with the ginger plant and its varieties such as turmeric, it also occurs in many other plants:

Alpha-elemene:

Chamaecyparis obtusa	Hinoki	Marjoram
Parsley	Rosa damascene	

Beta-elemene:

[1114] TGSC Information System - b-elemene, from: http://www.thegoodscentscompany.com/data/rw1053861.html Accessed January 10, 2020.

Allspice
Anise
Basil
Black pepper
Cajuput
Camel grass
Carrot
Cedrela
Cinnamon
Croton species
Curry
Ginger
Hyptis
Kachur
Lemongrass
Mandarin
Mikan
Orange
Palmarosa
Pepper tree
Phoebe
Potato
Rosemary
Sideritis species
Tamarind
Turmeric
Water mint
Zanthoxylum species

Amyris
Armoise
Bay
Burdock
Calamus
Cananga
Cascarilla
Celery
Citronella
Cumin
Dill
Grapefruit
Hyssop
Laurel
Lime
Manuka
Mint
Oregano
Parsley
Persea americana
Pine
Ravintsara
Salvia
Spearmint
Tansy
Valerian
Yarrow
Zinnia elegans

Angelica
Baccharis coridifolia
Bergamot
Cabocla
Callitris intratropica
Caraway
Cassis
Champaca
Clary sage
Curcuma amada
Galangal
Horsemint
Juniper
Lemon
Lovage
Marsypianthes chamaedrys
Occimum gratissimum
Orthodon dianthera
Patchouli
Phlomis
Polyalthia australis
Rose
Savin
Sugandha kokila
Thevetia peruviana
Vassoura
Yuzu

Gamma-elemene:

Angelica
Cajuput
Cassis
Galangal
Jambu
Lime
Parsley
Savin
Valerian

Anise
Canaga
Cedrela
Hinoki
Juniper
Morjoram
Pepper tree
Savory
Wormwood

Betel
Cascarilla
Curry
Hyptis
Kachur
Orange
Peppermint
Tagete

Delta-elemene:

Angelica
Bergamot
Cajuput
Cassis
Citronella
Cubeb
Ginger
Kachur

Anise
Black pepper
Calamus
Cedrela
Clary sage
Curcuma amada
Guava
Lavender

Basil
Boronia absolute
Cananga
Chamomile
Croton species
Curry
Jambu
Lime

Mandarin	Marjoram	Marsypianthes chamaedrys
Mikan	Mikania	Ocimum gratissimum
Orange	Patchouli	Phlomis
Phoebe	Piper cernum	Polyalthia australis
Ravintsara	Salvia	Savin
Sideritis species	Tagete	Valerian
Vassoura	Yarrow	Ylang ylang
Yuzu		

BIOLOGICAL ACTIVITY IN PLANTS

Little research has been undertaken to determine the biological functions of elemene in plants, however, some research appears to confirm roles in indirect defense, and as an allelopathic agent:

**Indirect Defense:* When attacked by the cotton ballworm (helicoverpa armigera), redroot pigweed (amaranthus retroflexus) significantly increases emissions of beta-elemene, which the endoparasitoid microplitis mediator uses to locate and prey upon cotton ballworm larvae[1115].

**Allelopathic Agent:* Mandarin orange trees have been shown to increase emissions of beta-elemene in response to herbivory by white-spotted longicorn beetles (anoplophora malasiaca), which causes males of the species to immediately reject conspecific females that have fed on this tree[1116].

USES IN INDUSTRY

Used almost exclusively as a research chemical, preparations of beta-elemene appear as clear, colorless, or yellowish liquids, which are generally extracted from curcuma species of plants like ginger and turmeric. This monocyclic sesquiterpene has also been efficiently manufactured from germacrone as isolated from geranium macrorrhizum[1117], and via a bioreactor that promotes the accumulation of beta-elemene in cells of the herb, rhizoma zedoariae[1118]. Research published in 2023 shows that beta-elemene can be produced by rewiring the cellular metabolism of Ogataea polymorpha[1119] (methylotrophic yeast).

POTENTIAL USES IN MEDICINE

Beta-elemene has strong potential and existing applications in cancer treatment, particularly as a cancer drug synergist, but also as a direct tumor reduction agent and other uses in oncology, among other potential medical use.

[1115] Yu, Huilin, Yongjun Zhang, Yunhe Li, Ziyun Lu, and Xiangju Li. Herbivore- and MeJA-induced volatile emissions from the redroot pigweed Amaranthus retroflexus Linnaeus: their roles in attracting Microplitis mediator (Haliday) parasitoids. Arthropod-Plant Interactions 12.4 (2018): 575-589.

[1116] Yasui, Hiroe, and Nao Fujiwara-Tsujii. Host plant affects the sexual attractiveness of the female white-spotted longicorn beetle, Anoplophora malasiaca. Scientific Reports 6.1 (2016): 1-9.

[1117] Barrero AF, Herrador MM, Quilez del Moral JF, Arteaga P, Meine N, Perez-Morales MC, Catalan JV. Efficient synthesis of the anticancer B-elemene and other bioactive elemanes from sustainable germacrone. Org Biomol Chem. 2011 Feb 21;9(4):1118-25.

[1118] Wang, Gui, and Hui Wang. Cell suspension culture of Rhizoma zedoariae in a two-stage perfusion bioreactor system for B-elemene production. In Vitro Cellular & Developmental Biology - Plant 55.2 (2018): 209-220.

[1119] Ye M, Gao J, Zhou YJ. Global metabolic rewiring of the nonconventional yeast Ogataea polymorpha for biosynthesis of the sesquiterpenoid β-elemene. Metab Eng. 2023 Mar;76:225-231.

Cancer Drug Synergist: Beta-elemene has been shown to substantially increase the suppressive effect of the cancer drug cisplatin on cell growth and proliferation in lung cancer cells[1120] and cisplatin-resistant ovarian cancer cells[1121], and as an individual compound was found to be a more potent inducer of apoptosis than cisplatin alone[1122]. Beta-elemene has also been shown to synergize the anticancer and anti-metastatic effects of the natural compound ligustrazine in bone cancer cells[1123], increase the antitumor activity of the chemotherapy drug etoposide phosphate (VP-16) in non-small cell lung carcinoma cells[1124], and strongly increases the cytotoxicity of the cancer drug doxorubicin in breast cancer cells[1125]. Finally, when combined with cancer treatment drug cisplatin, beta-elemene significantly suppressed cell proliferation, induced cell cycle arrest, promoted the apoptosis of Tca-8113-CDDP cells, and suppressed the activation of JAK2/STAT3 signaling pathway, thereby alleviating cellular resistance to cisplatin in oral squamous cell carcinoma[1126].

Cancer Treatment Synergist: Beta-elemene has been shown to significantly improve the radiosensitivity of lung cancer cells[1127], while also inducing apoptosis in A549 lung adenocarcinoma cells[1128]. Beta-elemene also synergized the effects of a modified dendritic cell vaccine in mice, significantly reducing the size and proliferation of pancreatic tumors[1129], and reduced morphine tolerance and nociception in bone cancer patients[1130].

Direct Cancer Treatment: Several beta-elemene piperazine (an organic molecule normally used to treat worm infestations) derivatives have been shown to induce apoptosis and activate both death-receptor and mitochondrial-mediated apoptotic pathways in human leukemia cells[1131]. Beta-elemene has also been shown to suppress cell viability via induction of apoptosis in gastric cancer cells[1132], and inhibits the cell migration, proliferation, and invasion of human cervical cancer cells[1133]. Other work has found that beta-elemene has a significant

---

[1120] Q. Quentin Li, Gangduo Wang, Furong Huang, Jueli M. Li, Christopher F. Cuff, Eddie Reed Sensitization of lung cancer cells to cisplatin by B-elemene is mediated through blockade of cell cycle progression: antitumor efficacies of B-elemene and its synthetic analogs Medical Oncology March 2013, 30:488.

[1121] Lee RX, Li QQ, Reed E. B-elemene effectively suppresses the growth and survival of both platinum-sensitive and platinum-resistant ovarian tumor cells. Anticancer Res. 2012 Aug;32(8):3103-13.

[1122] Li, Qingdi, Rebecca Lee, Huasheng Liang, Yuhua Zhong, and Eddie Reed. Enhancement of cisplatin-induced apoptosis by B-elemene in resistant human ovarian cancer cells. Medical Oncology 30.1 (2013): 1-13.

[1123] Min Fang, Xiaolong Mei, Hui Yao, Tao Zhang, Tao Zhang, Na Lu, Yanshi Liu, Wenyue Xu, and Chunyou Wan. B-elemene enhances anticancer and anti-metastatic effects of osteosarcoma of ligustrazine in vitro and in vivo Oncology Lett. 2018 Mar; 15(3): 3957-3964.

[1124] Zhang, Fan, Ling Xu, Xiujuan Qu, Mingfang Zhao, Bo Jin, Jian Kang, Yunpeng Liu, and Xuejun Hu. Synergistic antitumor effect of B-elemene and etoposide is mediated via induction of cell apoptosis and cell cycle arrest in non-small cell lung carcinoma cells. Molecular Medicine Reports 4.6 (2011): 1189-1193.

[1125] Xu, Hong-Bin, Ling Li, Jun Fu, Xia-Ping Mao, and Lu-Zhong Xu. Reversion of multidrug resistance in a chemoresistant human breast cancer cell line by B-elemene. Pharmacology 89.5-6 (2012): 303-312.

[1126] Wang H, Ma Y. β-Elemene alleviates cisplatin resistance in oral squamous cell carcinoma cell via inhibiting JAK2/STAT3 pathway in vitro and in vivo. Cancer Cell Int. 2022 Jul 31;22(1):244.

[1127] Li G, Xie B, Li X, Chen Y, Wang Q, Xu Y, Xu-Welliver M, Zou L. Down-regulation of survivin and hypoxia- inducible factor-1 a by B-elemene enhances the radiosensitivity of lung adenocarcinoma xenograft. Cancer Biother Radiopharm. 2012 Feb;27(1):56-64.

[1128] Jiang, Hao, Shenglin Ma, and Jianguo Feng. In vitro study of radiosensitization by B-Elemene in A549 cell line from adenocarcinoma of lung. The Chinese-German Journal of Clinical Oncology 8.1 (2009): 12-15.

[1129] Tan, Guang, Zhongyu Wang, Luanqing Che, and Shuo Yin. Immunotherapeutic effects on murine pancreatic carcinoma by B-elemene combined with dendritic cells modified with genes encoding interleukin-23. Frontiers of medicine in China 1.1 (2014): 41-45.

[1130] Zhou Q, Gong L, Bao G, Ding Q, Ji J. β-Elemene Improves Morphine Tolerance in Bone Cancer Pain via N-Methyl-D-Aspartate Receptor 2B Subunit-Mediated μ-Opioid Receptor. Comput Math Methods Med. 2022 Sep 17;2022:9897669.

[1131] Yu Z, Wang R, Xu L, Xie S, Dong J, Jing Y. B-Elemene piperazine derivatives induce apoptosis in human leukemia cells through downregulation of c-FLIP and generation of ROS. PLoS One. 2011 Jan 25;6(1):e15843.

[1132] Liu, Jing, Ye Zhang, Jinglei Qu, Ling Xu, Kezuo Hou, Jingdong Zhang, Xiujuan Qu, and Yunpeng Liu. B- Elemene-induced autophagy protects human gastric cancer cells from undergoing apoptosis. BMC Cancer 11.1 (2011): 1-10.

[1133] Wang L, Zhao Y, Wu Q, Guan Y, Wu X. Therapeutic effects of B-elemene via attenuation of the Wnt/B-catenin signaling pathway in cervical cancer cells.

antiproliferative effect on prostate cancer cells[1134], and human liver cancer cells[1135], increasing apoptosis in both cell lines. Intriguingly, beta-elemene has been shown to significantly inhibit the growth of glioma cells and markedly shrink brain cancer tumors in mice[1136], while other murine models have found that this monocyclic sesquiterpene inhibited melanoma growth and metastasis by reducing cell proliferation, shrinking tumor size, and suppressing angiogenesis[1137].

There are significant implications for the treatment of lung cancer after this compound inhibited the viability of non-small cell lung cancer cells via both apoptosis and protective autophagy[1138], and, as an injectable adjunctive treatment co-administered with chemotherapy, beta-elemene increased the survival time of patients suffering from lung cancer by more than 39%[1139], with researchers theorizing that the sesquiterpene works by causing lung cancer cell death through a mitochondrial release of the cytochrome c-mediated apoptotic pathway[1140]. Lung cancer research published in the summer of 2023 showed that beta-elemene exhibited significant inhibitory effects in lung adenocarcinoma cells by downregulating the glutathione system[1141].

One large review study found that the mechanisms of action of beta-elemene appears to be by the exertion of "therapeutic potential via modulation of core hallmark capabilities of cancer by suppressing proliferative signaling, such as MAPK and PI3K/Akt/mTOR pathway, inducing cell death, up-regulating growth suppressors, deactivating invasion and metastasis and interacting replicative immortality and attenuating angiogenesis[1142]." Other research has confirmed the significant anti-angiogenic effects of beta-elemene[1143], further cementing the concept that preventing or inhibiting the development of blood vessels to tumors is a critical step in preventing the spread and survival of the disease.

Recent research has shown that beta-elemene inhibits the migration of esophageal squamous cell carcinoma cells via the inhibition of multiple signaling pathways[1144], and inhibits cell division and cell proliferation while also modulating apoptosis in acute myeloid leukemia cells[1145]. Research published in 2022 showed that beta-elemene

---

Mol Med Rep. 2018 Mar;17(3):4299-4306.

[1134] Li QQ, Wang G, Huang F, Banda M, Reed E. Antineoplastic effect of beta-elemene on prostate cancer cells and other types of solid tumour cells. J Pharm Pharmacol. 2010 Aug;62(8):1018-27.

[1135] Dai, Zhi-Jun, Wei Tang, Wang-Feng Lu, Jie Gao, Hua-Feng Kang, Xiao-Bin Ma, Wei-Li Min, Xi-Jing Wang, and Wen-Ying Wu. Antiproliferative and apoptotic effects of B-elemene on human hepatoma HepG2 cells. Cancer Cell International 13.1 (2013): 1-10.

[1136] Zhu, Yongjian, Jue Hu, Fang Shen, Hong Shen, Weiguo Liu, and Jianmin Zhang. The cytotoxic effect of B- elemene against malignant glioma is enhanced by base-excision repair inhibitor methoxyamine. Journal of Neuro- Oncology 113.3 (2013): 375-384.

[1137] Chen, Wenxing, Yin Lu, Jiaming Wu, Ming Gao, Aiyun Wang, and Bo Xu. Beta-elemene inhibits melanoma growth and metastasis via suppressing vascular endothelial growth factor-mediated angiogenesis. Cancer Chemotherapy and Pharmacology 67.4 (2011): 799-808.

[1138] Liu J, Hu XJ, Jin B, Qu XJ, Hou KZ, Liu YP. B-Elemene induces apoptosis as well as protective autophagy in human non-small-cell lung cancer A549 cells. J Pharm Pharmacol. 2012 Jan;64(1):146-53.

[1139] Wang B, Peng XX, Sun R, Li J, Zhan XR, Wu LJ, Wang SL, Xie T. Systematic review of B-elemene injection as adjunctive treatment for lung cancer. Chin J Integr Med. 2012 Nov;18(11):813-23.

[1140] Wang, G., X. Li, F. Huang, J. Zhao, H. Ding, C. Cunningham, J. Coad, D. Flynn, E. Reed, and Q. Li. Antitumor effect of B-elemene in non-small-cell lung cancer cells is mediated via induction of cell cycle arrest and apoptotic cell death. Cellular and Molecular Life Sciences 62.8 (2005): 881-893.

[1141] Song GQ, Wu P, Dong XM, Cheng LH, Lu HQ, Lin YY, Tang WY, Xie T, Zhou JL. Elemene induces cell apoptosis via inhibiting glutathione synthesis in lung adenocarcinoma. J Ethnopharmacol. 2023 Jul 15;311:116409.

[1142] Jiang S, Ling C, Li W, Jiang H, Zhi Q, Jiang M. Molecular Mechanisms of Anti-cancer Activities of b-elemene: Targeting Hallmarks of Cancer. Anticancer Agents Med Chem. 2016;16(11):1426-1434.

[1143] Chen W, Lu Y, Wu J, Gao M, Wang A, Xu B. Beta-elemene inhibits melanoma growth and metastasis via suppressing vascular endothelial growth factor-mediated angiogenesis. Cancer Chemother Pharmacol. 2011 Apr; 67(4):799-808.

[1144] Liang Y, Li S. β-elemene Suppresses Migration of Esophageal Squamous Cell Carcinoma by Modulating Expression of MMP9 through the PI3K/Akt/NF-κB Pathway. Comb Chem High Throughput Screen. 2023;26(13):2304-2320.

[1145] Alafnan A, Dogan R, Bender O, Celik I, Mollica A, Malik JA, Rengasamy KRR, Break MKB, Khojali WMA, Alharby TN, Atalay A, Anwar S. Beta

inhibits cell proliferation, invasion, and metastasis by inhibiting the epithelial-mesenchymal transitions signaling pathway in HCT116 human colon cancer cell line[1146], while other work showed that beta-elemene inhibited the occurrence and development of colon cancer xenografts in mice, and significantly induced apoptosis and autophagy in colorectal cancer cells in vitro[1147]. Finally, beta-elemene inhibited triple negative breast cancer cell viability and proliferation in a concentration-dependent manner, inducing both apoptosis and senescence[1148].

Treatment for Traumatic Brain & Spinal Cord Injuries: Beta-elemene has been shown to significantly ameliorate neurological damage caused by traumatic brain injury in rats by attenuating alterations in the inflammatory pathway, while also synergizing the effects of hyperbaric oxygen treatment[1149]. Beta-elemene has also been found to reduce induced cellular death and apoptosis, while promoting neuron survival and locomotor behavioral recovery in a rat model of spinal cord injury[1150].

Treatment for Cardiovascular Disease: A derivative of beta-elemene, beta-elemene-13-yl, has been found to exhibit significant antioxidant activity in human umbilical vein endothelial cells after induced injuries with hydrogen peroxide, directly reducing ROS production and indirectly reducing nitric oxide production[1151], among other antioxidant activity.

Treatment for Ulcers: As a primary constituent in the essential oil of Pterodon emarginatus seeds, beta-elemene contributed to significant antiulcerogenic and anti-inflammatory activities against ethanol- and other induced ulcers[1152].

Treatment for Multiple Sclerosis (MS): Beta-elemene significantly delayed the onset of induced experimental autoimmune encephalomyelitis (a model of MS) and markedly suppressed myelin oligodendrocyte glycoprotein (MOG-specific) T cell proliferation in mice[1153].

Treatment for Rheumatoid Arthritis: Beta-elemene has been shown to significantly inhibit the viability of rheumatoid arthritis and fibroblast-like synoviocytes, causing a decline in mitochondrial membrane potential, an accumulation of cytochrome c in the cytosol, enhancement of ROS production[1154], and other anti-rheumatoid activities.

Elemene induces cytotoxic effects in FLT3 ITD-mutated acute myeloid leukemia by modulating apoptosis. Eur Rev Med Pharmacol Sci. 2023 Apr;27(8):3270-3287.

[1146] Deng H, Chen G, Zhang J. β-Elemene regulates epithelial-mesenchymal transformation and inhibits invasion and metastasis of colorectal cancer cells. J Complement Integr Med. 2022 Dec 9;20(2):425-430.

[1147] Wang GY, Zhang L, Geng YD, Wang B, Feng XJ, Chen ZL, Wei W, Jiang L. β-Elemene induces apoptosis and autophagy in colorectal cancer cells through regulating the ROS/AMPK/mTOR pathway. Chin J Nat Med. 2022 Jan;20(1):9-21.

[1148] Xie L, Zhang J, Yan H, Cai Y, Xu L. β-elemene induced apoptosis and senescence of triple-negative breast cancer cells through IGF1/IGF1R pathway. Tissue Cell. 2022 Dec;79:101914.

[1149] Xiangen Meng, Na Li, Yu Zhang, Danfeng Fan, Chen Yang, Hang Li, Dazhi Guo, and Shuyi Pan. Beneficial Effect of B-Elemene Alone and in Combination with Hyperbaric Oxygen in Traumatic Brain Injury by Inflammatory Pathway Translational Neuroscience 2018; 9: 33-37.

[1150] Wang J, Li H, Ren Y, Yao Y, Hu J, Zheng M, Ding Y, Chen YY, Shen Y, Wang LL, Zhu Y. Local Delivery of B- Elemene Improves Locomotor Functional Recovery by Alleviating Endoplasmic Reticulum Stress and Reducing Neuronal Apoptosis in Rats with Spinal Cord Injury. Cell Physiol Biochem. 2018;49(2):595-609.

[1151] Ahmad KA, Ze H, Chen J, Khan FU, Xuezhuo C, Xu J, Qilong D. The protective effects of a novel synthetic B- elemene derivative on human umbilical vein endothelial cells against oxidative stress-induced injury: Involvement of antioxidation and PI3k/Akt/eNOS/NO signaling pathways. Biomed Pharmacotherapy 2018 Oct;106:1734-1741.

[1152] Dutra, Rafael C., Marcelo B. Fava, Caio CS. Alves, Ana P. Ferreira, and Nadia R. Barbosa. Antiulcerogenic and anti-inflammatory activities of the essential oil from Pterodon emarginatus seeds. Journal of Pharmacy and Pharmacology: An International Journal of Pharmaceutical Science 61.2 (2009).

[1153] Zhang, Rongwei, Ayong Tian, Huimin Zhang, Zhike Zhou, Hongmei Yu, and Lei Chen. Amelioration of Experimental Autoimmune Encephalomyelitis by B-elemene Treatment is Associated with Th17 and Treg Cell Balance. Journal of Molecular Neuroscience 44.1 (2011): 31-40.

[1154] Zou, Shibo, Changyin Wang, Zhengjun Cui, Pengfei Guo, Qingnan Meng, Xun Shi, Ya Gao, Gaoyuan Yang, and Zhaofeng Han. B-Elemene induces

**Mosquitocidal Agent:* As a primary constituent in the essential oil made from the leaves of syzygium zeylanicum (an evergreen tree), beta-elemene contributed to significant toxic effects against larvae of several mosquitos: anopheles subpictus, aedes albopictus, and culex tritaeniorhynchus, while exhibiting little to no toxicity against non-target organisms[1155].

**Treatment for Hemangioma:* Although hemangiomas (vascular birthmarks) are not normally dangerous, there may be a novel option for people who want to reduce or control these marks, as beta-elemene combined with propranolol significantly inhibited proliferation, migration, and angiogenesis of hemangioma cells by synergistically down regulating the hypoxia inducible factor-1α/vascular endothelial growth factor signaling pathway[1156].

**Treatment for Chronic Pain:* As the primary component in an extract of Curcuma wenyujin, beta-elemene relieved spared nerve injury-induced mechanical allodynia, reversed spinal astrocytic ERK activation and subsequent upregulation of proinflammatory cytokines, and exerted antioxidative effects by increasing the levels of SOD and GSH-PX and decreasing the level of MDA[1157], producing remarkable analgesic effects on neuropathic pain.

IMPLICATIONS FOR HUMAN HEALTH & NUTRITION

Dietary consumption of elemene should be elementary for most people considering the wide variety of common foods, herbs, and spices that this monocyclic sesquiterpene is found in. For instance, when preparing, cooking, or serving foods, consider working with herbs and spices like allspice, anise, basil, black pepper, caraway, cinnamon, cumin, curry, dill, ginger, mint, oregano, parsley, peppermint, rosemary, spearmint, and turmeric. Meals or snacks that include vegetables like carrots, celery, ginger, or potatoes, or fruits like grapefruit, lemons, limes, mandarins, oranges, or tamarind are all high in elemene and other important nutrients. Lemongrass or mint teas are also likely to contain elemene, as well as common essential oils like those extracted from rose, patchouli, rosemary, and valerian root. As always, great care should be exercised when working with essential oils and extracts.

---

apoptosis of human rheumatoid arthritis fibroblast-like synoviocytes via reactive oxygen species-dependent activation of p38 mitogen-activated protein kinase. Pharmacological reports, PR 68.1 (2016): 7-11.

[1155] Govindarajan, Marimuthu, and Giovanni Benelli. a-Humulene and B-elemene from Syzygium zeylanicum (Myrtaceae) essential oil: highly effective and eco-friendly larvicides against Anopheles subpictus, Aedes albopictus, and Culex tritaeniorhynchus (Diptera: Culicidae). Parasitology Research 115.7 (2016): 2771-2778.

[1156] Wang Z, Chen Y, Yang L, Yao D, Shen Y. Combinative effects of β-elemene and propranolol on the proliferation, migration, and angiogenesis of hemangioma. PeerJ. 2023 Jul 12;11:e15643.

[1157] Ma LT, Bai Y, Cao P, Ren KX, Chen J, Zhang T, Fan BY, Qiao Y, Yan HY, Wang JJ, Li YQ, Zheng J. The analgesic effects of β-elemene in rats with neuropathic pain by inhibition of spinal astrocytic ERK activation. Mol Pain. 2022 Apr;18:17448069221121562.

Elemene Review

Answer the following questions to test your knowledge of this terpene:

Question #1: What type of terpene is elemene?

a. Monocyclic monoterpene
b. Monocyclic sesquiterpene
c. Bicyclic monoterpene
d. Acyclic monoterpene

Question #2: Which of the following best describes the primary flavor and aroma profile of elemene?

a. Citrus, fruity, dry, terpenic
b. Sweet, fir, turpentine, coniferous
c. Depends on the isomer
d. Herbal, waxy, fresh, green
e. Woody, carrot, spice, dry

Question #3: How common is elemene in cannabis?

a. Top ten
b. Top three
c. Top forty
d. Top one hundred

Question #4: What is the chemical formula for elemene?

a. $C_{10}H_{16}O$
b. $C_{15}H_{24}O_2$
c. $C_{10}H_{16}$
d. $C_{15}H_{24}$

Question #5: Name two uses of elemene in plants:

1 ______________________________ 2 ______________________________

Question #6: Name two potential medical uses of elemene:

1 ______________________________ 2 ______________________________

For the answer key to all individual compound chapters please go here: http://thebigbookofterps.com/bbot-quiz-answer-key/

AROMADENDRENE

Type: Tricyclic sesquiterpene
Chemical Formula: $C_{15}H_{24}$
Molecular Weight: 204.35 g/mol
Boiling Point: *Aromadendrene:* 258.10 °C. @ 760.00 mm Hg (estimated)[1158]
Alloaromadendrene: 257.00 to 258.00 °C. @ 760.00 mm Hg[1159]
D-aromadendrene: 258.00 to 259.00 °C. @ 760.00 mm Hg[1160]
L-aromadendrene: 260.00 to 265.00 °C. @ 760.00 mm Hg[1161]
Flash Point: *Aromadendrene:* 224.00 °F. TCC (106.50 °C.) (estimated by TGSC)
Alloaromadendrene: 248.00 °F. TCC (120.00 °C.)
D-aromadendrene: 223.00 °F. TCC (106.11 °C.)
L-aromadendrene: 223.00 °F. TCC (106.11 °C.)
Solubility: *D-aromadendrene:* Soluble in alcohol, DMF, DMSO, ethanol[1162], insoluble in water
Other isomers: Soluble in alcohol, insoluble in water
Biological Role: Allelopathic agent, insecticidal, acaricidal
Therapeutic Role: Antibacterial agent
Commercial Role: Research chemical
Occurrence in Cannabis: Top forty
Aroma/Flavor: *Aromadendrene:* Woody, earthy
Alloaromadendrene: Woody

Occurs in Cannabis Strains: Unknown

INTRODUCTION

Aromadendrene is a tricyclic sesquiterpene commonly found in the essential oils of Eucalyptus trees; in fact, this terpene is so abundant in eucalyptus that these trees were once referred to as aromadendron trees[1163]. Aromadendrene and alpha-cedrene are the only tricyclic sesquiterpenes thought to occur in cannabis within the top forty terpenes by concentration. However, the inclusion of this terpene within the top forty terpenes in cannabis is largely theoretical; during our review of hundreds of cannabis terpene laboratory tests, we did not find any occurrences of aromadendrene, nor are there any internet reported occurrences as of 2023. Nevertheless, many laboratories and companies within the cannabis industry report that this terpene is a common secondary compound in cannabis.

[1158] The Good Scents Company Data Sheet for Aromandendrene, from: http://www.thegoodscentscompany.com/data/rw1602521.html. Accessed September 10, 2023.

[1159] The Good Scents Company Data Sheet for Alloaromadendrene, from: http://www.thegoodscentscompany.com/data/rw1465731.html. Accessed September 10, 2023.

[1160] The Good Scents Company Data Sheet for D-aromadendrene, from: http://www.thegoodscentscompany.com/data/rw1465701.html. Accessed September 10, 2023.

[1161] The Good Scents Company Data Sheet for L-aromadendrene, from: http://www.thegoodscentscompany.com/data/rw1011191.html. Accessed September 10, 2023.

[1162] Cayman Chemical Data Sheet for Aromadendrene, from: https://www.caymanchem.com/product/33829/aromadendrene. Accessed September 10, 2023.

[1163] Yvonee M.A.W. Lamers Aromadendrene as a chiral starting material for the synthesis of fragrances and pheromones Thesis Wageningen University - ISBN 90-5808-833-2.

CHEMICAL STRUCTURE

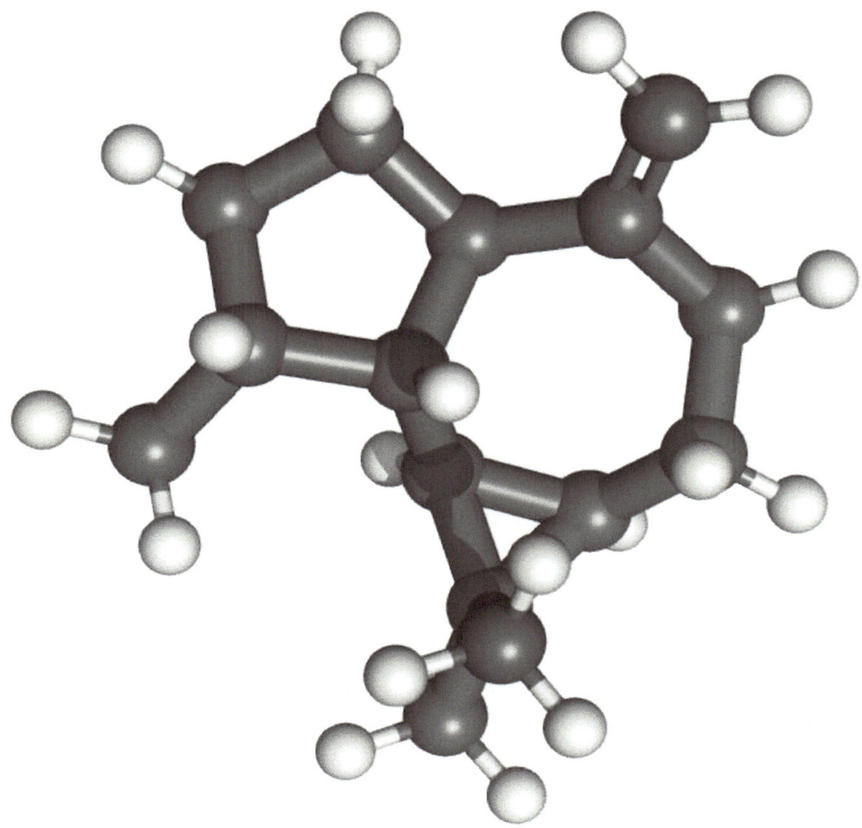

Formed in the mevalonate pathway in the cytosol of plants and derived from farnesyl pyrophosphate, aromadendrene is a sesquiterpene characterized by the standard fifteen carbon atoms and twenty-four hydrogen atoms, notated as $C_{15}H_{24}$. This compound contains three cyclopropane rings, one of which is strained, which further classifies this compound as tricyclic. The molecular skeleton of aromadendrene contains one set of exocyclic double bonds, lending the compound additional functionality. There are several isomers of aromadendrene, including a novel variation of the sesquiterpene recently discovered in the essential oil of Campomanesia lineatifolia that researchers named Aromadendrene 2[1164], and at least the following:

Alpha-aromadendrene (alloaromadendrene)
Beta-aromadendrene (is known to exist but no data has been reported)
(+)-aromadendrene (D-aromadendrene)
(-)-aromadendrene (L-aromadendrene)
allo-9-aromadendrene (is known to exist but no data has been reported)

[1164] Neves NCV, de Mello MP, Smith SM, Boylan F, Caliari MV, Castilho RO. Chemical Composition and In Vitro Anti-Helicobacter pylori Activity of Campomanesia lineatifolia Ruiz & Pavón (Myrtaceae) Essential Oil. Plants (Basel). 2022 Jul 27;11(15):1945.

OCCURRENCE IN PLANTS

Aromadendrene is known to occur in a moderate number of plants, but the compound is typically extracted from eucalyptus varieties. A study published in 2023 showed that aromadendrene and alloaromadendrene occurred in several samples of wild hop plants[1165] (Humulus lupulus), while other work published in 2022 showed that researchers were surprised when they found alloaromadendrene in the essential oil of a Centaurea variety (C. scabiosa) for the first time[1166].

Aromadendrene:

Artemisia species
Champaca
Curry
Hyptis
Lemon
Manuka
Microstrobos
Oregano
Prangos
Sugandha kokila
Thyme
Bay
Clary sage
Eucalyptus globulus
Juniper
Lime
Marjoram
Mikania
Phoebe
Sage
Tagete
Vassoura
Cajuput
Croton
Ginger
Lavender
Lippia
Mastic
Myrtle
Polyalthia australis
Salvia
Tea tree
Zanthoxylum species

Alloaromadendrene:

Allspice
Backhousia citriodora
Cassis
Clove
Eucalyptus species
Horsemint
Lemon
Myrtle
Oregano
Polyalthia australis
Sage
Stachys recta
Tea tree
Water mint
Zanthoxylum species
Anise
Bay
Cedrela
Croton species
Ginger
Hyssop
Manuka
Niaouli
Phoebe
Rosemary
Salvia
Sundacarpus amarus
Thyme
Wormseed
Baccharis coridifolia
Cascarilla
Cistus
Erigeron garveolens
Guava
Labdanum
Marjoram
Occimum gratissimum
Pimenta
Safflower
Santolina
Tansy
Vassoura
Wormwood

D-aromadendrene:

Allspice
Ginger
Guava

[1165] Paguet AS, Siah A, Lefèvre G, Moureu S, Cadalen T, Samaillie J, Michels F, Deracinois B, Flahaut C, Alves Dos Santos H, Etienne-Debaecker A, Rambaud C, Chollet S, Molinié R, Fontaine JX, Waterlot C, Fauconnier ML, Sahpaz S, Rivière C. Multivariate analysis of chemical and genetic diversity of wild Humulus lupulus L. (hop) collected in situ in northern France. Phytochemistry. 2023 Jan;205:113508.

[1166] Carev I, Golemac A, Siljak-Yakovlev S, Pellay FX, Politeo O. Volatile Oil Chemical Composition of Wild, Edible Centaurea scabiosa L. and Its Cytotoxic Activity. Plants (Basel). 2022 Nov 28;11(23):3267.

Oregano	Sage	Salvia
Savory	St. John's Wort	Tamarind

L-aromadendrene:

Annatto seed	Eucalyptus globulus	Eucalyptus rariflora

BIOLOGICAL ACTIVITY IN PLANTS

Little research has been published on the biological roles of aromadendrene in plants, however, some studies indicate allelopathic, insecticidal, and acaricidal functions:

**Allelopathic Agent:* Eucalyptus benthamii plants have been shown to significantly increase volatile emissions of aromadendrene (along with alpha-pinene and globulol) in response to herbivory by bronze bugs (thaumastocoris peregrinus), which deterred feeding by mated conspecific females[1167].

**Insecticidal Agent:* As a primary constituent in the essential oil of spiked pepper (piper aduncum), aromadendrene contributed to contact toxicity against the brown stink bug (euschistus heros) by causing body fat cellular stress[1168], which scientists theorized compromises the insect's survival processes.

**Acaricidal Agent:* As a primary constituent in the essential oils made from the leaves of ocotea duckei and ocotea glomerate, aromadendrene contributed to significant contact and fumigant toxicity against the two-spotted spider mite (tetranychus urticae)[1169]. When the same mites infested cassava, the plants significantly increased emissions of alloaromadendrene – in addition to several other terpenes including variations of ocimene and caryophyllene – when the mites reached about 200 in number[1170].

USES IN INDUSTRY

Aromadendrene has no reported industrial or commercial uses in the literature and appears to be for research purposes only.

POTENTIAL USES IN MEDICINE

The primary medical potential offered by aromadendrene is as an antibacterial agent, particularly against multi-drug resistant bacteria:

**Antibacterial Agent:* As the primary constituent in the essential oil of Eucalyptus globulus, aromadendrene was shown to exhibit marked inhibition of multi-drug resistant bacteria including methicillin-resistant staphylococcus aureus (MRSA) and vancomycin-resistant enterococci (VRE) enterococcus faecalis[1171]. Similar work but with the

---

[1167] Martins CB, Zarbin PH. Volatile organic compounds of conspecific-damaged Eucalyptus benthamii influence responses of mated females of Thaumastocoris peregrinus. J Chem Ecol. 2013 May;39(5):602-11.

[1168] Cossolin, Jamile, Monica Pereira, Luis Martinez, Leonardo Turchen, Muhammad Fiaz, Hakan Bozdogan, and Jose Serrao. Cytotoxicity of Piper aduncum (Piperaceae) essential oil in brown stink bug Euschistus heros (Heteroptera: Pentatomidae). Ecotoxicology 28.7 (2019): 763-770.

[1169] Moraes, Marcilio M DE, Claudio A G DA Camara, and Milena M C DA Silva. Comparative toxicity of essential oil and blends of selected terpenes of Ocotea species from Pernambuco, Brazil, against Tetranychus urticae Koch. Anais da Academia Brasileira de Ciencias 89.3 (2018): 1417-1429.

[1170] Pinto-Zevallos DM, Bezerra RHS, Souza SR, Ambrogi BG. Species- and density-dependent induction of volatile organic compounds by three mite species in cassava and their role in the attraction of a natural enemy. Exp Appl Acarol. 2018 Mar;74(3):261-274.

[1171] Mulyaningsih S, Sporer F, Zimmermann S, Reichling J, Wink M. Synergistic properties of the terpenoids aromadendrene and 1,8-cineole from the essential

essential oils made from Eucalyptus sargentii showed that aromadendrene contributed to moderate to high activity against bacillus subtilis, staphylococcus aureus, staphylococcus epidermidis, Escherichia coli, paeruginosa, klebsiella pneumoniae, proteus vulgaris, and shigella dysenteriae bacteria[1172].

IMPLICATIONS FOR HUMAN HEALTH & NUTRITION

Dietary intake of aromadendrene can be accomplished by working with spices and herbs like allspice, bay leaf, clary sage, curry, ginger, marjoram, oregano, sage, or thyme when preparing, cooking, or serving food. Lemons and limes both contain this tricyclic sesquiterpene, as well as the essential oils made from eucalyptus and tea tree.

oil of Eucalyptus globulus against antibiotic-susceptible and antibiotic-resistant pathogens. Phytomedicine. 2010 Nov;17(13):1061-6.

[1172] Javad Safaei-Ghomi, Hossein Batooli Chemical Composition and Antimicrobial Activity of the Volatile Oil of Eucalyptus Sargentii Maiden Cultivated in Central Iran International Journal of Green Pharmacy Vol 4, No 3 (2010).

Aromadendrene Review

Answer the following questions to test your knowledge of this terpene:

Question #1: What type of terpene is aromadendrene?

a. Acyclic sesquiterpene
b. Bicyclic sesquiterpene
c. Tricyclic sesquiterpene
d. Trescyclic sesquiterpene

Question #2: Which of the following best describes the primary flavor & aroma profile of aromadendrene?

a. Woody, cedar
b. Citrus, pepper
c. Lemon, sharp
d. Woody, earthy

Question #3: How common is aromdendrene in cannabis?

a. Top ten
b. Top three
c. Top fifteen
d. Top forty

Question #4: What is the chemical formula for aromadendrene?

a. $C_{10}H_{16}O$
b. $C_{15}H_{24}O_2$
c. $C_{15}H_{24}$
d. $C_{20}H_{32}O$

Question #5: What feature makes aromadendrene a tricyclic sesquiterpene?

____

Question #6: Aromadendron trees are now known as what type of tree:

____

For the answer key to all individual compound chapters, please go here: http://thebigbookofterps.com/bbot-quiz-answer-key/

CUBEBENE

Type: Tricyclic sesquiterpene
Chemical Formula: $C_{15}H_{24}$
Molecular Weight: 204.35 g/mol
Boiling Point: *Alpha-cubebene:* 245.00 to 246.00 °C. @ 760.00 mm Hg[1173]
Beta-cubebene: 283.00 to 285.00 °C. @ 760.00 mm Hg[1174]
Flash Point: *Alpha-cubebene:* >212.00 °F. TCC (>100.00 °C.)
Beta-cubebene: 235.00 °F. TCC (112.78 °C.)
Solubility: *Alpha-cubebene:* Soluble in alcohol, insoluble in water
Beta-cubebene: Soluble in alcohol, insoluble in water
Oral LD50: *Alpha-cubebene:* 0.45565 g/kg (rat)[1175]
Beta-cubebene: 1.04383 g/kg (rat)[1176]
Biological Role: Signaling molecule
Commercial Role: Flavoring agent, research chemical
Therapeutic Role: Anticancer, anti-inflammatory, anti-fungal, neuroprotective agent
Occurrence in Cannabis: Unknown, probably Top 40
Aroma/Flavor: *Alpha-cubebene:* Herbal, waxy, warm, woody, camphoraceous.
Beta-cubebene: Citrusy, fruity, radish.

Occurs in Cannabis Strains: At the time of publication of this text, there were no available Certificates of Analysis or other commercial or private cannabis testing data available showing cubebene as a constituent of cannabis or cannabis products. However, research conducted in 2019 analyzed the phytochemical content of 33 chemovars, with beta-cubebene appearing in all the chemovars except the low-THC, high-CBD varieties[1177]. Additionally, a white paper published by the LECO corporation showed alpha-cubebene in an "indica-dominant" sample[1178], while researchers in 2020 discovered alpha-cubebene in a hemp chemotype known as "Futura"[1179]. Israeli researchers have found alpha-cubebene in samples of Moroccan and Lebanese hash[1180], and the University of Cambridge found the alpha isomer of this compound in the particulate phase fraction of mainstream cannabis smoke.[1181]

[1173] The Good Scents Company Data Sheet for Alpha-cubebene, from: http://www.thegoodscentscompany.com/data/rw1054531.html. Accessed September 10, 2023.

[1174] The Good Scents Company Data Sheet for Beta-cubebene, from: http://www.thegoodscentscompany.com/data/rw1454791.html. Accessed September 10, 2023.

[1175] AromaDb Database Specification Sheet for Alpha-cubebene, from: http://bioinfo.cimap.res.in/aromadb/web_compound_detail.php?id=CRMOL591. Accessed September 10, 2023.

[1176] AromaDb Database Specification Sheet for Beta-cubebene, from: http://bioinfo.cimap.res.in/aromadb/web_compound_detail.php?id=CRMOL530. Accessed September 10, 2023.

[1177] Mudge EM, Brown PN, Murch SJ. The Terroir of Cannabis: Terpene Metabolomics as a Tool to Understand Cannabis sativa Selections. Planta Med. 2019 Jul;85(9-10):781-796.

[1178] Determining Terpene Profiles of Cannabis Strains Using GC and GCxGC with High Performance TOFMS - Instrument: Pegasus BT 4D. LECO Corporation white paper.

[1179] Hanuš LO, Hod Y. Terpenes/Terpenoids in Cannabis: Are They Important? Med Cannabis Cannabinoids. 2020 Aug 10;3(1):25-60.

[1180] Lumír O Hanuš, Rina Levy, Dafna De La Vega, Limor Katz, Michael Roman, Pavel Tomíček. The main cannabinoids content in hashish samples seized in Israel and Czech Republic 2 July, 2016 Israel Journal of Plant Sciences Volume 63 Issue 3 Pages 182-190. (From: https://lumirlab.com/publication/the-main-cannabinoids-content-in-hashish-samples-seized-in-israel-and-czech-republic/, Accessed: June 19, 2023.

[1181] Brian M. Graves, Tyler J. Johnson, Robert T. Nishida, Ryan P. Dias, Benjamin Savareear, James J. Harynuk, Mohsen Kazemimanesh, Jason S. Olfert, and Adam M. Boies..Supplementary Information: Comprehensive characterization of mainstream marijuana and tobacco smoke. University of Cambridge, from: https://api.repository.cam.ac.uk/server/api/core/bitstreams/a6c1b82c-e9d0-45ad-97c8-565728882d17/content, Accessed: June 19, 2023.

INTRODUCTION

A lesser-known terpene that is likely present in many cannabis varieties, cubebene is a tricyclic sesquiterpene that occurs as two isomers, the alpha- and beta- versions. Although the alpha version of this compound is much more common in the literature than the beta version, both occur naturally, with some plants producing the racemic mixture of the two isomers. Each isomer has its own unique flavor and aroma profile, with the alpha version generally being described as herbaceous, waxy, and woody, and the beta version as citrusy, fruity, and with a hint of radish.

Interestingly, stingless bees of the species Frieseomelitta have been shown to collect sesquiterpenes from plants – notably alpha-cubebene – which is then transported to the nest and used as construction adhesive and structural compounds[1182]. Antenna of two types of stingless bees in the same species have also been shown to respond to the pure compound, alpha-cubebene[1183].

At the microbial level, alpha-cubebene is produced in significant quantities by the phytopathogen Fusarium solani when exposed to plant-growth-promoting bacteria, with researchers theorizing that F. solani likely uses the phytochemical as a method of chemical communication with the bacteria[1184].

Finally, of particular interest to cannabis cultivators is research carried out by Saloner and Bernstein, which showed that the production of alpha-cubebene in the apical inflorescences of medical cannabis plants was reduced in response to an increase in nitrogen[1185].

[1182] Patricio EF, Cruz-López L, Maile R, Tentschert J, Jones GR, Morgan ED. The propolis of stingless bees: terpenes from the tibia of three Frieseomelitta species. J Insect Physiol. 2002 Feb;48(2):249-254.

[1183] Patricio EF, Cruz-López L, Morgan ED. Electroantennography in the study of two stingless bee species (Hymenoptera: Meliponini). Braz J Biol. 2004 Nov;64(4):827-31.

[1184] Ana, A. Gutiérrez-Santa, H. A. Carrillo-Cerda, J. Rodriguez-Campos, J. B. Velázquez-Fernández, O. A. Patrón-Soberano, and S. M. Contreras-Ramos. "Dynamics of volatilomes emitted during cross-talking of plant-growth-promoting bacteria and the phytopathogen, Fusarium solani." World Journal of Microbiology and Biotechnology 36.10 (2020).

[1185] Avia Saloner, Nirit Bernstein Nitrogen supply affects cannabinoid and terpenoid profile in medical cannabis (Cannabis sativa L.), Industrial Crops and Products, Volume 167, 2021,
113516, ISSN 0926-6690.

CHEMICAL STRUCTURE

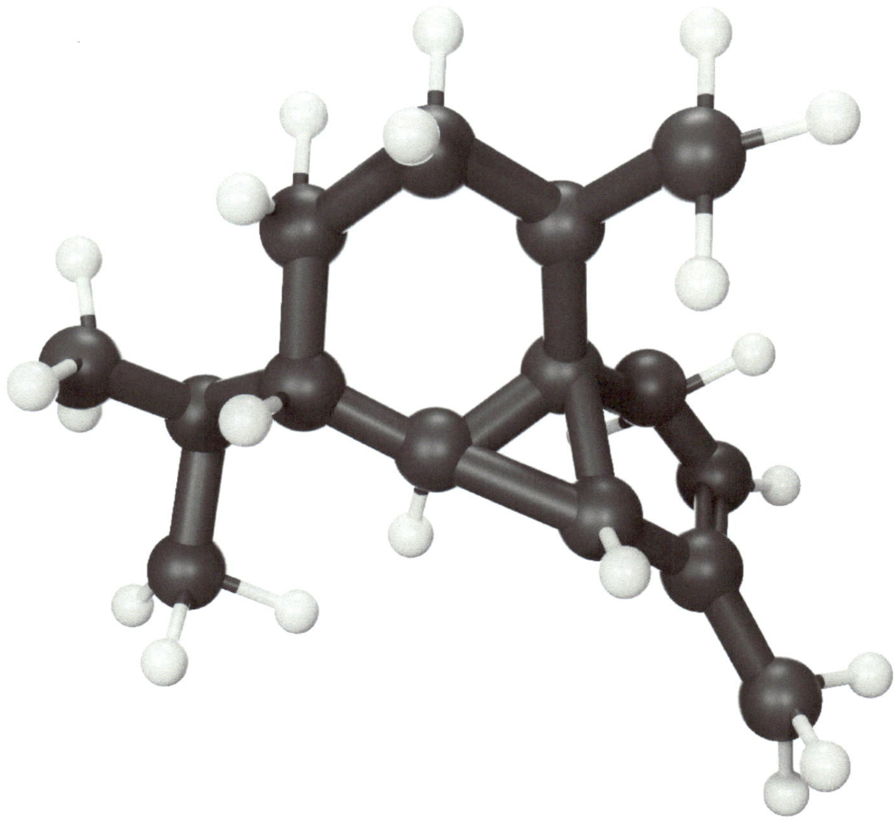

Cubebene is derived from farnesyl pyrophosphate in the mevalonate pathway of plants. This terpene consists of 3 linked units of isoprene forming the classic sesquiterpene skeleton containing 15 carbon atoms and 24 hydrogen atoms, notated as $C_{15}H_{24}$. This compound contains 3 rings, denoting it as a tricyclic sesquiterpene. Cubebene features 2 isomers; the alpha and beta versions, where the solitary double carbon bond in the alpha version is endocyclic, and the double carbon bond in the beta version is exocyclic.

OCCURRENCE IN PLANTS

Found in a wide variety of plants, fruits, and herbs, cubebene is thought to be an active ingredient in some traditional folk medicines. In Hungary, for instance, researchers studying the principal components of Garcinia zeylanica (chamomile) found pharmacologically active alpha-cubebene as a major constituent in the leaves, which are currently considered agricultural waste[1186]. Similarly, researchers studying the active components of the essential oil of Phlomis bruguieri (referred to as "calba" in Turkish folk medicine) found alpha-cubebene as a primary constituent[1187].

Alpha-cubebene:

Angelica Anise Ayou

[1186] Kokilananthan S, Bulugahapitiya VP, Manawadu H, Gangabadage CS. Investigations of chemical compositions and antioxidative potential of essential oils isolated from the leaves of two Garcinia species. J Adv Pharm Technol Res. 2023 Jan-Mar;14(1):12-17.

[1187] Akman F, Demirpolat A, Kazachenko AS, Kazachenko AS, Issaoui N, Al-Dossary O. Molecular Structure, Electronic Properties, Reactivity (ELF, LOL, and Fukui), and NCI-RDG Studies of the Binary Mixture of Water and Essential Oil of Phlomis bruguieri. Molecules. 2023 Mar 16;28(6):2684.

Basil
Bay
Black pepper
Cascarilla
Cedrela
Chamomile
Clary sage
Clove
Croton species
Cubeb
Curry
Cypress
Dill
Eucalyptus globulus
Galangal
Geranium
Ginger
Guava
Hyptis
Juniper
Labdanum
Lemon
Lime
Mandarin
Manuka
Melissa
Mikania
Nutmeg
Orange
Oregano
Parsley
Pepper tree
Peppermint
Phlomis
Phoebe
Pine
Piper cernum
Polyalthia australis
Ravintsara
Rosemary
Sage
Salvia
Santolina
Sugandha kokila
Tea tree
Wormwood
Zanthoxylum nitidum

Beta-cubebene:

Allamanda cathartica
Basil
Black pepper
Calamus
Cananga
Carrot
Cascarilla
Cedrela
Chamomile
Citronella
Clary sage
Costmary
Cubeb
Curry
Eucalyptus globulus
Fleabane
Grapefruit
Guava
Hinoki
Hyptis
Labdanum
Lemon
Lime
Litsea cubeba
Lovage
Mace
Mandarin
Michelia alba
Myristica fragrans
Nutmeg
Orange
Oregano
Persea americana
Phlomis
Polyalthia australis
Prangos
Protium heptaphyllum
Ravintsara
Salvia
Stachys recta
Sundacarpus amarus
Water mint
Wormwood
Yarrow
Ylang ylang

BIOLOGICAL ACTIVITY IN PLANTS

Little is known about how plants use cubebene, however, some research indicates that this sesquiterpene may be produced by plants in response to certain types of pathogenic infection, while some pests use emissions of the compound to locate their host. For instance, poplar trees significantly increase emissions of alpha-cubebene in response to infection by Phytophthora cactorum and Phytophthora plurivora, using the terpene as a signaling molecule to suppress Phytophthora hyphae spread from infected sites to healthy tissues[1188]. Meanwhile, Agrilus planipennis (emerald ash borer) uses emissions of bark volatiles from Fraxinus pennsylvanica (green ash) to locate its host, where alpha-cubebene is the primary constituent of the volatiles[1189]. Another pest, the elm bark beetle,

[1188] Ďurkovič J, Bubeníková T, Gužmerová A, Fleischer P, Kurjak D, Čaňová I, Lukáčik I, Dvořák M, Milenković I. Effects of Phytophthora Inoculations on Photosynthetic Behaviour and Induced Defence Responses of Plant Volatiles in Field-Grown Hybrid Poplar Tolerant to Bark Canker Disease. J Fungi (Basel). 2021 Nov 15;7(11):969.

[1189] Crook DJ, Khrimian A, Francese JA, Fraser I, Poland TM, Sawyer AJ, Mastro VC. Development of a host-based semiochemical lure for trapping emerald ash borer Agrilus planipennis (Coleoptera: Buprestidae). Environ Entomol. 2008 Apr;37(2):356-65.

hijacks the phytochemical production of the Dutch elm tree to cause a significant increase in the emissions of alpha-cubebene and 3 other compounds, which then increases the attractiveness of the tree to other beetles[1190]. The elm bark beetle is a known carrier of the pathogen Ophiostoma novo-ulmi, which causes the deadly Dutch elm disease.

USES IN INDUSTRY

Commercial preparations of cubebene are generally available for research purposes only, although the compound is also used in some food items including dairy products, sherbet and sorbet, processed fruit and vegetables, confectionary items, bakery items, processed meats and fish, spices, soups, sauces, beverages, and prepared foods. This compound can appear as a colorless clear liquid, a pale green or blue-yellow viscous liquid, or a yellowish oil, and is often isolated from Piper cubeba berries.

Some research has looked at alpha-cubebene as a potential insecticide. Scientists found that alpha-cubebene was a major constituent of croton seed oil, contributing to significant insecticidal activity including 80% of mosquito deaths and 71.8% of cockroach deaths within 4 hours of exposure[1191].

POTENTIAL USES IN MEDICINE

Much of the medical research regarding cubebene focuses on using the detection of the compound as a type of biological marker in the detection and analysis of several diseases, most notably celiac disease and other gastrointestinal diseases. The sesquiterpene may also have use in the treatment of cancer, and as an anti-inflammatory or antifungal agent.

**Anticancer Agent:* As the primary constituent in the essential oil of Etlingera velutina (a plant in the ginger family), alpha-cubebene contributed to a dose-dependent reduction in the viability of NCI-N87 (gastric carcinoma) and MKN45 cell lines (gastric adenocarcinoma)[1192]. Additionally, US Patent #US20220401504A1 consists of a proprietary extract of Magnolia figo - which is high in alpha-cubebene - used as an inhibitor of lung cancer cells[1193].

**Anti-inflammatory:* As a primary constituent in the essential oil of Piper miniatum, alpha-cubebene contributed to significant anti-inflammatory effects, reducing inflammation by nearly 86% in a TPA-induced mouse ear edema model[1194].

**Antifungal:* As a major constituent in the essential oil of Matricaria chamomilla L., alpha-cubebene contributed to potent effects against Aspergillus niger (a type of mold that can sometimes cause disease in humans), causing, among other effects, the disruption of cytoplasmic membranes and intracellular organelles, detachment of plasma membrane from the cell wall, cytoplasm depletion, complete disorganization of hyphal compartments, swelling and deformation of hyphal tips, formation of short branches, and collapse of entire hyphae[1195]. As a major constituent in

---

[1190] McLeod G, Gries R, von Reuss SH, Rahe JE, McIntosh R, König WA, Gries G. The pathogen causing Dutch elm disease makes host trees attract insect vectors. Proc Biol Sci. 2005 Dec 7;272(1580):2499-503.

[1191] Endris YA, Mekonnen KD. Formulation of Neem Leaf and Croton Seed Essential Oils as a Natural Insecticide Tested on Mosquitoes and Cockroaches. ACS Omega. 2023 Apr 24;8(17):15052-15061.

[1192] Na Wei, Zhiqiang Hu, Jianfeng Xue, Jianxia Wen, Chun Yang & Huaping Gong (2022) Introducing a Modern Chemotherapeutic Drug Formulated by Etlingera velutina Essential Oil for the Treatment of Human Gastric Cancer, Journal of Essential Oil Bearing Plants, 25:5, 967-975.

[1193] https://patents.justia.com/patent/20220401504, Accessed June 20, 2023.

[1194] Salleh WM, Kammil MF, Ahmad F, Sirat HM. Antioxidant and Anti-inflammatory Activities of Essential Oil and Extracts of Piper miniatum. Nat Prod Commun. 2015 Nov;10(11):2005-8.

[1195] Tolouee, Marziyeh, Soheil Alinezhad, Reza Saberi, Ali Eslamifar, Seyed Javad Zad, Kamkar Jaimand, Jaleh Taeb, Mohammad-Bagher Rezaee, Masanobu Kawachi, Masoomeh Shams-Ghahfarokhi, and Mehdi Razzaghi-Abyaneh. "Effect of Matricaria chamomilla L. flower essential oil on the growth and ultrastructure of Aspergillus niger van Tieghem.." International journal of food microbiology 139.3 (2010): 127-133.

the essential oil made from the leaves of Ocimum basilicum L. (basil), alpha-cubebene contributed to the complete inhibition of A. flavus growth at 1000 ppm oil concentration, and markedly inhibited aflatoxin B1 at all oil concentrations tested[1196].

IMPLICATIONS FOR HUMAN HEALTH & NUTRITION

To include cubebene as part of a healthy diet, consider preparing, cooking, or serving foods with spices and herbs such as anise, basil, black pepper, cinnamon, clary sage, clove, dill, ginger, nutmeg, oregano, parsley, peppermint, rosemary, and sage. Snacking on fruits including grapefruit, guava, lemon, lime, mandarin, and orange may help increase intake of cubebene, while teas containing chamomile, eucalyptus, anise, and geranium may also contain this terpene.

Essential oils derived from any of the plants in the "Occurrence in Plants" section above are likely to contain one or both isomers of cubebene, although caution is advised in the use of essential oils, which can also contain toxic substances. Professional advice or instruction should be sought. These oils can be taken directly, added to products or food, or used in diffusers or candles.

At this time, the only known tested cannabis variety found to contain alpha-cubebene is the Futura hemp variety, although the author expects that many more varieties will be found that contain this tricyclic sesquiterpene as laboratories begin to include this in their phytochemical testing standards, particularly within the cannabis industry.

[1196] El-Soud NH, Deabes M, El-Kassem LA, Khalil M. Chemical Composition and Antifungal Activity of Ocimum basilicum L. Essential Oil. Open Access Maced J Med Sci. 2015 Sep 15;3(3):374-9.

Cubebene Review

Answer the following questions to test your knowledge of this terpene:

Question #1: What type of terpene is cubebene?

a. Monocyclic monoterpene
b. Linear sesquiterpene
c. Bicyclic monoterpene
d. Tricyclic sesquiterpene

Question #2: Which of the following best describes the primary flavor & aroma profile of cubebene?

a. Spicy, turpentine, dry
b. Hops, woody, spicy, clove, bitter
c. Depends on the isomer
d. Depends on pentanyl framing

Question #3: How common is cubebene in cannabis?

a. Top three
b. Top ten
c. Top forty
d. Less common

Question #4: Which plant is the most common source of cubebene?

a. Piter cruella berries
b. Piper cubeba berries
c. Hyssop culpeper berries
d. Piper culpeper berries

Question #5: Describe the difference between the alpha- and beta- isomers of cubebene:

____
____

Question #6: Name two potential medical uses of cubebene:

1 ______________________________ 2 ______________________________

For the answer key to all individual compound chapters please go here: http://thebigbookofterps.com/bbot-quiz-answer-key/

COPAENE

Type: Tricyclic sesquiterpene
Chemical Formula: $C_{15}H_{24}$
Molecular Weight: 204.35 g/mol
Boiling Point: *Alpha-copaene:* 246.00 to 251.00 °C. @ 760.00 mm Hg[1197]
Beta-copaene: 254.00 to 256.00 °C. @ 760.00 mm Hg[1198]
Flash Point: *Alpha-copaene:* 109.00 °F. TCC (42.78 °C.)
Beta-copaene: 221.00 °F. TCC (105.00 °C.)
Solubility: *Alpha-copaene:* Soluble in alcohol, slightly soluble in water, chloroform, methanol
Biological Role: Signaling agent, antimicrobial, allelopathic agent
Therapeutic Role: Anticancer, antibacterial, antioxidant
Commercial Role: Research chemical
Occurrence in Cannabis: Unknown, but probably Top 40-50
Aroma/Flavor: *Alpha-copaene:* Woody, spicy, honey

Occurs in Cannabis Strains: At the time of publication of this text, there were no publicly available COAs or published testing laboratory analysis of commercial medical or recreational cannabis varieties that contained copaene. However, alpha-copaene was initially reported in the essential oil of cannabis sativa grown in Mexico in 1974[1199]. In 2014, researchers found both alpha-copaene and beta-copaene in wild cannabis sativa growing in Nepal[1200]. Research edited by ElSohly – a scientist who worked closely with Mechoulam – found only trace amounts of alpha-copaene in 5 drug-type cultivars from the European Union[1201]. Alpha-copaene has also been found in extracts made from the dried flowering tops of cannabis sativa L. grown in Italy[1202], and in nearly all of 33 chemovars tested from licensed Canadian producers[1203]. This compound has also been found in dried cannabis supplied by the United States National Institute on Drug Abuse at the National Institute of Health[1204], and in the essential oils of non-psychoactive chemotypes of cannabis[1205]. Additionally, cannabis industry technical and scientific equipment providers have found beta-copaene in 5% CBD in organic hemp oil[1206], and alpha-copaene in an unidentified indica dominant cannabis sample[1207]. Also noteworthy is a terpene blend product sold by Abstrax

---

1197 The Good Scents Company Data Sheet for Alpha-copaene, from: http://www.thegoodscentscompany.com/data/rw1053841.html. Accessed September 10, 2023.

1198 The Good Scents Company Data Sheet for Beta-copaene, from: http://www.thegoodscentscompany.com/data/rw1561521.html. Accessed September 10, 2023.

1199 Radwan MM, Chandra S, Gul S, ElSohly MA. Cannabinoids, Phenolics, Terpenes and Alkaloids of Cannabis. Molecules. 2021 May 8;26(9):2774.

1200 Satyal, Prabodh and William N. Setzer. 2014. Chemotyping and Determination of Antimicrobial, Insecticidal, and Cytotoxic Properties of Wild-Grown Cannabis sativa from Nepal. Journal of Medicinally Active Plants 3, (1):9-16.

1201 Brenneisen, Rudolf; ElSohly, Mahmoud; Marijuana and the Cannabinoids. Forensic Science and Medicine, Chapter 2: Chemistry and Analysis of Phytocannabinoids and Other Cannabis Constituents Pg. 28, 1.2.1.

1202 Muscarà C, Smeriglio A, Trombetta D, Mandalari G, La Camera E, Grassi G, Circosta C. Phytochemical characterization and biological properties of two standardized extracts from a non-psychotropic Cannabis sativa L. cannabidiol (CBD)-chemotype. Phytother Res. 2021 Sep;35(9):5269-5281.

1203 Mudge EM, Brown PN, Murch SJ. The Terroir of Cannabis: Terpene Metabolomics as a Tool to Understand Cannabis sativa Selections. Planta Med. 2019 Jul;85(9-10):781-796.

1204 Stenerson, Katherine. Headspace SPME-GC/MS Analysis of Terpenes in Hops and Cannabis. Analytix Reporter Issue 3.

1205 Ruth Gallily, Zhannah Yekhtin, and Lumír Ondřej Hanuš. The Anti-Inflammatory Properties of Terpenoids from Cannabis. Cannabis and Cannabinoid Research.Dec 2018.282-290.

1206 Sam, Karen; Application Note: Cannabidiol Oil Analysis with the Pyroprobe. Thermal Analysis of CBD oil using a CDS Model 6150 Pyroprobe. CDS Analytical. From: https://www.go-jsb.com/resources/media/application_notes/195_TD_Py_Cannabidiol_Oil_Analysis_with_the_Pyroprobe.pdf, accessed June 22, 2023.

1207 Application Note for Pegasus BT 4D Instrument: Determining Terpene Profiles of Cannabis Strains Using GC and GCxGC with High Performance TOFMS. LECO Corporation; Saint Joseph, Michigan USA 2019.

Tech that is purported to recreate the aromas of the popular strain, Ghost Train Haze. In this product, "copaene" is listed in the ingredients as the 23rd of 27 ingredients[1208].

INTRODUCTION

Although two isomers of copaene exist – the alpha and beta versions – most references in the scientific literature are to the alpha version. Also known as 'aglaiene,' copaene earned its name after it was isolated for the first time from Copaifera langsdorffii (copaiba tree). This compound is strongly attracting to several economically and agriculturally important insect pests as detailed below. Interestingly, alpha-copaene has also been found to be produced by the bacteria Streptomyces griseus.

Readers should use caution when working with this pure or diluted compound, as it can be irritating to skin and presents a significant hazard if aspirated.

CHEMICAL STRUCTURE

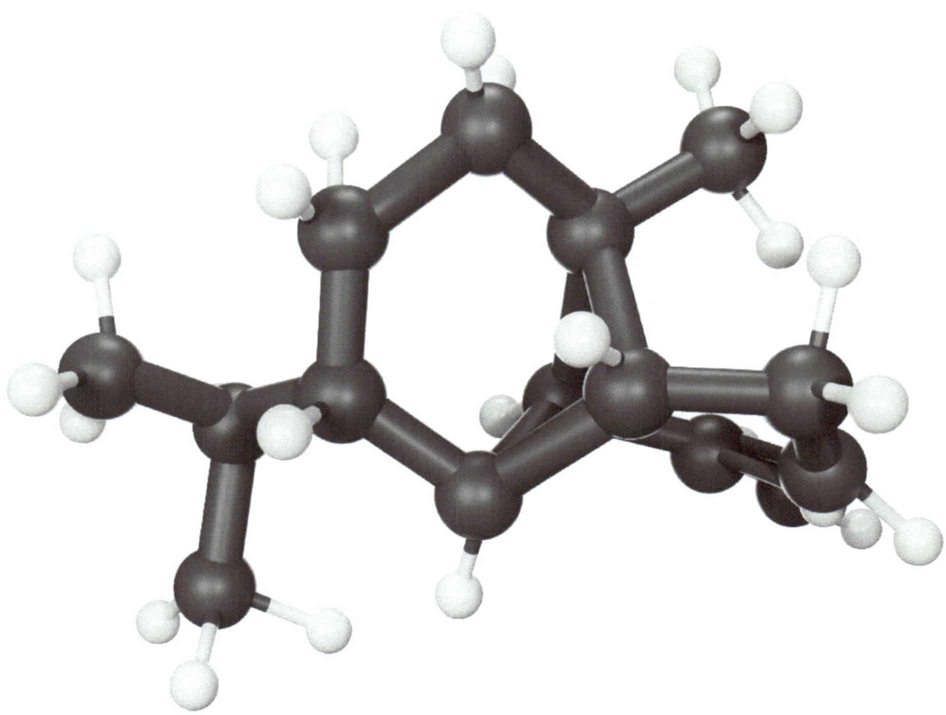

Formed in the cytosol of plants via the mevalonate pathway from isoprene and then farnesyl diphosphate, copaene is a sesquiterpene containing 15 carbon atoms and 24 hydrogen atoms, noted as $C_{15}H_{24}$. While alpha-copaene contains an isopropyl substituent at position 8 and two methyl substituents at positions 1 and 3 with no double bond, beta-copaene differs in that it features a double carbon bond and an exocyclic methylene group.

OCCURENCE IN PLANTS

Copaene is found in many plants, which can include both isomers and the racemic mixture of the two.

[1208] Ghost Train Haze Terpene Profile, From: https://www.leafly.com/brands/abstrax-tech/products/abstrax-tech-ghost-train-haze-terpene-profile-terpenes, accessed June 22, 2023.

Alpha-copaene:

Allspice fruit
Anise star
Ayou wood
Bell pepper
Blood orange
Calamus
Carrot
Cassinia laevis
Chamomile
Clary sage
Costmary
Cumin
Eucalyptus globulus
Geranium
Guava
Juniper
Lavender
Lovage
Marjoram
Melissa
Nepeta
Oregano
Pepper tree
Phlomis fruticosa
Piper cernum vell.
Ravintsara
Sage
Satureja viminea l.
Spearmint
Tamarind
Tea Tree
Water mint
Yarrow
Zingiber officinale
Alpinia galanga oil
Artemisia campestris spp.
Basil
Black pepper
Cabocla
Cannabis
Cascarilla bark
Cedrela wood
Cinnamon
Clementine
Croton flavens
Curry
Featherfew
Ginger
Ho leaf
Labdanum
Lemon
Mandarin
Marsypianthes chamaedrys
Mikan
Nutmeg
Parsley
Peppermint
Phoebe
Polyalthia australis
Rosemary
Salvia
Savory
Sugandha kokila berry
Tangerine
Thyme
Wormwood
Ylang ylang
Zinnia elegans jacq.
Angelica
Asarum canadense
Bay laurel
Bergamot
Cajuput
Cardamom
Cassia
Chamaecyparis obtusa
Citronella
Clove
Cubeb
Cypress
Fleabane
Grapefruit
Jambu
Laurel leaf
Lime
Manuka
Mastic
Mikania
Orange
Patchouli
Petitgrain
Pine needle
Prangos varieties
Safflower
Santolina
Sidertis varieties
Sundacarpus amarus
Tansy
Valerian root
xylopia aromatica
Zanthoxylum ovalifolium l.

Beta-copaene:

Allspice
Basil
Corn
Guava
Manuka
Peppermint
Sage
Angelica
Cassinia
Galangal
Horsemint
Mikania
Pine
Tansy
Anise
Chamomile
Grapefruit
Jambu
Orange
Ravintsara
Wormseed

BIOLOGICAL ACTIVITY IN PLANTS

Although scant work exists that explores the biological uses of copaene in plants, there is some evidence to suggest that the compound may be used as an allelopathic agent, and as an antibacterial agent. However, much of the scientific body of work in this regard has found that this sesquiterpene appears to be a signaling molecule that plants release in response to insect predation, and, more importantly, that insects then use to locate their hosts.

**Allelopathic agent:* As a major constituent in the essential oil derived from the leaves of Unonopsis guatterioides (a species of evergreen tree in the family Annonaceae), alpha-copaene contributed to the inhibition of seed germination and growth in Allium cepa (onion) and Lactuca sativa (lettuce)[1209].

**Antimicrobial agent:* As a primary component of essential oils obtained by hydro distillation of fresh and dry leaves of Eremanthus erythropappus (a type of deciduous tree), alpha-copaene contributed to antibacterial activity against Staphylococcus aureus, Streptococcus pyogenes, and Escherichia coli, as well as the fungi Candida albicans and C. tropicalis[1210].

**Signaling molecule:* Copaene acts as a signaling molecule that various pests use to locate the host emitting the volatile compound. For instance, electrophysiological responses of the Naupactus bipes beetle to essential oils from Piperaceae species – where alpha-copaene is a primary constituent – suggest that the beetles are using the sesquiterpene to locate their host[1211], the pepper plant, an important agricultural crop. Similarly, emissions of a chemical blend where alpha-copaene is a major compound are used by the redbay ambrosia beetle to locate its host[1212], while both alpha-copaene and beta-copaene as produced by Olea europaea L. (olive tree) are ovipositional attractants to Bactrocera oleae[1213] (fruit fly). Finally, tomato plants have been shown to significantly increase emissions of alpha-copaene in response to botrytis cinerea infection in the leaves of the plant[1214], while Artemisia ordosica produces the compound in response to predation by weevils[1215].

USES IN INDUSTRY

Commercial preparations of copaene generally appear as a clear viscous liquid, with little to no use of this compound as a fragrance or flavor additive. In fact, The Good Scents Company reports that both isomers of the compound are specifically for research purposes only.

Some work has explored the use of alpha-copaene as an attractant or lure for Euwallacea nr. Fornicates (tea shot hole borer), an agriculturally important pest of tea plants[1216]. Researchers have produced beta-copaene in

---

[1209] Yoshida NC, Saffran FP, Lima WG, Freire TV, de Siqueira JM, Garcez WS. Chemical characterization and bioherbicidal potential of the essential oil from the leaves of Unonopsis guatterioides (A.DC.) R.E.Fr. (Annonaceae). Nat Prod Res. 2019 Nov;33(22):3312-3316.

[1210] Silvério MS, Del-Vechio-Vieira G, Pinto MA, Alves MS, Sousa OV. Chemical composition and biological activities of essential oils of Eremanthus erythropappus (DC) McLeisch (Asteraceae). Molecules. 2013 Aug 16;18(8):9785-96.

[1211] Ramos CS, Soares MG, da Silva AM, Batista-Pereira LG, Corrêa AG, Kato MJ. Electrophysiological responses of the Naupactus bipes beetle to essential oils from Piperaceae species. Nat Prod Commun. 2012 Aug;7(8):1103-6.

[1212] Kendra, P.E., Montgomery, W.S., Niogret, J. et al. Attraction of the Redbay Ambrosia Beetle, Xyleborus glabratus, to Avocado, Lychee, and Essential Oil Lures. J Chem Ecol 37, 932–942 (2011).

[1213] de Alfonso I, Vacas S, Primo J. Role of α-copaene in the susceptibility of olive fruits to Bactrocera oleae (Rossi). J Agric Food Chem. 2014 Dec 10;62(49):11976-9.

[1214] Thelen, John, Jeremy Harbinson, Roel Jansen, Gerrit Van Straten, Maarten A. Posthumus, Ernst J. Woltering, and Harro J. Bouwmeester. "The sesquiterpene α-copaene is induced in tomato leaves infected by Botrytis cinerea." Journal of Plant Interactions 1.3 (2005): 163-170.

[1215] Zhang H, Zhou D, Luo Y, Wang J, Zong S. Identification of volatile compounds emitted by Artemisia ordosica (Artemisia, Asteraceae) and changes due to mechanical damage and weevil infestation. Z Naturforsch C J Biosci. 2013 Jul-Aug;68(7-8):313-7.

[1216] Kendra PE, Owens D, Montgomery WS, Narvaez TI, Bauchan GR, Schnell EQ, et al. (2017)
α-Copaene is an attractant, synergistic with quercivorol, for improved detection of Euwallacea nr. fornicatus (Coleoptera: Curculionidae: Scolytinae). PLoS

Escherichia coli[1217], and increased the production of alpha-copaene in potato plants by 15-fold compared to controls[1218].

POTENTIAL USES IN MEDICINE

Most medical research focuses on the alpha version of copaene. The compound has potential as a potent inhibitor of various types of cancer, an antibacterial agent, antioxidant, larvicidal agent, and treatment for trichomoniasis - a sexually transmitted parasitic disease.

**Anticancer agent:* Copaene exhibits a wide range of anticancer activities. For instance, researchers studying the essential oil of Cedrelopsis grevei leaves – widely used in folk medicine in Madagascar – discovered that alpha-copaene was a top 3 compound in the oil, which was active against the MCF7 cell line of human breast adenocarcinoma[1219]. As a major constituent in an essential oil made from the leaves of Xylopia frutescens (a type of evergreen tree known locally as embira), alpha-copaene likely contributed to up to 37.5% inhibition of OVCAR-8 (ovarian adenocarcinoma), NCI-H358M (bronchoalveolar lung carcinoma) and PC-3M (metastatic prostate carcinoma) human tumor cell lines with no expressive toxicity[1220]. These tumor inhibition effects were observed both in vitro and in vivo. Similar work showed that alpha-copaene was a major component of the essential oil from leaves of Xylopia laevigata (known locally as meiu), which inhibited tumor growth rates by up to 42.5% in HL-60 (promyelocytic leukemia), OVACAR-8 (ovarian carcinoma), SF295 (glioblastoma), and HCT-116 (colon carcinoma) human cancer cell lines[1221]. As a primary constituent in the essential oil made from the leaves of two Annonaceae species, alpha-copaene contributed to potent antitumor activities against glioma, melanoma, breast, lung, kidney, and colon cancer cell lines[1222].

**Treatment for Prostate Cancer:* Used in folk medicine of Africa for centuries, scientists found that an extract of guava leaf with alpha-copaene as a primary constituent was a potent inducer of cytotoxic and apoptotic effects in PC-3 human prostate cancer cells[1223]. These actions were attributed by the researchers to suppression of the AKT/mTOR/S6K1 and MAPK signaling pathways.

**Treatment for Breast Cancer:* As the primary constituent in the essential oil from Garcinia celebica L., (known locally in Malaysia as manggis hutan), alpha-copaene displayed significant anti-proliferative effects against MCF-7 human breast cancer cells[1224].

ONE 12(6): e0179416.

[1217] Mischko W, Hirte M, Fuchs M, Mehlmer N, Brück TB. Identification of sesquiterpene synthases from the Basidiomycota Coniophora puteana for the efficient and highly selective β-copaene and cubebol production in E. coli. Microb Cell Fact. 2018 Oct 22;17(1):164.

[1218] Morris WL, Ducreux LJ, Shepherd T, Lewinsohn E, Davidovich-Rikanati R, Sitrit Y, Taylor MA. Utilisation of the MVA pathway to produce elevated levels of the sesquiterpene α-copaene in potato tubers. Phytochemistry. 2011 Dec;72(18):2288-93.

[1219] Afoulous, Samia, Hicham Ferhout, Emmanuel Guy Raoelison, Alexis Valentin, Béatrice Moukarzel, François Couderc, and Jalloul Bouajila. "Chemical composition and anticancer, antiinflammatory, antioxidant and antimalarial activities of leaves essential oil of Cedrelopsis grevei." Food and chemical toxicology : an international journal published for the British Industrial Biological Research Association 56 (2014): 352-362.

[1220] Rosana P.C. Ferraz, Gabriella M.B. Cardoso, Thanany B. da Silva, José Eraldo do N. Fontes, Ana Paula do N. Prata, Adriana A. Carvalho, Manoel O. Moraes, Claudia Pessoa, Emmanoel V. Costa, Daniel P. Bezerra. Antitumour properties of the leaf essential oil of Xylopia frutescens Aubl. (Annonaceae), Food Chemistry, Volume 141, Issue 1, 2013, Pages 196-200, ISSN 0308-8146.

[1221] Quintans Jde S, Soares BM, Ferraz RP, Oliveira AC, da Silva TB, Menezes LR, Sampaio MF, Prata AP, Moraes MO, Pessoa C, Antoniolli AR, Costa EV, Bezerra DP. Chemical constituents and anticancer effects of the essential oil from leaves of Xylopia laevigata. Planta Med. 2013 Jan;79(2):123-30.

[1222] Costa EV, Dutra LM, Salvador MJ, Ribeiro LH, Gadelha FR, de Carvalho JE. Chemical composition of the essential oils of Annona pickelii and Annona salzmannii (Annonaceae), and their antitumour and trypanocidal activities. Nat Prod Res. 2013;27(11):997-1001.

[1223] Ryu NH, Park KR, Kim SM, Yun HM, Nam D, Lee SG, Jang HJ, Ahn KS, Kim SH, Shim BS, Choi SH, Mosaddik A, Cho SK, Ahn KS. A hexane fraction of guava Leaves (Psidium guajava L.) induces anticancer activity by suppressing AKT/mammalian target of rapamycin/ribosomal p70 S6 kinase in human prostate cancer cells. J Med Food. 2012 Mar;15(3):231-41.

[1224] Tan WN, Tan ZH, Zulkifli NI, Nik Mohamed Kamal NNS, Rozman NAS, Tong WY, Leong CR, Lim JW. Sesquiterpenes rich essential oil from Garcinia

Antibacterial Agent: As the primary constituent in the essential oil made from the leaves of Murraya koenigii L. (curry tree), alpha-copaene contributed to the complete inhibition of growth of Listeria innocua[1225], which causes the potentially fatal Listeria illness. As a primary constituent in the essential oil made from the leaves of Psidium myrtoides (purple guava), alpha-copaene contributed to moderate activity against Streptococcus mitis, S. sanguinis, S. sobrinus, and S. salivarius, and strong activity against S. mutans[1226].

Antioxidant: As a major component in the essential oil made from the leaves of two Annonaceae species, alpha-copaene contributed to significant antioxidant activity[1227]. The author who led this study also led the study referenced in citation #1222 above, which found significant anticancer properties.

Larvicidal Agent: Researchers around the world are constantly studying new mosquito control molecules and protocols, and some work has shown that copaene might be useful in preventing mosquito-borne illnesses. As the primary constituent in the essential oil made from the peel of ripe and unripe fruits of Hymenaea courbaril, (West India locust), alpha-copaene contributed to significant larvicidal activity against third instar larvae of Aedes aegypti[1228], a prolific disease vector mosquito.

Treatment for Trichomoniasis: As the primary constituent in the essential oil from Dracocephalum kotschyi (known locally in Iran as Badrandjboie-Dennaie), alpha-copaene contributed to the inducement of apoptosis in Trichomonas vaginalis (a sexually transmitted parasitic disease) without toxicity in haematopoietic cells[1229] (an immature or stem blood cell).

IMPLICATIONS FOR HUMAN HEALTH & NUTRITION

Although it's relatively unknown how the human body processes copaene, and particularly whether this compound is useful when ingested, there are plenty of ways to obtain this compound dietarily. For instance, when preparing, cooking, or serving food, consider working with spices and herbs such as allspice, angelica, anise, basil, black pepper, bergamot, cajuput, chamomile, cinnamon, clary sage, clove, cumin, curry, ginger, lavender, marjoram, nutmeg, oregano, parsley, peppermint, rosemary, safflower, sage, savory, thyme, or water mint. Eat vegetables like bell peppers and carrots, and eat fruits like blood oranges, grapefruit, guava, lemons, limes, mandarin, oranges, and tangerines. Teas containing anise, ginger, spearmint, or valerian root may also contain copaene.

Finally, essential oils from any of the plants listed in the "Occurrence in Plants" section – including cannabis – may contain copaene, but readers should exercise extreme caution when working with any EO or pure compound.

---

celebica L. and its cytotoxic and antimicrobial activities. Nat Prod Res. 2020 Dec;34(23):3404-3408.

[1225] Erkan N, Tao Z, Rupasinghe HP, Uysal B, Oksal BS. Antibacterial activities of essential oils extracted from leaves of Murraya koenigii by solvent-free microwave extraction and hydro-distillation. Nat Prod Commun. 2012 Jan;7(1):121-4.

[1226] Dias ALB, Batista HRF, Estevam EBB, Alves CCF, Forim MR, Nicolella HD, Furtado RA, Tavares DC, Silva TS, Martins CHG, Miranda MLD. Chemical composition and in vitro antibacterial and antiproliferative activities of the essential oil from the leaves of Psidium myrtoides O. Berg (Myrtaceae). Nat Prod Res. 2019 Sep;33(17):2566-2570.

[1227] Costa EV, Dutra LM, de Jesus HC, Nogueira PC, Moraes VR, Salvador MJ, Cavalcanti SC, dos Santos RL, Prata AP. Chemical composition and antioxidant, antimicrobial, and larvicidal activities of the essential oils of Annona salzmannii and A. pickelii (Annonaceae). Nat Prod Commun. 2011 Jun;6(6):907-12.

[1228] Aguiar, José Cláudio D., Gilvandete M. P. Santiago, Patrícia L. Lavor, Helenicy N. H. Veras, Yana S. Ferreira, Michele A. A. Lima, Ângela M. C. Arriaga, Telma L. G. Lemos, Jefferson Q. Lima, Hugo C. R. de Jesus, Péricles B. Alves, and Raimundo Braz-Filho. "Chemical Constituents and Larvicidal Activity of Hymenaea courbaril Fruit Peel." Natural Product Communications 5.12 (2010)1.

[1229] Chelgerdi Dehkordi B, Esmaeilifallah M, Kalantari R, Benchimol M, Khamesipour F. Induction of apoptosis by essential oil of Dracocephalum kotschyi on Trichomonas vaginalis. Vet Med Sci. 2023 Mar;9(2):1008-1016.

Copaene Review

Answer the following questions to test your knowledge of this terpene:

Question #1: What type of terpene is copaene?

a. Monocyclic monoterpene
b. Tricyclic sesquiterpene
c. Bicyclic monoterpene
d. Linear sesquiterpene

Question #2: Which of the following best describes the primary flavor & aroma profile of copaene?

a. Spicy, turpentine, dry
b. Woody, spicy, honey
c. Depends on the isomer
d. Depends on pentanyl framing

Question #3: How common is copaene in cannabis?

a. Top three
b. Top ten
c. Top twenty
d. Less common

Question #4: Which plant was copaene first isolated from?

a. Copenbe bush
b. Coreol tea flower
c. Copaiba tree
d. Claston disatrom F.

Question #5: Describe the difference between the alpha- and beta- isomers of copaene:

____
____

Question #6: Name two potential medical uses of alpha-copaene:

1 ________________________________ 2 ________________________________

For the answer key to all individual compound chapters please go here: http://thebigbookofterps.com/bbot-quiz-answer-key/

CADINENE

Type: Bicyclic sesquiterpene
Chemical Formula: $C_{15}H_{24}$
Molecular Weight: 204.356 g/mol
Boiling Point: *Alpha-cadinene:* 271.00 to 272.00 °C. @ 760.00 mm Hg[1230]
Beta-cadinene: 273.00 to 276.00 °C. @ 760.00 mm Hg[1231]
Gamma-cadinene: 271.00 to 276.00 °C. @ 760.00 mm Hg[1232]
Delta-cadinene: 279.00 to 280.00 °C. @ 760.00 mm Hg[1233]
Flash Point: *Alpha-cadinene:* 224.00 °F. TCC (106.67 °C.)
Beta-cadinene: 224.00 °F. TCC (106.67 °C.)
Gamma-cadinene: 224.00 °F. TCC (106.67 °C.)
Delta-cadinene: 231.00 °F. TCC (110.56 °C.)
Solubility: *Alpha-cadinene:* Soluble in alcohol, insoluble in water
Beta-cadinene: Soluble in acetone, petroleum ether, chloroform, slightly soluble in alcohol, insoluble in water
Gamma-cadinene: Soluble in alcohol, insoluble in water
Delta-cadinene: Soluble in alcohol, insoluble in water
Biological Role: Allelopathic agent, antifeedant
Therapeutic Role: Antimicrobial, antioxidant, insecticidal
Commercial Role: Research chemical, former flavoring and fragrance agent
Occurrence in Cannabis: Unknown, probably Top 50
Aroma/Flavor: *Alpha-cadinene:* Woody, dry
Beta-cadinene: Green, woody
Gamma-cadinene: No descriptors found
Delta-cadinene: Thyme, lemon, herbal, woody, dry

Occurs in Cannabis Varieties: Edom, Futura hemp, Pandora's Box

At the time of publication of this text, there were no publicly available commercial laboratory testing analyses or COAs available that showed cannabis or cannabis products containing cadinene. However, a 1975 study that examined the essential oil of cannabis sativa found delta, gamma, and alpha isomers of cadinene in low concentrations[1234]. A 2016 study found cadinene in both "indica" and "sativa" drug varieties of cannabis[1235], and in 2020 scientists found delta-cadinene in a sample of a beta-caryophyllene-dominant chemotype called Edom, alpha-cadinene in a sample of Pandora's Box, and delta-cadinene in a sample of Futura hemp[1236].

1230 The Good Scents Company Data Sheet for Alpha-cadinene, from: http://www.thegoodscentscompany.com/data/rw1443261.html. Accessed September 11, 2023.

1231 The Good Scents Company Data Sheet for Beta-cadinene, from: http://www.thegoodscentscompany.com/data/rw1054311.html. Accessed September 11, 2023.

1232 The Good Scents Company Data Sheet for Gamma-cadinene, from: http://www.thegoodscentscompany.com/data/rw1595501.html. Accessed September 11, 2023.

1233 The Good Scents Company Data Sheet for Delta-cadinene, from: http://www.thegoodscentscompany.com/data/rw1054621.html. Accessed September 11, 2023.

1234 Hendriks, Henk, Theo M. Malingré, Sieb Batterman, Rein Bos. "Mono- and sesqui-terpene hydrocarbons of the essential oil of Cannabis sativa." Phytochemistry 14.3 (1975): 814-815.

1235 Hazekamp, Arno et al. "Cannabis: From Cultivar to Chemovar II—A Metabolomics Approach to Cannabis Classification." (2016).

1236 Hanuš LO, Hod Y. Terpenes/Terpenoids in Cannabis: Are They Important? Med Cannabis Cannabinoids. 2020 Aug 10;3(1):25-60.

INTRODUCTION

Cadinene is a sesquiterpene that can appear somewhat confusing because this compound occurs in 9 different isomers, each with a unique configuration and aroma and/or flavor profile. Understanding how cadinene might impact the bouquet of cannabis is difficult without knowing which isomer/s is/are present, but it's likely that the primary trait imparted by this compound will be a woody aroma and/or flavor. However, in general cadinene is not used in fragrances or as a flavoring agent, and instead has potential use as an antimicrobial agent and an antioxidant.

CHEMICAL STRUCTURE

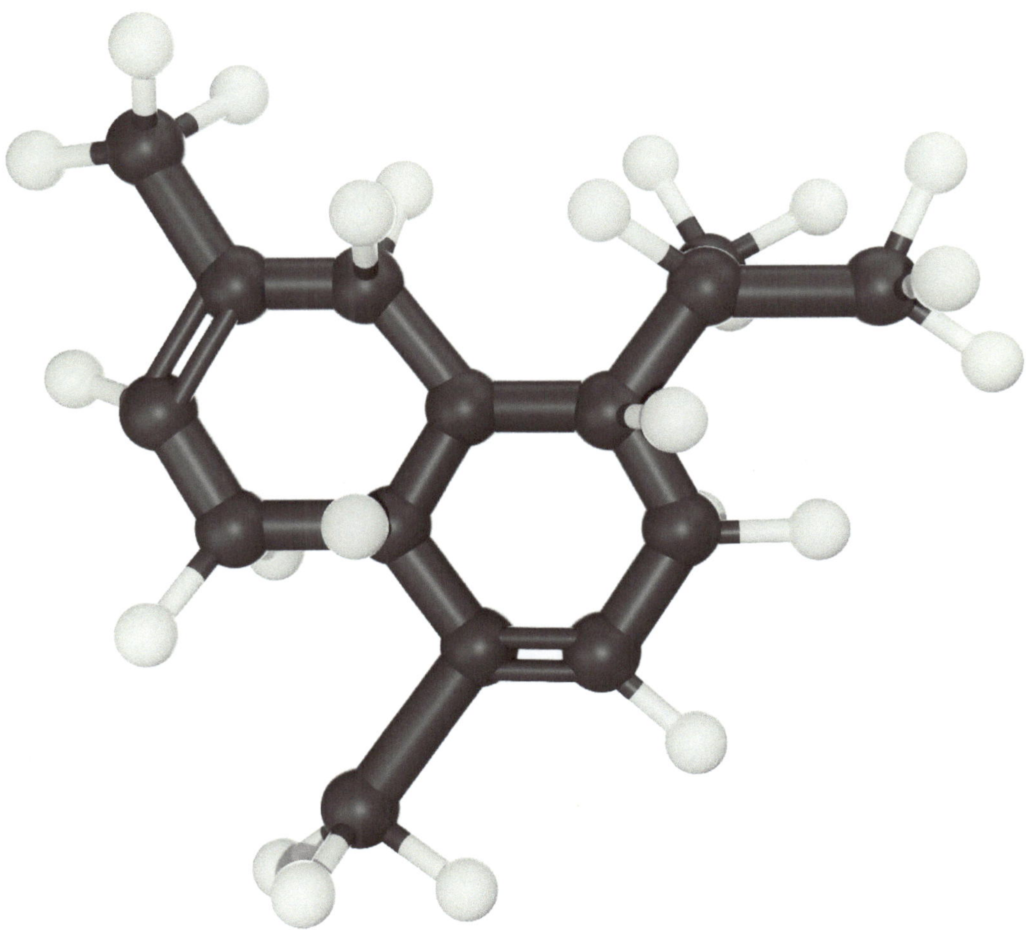

Cadinene is formed in the cytosol of plants via the mevalonic acid (MVA) pathway, with farnesyl diphosphate (FPP) as the key intermediate. The cadinene molecule features two rings – often with endocyclic double bonds – which classifies this compound as bicyclic. Comprised of three units of isoprene, cadinene contains 15 carbon atoms and 24 hydrogen atoms, notated as $C_{15}H_{24}$. This sesquiterpene may occur as one of nine different isomers, with the beta isomer being common generally in nature, and the delta version seemingly common in cannabis, although more research is needed to quantify this compound in both drug and hemp varieties of cannabis. The primary difference in the various isomers of cadinene is the arrangement of the molecule and the position of the double bonds.

OCCURRENCE IN PLANTS

Cadinene occurs in many plants overall, but sometimes plants that produce this terpene only produce certain isomers rather than racemic mixtures. This list is general and includes the primary isomers alpha, beta, delta, and gamma.

Alpha-cadinene:

Allspice
Basil
Cedrela
Chamomile
Clary sage
Clove
Dacrydium colensol
Dill
Ginger
Hop
Hyssop
Juniper
Lemon
Peppermint
Pine
Pteronia
Rosemary

Beta-cadinene:

Basil
Bay
Calamus
Cananga
Dacrydium colensol
Ginger
Lavender
Oregano
Thyme
Wormwood
Zingiber officinale

Gamma-cadinene:

Allspice
Angelica
Armoise
Ayou
Basil
Bay
Black pepper
Blood orange
Cajuput
Calamus
Cananga
Carrot
Cascarilla
Cedrela
Cinnamon
Citronella
Clary sage
Clove4Croton
Curry
Davana
Dill
Erigeron graveolens
Eucalyptus globulus
Galngal
Geranium
Ginger
Hinoki
Horsemint
Hyssop
Jambu
Juniper
Labdanum
Laurel
Lavandin
Lavender
Lemon
Lovage
Mandarin
Manuka species
Marjoram
Mentha longifolia
Mikania
Myrrhinium atropurpureum
Orange
Oregano
Parsley
Peppermint
Phoebe
Pimento
Pine
Polyalthia australis
Ravintsara
Rosemary
Sage
Salvia
Satureja viminea
Savin
Savory
Sundacarpus amarus
Tagete
Tamarind
Tansy
Thyme
Vassoura
Yarrow
Ylang ylang
Zanthoxylum species

Delta-cadinene:

Allspice
Amyris
Angelica
Anise
Armoise
Artemisia species
Baccharis coridifolia
Basil
Bay
Bergamot
Betel
Black pepper
Blood orange
Cabocla
Cajuput
Calamus
Cananga
Cascarilla
Cassinia
Cassis
Cedarwood
Cedrela
Chamomile
Cinnamon
Cistus
Citronella
Clary sage
Clove
Corn
Croton
Cubeb
Curry
Cypress
Dill
Erigeron graveolens
Eucalyptus species
Featherfew
Fennel
Ferula
Feverfew
Fir
Fleabane
Galangal
Geranium
Ginger
Grapefruit
Guava
Helichrysum arenarium
Herniaria incana
Hinoki
Horsemint
Hyptis
Hyssop
Jambu
Juniper
Kewda
Labdanum
Laurel
Layana
Lemon
Lemongrass
Lime
Lovage
Mace
Mandarin
Marjoram
Mastic
Mentha longifolia
Microstrobos
Mikan
Mikania
Myrrhinium atropurpureum
Nutmeg
Orange
Oregano
Parsley
Pennyroyal
Pepper tree
Peppermint
Phlomis
Phoebe
Pimenta
Pimento
Pine
Piper cernum
Polyalthia australis
Prangos species
Pteronia
Ravintsara
Rosemary
Rue
Rugula
Safflower
Sage
Salvia
Santolina
Satureja viminea
Savory
Sideritis species
Snake root
Spearmint
Stachy recta
Sugandha kokila
Sundacarpus amarus
Tansy
Taxodium distichum
Tea tree
Thevetia peruviana
Thyme
Valerian
Vassoura
Witch hazel
Wormseed
Wormwood
Xylopia aromatica
Yarrow
Ylang ylang
Zanthoxylum species

BIOLOGICAL ACTIVITY IN PLANTS

Although little published information exists that describes the biological activity of cadinene in plants, some work has shown that this compound can serve as an allelopathic agent, and as an antifeedant. Interestingly, research that studied the volatile emissions of plants exposed to greenhouse gases showed that Astronium graveolens (glassywood) in Brazil significantly increased emissions of beta-cadinene in response to O_3 exposure[1237], although it's unclear what biological purpose this might serve.

[1237] Moura BB, Bolsoni VP, de Paula MD, Dias GM, de Souza SR. Ozone Impact on Emission of Biogenic Volatile Organic Compounds in Three Tropical

**Allelopathic agent:* As primary constituents in an extract made from the aerial parts of Eupatorium adenophorum Spreng, several different cadinene isomers contributed to the significant inhibition of the germination of seeds and seedling growth in Allium cepa (onion), Raphanus sativus (radish), and Cucumis sativus (cucumber)[1238].

**Antifeedant (cadinenes in general):* As the major constitutive and inducible chemicals of Eupatorium adenophorum – a plant to which researchers attributed "extraordinary defense potency" – cadinenes exhibited potent antifeedant activity against Spodoptera exigua[1239] (beet armyworm).

USES IN INDUSTRY

Cadinene is rarely used in modern formulations of fragrances of food items. Normally appearing as a colorless viscous oil, much of the use of this compound occurred between the 1950s and 1980s, with estimates in 1979 of total yearly use in the United States of less than 4,000 pounds[1240], with much of that amount earmarked for research purposes. Various isomers of cadinene have been used in soaps, detergents, creams, lotions, and perfumes.

POTENTIAL USES IN MEDICINE

Cadinene is not well-represented in the medical literature, however, some research shows that this bicyclic sesquiterpene may have anticancer potential, as well as potential as an antimicrobial agent, an insecticidal agent, and as an antioxidant.

**Treatment for Ovarian Cancer:* As an isolated pure compound, delta-cadinene exhibited dose-dependent anticancer effects against OVCAR-3 (human ovary cancer) cells, inducing cell shrinkage, chromatin condensation, nuclear membrane rupture, cell cycle arrest in the sub-G1 phase, and caspase activation[1241].

**Antioxidant:* Researchers studying the medicinal effects of the essential oil of Commiphora gileadensis found that beta-cadinene was a major component, which likely contributed to higher antioxidant activity against cyclooxygenase, xanthine oxidase, and protein denaturation than that of ascorbic acid[1242], a well-known and prolifically used antioxidant.

**Antimicrobial Agent:* As an isolated component from an extraction made from Acorus calamus (calamus root), beta-cadinene exhibited antifungal and antibacterial activity by fitting in the binding pockets of the microbes, ultimately exhibiting the best binding interaction among a selection of tested phytochemicals from calamus[1243].

Tree Species From the Atlantic Forest Remnants in Southeast Brazil. Front Plant Sci. 2022 Jun 24;13:879039.

1238 Baruah, Narayan, Jadab Sarma, Soneswar Sarma, and Ram Sharma. "Seed germination and growth inhibitory cadinenes from Eupatorium adenophorum spreng." Journal of Chemical Ecology 20.8 (2005): 1885-1892.

1239 Liu, Yan, Shi-Hong Luo, Juan Hua, De-Sen Li, Yi Ling, Qian Luo, and Sheng-Hong Li. "Characterization of defensive cadinenes and a novel sesquiterpene synthase responsible for their biosynthesis from the invasive Eupatorium adenophorum." New Phytologist 229.3 (2021): 1740-1754.

1240 Opdyke, D.L.J. (ed.). Monographs on Fragrance Raw Materials. New York: Pergamon Press, 1979., p. 164.

1241 Hui LM, Zhao GD, Zhao JJ. δ-Cadinene inhibits the growth of ovarian cancer cells via caspase-dependent apoptosis and cell cycle arrest. Int J Clin Exp Pathol. 2015 Jun 1;8(6):6046-56.

1242 Shadid KA, Shakya AK, Naik RR, Al-Qaisi TS, Oriquat GA, Atoom AM, Farah HS. Exploring the Chemical Constituents, Antioxidant, Xanthine Oxidase and COX Inhibitory Activity of Commiphora gileadensis Commonly Grown Wild in Saudi Arabia. Molecules. 2023 Mar 2;28(5):2321.

1243 Shalini K, Guleria S, Salaria D, Rolta R, Fadare OA, Mehta J, Awofisayo O, Mandyal P, Shandilya P, Kaushik N, Choi EH, Chandel SR, Kaushik NK. Antimicrobial potential of phytocompounds of Acorus calamus: in silico approach. J Biomol Struct Dyn. 2023 May 12:1-12.

Insecticidal Agent against Mosquitoes: As major constituents in an essential oil made from Schinus molle L. leaves, gamma-cadinene and beta-cadinene both exhibited binding energy and insecticidal activity against Aedes aegypti (yellow fever mosquito), with the activity of beta-cadinene like those of the insecticide, pyriproxyfen[1244].

IMPLICATIONS FOR HUMAN HEALTH & NUTRITION

While the effects of directly inhaling or ingesting cadinene are not well understood, it may be possible to acquire this bicyclic sesquiterpene in certain food items, and possibly in small amounts in cannabis. When preparing, cooking, or serving foods, consider working with spices or herbs like allspice, angelica, basil, cajuput, cinnamon, clary sage, dill, ginger, marjoram, oregano, parsley, rosemary, sage, or thyme. Foods that you can eat directly that might contain cadinene include fruits such as blood oranges, mandarins, oranges, and tamarind, as well as vegetables like carrots and ginger. Teas made with chamomile, clove, eucalyptus, geranium, peppermint, or spearmint may contain various isomers of cadinene.

Essential oils made from the plants detailed in the "Occurrence in Plants" section should be tested to determine cadinene content, and subsequently used with care due to the potentially toxic nature of some EO. This includes EO made from cannabis and may also extend to cannabis products that are smoked or vaped, although it's likely that cadinene will only appear in these cases in trace amounts.

[1244] Herrera-Calderon O, Chavez H, Enciso-Roca EC, Común-Ventura PW, Hañari-Quispe RD, Figueroa-Salvador L, Loyola-Gonzales EL, Pari-Olarte JB, Aljarba NH, Alkahtani S, Batiha GE. GC-MS Profile, Antioxidant Activity, and In Silico Study of the Essential Oil from Schinus molle L. Leaves in the Presence of Mosquito Juvenile Hormone-Binding Protein (mJHBP) from Aedes aegypti. Biomed Res Int. 2022 May 16;2022:5601531.

Cadinene Review

Answer the following questions to test your knowledge of this terpene:

Question #1: What type of terpene is cadinene?

a. Bicyclic hemiterpene
b. Bricyclic sesquiterpene
c. Bicyclic monoterpene
d. Bicyclic sesquiterpene

Question #2: Which of the following best describes the primary flavor & aroma profile of cadinene?

a. Spicy, turpentine, dry
b. Woody, dry, green
c. Depends on the iso version
d. Depends on the isomer

Question #3: How common is cadinene in cannabis?

a. Top three
b. Top ten
c. Top twenty
d. Less common

Question #4: How many known isomers of cadinene exist?

a. 5
b. 7
c. 9
d. 11

Question #5: Describe the primary difference/s between the isomers of cadinene:

____
____

Question #6: Name two potential medical uses of cadinene:

1 ________________________________ 2 ________________________________

For the answer key to all individual compound chapters please go here: http://thebigbookofterps.com/bbot-quiz-answer-key/

SELINENE

Type: Bicyclic Sesquiterpene
Chemical Formula: $C_{15}H_{24}$
Molecular Weight: 204.357 g/mol
Boiling Point: *Alpha-selinene:* 270.00 °C. @ 760.00 mm Hg (estimated)[1245]
Beta-selinene: 260.00 to 263.00 °C. @ 760.00 mm Hg[1246]
Gamma-selinene: 269.00 to 270.00 °C. @ 760.00 mm Hg[1247]
Delta-selinene: 282.00 to 283.00 °C. @ 760.00 mm Hg[1248]
Flash Point: *Alpha-selinene:* 216.00 °F. TCC (102.00 °C.) (estimated by TGSC)
Beta-selinene: 216.00 °F. TCC (102.22 °C.)
Gamma-selinene: 222.00 °F. TCC (105.56 °C.)
Delta-selinene: 234.00 °F. TCC (112.22 °C.)
Solubility: *Alpha-selinene:* Soluble in alcohol, insoluble in water
Beta-selinene: Soluble in alcohol, insoluble in water
Gamma-selinene: Insoluble in water, paraffin oil
Biological Role: Antifeedant
Therapeutic Role: Anticancer agent, antimicrobial, antioxidant
Commercial Role: Research chemical
Occurrence in Cannabis: Unknown, probably top 50
Aroma/Flavor: *Alpha:* Amber, woody
Beta: Herbal
Gamma: Woody

Occurs in Cannabis Strains: Cali Pink Diamonds[1249], Nurse Jackie (hemp), Jilly Bean (hemp).

There were virtually no publicly available reports of commercial testing of medical, recreational, or hemp varieties of cannabis showing selinene as a constituent at the time of publication of this text. However, in 2022 researchers found beta-selinene and alpha-selinene in the essential oil of cannabis sativa L.[1250] (hemp), and the same year other researchers quantified beta-selinene in samples of Nurse Jackie and Jilly Bean hemp varieties[1251].

INTRODUCTION

Selinene is a bicyclic sesquiterpene that occurs as several different isomers, each with an individual aroma and flavor profile. Occurring in many plants as individual versions of selinene and as racemic mixtures, the alpha and

---

1245 The Good Scents Company Data Sheet for Alpha-selinene, from: http://www.thegoodscentscompany.com/data/rw1511191.html. Accessed September 11, 2023.

1246 The Good Scents Company Data Sheet for Beta-selinene, from: http://www.thegoodscentscompany.com/data/rw1053721.html. Accessed September 11, 2023.

1247 The Good Scents Company Data Sheet for Gamma-selinene, from: http://www.thegoodscentscompany.com/data/rw1104381.html. Accessed September 11, 2023.

1248 The Good Scents Company Data Sheet for Delta-selinene, from: http://www.thegoodscentscompany.com/data/rw1562081.html. Accessed September 11, 2023.

1249 From: https://merryjade.com/cali-pink-diamonds-review-cannabis-photos/. Accessed: June 30, 2023.

1250 Zheljazkov VD, Noller JS, Maggi F, Dale R. Terpenes and Cannabinoids Yields and Profile from Direct-Seeded and Transplanted CBD-Cannabis sativa. J Agric Food Chem. 2022 Aug 31;70(34):10417-10428.

1251 Janatová A, Doskočil I, Božik M, Fraňková A, Tlustoš P, Klouček P. The chemical composition of ethanolic extracts from six genotypes of medical cannabis (Cannabis sativa L.) and their selective cytotoxic activity. Chem Biol Interact. 2022 Feb 1;353:109800.

beta versions of this compound are the most prevalent, particularly in cannabis, although delta, gamma, and other versions are known to occur in plants as well.

Interestingly, Ceroplastes rubens (pink scale wax, an insect) has been shown to emit large quantities of beta-selinene[1252] for unknown purposes, while Reticulitermes speratus (a type of subterranean termite) uses the same compound as an aggregation pheromone, and – in a surprising turning of the tables – the termites also use this same compound to hunt Lasius japonicus, a predator of the termite that uses their emissions of beta-selinene to locate its prey[1253].

CHEMICAL STRUCTURE

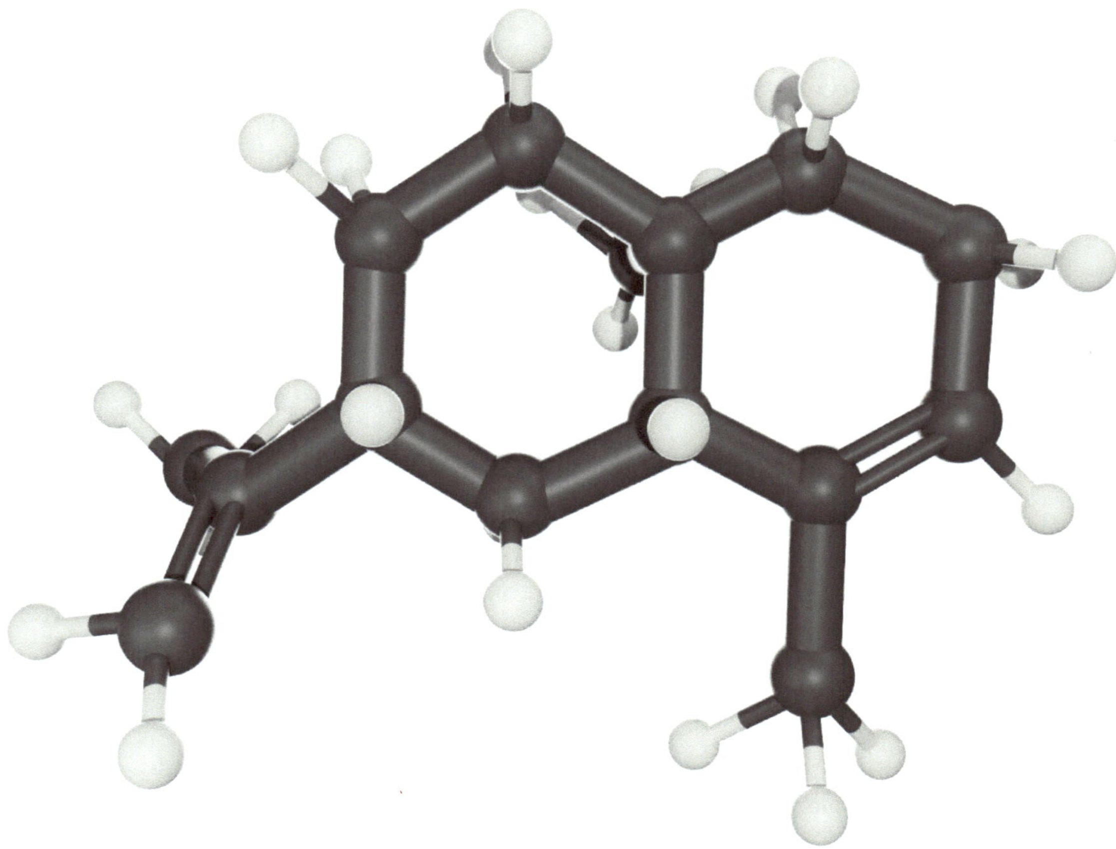

Formed in the cytosol of plants in the MVA pathway and derived from farnesyl diphosphate, selinene contains 15 carbon atoms and 24 hydrogen atoms, notated as $C_{15}H_{24}$. The presence of two rings makes this a bicyclic compound, and its basis on 3 units of isoprene classifies it as a sesquiterpene. There are numerous isomers of selinene, including alpha, beta, delta, gamma, 7-epi-selinene, and 6-epi-selinene, where the arrangement of these molecules determines their aroma, flavor, and function. For instance, the delta version of selinene contains a trans-fused decalin ring and a six-membered ring[1254], the alpha version features an endocyclic double bond in the

---

[1252] Sakurai K, Tomiyama K, Yaguchi Y, Asakawa Y. The characteristic smell emitted from two scale insects, Ceroplastes japonicus and Ceroplastes rubens. Biosci Biotechnol Biochem. 2020 Aug;84(8):1541-1545.

[1253] Nguyen TT, Kanaori K, Hojo MK, Kawada T, Yamaoka R, Akino T. Chemical identification and ethological function of soldier-specific secretion in Japanese subterranean termite Reticulitermes speratus (Rhinotermitidae). Biosci Biotechnol Biochem. 2011;75(9):1818-22.

[1254] From: https://www.thebiotek.com/product/others/bt-1587183. Accessed: July 1, 2023.

octahydronaphthalene ring[1255], and the decahydronaphthalene beta version is substituted by an isopropenyl group at position 7, a methyl group at position 4a, and a methylidene group at position 1[1256].

OCCURRENCE IN PLANTS

Selinene is known to be produced by a moderate variety of plants, which are listed here according to specific isomers:

Alpha-selinene:

Allspice
Amyris
Basil
Bay
Black pepper
Cajuput
Calamus
Callitris species
Carrot
Cedarwood
Celery
Curcuma amada
Curry
Eucalyptus globulus
Galangal
Geranium
Ginger
Guava
Kachur
Laurel
Lemon
Lime
Lovage
Mandarin
Manuka
Mikan
Oregano
Palmarosa
Phoebe
Pimento
Polyalthia australis
Rosemary
Salvia
Sugandha kokila
Sundacarpus amarus
Zanthoxylum nitidum

Beta-selinene:

Ajowan
Allspice
Anise
Baccharis coridifolia
Basil
Bay
Black pepper
Cajuput
Callitris species
Carrot
Cascarilla
Cassia
Cassinia
Cedarwood
Celery
Chamomile
Cinnamon
Clove
Curcuma amada
Curry
Cypress
Erigeron garveolens
Eucalyptus globulus
Geranium
Ginger
Guava
Kachur
Laurel
Lemon
Lime
Lovage
Mandarin
Manuka species
Mikan
Parsley
Phoebe
Pimento
Polyalthia australis
Prangos
Protium haptaphyllum
Ravintsara
Rose
Safflower
Sage
Salvia
Savin
Sugandha kokila
Sundacarpus amarus
Zanthoxylum nitidum

Gamma-selinene:

Bois de rose
Celery
Eucalyptus globulus

[1255] From: https://pubchem.ncbi.nlm.nih.gov/compound/alpha-Selinene. Accessed: July 1, 2023.

[1256] From: https://contaminantdb.ca/contaminants/CHEM022789. Accessed: July 1, 2023.

Ginger	Hop	Laurel
Lime	Lovage	Mandarin
Patchouli	Savin	Wormwood

Delta-selinene:

Allspice	Celery	Citronella
Eucalyptus globulus	Lime	Lovage

BIOLOGICAL ACTIVITY IN PLANTS

The biological uses of selinene by plants are not well-studied, with virtually no mention of this subject in the literature available by the summer of 2023. However, the compound is widely regard as an obvious antifeedant, with beta-selinene as extracted from the essential oil of Artemisia nakaii having been shown to exhibit potent antifeedant activity against third instar larvae of Spodoptera litura[1257] (tobacco cutworm).

USES IN INDUSTRY

Although some reports exist of the use of selinene in perfumes and cosmetics, the author could find no evidence of such use within the last decade. In general, the various isomers of selinene are available commercially for research purposes only, appearing as a colorless liquid at room temperature. Synthesized from a variety of plants, the most common selinene isomers – particularly the alpha version – are sourced from certain species of the Asteraceae family, such as chamomile and yarrow. Selinene can also be synthesized from other terpenes including geraniol.

POTENTIAL USES IN MEDICINE

Selinene is not well-represented in the medical literature, however, promising work has shown that this bicyclic sesquiterpene may be useful in the treatment of breast cancer, and potential as an anti-inflammatory, an antimicrobial agent, an insecticidal agent, or as an antioxidant[1258].

Treatment for Breast Cancer: As a primary constituent in the essential oil made from the stems and leaves of Blepharispermum hirtum Oliver – a small tree native to Oman – beta-selinene contributed to considerable cytotoxic effects against MDA-MB-231 (human breast cancer) cells, while leaving healthy MCF-10A human breast epithelial cells mostly undamaged[1259].

Anti-inflammatory Agent: Researchers studying the therapeutic activity of the essential oil made from the leaves of Piper gaudichaudianum Kunth (known as "pariparoba" locally) found that beta-selinene was the primary constituent, which inhibited neutrophil chemotaxis (essentially, the movement of white blood cells in response to a chemical) in a dose-dependent manner, preventing inflammation in injury sites[1260].

[1257] Liu J, Hua J, Qu B, Guo X, Wang Y, Shao M, Luo S. Insecticidal Terpenes From the Essential Oils of Artemisia nakaii and Their Inhibitory Effects on Acetylcholinesterase. Front Plant Sci. 2021 Aug 12;12:720816.

[1258] Araujo CA, da Camara CAG, de Moraes MM, de Vasconcelos GJN, Pereira MR, Zartman CE. Chemical composition of essential oils from four Piper species, differentiation using multivariate analysis and antioxidant activity. Nat Prod Res. 2022 Jan;36(1):436-439.

[1259] Shah M, Al-Housni SK, Khan F, Ullah S, Al-Sabahi JN, Khan A, Al-Yahyaei BEM, Al-Ruqaishi H, Rehman NU, Al-Harrasi A. First Report on Comparative Essential Oil Profile of Stem and Leaves of Blepharispermum hirtum Oliver and Their Antidiabetic and Anticancer Effects. Metabolites. 2022 Sep 26;12(10):907.

[1260] Soares KD, Bordignon SAL, Apel MA. Chemical composition and anti-inflammatory activity of the essential oils of Piper gaudichaudianum and Piper mikanianum. J Ethnopharmacol. 2022 Oct 28;297:115533.

***Insecticidal Agent:** As primary constituents in a hexane extract made from Apium graveolens (celery), beta-selinene and gamma-selinene contributed to remarkable repellency against female Aedes aegypti mosquitos, with a mean protection time of 4.5 hours, which is comparable to one of the most widely used commercial repellants, 28.5% DEET solution, with no skin toxicity noted among test subjects[1261]. As a major component in the essential oil made from Arisaema fargesii (Chinese cobra lily), beta-selinene contributed to larvicidal activity against Aedes aegypti and Aedes albopictus by inhibiting the larvae's defensive enzymes[1262].

***Antimicrobial:** As the primary constituent in the essential oil of Psidium guajava (guava), beta-selinene exhibited remarkable activity against Salmonella Enteritidis, with researchers noting that the effects of the EO were comparable to commercial antibiotics[1263]. As a major component in the essential oil of Lindera nacusua (a subtropical shrub or tree), beta-selinene contributed to potent antibacterial activity and potential high selectivity against Staphylococcus aureus and Candida albicans[1264] – two extremely common microbes that cause disease in humans.

IMPLICATIONS FOR HUMAN HEALTH & NUTRITION

While little is known about the biological availability of selinene once ingested, dietary inclusion of this sesquiterpene is moderately simple. For instance, when preparing, cooking, or serving food, consider working with spices or herbs such as allspice, anise, basil, black pepper, cajuput, cinnamon, clove, oregano, parsley, rosemary, safflower, or sage. Add selinene-rich celery or carrots to vegetable dishes, or eat raw ginger with sushi or similar meals. Snacking on fruits such as guava, lemons, limes, or mandarin could boost levels of this compound, as well as drinking teas made from anise, chamomile, geranium, or ginger.

Essential oils made from the plants listed in the "Occurrences in Plants" section above are likely to contain selinene, but extreme care should be taken when working with essential oils, as these complex mixtures can be toxic even in small amounts.

Cannabis varieties thought to contain selinene include Cali Pink Diamonds, and the Nurse Jackie and Jilly Bean hemp varietals. EO made from these varieties may contain selinene, and, though less likely, side-stream smoke or vapors from these or other selinene-containing varieties may result in the ingestion of this bicyclic sesquiterpene.

1261 Tuetun B, Choochote W, Pongpaibul Y, Junkum A, Kanjanapothi D, Chaithong U, Jitpakdi A, Riyong D, Pitasawat B. Celery-based topical repellents as a potential natural alternative for personal protection against mosquitoes. Parasitol Res. 2008 Dec;104(1):107-15.

1262 Huang Y, Lin M, Jia M, Hu J, Zhu L. Chemical composition and larvicidal activity against Aedes mosquitoes of essential oils from Arisaema fargesii. Pest Manag Sci. 2020 Feb;76(2):534-542.

1263 de Souza WFC, de Lucena FA, de Castro RJS, de Oliveira CP, Quirino MR, Martins LP. Exploiting the chemical composition of essential oils from Psidium cattleianum and Psidium guajava and its antimicrobial and antioxidant properties. J Food Sci. 2021 Oct;86(10):4637-4649.

1264 Wei G, Kong L, Zhang J, Ma C, Wu X, Li X, Jiang H. Essential oil composition and antibacterial activity of Lindera nacusua (D. Don) Merr. Nat Prod Res. 2016 Dec;30(23):2704-2706.

Selinene Review

Answer the following questions to test your knowledge of this terpene:

Question #1: What type of terpene is selinene?

a. Bitcyclic hesterterpene
b. Bricyclic sequioaterpene
c. Bicyclic sesquiterpene
d. Bicyclic monoterpene

Question #2: Which of the following best describes the primary flavor & aroma profile of selinene?

a. Depends on the isometer
b. Depends on the isomer
c. Depends on the iso version
d. Fruity, citrus, spice

Question #3: How common is selinene in cannabis?

a. Top 10
b. Top 20
c. Top 40
d. Top 50

Question #4: Name 4 isomers of selinene:

1 ______________________ 2 ______________________

3 ______________________ 4 ______________________

Question #5: Describe the primary known biological use of selinene in plants:

____

Question #6: Name two potential medical uses of selinene:

1 ______________________ 2 ______________________

For the answer key to all individual compound chapters please go here: http://thebigbookofterps.com/bbot-quiz-answer-key/

EPIZONARENE

Type: Bicyclic sesquiterpene
Chemical Formula: $C_{15}H_{24}$
Molecular Weight: 204.35 g/mol
Boiling Point: Unknown
Flash Point: Unknown
Solubility: Soluble in water
Biological Role: Unknown
Therapeutic Role: Unknown
Commercial Role: Unknown
Occurrence in Cannabis: Unknown, probably Top 50
Aroma/Flavor: Unknown

Occurs in Cannabis Strains: At the time of publication of this text, there were no publicly available COAs or other testing data for commercial cannabis varieties of the drug type that contained epizonarene. However, epizonarene has been quantified in several varieties of hemp, including Futura, Carmagnola, Eletta Campana, and Fibranova varieties[1265].

INTRODUCTION

Epizonarene rarely appears in published scientific and medical literature. Therefore, many details normally included herein are absent for this compound. For instance, there are no known aroma or flavor descriptors, no known biological roles, and nearly no potential medical or therapeutic use information available. The author has set up alerts for scientific research on this sesquiterpene and will update this chapter in a subsequent edition.

CHEMICAL STRUCTURE

[1265] Pieracci Y, Ascrizzi R, Terreni V, Pistelli L, Flamini G, Bassolino L, Fulvio F, Montanari M, Paris R. Essential Oil of Cannabis sativa L: Comparison of Yield and Chemical Composition of 11 Hemp Genotypes. Molecules. 2021 Jul 3;26(13):4080.

Epizonarene is a bicyclic sesquiterpene containing 15 carbon atoms and 24 hydrogen atoms notated as $C_{15}H_{24}$. With a molecular skeleton containing 2 rings with 2 sets of endocyclic double bonds, this compound is substituted at positions 2 and 5 by methyl groups, and at position 8 by an isopropyl group.

OCCURRENCE IN PLANTS

Epizonarene is known to occur in allspice and ginger, and has also been found in the leaf oil of Cupressus cashmeriana[1266] (Kashmir cypress), the leaves of Opisthopappus taihangensis[1267] (Taihang Mingzhu), and as a trace compound in the essential oil of Cymbopogon flexuosus[1268] (cochin grass).

BIOLOGICAL ACTIVITY IN PLANTS

Unknown.

USES IN INDUSTRY

Unknown.

POTENTIAL USES IN MEDICINE

Only one published study could be found on the potential medical or therapeutic uses of epizonarene, which suggests multiple uses including anticancer, antioxidant, and antimicrobial. As a major compound in the essential oil made from the leaves of Stachys viticina Boiss, epizonarene likely contributed to a number of effects, including; antimicrobial activity against methicillin-resistant Staphylococcus aureus (MRSA), Escherichia coli and Epidermophyton floccosum; cytotoxicity and 95% inhibition rates against HeLa (cervical adenocarcinoma) cells and 90% inhibition against Colo-205 (colon) cancer cell lines, and; high cyclooxygenase inhibitory activity against COX-1 and COX-2[1269].

IMPLICATIONS FOR HUMAN HEALTH & NUTRITION

Obtaining epizonarene via a healthy diet is unlikely, but not impossible. Foods or teas that contain allspice or ginger may feature this compound, while essential oils made from allspice, ginger, Kashmir cypress, cochin grass, or Taihang Mingzhu may also contain epizonarene.

Cannabis oils or other products made from the hemp varieties Futura, Carmagnola, Eletta Campana, and Fibranova varieties may also contain epizonarene.

As always, extreme care should be taken when working with any type of essential oil, as these complex mixtures (and pure compounds) can be toxic in some situations.

[1266] Su YC, Hsu KP, Hua KF, Ho CL. Composition, in vitro Anti-inflammatory, Antioxidant and Antimicrobial Activities of Essential Oils from Leaf and Twig Parts of Cupressus cashmeriana. Nat Prod Commun. 2015 Aug;10(8):1461-4.

[1267] Liu, Hua, Wendan Chen, Yuhong Chai, Wenchao Liu, Haixia Chen, Lei Sun, Xiaowei Tang, Chang Luo, Dongliang Chen, Xi Cheng, Fengjun Wang, Xiaohuan Yuan, Conglin Huang, and . "Terpenoids and their gene regulatory networks in Opisthopappus taihangensis 'Taihang Mingzhu' as detected by transcriptome and metabolome analyses." Frontiers in Plant Science (2022).

[1268] Pathania, D., Sharma, M., Thakur, P. et al. Exploring phytochemical composition, photocatalytic, antibacterial, and antifungal efficacies of Au NPs supported by Cymbopogon flexuosus essential oil. Sci Rep 12, 14249 (2022).

[1269] Jaradat N, Al-Maharik N. Fingerprinting, Antimicrobial, Antioxidant, Anticancer, Cyclooxygenase and Metabolic Enzymes Inhibitory Characteristic Evaluations of Stachys viticina Boiss. Essential Oil. Molecules. 2019 Oct 28;24(21):3880.

Epizonarene Review

Answer the following questions to test your knowledge of this terpene:

Question #1: What type of terpene is epizonarene?

a. Bicyclic sesquiterpene
b. Bricyclic sequioaterpene
c. Bitcyclic hesterterpene
d. Bicyclic monoterpene

Question #2: What is the chemical formula for epizonarene?

a. C_5H_{24}
b. Depends on the isomer
c. $C_{15}H_{24}$
d. $C_{24}H_{15}$

Question #3: How common is epizonarene in cannabis?

a. Top 10
b. Top 20
c. Top 40
d. Top 50

Question #4: What is the position of the two double bonds in epizonarene?

a. Exocyclic
b. Extencyclic
c. Endogenous
d. Endocyclic

Question #5: Name the two common plants/foods that epizonarene is found in:

1 ______________________________ 2 ______________________________

Question #6: Name two hemp varieties that are known to produce epizonarene:

1 ______________________________ 2 ______________________________

For the answer key to all individual compound chapters please go here: http://thebigbookofterps.com/bbot-quiz-answer-key/

EUDESMOL

Type: Bicyclic sesquiterpenoid
Chemical Formula: $C_{15}H_{26}O$
Molecular Weight: 222.37 g/mol
Boiling Point: *Alpha-eudesmol:* 299.00 to 302.00 °C. @ 760.00 mm Hg[1270]
Beta-eudesmol: 301.00 to 302.00 °C. @ 760.00 mm Hg[1271]
Gamma-eudesmol: 301.00 to 302.00 °C. @ 760.00 mm Hg[1272]
Flash Point: *All isomers:* 227.00 °F. TCC (108.33 °C.)
Melting Point: *Alpha-eudesmol:* 81.00 to 82.00 °C. @ 760.00 mm Hg
Beta-eudesmol: 72.00 to 74.00 °C. @ 760.00 mm Hg
Solubility: *Alpha-eudesmol:* Soluble in alcohol, insoluble in water
Beta-eudesmol: Soluble in alcohol, slightly soluble in water
Gamma-eudesmol: Slightly soluble in water, insoluble in paraffin oil
Biological Role: Antifungal, signaling molecule, insecticidal agent
Therapeutic Role: Anticancer, anti-angiogenesis, anti-nociceptive
Commercial Role: Cosmetic ingredient
Occurrence in Cannabis: Unknown, probably top 50
Aroma/Flavor: *Alpha:* sweet, woody, yuzu
Beta: woody, green, clove, turpentine
Gamma: waxy, sweet, yuzu

Occurs in Cannabis Varieties: According to Ascendancy Strains, eudesmol is found in Haze 20, Moby Dick CBD, Purple Orange CBD, and Critical 20[1273], however, no COA is provided on the company website. Otherwise, no other publicly available information or commercial testing laboratory analyses were found describing cannabis varieties that contain this sesquiterpene.

[1270] The Good Scents Company Data Sheet for Alpha-eudesmol, from: http://www.thegoodscentscompany.com/data/rw1023401.html. Accessed September 11, 2023.

[1271] The Good Scents Company Data Sheet for Beta-eudesmol, from: http://www.thegoodscentscompany.com/data/rw1054041.html. Accessed September 11, 2023.

[1272] The Good Scents Company Data Sheet for Gamma-eudesmol, from: http://www.thegoodscentscompany.com/data/rw1104361.html. Accessed September 11, 2023.

[1273] From: https://ascendancystrains.com/eudesmol-cannabis-terpene-profile. Accessed July 4, 2023.

Eudesmol has been found to be associated with CBD dominant cannabis varieties so reliably that the compound can be used as a chemotypic marker of CBD strains[1274]. In 2005, Hillig quantified the compound in wild wide-leaflet drug biotypes so consistently that he suggested a taxonomic revision of cannabis based on his findings of elevated levels of guaiol and isomers of eudesmol[1275], findings which were echoed 18 years later when researchers found a positive correlation between gamma- and beta-eudesmol and "indica" labelled varieties[1276]. These positive correlations were also found with the acid version of cannabidiol, CBDA, with researchers writing that gamma-eudesmol "showed absolute and unique correlation with CBD"[1277], while within the last year scientists have found beta- and gamma-eudesmol in the essential oil from an all-female, clonal transplant production system of a CBD variety[1278].

INTRODUCTION

Eudesmol is a bicyclic sesquiterpenoid that occurs as multiple isomers, each with a unique aroma and flavor profile. This compound is well-represented in the scientific literature, with many potential uses in the treatment of a wide variety of diseases and conditions. Readers are advised that beta-eudesmol – and possibly other isomers – are moderately toxic and may cause skin irritation and inflammation.

Interestingly, beta-eudesmol may affect the appetite of mammals that consume plants or other food sources containing this compound. Scientists seeking to understand why humans prefer to consume foods containing TRPA1 (transient receptor potential ankyrin 1) channel activators – which normally is activated by noxious substances – discovered that beta-eudesmol activates TRPA1 in rats, but instead of a noxious effect, an increase in appetite and food intake was elicited, which researchers attributed to a significant increase in gastric vagal nerve activity and elevated plasma ghrelin levels[1279] (a hormone that increases significantly immediately prior to eating food).

Researchers in Taiwan found that the essential oil of Taiwania cryptomerioides – a valuable raw wood material – continued to emit volatile organic chemicals including beta-eudesmol long after installation, with the chemical blend from the wood eliciting feelings of relaxation, alleviating negative feelings, and lowering the heart rate of study participants[1280]. Ongoing emissions or off-gassing of natural products used in construction is a serious but understudied issue.

[1274] Jin D, Henry P, Shan J, Chen J. Identification of Chemotypic Markers in Three Chemotype Categories of Cannabis Using Secondary Metabolites Profiled in Inflorescences, Leaves, Stem Bark, and Roots. Front Plant Sci. 2021 Jul 1;12:699530.

[1275] Hillig K. W. (2005a). A Systematic Investigation of Cannabis. Ph.D. thesis. Bloomington, IN: Indiana University.

[1276] Watts S, McElroy M, Migicovsky Z, Maassen H, van Velzen R, Myles S. Cannabis labelling is associated with genetic variation in terpene synthase genes. Nat Plants. 2021 Oct;7(10):1330-1334.

[1277] Namdar D, Voet H, Ajjampura V, Nadarajan S, Mayzlish-Gati E, Mazuz M, Shalev N, Koltai H. Terpenoids and Phytocannabinoids Co-Produced in Cannabis Sativa Strains Show Specific Interaction for Cell Cytotoxic Activity. Molecules. 2019; 24(17):3031.

[1278] Zheljazkov VD, Noller JS, Maggi F, Dale R. Terpenes and Cannabinoids Yields and Profile from Direct-Seeded and Transplanted CBD-Cannabis sativa. J Agric Food Chem. 2022 Aug 31;70(34):10417-10428.

[1279] Ohara K, Fukuda T, Ishida Y, Takahashi C, Ohya R, Katayama M, Uchida K, Tominaga M, Nagai K. β-Eudesmol, an oxygenized sesquiterpene, stimulates appetite via TRPA1 and the autonomic nervous system. Sci Rep. 2017 Nov 17;7(1):15785.

[1280] Yu, CP., Weng, WC., Ramanpong, J. et al. Physiological and psychological responses to olfactory simulation by Taiwania (Taiwania cryptomerioides) essential oil and the influence of cognitive bias. J Wood Sci 68, 12 (2022).

CHEMICAL STRUCTURE

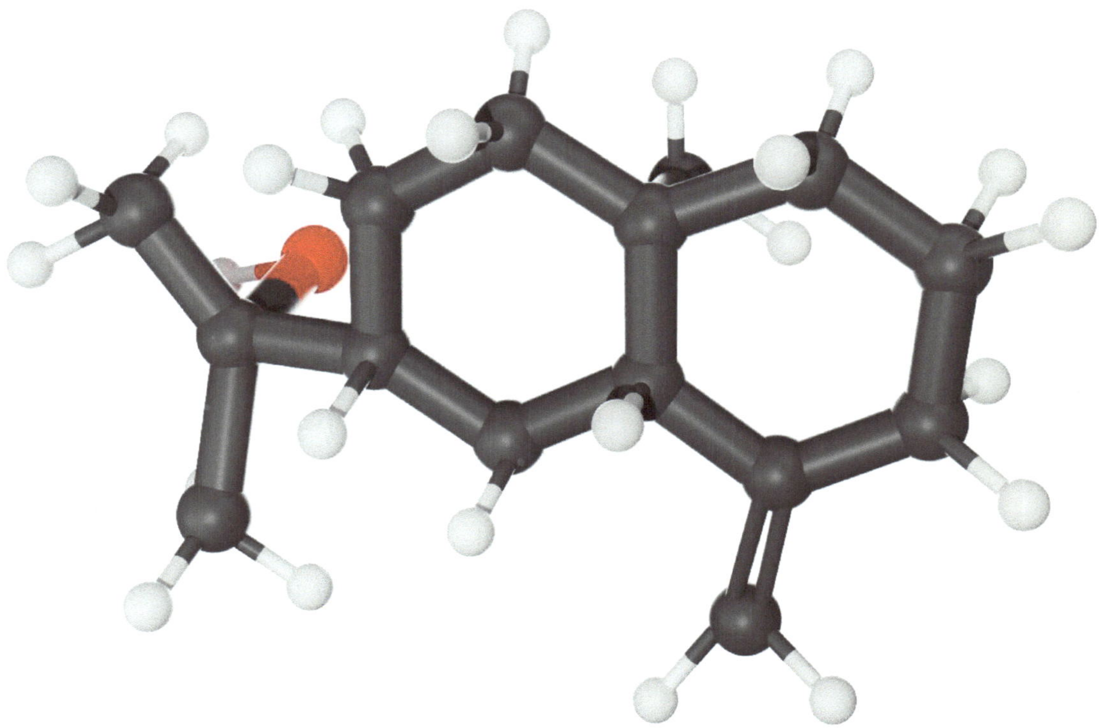

Eudesmol is biosynthesized in the mevalonate acid (MVA) pathway via farnesyl diphosphate as the derivative compound. Based on three units of isoprene, eudesmol contains 15 carbon atoms, 26 hydrogen atoms, and 1 oxygen atom notated as $C_{15}H_{26}O$. The eudesmol molecule features two rings, which categorizes this compound as bicyclic.

Eudesmol occurs as several isomers, with the alpha version containing an endocyclic double bond between C-3 and C-4 and a hydroxy substituent at C-11, while beta-eudesmol features an exocyclic double bond, with a substitute at position 2 of 2-hydroxypropan-2-yl, a methyl substituent at position 4a, and methylidene at position 8. Gamma-eudesmol is similar to the alpha isomer, also featuring a hydroxy substituent at C-11 and an endocyclic double bond, but between C-4 and C-5.

OCCURRENCE IN PLANTS

Here we list the known occurrences in plants of 3 isomers of eudesmol, although other isomers are known to occur in some plants, and many produce racemic mixtures that often contain alpha-, beta-, and gamma-eudesmol:

Alpha:

Black pepper
Camel grass
Helichrysum italicum
Laurel leaf
Nepeta betonicifolia
Pimenta
Sage
Bombycilaena erecta
Celery seed
Hinoki
Mauka species
Pepper tree
Ravintsara
Water mint
Callitris species
Eucalyptus
Juniperus virginiana
Microstrobos fitzgeraldii
Phoebe
Rue oil
Witch hazel

Beta:

Amyris wood
Angelica
Basil
Bay laurel
Burdock
Callitris species
Camel grass
Cascarilla
Celery seed
Cistus oil
Citronella
Clary sage
Curry
Dill
Eucalyptus
Ferula cupularis
Geranium
Ginger
Guava
Helichrysum italicum
Hinoki
Ho leaf
Hyssop
Labdanum
Manuka species
Mayur pankh
Microstrobos fitzgeraldii
Oregano
Otanthus maritimus
Pepper tree
Phoebe
Pimenta
Piper cernum vell.
Ravintsara
Rue oil
Sage
Salvia
Tanacetum annum l.
Tansy
Valerian root
Walnut (English)
Water mint
Yuzu

Gamma:

Cajuput
Callitris species
Camel grass
Celery seed
Cistus
Eucalyptus species
Ginger
Ginko biloba
Guava
Hinoki
Laurel leaf
Lime
Mango
Manuka species
Pepper tree
Phoebe
Pineapple
Piper cernum vell.
Ravintsara
Rosemary
Rue oil
Salvia
Tagete
Thyme
Witch hazel

BIOLOGICAL ACTIVITY IN PLANTS

The primary biological activity of eudesmol in plants is likely as an antimicrobial agent and insecticide. Cannabis cultivators might be interested to know that low pH has been shown to increase production of beta-eudesmol in Atractylodes lancea, a traditional Chinese medicinal plant, where the sesquiterpenoid is the primary constituent. The various isomers of eudesmol also appear to serve multiple other roles in plants as detailed below:

Antimicrobial Agent: As primary constituents in the essential oil of Salvia microphylla (baby sage), beta-eudesmol and gamma-eudesmol showed significant bactericidal activity against multiple organisms[1281], and as the primary constituents in the essential oil of Guatteriopsis friesiana, the same compounds exhibited significant antimicrobial activity against 11 different organisms[1282].

Antifungal Agent: As the primary constituent in the essential oil made from the aerial parts of Warionia saharae (known locally as afessas), beta-eudesmol significantly inhibited the mycelial growth of 3 phytopathogenic strains

[1281] Lima, Rafaela K., Maria das Graças Cardoso, Milene A. Andrade, Paula L. Guimarães, Luís R. Batista, and David L. Nelson. "Bactericidal and Antioxidant Activity of Essential Oils from Myristica fragrans Houtt and Salvia microphylla H.B.K." Journal of the American Cil Chemists' Society 89.3 (2012): 523-528.

[1282] Costa EV, Teixeira SD, Marques FA, Duarte MC, Delarmelina C, Pinheiro ML, Trigo JR, Sales Maia BH. Chemical composition and antimicrobial activity of the essential oils of the Amazon Guatteriopsis species. Phytochemistry. 2008 Jun;69(9):1895-9.

of bacteria, with fungal spore production completely inhibited in Alternaria sp., Penicillium expansum, and Rhizopus stolonifer[1283]. As major components in the essential oil made from the leaves of Sabina chinensis (blue point juniper), alpha-eudesmol and gamma-eudesmol contributed to high and long-term antifungal activity against Fusarium oxysporum and Fusarium incarnatum, reaching approximately 76% and 90% growth inhibition respectively[1284].

Signaling Molecule: As a primary constituent in a hexane extract made from Eucalyptus maculate, beta-eudesmol contributed to the interference of recognition mechanisms in Atta sexdens rubropilosa Forel (a type of ant) workers, with researchers theorizing that this may explain the resistance of the plant species to ant attack or infestation[1285]. Other work has shown that the same species of ant becomes aggressive among nestmates when in the presence of beta-eudesmol as extracted from the leaves of Eucalyptus maculate, with aggression attributed to interference with ant-to-ant recognition[1286].

Insecticidal Agent: As the primary constituents in the essential oil made from the leaves and shoots of Drimys winteri (also known as canelo, an evergreen tree or shrub), gamma-, beta-, and alpha-eudesmol contributed to a 94% mortality rate in Acanthoscelides obtectus (bean weevil, an important post-harvest pest of beans), as well as repellency effects against Aegorhinus superciliosus[1287]. As a major compound in the essential oil made from the leaves of Eucalyptus torquata, beta-eudesmol contributed to moderate insecticidal activities against Ceratitis capitata medflies (Mediterranean fruit flies)[1288], while alpha-, beta-, and gamma-eudesmol as the major compounds of the essential oil made from the heartwood of Callitris glaucophylla (white cypress pine) contributed to repellency against Coptotermes formosanus (Formosan subterranean termite) workers[1289]. Finally, as the principal constituent in the essential oil made from the rhizomes of Rhynchanthus beesianus (a plant in the ginger family), beta-eudesmol contributed to fumigant and contact toxicity against Liposcelis entomophila (booklouse) and Tribolium castaneum (red flour beetle)[1290].

USES IN INDUSTRY

Industrial uses of eudesmol are relatively unknown, although some research has been carried out that seeks to use this compound in cosmetics for its antioxidant, anti-inflammatory, and cell preservation effects[1291]. Commercial preparations of eudesmol are typically for research purposes only, appearing as a pale beige powder or colorless to pale yellow oily liquid, often isolated from the rhizomes of Atractylodes lancea. Research has also been carried out

---

[1283] Znini, M., Cristofari, G., Majidi, L. et al. In vitro antifungal activity and chemical composition of Warionia saharae essential oil against 3 apple phytopathogenic fungi. Food Sci Biotechnol 22 (Suppl 1), 113–119 (2013).

[1284] Zhang J, Zhao Z, Liang W, Bi J, Zheng Y, Gu X, Fang H. Essential oil from Sabina chinensis leaves: A promising green control agent against Fusarium sp. Front Plant Sci. 2022 Nov 11;13:1006303.

[1285] Marsaro, A.L., Souza, R.C., Della Lucia, T.M.C. et al. Behavioral Changes in Workers of the Leaf-Cutting Ant Atta sexdens rubropilosa Induced by Chemical Components of Eucalyptus maculata Leaves. J Chem Ecol 30, 1771–1780 (2004).

[1286] Marinho CG, Della Lucia TM, Ribeiro MM, Magalhães ST, Guedes RN, Jham GN. Interference of beta-eudesmol in nestmate recognition in Atta sexdens rubropilosa (Hymenoptera: Formicidae). Bull Entomol Res. 2008 Oct;98(5):467-73.

[1287] Tampe J, Espinoza J, Chacón-Fuentes M, Quiroz A, Rubilar M. Evaluation of Drimys winteri (Canelo) Essential Oil as Insecticide against Acanthoscelides obtectus (Coleoptera: Bruchidae) and Aegorhinus superciliosus (Coleoptera: Curculionidae). Insects. 2020 May 30;11(6):335.

[1288] Dhaouadi, F., Bargougui, A., Maamer, S. et al. Chemical composition and insecticidal activity of two Eucalyptus essential oils against the Mediterranean fruit fly, Ceratitis capitata (Wiedemann) (Diptera: Tephritidae). J Plant Dis Prot 130, 483–493 (2023).

[1289] Watanabe, Y., Mihara, R., Mitsunaga, T. et al. Termite repellent sesquiterpenoids from Callitris glaucophylla heartwood. J Wood Sci 51, 514–519 (2005).

[1290] Pan X, Xiao H, Hu X, Liu ZL. Insecticidal activities of the essential oil of Rhynchanthus beesianus rhizomes and its constituents against two species of grain storage insects. Z Naturforsch C J Biosci. 2022 Sep 22;78(1-2):83-89.

[1291] Kim, K.Y. Anti-inflammatory and ECM gene expression modulations of β-eudesmol via NF-κB signaling pathway in normal human dermal fibroblasts. biomed dermatol 2, 3 (2018).

that attempts to produce beta-eudesmol using engineered Escherichia coli bacteria[1292], as well as work that seeks to produce beta-eudesmol in-vitro in the roots of Atractylodes chinensis[1293], and research that attempted the artificial inducement of agarwood resin formation – where gamma-eudesmol is the principal constituent – in Gyrinops walla (a tree native to Sri Lanka) by exposing the plant to Aspergillus niger and Fusarium solani fungi[1294].

POTENTIAL USES IN MEDICINE

Eudesmol offers significant potential in the treatment of a variety of cancers, particularly bile duct cancer, where much of the research into this compound has been focused, and the anti-angiogenesis properties of this sesquiterpene adds potently to these anticancer effects. This compound may also be useful in the treatment of tuberculosis, migraine, brain injury, pancreatitis, and even treatment for COVID-19, among other medical and therapeutic potential as detailed below:

**Anticancer Agent (breast, lung, prostate)*: As the primary constituents in the essential oil made from the leaves of Anaxagorea brevipes (an evergreen shrub or tree, native to South America), beta-eudesmol, alpha-eudesmol, and gamma-eudesmol exhibited significant antiproliferative effects against MCF-7 (breast cancer), NCI-H460 (lung cancer), and PC-3 (prostate cancer) cell lines[1295]. As the primary constituents in the essential oil made from the leaves of Guatteria friesiana, the same compounds displayed cytotoxicity against HCT-8 (intestinal adenocarcinoma), HL-60 (leukemia), and MDA-MB-435 (breast cancer) cell lines in mice[1296].

****Treatment for Breast Cancer (estrogen-receptor negative tumors):*** As major components in the essential oil made from the leaves of Blepharocalyx salicifolius (known locally in Brazil as "murta"), gamma-eudesmol and alpha-eudesmol contributed to impairment of cellular metabolism and other cytotoxic effects against the MDA-MB-231 human breast cancer cell line[1297].

****Treatment for Cervical Cancer:*** As the primary constituent in the essential oil made from the aerial parts of Chenopodium botrys (Jerusalem-oak), alpha-eudesmol inhibited the growth of HeLa (cervical cancer) cells by inducing cell cycle arrest via the p53/p21 pathway, increasing the sub-G1 DNA content, and inducing apoptosis[1298].

****Treatment for Cholangiocarcinoma (Bile Duct Cancer)***: Much of the scientific research regarding eudesmol has focused on use of this compound in the prevention and treatment of cholangiocarcinoma. As a pure compound derived from Atractylodes lancea, beta-eudesmol moderately inhibited HuCCT-1 (human cholangiocellular carcinoma) cell growth, significantly downregulated gene expression levels, and downregulated the expression of mTOR gene and protein expression levels, with the result being the downregulation of the Notch signaling pathway[1299], which is implicated in the progression of cholangiocarcinoma. Beta-eudesmol has also been shown to

---

1292 Yu F, Harada H, Yamasaki K, Okamoto S, Hirase S, Tanaka Y, Misawa N, Utsumi R. Isolation and functional characterization of a beta-eudesmol synthase, a new sesquiterpene synthase from Zingiber zerumbet Smith. FEBS Lett. 2008 Mar 5;582(5):565-72.

1293 Yun C, Zhao Z, Gu L, Zhang Z, Wang S, Shi Y, Miao N, Ri I, Wang W, Wang H. In vitro production of atractylon and β-eudesmol from Atractylodes chinensis by adventitious root culture. Appl Microbiol Biotechnol. 2022 Nov;106(21):7027-7037.

1294 Subasinghe, S.M.C.U.P., Hitihamu, H.I.D. & Fernando, K.M.E.P. Use of two fungal species to induce agarwood resin formation in Gyrinops walla. J. For. Res. 30, 721–726 (2019).

1295 de Alencar DC, Pinheiro ML, Pereira JL, de Carvalho JE, Campos FR, Serain AF, Tirico RB, Hernández-Tasco AJ, Costa EV, Salvador MJ. Chemical composition of the essential oil from the leaves of Anaxagorea brevipes (Annonaceae) and evaluation of its bioactivity. Nat Prod Res. 2016;30(9):1088-92.

1296 Britto AC, de Oliveira AC, Henriques RM, Cardoso GM, Bomfim DS, Carvalho AA, Moraes MO, Pessoa C, Pinheiro ML, Costa EV, Bezerra DP. In vitro and in vivo antitumor effects of the essential oil from the leaves of Guatteria friesiana. Planta Med. 2012 Mar;78(5):409-14.

1297 Furtado FB, Borges BC, Teixeira TL, Garces HG, Almeida Junior LD, Alves FCB, Silva CVD, Fernandes Junior A. Chemical Composition and Bioactivity of Essential Oil from Blepharocalyx salicifolius. Int J Mol Sci. 2018 Jan 4;19(1):33.

1298 Rezaieseresht H, Shobeiri SS, Kaskani A. Chenopodium Botrys Essential Oil as A Source of Sesquiterpenes to Induce Apoptosis and G1 Cell Cycle Arrest in Cervical Cancer Cells. Iran J Pharm Res. 2020 Spring;19(2):341-351.

1299 Vanaroj P, Chaijaroenkul W, Na-Bangchang K. Atractylodin and β-eudesmol from Atractylodes lancea (Thunb.) DC. Inhibit Cholangiocarcinoma Cell

exhibit potent antiproliferative activity and to significantly reduce cholangiocarcinoma cell migration and invasion, while also significantly up-regulating the expression of epithelial marker E-cadherin and down-regulating the expression of mesenchymal markers-vimentin and snail-1, and inhibiting PI3K and AKT phosphorylation while activating p38MAPK activity[1300]. Beta-eudesmol has also been shown to help sensitize cholangiocarcinoma cells to chemotherapeutic agents[1301], although in the last year researchers issued a warning against clinical use of beta-eudesmol in cholangiocarcinoma, stating that; "Results revealed some adverse effects on the functions of the Na/K-ATPase-related genes, which could subsequently affect the functioning of the heart. Similar adverse effects were also observed in the liver, whereby the biomarkers for liver toxicity were triggered."[1302]

Treatment for Hepatocellular Carcinoma (liver cancer): As pure compounds, alpha-, beta-, and gamma-eudesmol exhibited significant cytotoxicity against HepG2 (liver cancer) cells, reducing cell proliferation and inducing tumor cell death by caspase-mediated apoptosis pathways, however, the sesquiterpenes were also cytotoxic to normal cells[1303].

Anti-angiogenesis Agent: As a pure compound, beta-eudesmol as extracted from the rhizome of Atractylodes lancea inhibited vessel development at higher concentrations, significantly downregulated the expression of the Vegfaa gene and its receptor Vegfr2, and interfered with the expression of Vegfaa protein[1304]. Other research has found that beta-eudesmol inhibits angiogenesis by blocking the ERK signaling pathway and inhibiting HUVEC migration and tube formation, with the compound inhibiting angiogenesis both in-vitro and in-vivo[1305].

Antinociceptive Agent: As the primary constituents in the essential oil made from the leaves of Guatteria friesiana (a small South American tree called envira), beta-, alpha-, and gamma-eudesmol, along with a minor compound that appeared as a new alpha-eudesmol derivative, 5-hydroxy-α-eudesmol, the eudesmol mixture contributed to significant antinociceptive properties in male Swiss albino mice[1306].

Anti-tuberculosis Agent: Researchers working on an inhalation therapy for tuberculosis found that the essential oil of Eucalyptus citriodora – where beta-eudesmol was a major compound – exhibited airborne anti-tuberculosis effects[1307].

Anti-inflammatory Agent: As a major component in Agarwood oil, a seemingly rare isomer of eudesmol, 10-epi-γ-Eudesmol, contributed to the significant reduction in skin thickness, ear weight, oxidative stress, and pro-

---

Proliferation by Downregulating the Notch Signaling Pathway. Asian Pac J Cancer Prev. 2023 Feb 1;24(2):551-558.

[1300] Acharya B, Chajaroenkul W, Na-Bangchang K. β-Eudesmol Inhibits the Migration of Cholangiocarcinoma Cells by Suppressing Epithelial-Mesenchymal Transition via PI3K/AKT and p38MAPK Modulation. Asian Pac J Cancer Prev. 2022 Aug 1;23(8):2573-2581.

[1301] Srijiwangsa, P., Ponnikorn, S. & Na-Bangchang, K. Effect of β-Eudesmol on NQO1 suppression-enhanced sensitivity of cholangiocarcinoma cells to chemotherapeutic agents. BMC Pharmacol Toxicol 19, 32 (2018).

[1302] Tshering G, Pimtong W, Plengsuriyakarn T, Na-Bangchang K. Effects of β-eudesmol and atractylodin on target genes and hormone related to cardiotoxicity, hepatotoxicity, and endocrine disruption in developing zebrafish embryos. Sci Prog. 2022 Oct-Dec;105(4):368504221137458.

[1303] Bomfim DS, Ferraz RP, Carvalho NC, Soares MB, Pinheiro ML, Costa EV, Bezerra DP. Eudesmol isomers induce caspase-mediated apoptosis in human hepatocellular carcinoma HepG2 cells. Basic Clin Pharmacol Toxicol. 2013 Nov;113(5):300-6.

[1304] Tshering G, Pimtong W, Plengsuriyakarn T, Na-Bangchang K. Anti-angiogenic effects of beta-eudesmol and atractylodin in developing zebrafish embryos. Comp Biochem Physiol C Toxicol Pharmacol. 2021 May;243:108980.

[1305] Tsuneki H, Ma EL, Kobayashi S, Sekizaki N, Maekawa K, Sasaoka T, Wang MW, Kimura I. Antiangiogenic activity of beta-eudesmol in vitro and in vivo. Eur J Pharmacol. 2005 Apr 11;512(2-3):105-15.

[1306] Costa, Emmanoel V., Leociley R. Alencar Menezes, Lívia M. Dutra, Maria Lúcia B. Pinheiro, Érica M. Lavor, Mariana G. Silva, Cristiane dos S. C. Alves, Jackson R. G. S. Almeida, Felipe Moura A. da Silva, Hector H. F. Koolen, and Andersson Barison. "A novel eudesmol derivative from the leaf essential oil of Guatteria friesiana (Annonaceae) and evaluation of the antinociceptive activity." Zeitschrift für Naturforschung C: A Journal of Biosciences 78.5-6 (2023): 169-177.

[1307] Ramos Alvarenga RF, Wan B, Inui T, Franzblau SG, Pauli GF, Jaki BU. Airborne antituberculosis activity of Eucalyptus citriodora essential oil. J Nat Prod. 2014 Mar 28;77(3):603-10.

inflammatory cytokines production in a 12-O-tetradecanoylphorobol-13-acetate-induced mouse ear inflammation model[1308].

**Treatment for Brain Disorders:* As a primary constituent in the essential oil made from the leaves of Eucalyptus citriodora, alpha-eudesmol potently inhibited the presynaptic omega-agatoxin IVA-sensitive (P/Q-type) Ca(2+) channel after electrical stimulation in rats with neurogenic inflammation[1309]. Other research has shown that alpha-eudesmol inhibited glutamate release from brain synaptosomes, significantly attenuated the post-ischemic increase in brain water content, significantly reduced the size of the infarct area, and inhibited the elevation of the extracellular concentration of glutamate during focal ischemia in rats[1310].

**Treatment for Pancreatitis:* As major components in the essential oil of Magnoliae officinalis (magnolia), several eudesmols were shown to have high binding activity to select acute pancreatitis therapeutic targets, where alpha-eudesmol was associated with TNF (tumor necrosis factor), gamma-eudesmol was associated with IL-1B (interleukin 1-beta), and beta-eudesmol was associated with ESR1 (Estrogen Receptor Alpha Gene)[1311].

**Microbiostatic Agent:* Microbiostatic agents inhibit the growth or ability of an organism to reproduce. As primary constituents in the essential oil of Algerian propolis, beta- and alpha-eudesmol contributed – along with cedrol – to bacteriostatic effects against Escherichia coli and Pseudomonas aeruginosa, and fungistatic effects against Candida albicans[1312], three common pathogens that can cause infections in humans.

**Treatment/Preventative for COVID-19:* As major components in the essential oil made from the leaves of Melaleuca cajuputi (cajuput), alpha- and gamma-eudesmol contributed to excellent inhibition of ACE2 (Angiotensin-Converting Enzyme 2) and PDB6LU7 (primary protease) proteins, with researchers speculating that; "Melaleuca cajuputi essential oil is considered as a valuable resource for preventing SARS-CoV-2 invasion into [the] human body."[1313]

**Antimicrobial Agent:* As primary constituents in the hydro distillated leaf essential oil of Litsea kostermansii, beta-, gamma-, and alpha-eudesmol contributed to particularly strong biological activity against Bacillus cereus, Staphylococcus aureus, and Staphylococcus epidermidis[1314].

**Treatment for Gastrointestinal Disease:* As an isolated compound from an extract of Atractylodes lancea, beta-eudesmol stimulated small intestinal motility, inhibited reductions in gastric emptying and gastrointestinal motility induced by dopamine, and inhibited the atropine-induced decrease in small intestinal motility in mice[1315].

[1308] Yadav DK, Mudgal V, Agrawal J, Maurya AK, Bawankule DU, Chanotiya CS, Khan F, Thul ST. Molecular docking and ADME studies of natural compounds of Agarwood oil for topical anti-inflammatory activity. Curr Comput Aided Drug Des. 2013 Sep;9(3):360-70.

[1309] Batish DR, Singh HP, Setia N, Kaur S, Kohli RK. Chemical composition and phytotoxicity of volatile essential oil from intact and fallen leaves of Eucalyptus citriodora. Z Naturforsch C J Biosci. 2006 Jul-Aug;61(7-8):465-71.

[1310] Asakura K, Matsuo Y, Oshima T, Kihara T, Minagawa K, Araki Y, Kagawa K, Kanemasa T, Ninomiya M. omega-agatoxin IVA-sensitive Ca(2+) channel blocker, alpha-eudesmol, protects against brain injury after focal ischemia in rats. Eur J Pharmacol. 2000 Apr 7;394(1):57-65.

[1311] Li S, Huang Y, Liu L, Zhang F, Ao H, Luo Y. Mechanism of Magnolia Volatile Oil in the Treatment of Acute Pancreatitis Based on GC-MS, Network Pharmacology, and Molecular Docking. Evid Based Complement Alternat Med. 2023 Feb 7;2023:3503888.

[1312] Boulechfar S, Zellagui A, Asan-Ozusaglam M, Bensouici C, Erenler R, Yildiz İ, Tacer S, Boural H, Demirtas I. Chemical composition, antioxidant, and antimicrobial activities of two essential oils from Algerian propolis. Z Naturforsch C J Biosci. 2021 Aug 16;77(3-4):105-112.

[1313] My TTA, Loan HTP, Hai NTT, Hieu LT, Hoa TT, Thuy BTP, Quang DT, Triet NT, Anh TTV, Dieu NTX, Trung NT, Hue NV, Tat PV, Tung VT, Nhung NTA. Evaluation of the Inhibitory Activities of COVID-19 of Melaleuca cajuputi Oil Using Docking Simulation. ChemistrySelect. 2020 Jun 8;5(21):6312-6320.

[1314] Ho CL, Wang EI, Hsu KP, Lee PY, Su YC. Composition and antimicrobial activity of the leaf essential oil of Litsea kostermansii from Taiwan. Nat Prod Commun. 2009 Aug;4(8):1123-6.

[1315] Kimura Y, Sumiyoshi M. Effects of an Atractylodes lancea rhizome extract and a volatile component β-eudesmol on gastrointestinal motility in mice. J Ethnopharmacol. 2012 May 7;141(1):530-6.

Antiprotozoal Agent: As primary constituents in the essential oil made from the leaves of Eugenia stipitate, beta-eudesmol and gamma-eudesmol were shown to be highly selective against Leishmania braziliensis and L. infantum promastigotes,[1316] which could be useful in the prevention of diseases like Leishmaniasis.

IMPLICATIONS FOR HUMAN HEALTH & NUTRITION

To increase intake of eudesmol, consumption or ingestion of the spices, herbs, fruits, vegetables, teas, and essential oils listed below might be effective. When preparing, cooking, or serving food, consider working with herbs and spices such as angelica, basil, black pepper, cajuput, celery seed, clary sage, curry, dill, laurel leaf, oregano, sage, thyme, or water mint. Snacking on fruits such as guava, lime, mango, pineapple, or yuzu might increase dietary intake of eudesmol, and walnuts might help in this effort as well. Teas containing eucalyptus, phoebe, geranium, ginger, valerian root, or ginkgo biloba may contain isomers of eudesmol, and eating raw ginger may also be effective.

Essential oils derived from any of the plants listed in the "Occurrence in Plants" section may contain isomers of eudesmol or racemic mixtures. Use of any EO should be undertaken with great care as these complex mixtures can be toxic in some situations.

Cannabis plants that are high in CBD are likely to contain at least trace amounts of eudesmol, however, it's unclear if concentrations of this compound are sufficient in cannabis or cannabis products to elicit biological effects.

[1316] Dos Santos CRB, Sampaio MGV, Vandesmet LCS, Dos Santos BS, de Menezes SA, Portela BYM, Gomes DWR, Correia MTS, Gomez MCV, de Alencar Menezes IR, da Silva MV. Chemical composition and biological activities of the essential oil from Eugenia stipitata McVaugh leaves. Nat Prod Res. 2022 Dec 5:1-7.

Eudesmol Review

Answer the following questions to test your knowledge of this terpene:

Question #1: What type of terpene is eudesmol?

a. Bicyclic sesquiterpene
b. Linear sesquiterpene
c. Bicyclic monoterpene
d. Tricyclic sesquiterpene

Question #2: Which of the following best describes the primary flavor & aroma profile of alpha-eudesmol?

a. Spicy, turpentine, dry
b. Sweet, woody, yuzu
c. Depends on the isomer
d. Depends on pentanyl framing

Question #3: How common is eudesmol in cannabis?

a. Top three
b. Top ten
c. Top forty
d. Top fifty

Question #4: Which plant is the most common source of eudesmol?

a. Attractivicus hoti
b. Atractylodes lancea
c. Atractylodes cornae
d. Rhizome EO

Question #5: Describe the connection between eudesmol and CBD:

____

____

Question #6: Name two potential medical uses of eudesmol:

1 ______________________________________ 2 _____________________________________

For the answer key to all individual compound chapters please go here: http://thebigbookofterps.com/bbot-quiz-answer-key/

GURJUNENE

Type: *Alpha:* Tricyclic sesquiterpene
Beta: Tricyclic sesquiterpene
Gamma: Bicyclic sesquiterpene
Chemical Formula: $C_{15}H_{24}$
Molecular Weight: 204.356 g/mol
Boiling Point: *Alpha:* 262.00 to 263.00 °C. at 760.00 mm Hg
Beta: 257.00 to 258.00 °C. at 760.00 mm Hg
Gamma: 276.00 to 277.00 °C. at 760.00 mm Hg
Flash Point: *Alpha:* > 200.00 °F. TCC (> 93.33 °C.)
Beta: 223.00 °F. TCC (106.11 °C.)
Gamma: 228.00 °F. TCC (108.89 °C.)
Solubility: *Alpha:* soluble in alcohol, water
Beta: soluble in paraffin oil, water
Gamma: soluble in alcohol, water
Biological Role: Insecticidal agent
Therapeutic Role: Anticancer, sedative, antimicrobial
Commercial Role: Alpha-gurjunene; perfuming agent
Occurrence in Cannabis: Unknown, probably top 50
Aroma/Flavor: *Alpha:* woody, balsamic, earthy
Beta: no aroma or flavor descriptors found
Gamma: musty

Occurs in Cannabis Varieties: At the time of printing of this text, there were no publicly available COAs or other laboratory testing analyses or other data that showed gurjunene in commercially available samples of cannabis or cannabis products. However, alpha-gurjunene has been found in a mid-level THC-dominant chemovar[1317], a chemotype I high-THCA variety[1318], and in the hexane extract of a high-CBD fiber type variety (Cannabis sativa var. fibrante)[1319]. Alpha-gurjunene has also been found in illicit Egyptian hash[1320], and in illicit hash confiscated by Italian authorities[1321].

INTRODUCTION

Originally isolated by Palmade et al. in 1963, Gurjunene is a sesquiterpene that occurs as three primary isomers – the alpha-, beta- and gamma- versions, although alpha-gurjunene appears much more frequently in plants than the other isomers. This compound has a woody, balsamic, and earthy aroma and flavor profile as the alpha isomer, and a musty profile as the gamma isomer.

[1317] Mudge EM, Brown PN, Murch SJ. The Terroir of Cannabis: Terpene Metabolomics as a Tool to Understand Cannabis sativa Selections. Planta Med. 2019 Jul;85(9-10):781-796.

[1318] Chacon FT, Raup-Konsavage WM, Vrana KE, Kellogg JJ. Secondary Terpenes in Cannabis sativa L.: Synthesis and Synergy. Biomedicines. 2022; 10(12):3142.

[1319] Muscarà C, Smeriglio A, Trombetta D, Mandalari G, La Camera E, Grassi G, Circosta C. Phytochemical characterization and biological properties of two standardized extracts from a non-psychotropic Cannabis sativa L. cannabidiol (CBD)-chemotype. Phytother Res. 2021 Sep;35(9):5269-5281.

[1320] Hassan, Saad S.M., Ayman H. Kamel, Nasser S. Awwad, Awaad H.A. Aboterika, and "Characterization of Some "Hashish" Samples in the Egyptian Illicit Trafficking Market Using a Thermal Separation Probe and Gas Chromatography–Mass Spectrometry." ACS Omega (2023).

[1321] Arnoldi, Sebastiano, Gabriella Roda, Eleonora Casagni, Lucia Dell Acqua, Michele Dei Cas, Fiorenza Fare, Chiara Rusconi, Giacomo Luca Visconti, Veniero Gambaro, and . "Characterization of the Volatile Components of Cannabis Preparations by Solid-Phase Microextraction Coupled to Headspace-Gas Chromatography with Mass Detector (SPME-HSGC/MS)." Journal of Chromatography & Separation Techniques (2017).

Interestingly, scientists studying volatile organic compounds in soil inadvertently discovered that the high sesquiterpene content of the environment above the forest floor – which included alpha-gurjunene as a major compound – was being emitted from plant roots and root-associated microbes, as opposed to leaf litter or emissions from inflorescences[1322]. Other intriguing research has shown that the fungi Aspergillus flavus produces alpha-gurjunene, with scientists finding an apparent correlation between the release of sesquiterpene volatile compounds and the initiation of aflatoxin (a fungal toxin) biosynthesis[1323].

CHEMICAL STRUCTURE

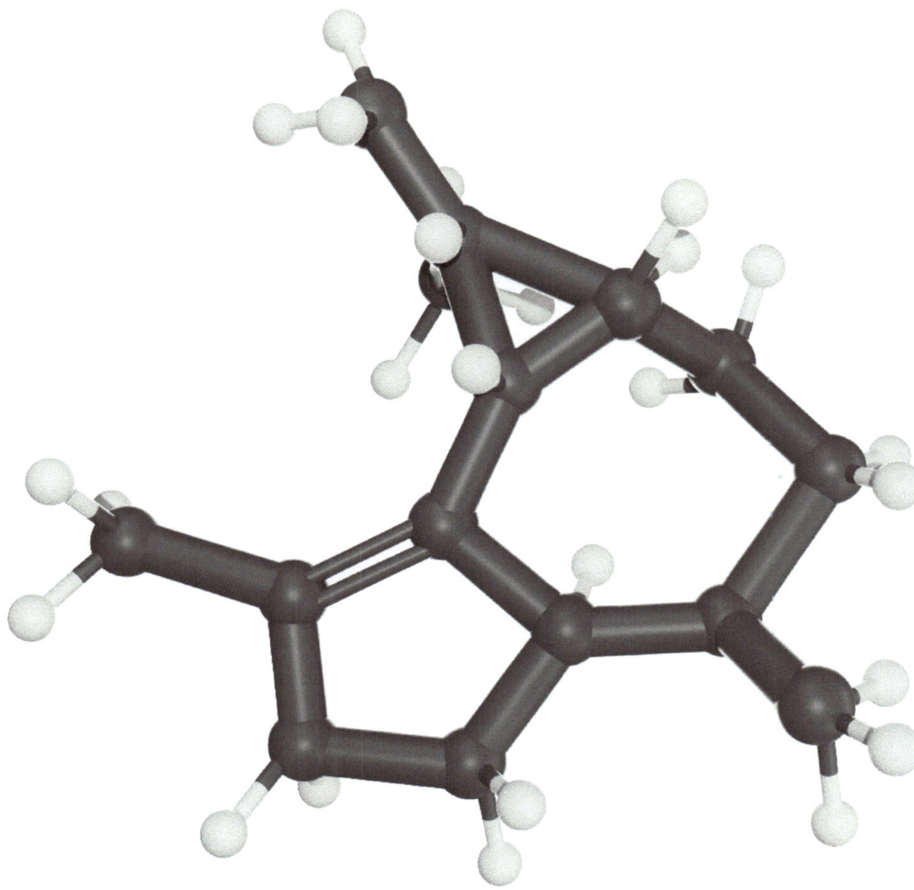

Gurjunene is a sesquiterpene produced from farnesyl diphosphate in the mevalonate pathway of plants. Containing 15 carbon atoms and 24 hydrogen atoms, gurjunene is noted as $C_{15}H_{24}$. Occurring as 3 isomers, each isomer features a molecular skeleton that has undergone various re-arrangements:

Alpha-gurjunene contains 3 rings, making it a tricyclic sesquiterpene, and features one endocyclic double bond. Alpha-gurjunene is substituted by methyl groups at positions 1, 4, and 7.

[1322] Mäki, M., Aaltonen, H., Heinonsalo, J. et al. Boreal forest soil is a significant and diverse source of volatile organic compounds. Plant Soil 441, 89–110 (2019).

[1323] Zeringue HJ, Bhatnagar D, Cleveland TE. C(15)H(24) Volatile Compounds Unique to Aflatoxigenic Strains of Aspergillus flavus. Appl Environ Microbiol. 1993 Jul;59(7):2264-70.

Beta-gurjunene contains 3 rings and features an exocyclic double bond, and is substituted by methyl groups at positions 1 and 4, and by a methylidene group at position 7.

Gamma-gurjunene contains 2 rings and is therefore classified as a bicyclic sesquiterpene. This compound contains both an exocyclic and an endocyclic double bond.

Interestingly, scientists have commented that nerolidol may be confused in some cases with alpha-gurjunene, with cannabis researchers in 2016 concluding that; "nerolidol may in fact be alpha-gurjunene.[1324]"

OCCURRENCE IN PLANTS

Alpha:

Allspice
Bay laurel
Cajuput
Eucalyptus
Horsemint
Labdanum
Parsley
Polyalthia australis
Sage
Spikenard
Zanthoxylum bracc.

Amyris wood
Bergamot mint
Carrot
Guava
Hyssop
Manuka scoparium
Pepper tree
Protium heptaphyllum
Salvia
Tea tree

Anaphalis nubigena
Black pepper
Croton flavens
Gurju balsam
Jatropha ribifolia
Laurel leaf
Peppermint
Safflower
Satureja viminea
Water mint

Beta:

Acorus calamus
Calamus
Clary sage
Eucalyptus
Labdanum
Patchouli
Plectranthus glandulosus
Spikenard
Valerian root

Baccharis coridifolia
Carrot seed
Croton cajucara
Ginger
Lantana camara
Peppermint
Rosemary
Tea tree

Basil
Cedarwood
Curry
Helichrysum odoratissimum
Mandarin
Pinus peuce
Santolina
Thyme

Gamma:

Basil
Guava
Nasutitermes nigriceps
Salvia

Curry
Lantana camara
Oregano
Tea tree

Eucalyptus
Lime
Phoebe
Xylopia aromatica

[1324] Arno Hazekamp, Katerina Tejkalová, and Stelios Papadimitriou. Cannabis: From Cultivar to Chemovar II—A Metabolomics Approach to Cannabis Classification. Cannabis and Cannabinoid Research. Dec 2016.202-215.

BIOLOGICAL ACTIVITY IN PLANTS

Virtually no research has been published on the biological activity of gurjunene in plants. However, one 2005 study found that, as the primary constituent in the resin produced by Dipterocarpus kerrii (a hardwood tree that produces copious amounts of resin from natural or artificial wounds), alpha-gurjunene strongly inhibited and caused substantial mortality to attacking insects[1325]. This raises the possibility that alpha-gurjunene serves as an insecticidal agent in at least some plants.

USES IN INDUSTRY

Commercial preparations of alpha-gurjunene appear as colorless clear liquid and are sometimes used in fragrances or as a perfuming agent in personal care products. No commercial use of beta-gurjunene or gamma-gurjunene appears in the literature outside of research purposes.

POTENTIAL USES IN MEDICINE

Gurjunene has significant potential as an anticancer agent used in the treatment or prevention of several different types of cancer, as well as strong sedative properties, and possible use as an antimicrobial agent as described below:

****Treatment for Liver Cancer:*** As the primary constituent in the essential oil made from the leaves of Duguetia pycnastera (a tropical tree), alpha-gurjunene contributed to cytotoxicity in SCC4 (squamous cell carcinoma) and CAL27 cancer cell lines, and inhibited tumor mass by 47% in HepG2 (liver cancer) cells in tumor-bearing mice[1326].

****Treatment for Colon Cancer:*** As a primary constituent in the essential oil made from Virola surinamensis (known in Brazil as mucuíba), alpha-gurjunene contributed to cell death in HCT116 (colon cancer) cells, with tumor inhibition rates approaching 58%[1327].

****Treatment for Lymphoma:*** As the primary constituent in the oleo resin from Dipterocarpus alatus (the resin tree), alpha-gurjunene (and with gamma-gurjunene as a major compound) contributed to the significant inhibition of U937 (lymphoma) cells without affecting normal cells, as well as cytotoxicity greater than that of melphalan[1328], a common chemotherapy drug. As a primary constituent in the leaf oil made from Melicope denhamii, alpha-gurjunene contributed to significant activity against Dalton's lymphoma ascites cells, inducing morphological changes, nuclear damage, and apoptosis[1329].

****Sedative:*** As the primary constituent in agarwood oil, alpha-gurjunene contributed to significant sedative effects in mice when administered as a vapor[1330].

[1325] Richardson, D., A. Messer, S. Greenberg, H. Hagedorn, and J. Meinwald. "Defensive sesquiterpenoids from a dipterocarp (Dipterocarpus kerrii)." Journal of Chemical Ecology 15.2 (2005): 731-747.

[1326] Costa EV, de Souza CAS, Galvão AFC, Silva VR, Santos LS, Dias RB, Rocha CAG, Soares MBP, da Silva FMA, Koolen HHF, Bezerra DP. Duguetia pycnastera Sandwith (Annonaceae) Leaf Essential Oil Inhibits HepG2 Cell Growth In Vitro and In Vivo. Molecules. 2022 Sep 2;27(17):5664.

[1327] Anunciação TAD, Costa RGA, Lima EJSP, Silva VR, Santos LS, Soares MBP, Dias RB, Rocha CAG, Costa EV, Silva FMAD, Koolen HHF, Bezerra DP. In vitro and in vivo inhibition of HCT116 cells by essential oils from bark and leaves of Virola surinamensis (Rol. ex Rottb.) Warb. (Myristicaceae). J Ethnopharmacol. 2020 Nov 15;262:113166.

[1328] Yongram C, Sungthong B, Puthongking P, Weerapreeyakul N. Chemical Composition, Antioxidant and Cytotoxicity Activities of Leaves, Bark, Twigs and Oleo-Resin of Dipterocarpus alatus. Molecules. 2019 Aug 25;24(17):3083.

[1329] George S, Nair SA, Venkataraman R, Baby S. Chemical composition, antibacterial and anticancer activities of volatile oil of Melicope denhamii leaves. Nat Prod Res. 2015;29(20):1959-62.

[1330] Takemoto, H., Ito, M., Shiraki, T. et al. Sedative effects of vapor inhalation of agarwood oil and spikenard extract and identification of their active components. J Nat Med 62, 41–46 (2008).

**Antibacterial Agent:* As a primary constituent in a methanol extract made from Porella arboris-vitae, alpha-gurjunene contributed to significant morphological and ultrastructural alterations in Salmonella Enteritidis[1331]. As a primary constituent in the essential oil made from the leaves of Murraya koenigii, beta-gurjunene contributed to the complete inhibition of growth of Listeria innocua[1332].

IMPLICATIONS FOR HUMAN HEALTH & NUTRITION

The effects of ingesting gurjunene are relatively unknown. To increase dietary consumption of this sesquiterpene, consider the following:

When preparing, cooking, or serving foods, try working with spices or herbs such as allspice, basil, bergamot, black pepper, cajuput, curry, parsley, peppermint, rosemary, safflower, sage, thyme, or water mint. Snack on fruits including guava and mandarin, and vegetables like carrots and ginger. Teas that are made with eucalyptus, peppermint, ginger, or valerian root may contain gurjunene, as well as essential oils made from any of the plants in the "Occurrences in Plants" section of this chapter. As always, extreme care should be taken when working with any EO, as these complex mixtures can be toxic in some situations.

Cannabis varieties that contain gurjunene could include high-CBD and fiber types, as well as traditionally produced hashish. Drug type varieties might contain gurjunene, however, most laboratories are not testing for this sesquiterpene as of 2023.

[1331] Kumar Tyagi A, Bukvicki D, Gottardi D, Veljic M, Guerzoni ME, Malik A, Marin PD. Antimicrobial Potential and Chemical Characterization of Serbian Liverwort (Porella arboris-vitae): SEM and TEM Observations. Evid Based Complement Alternat Med. 2013;2013:382927.

[1332] Erkan N, Tao Z, Rupasinghe HP, Uysal B, Oksal BS. Antibacterial activities of essential oils extracted from leaves of Murraya koenigii by solvent-free microwave extraction and hydro-distillation. Nat Prod Commun. 2012 Jan;7(1):121-4.

Gurjunene Review

Answer the following questions to test your knowledge of this terpene:

Question #1: What type of terpene is gurjunene?

a. Tricyclic sesquiterpene
b. Depends on the isomer
c. Bicyclic monoterpene
d. Bicyclic sesquiterpene

Question #2: Which of the following best describes the primary flavor & aroma profile of gamma-gurjunene?

a. Spicy, turpentine, dry
b. Sweet, woody, yuzu
c. Depends on the isomer
d. Musty

Question #3: How common is gurjunene in cannabis?

a. Top 40
b. Top 50
c. Top 60
d. Top 25

Question #4: Which gurjunene isomer contains 2 double bonds?

a. Alpha
b. Beta
c. Gamma
d. Delta

Question #5: Name 6 plants that are known to contain gurjunene:

____

____

Question #6: Name 2 potential medical uses of gurjunene:

1 ____________________ 2 ____________________

For the answer key to all individual compound chapters please go here: http://thebigbookofterps.com/bbot-quiz-answer-key/

HASHISHENE

Type: Bicyclic monoterpene
Chemical Formula: $C_{10}H_{16}$
Molecular Weight: 136.234 g/mol
Boiling Point: 161°C (321°F)
Solubility: Slightly soluble in chloroform, methanol
Biological Role: Unknown
Therapeutic Role: Unknown
Commercial Role: Hash biomarker
Occurrence in Cannabis: Trace or minor in plant parts, major constituent in hashish
Aroma/Flavor: No descriptors available

Occurs in Cannabis Varieties: No data available as of July 2023.

INTRODUCTION

Hashishene was recently re-discovered when a group of researchers set out to find a volatile biomarker in hashish that could be used to develop a remote detection system for law enforcement and interdiction purposes. The researchers hit gold when they stumbled upon a compound technically called 5,5-dimethyl-1-vinylbicyclo[2.1.1]hexane that appeared as a major constituent in hash samples, but only in trace or low amounts in samples of fresh and dried cannabis inflorescences[1333]. All samples were sourced from seizures carried out by police and stored at the French Police Forensic Institute, with most of the hash presumed to have originated in Morocco. This research was so successful that those involved immediately filed a patent to use hashishene as a volatile marker of hashish, filing patent application EP 14305628.1 on April 28, 2014. Originally discovered in 1965, there exists little research on this compound, and many of the details normally provided in this text for each compound are missing here. Subsequent editions of this book will be updated with emerging research.

[1333] Marchini M, Charvoz C, Dujourdy L, Baldovini N, Filippi JJ. Multidimensional analysis of cannabis volatile constituents: identification of 5,5-dimethyl-1-vinylbicyclo[2.1.1]hexane as a volatile marker of hashish, the resin of Cannabis sativa L. J Chromatogr A. 2014 Nov 28;1370:200-15.

CHEMICAL STRUCTURE

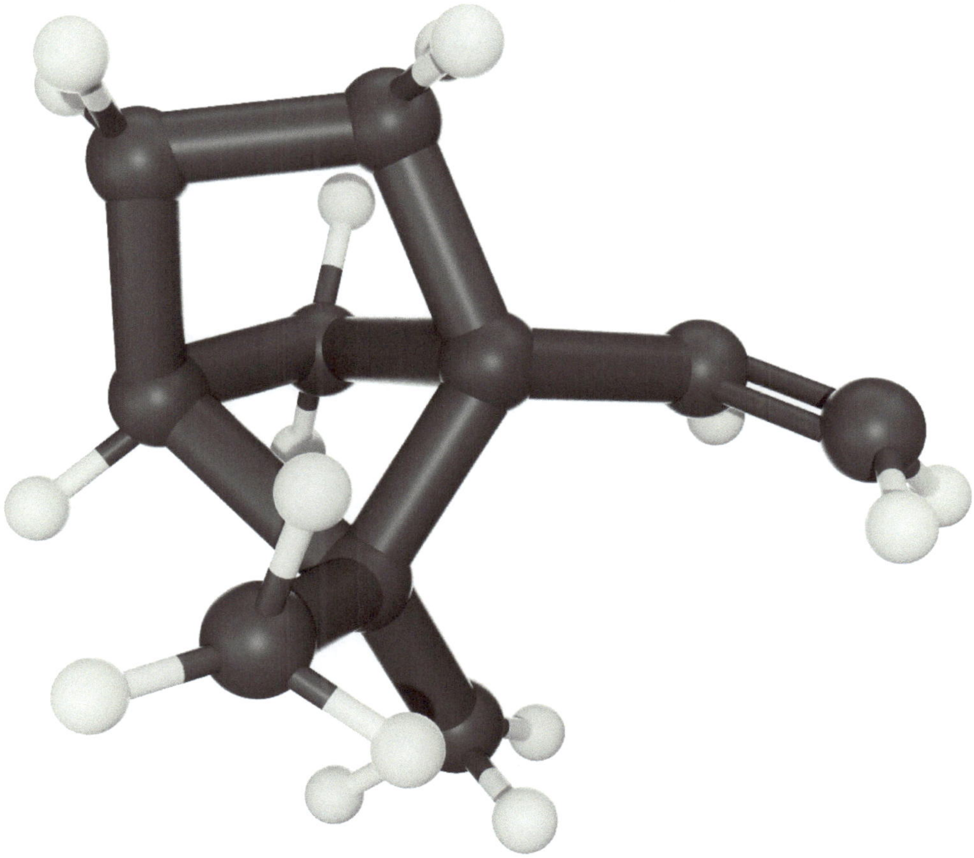

Hashishene is a monoterpene based on two units of isoprene, containing 10 carbon atoms and 16 hydrogen atoms notated as $C_{10}H_{16}$. This compound features one double bond, one strained 4-membered ring, and one 5-membered ring, the latter two of which classify it as a bicyclic monoterpene. Found to occur as a racemic mixture of two stereoisomers, researchers were unsuccessful in isolating this compound, instead synthesizing it from beta-myrcene based on earlier work by Liu et al., 1965.

Hashishene is likely formed by the photolytic rearrangement of beta-myrcene[1334], where light photons gradually transform the compound without the process of oxidation being involved. In the case of the work by Marchini et al., the hash samples supposedly originating in Morocco were likely exposed to sunlight, which began the photolytic transformation process.

OCCURRENCE IN PLANTS

To date, hashishene has only been found in low amounts in cannabis inflorescences, and in the essential oil of Mentha cardiaca L., a hybrid mint species.

[1334] Robert S. H. Liu and George S. Hammond; Photosensitized internal addition of dienes to olefins. J. Am. Chem. Soc. 1967, 89, 19, 4936–4944. Publication Date: September 1, 1967.

USES IN INDUSTRY

Hashishene appears as a colorless oil, with no current uses in industry other than the patent referred to earlier, which seeks to use the compound as a volatile marker for detection of hash. This compound is also available as a research chemical.

IMPLICATIONS FOR HUMAN HEALTH & NUTRITION

Until further research is carried out on hashishene, the implications of consuming hash containing this compound are unknown.

Hashishene Review

Answer the following questions to test your knowledge of this terpene:

Question #1: What type of terpene is hashishene?

a. Tricyclic monoterpenoid
b. Bicyclic monoterpene
c. Acyclic monoterpene
d. Bicyclic sesquiterpene

Question #2: Which of the following describe hashishene's two rings:

a. One 3 member ring and one 5 member ring
b. One 5 member ring and one 2 member ring
c. One 4 member ring and one 5 member ring
d. One 4 member ring and one 3 member ring

Question #3: Which two plants has hashishene been found in so far?

a. Cannabis and rosemary
b. Cannabis and melangtholia
c. Cannabis and marjoram
d. Cannabis and mint

Question #4: Hashishene is formed by:

a. Pyrolysis of alpha-myrcene
b. Photolytic rearrangement of beta-myrcene
c. Photo oxidation of beta-myrcene
d. Photon induction rearrangement of beta-myrcene

Question #5: Describe the original intent of researchers who rediscovered hashishene in 2014:

____
____

Question #6: Name the two current uses of hashishene:

1 ____________________________ 2 ____________________________

For the answer key to all individual compound chapters please go here: http://thebigbookofterps.com/bbot-quiz-answer-key/

MANIPULATION OF PHYTOCHEMICAL CONTENT

This chapter is mostly theoretical and discusses general experimental concepts in the manipulation of cannabis terpene and flavonoid profiles. The majority of information in this section is derived, like much of this book, from studies and research conducted with plants other than cannabis. This chapter and all other relevant sections of this book will be updated as information specific to cannabis becomes available.

Readers are encouraged to report their experimentation and research efforts to the author and moderators at TheBigBookOfTerps.com.

[NOTE: Cannabis plants that have been part of experimental programs where chemical, pathogen, or insect stressors have been added should be composted or otherwise rendered unusable.]

GENETICS

The most crucial factor in determining the terpene and flavonoid profiles of a particular cannabis plant is genetics[1335]. Inherited genetic traits will largely determine the total maximum *potential* terpene and other phytochemical content of a plant grown under optimal conditions. Therefore, the primary method of manipulating cannabis for terpene or flavonoid profiles is through selective breeding and propagation.

However, as this chapter will reveal, even stabilized genetics do not necessarily mean that a grower will achieve the maximum potential phytochemical content softly promised by DNA; nor that every grower will want exactly that theoretical genetic profile. From seed germination to curing, every step of the cannabis harvesting process alters this potential profile. Growers can capitalize on this sensitivity to experiment and potentially develop and produce cannabis varieties with unique terpene and flavonoid profiles.

HARVEST TIMING

Daily

For volatile molecules like terpenes, the aging of each day results in significantly increased vaporization as daytime temperatures climb. Therefore, the timing of harvest is critical. This is especially true for volatile monoterpene content such as pinene, myrcene, limonene, and linalool, which are not as stable as sesquiterpenes like humulene, caryophyllene, and bisabolol. Cannabis plants that are harvested in the middle or later part of the day are likely to contain far less monoterpene content.

To retain monoterpenes, harvest in the early part of the day, before the temperature starts to rise, and keep harvested materials below the point of vaporization (see individual compound chapters for specifics), or generally below 21°C. Harvesting at dawn before the sun has crested the horizon is the ideal time to avoid both heat and light.

Seasonal

Most experienced cannabis growers know that the time of season affects terpene and flavonoid content. Growers know to look for the ripeness of trichomes, which is where the majority of phytochemical content like terpenes, flavonoids, and cannabinoids are produced and stored in the cannabis plant. Selectively harvesting based on this ripeness is a common practice in the cannabis industry, where clear to cloudy trichomes are harvested for one therapeutic or psychoactive effect, while amber or dark trichomes are harvested for a different effect. These subjective yet consistently reported effects are almost certainly due to differences in phytochemical content based on the timing of harvest.

[1335] Franz, Chlodwig & Novak, Johannes. (2009). Sources of Essential Oils.

Growers and researchers are encouraged to test for phytochemical content at different plant life cycle stages to determine what part of the season is the ideal time to harvest an individual strain or variety, and for what purpose. For best results, testing and harvesting solely for maximum cannabinoid content must be avoided.

DISEASE STRESS

Many plants respond to disease stress by increasing production of beneficial phytochemicals, most notably, terpenes. For example, when attacked by a common bacterial pathogen, thale cress greatly increases production of Beta-caryophyllene to inhibit the growth of the pathogen[1336]. When infected by the fungus S. cardinale, Italian cypress markedly increases production of terpinolene, limonene, and alpha-terpinene, which allow the tree to resist the attacker[1337].

Cannabis plants will also respond to infection by pathogens by increasing emissions and total content of various terpenes and other phytochemicals. Growers can manipulate cannabis by deliberately exposing the plant to bacteria, fungi, or viruses, with a remedy prepared in advance.

For instance, the cannabis strain Black Widow, which is typically low in Beta-caryophyllene content, could be induced to increase production of this sesquiterpene by exposing it to a bacterial pathogen similar to the one that elicits an increase of Beta-caryophyllene in thale cress, presuming the plants use this chemical in a similar manner. Meanwhile, haze strains could be induced to increase production of terpinolene - the signature terpene of the hazes - for a 'super haze' variety by exposing the plants to a fungus like seiridium cardinale. Join our discussion around other examples of disease stress, and updates to the topic at TheBigBookofTerps.com.

INSECT STRESS

Exposing cannabis plants to insect stressors is another experimental method of manipulating terpene and other phytochemical content. For example, if cannabis responds to bore weevils similarly to the Chinese plant curcuma wenyujin, it will significantly increase the amount of 1,8-cineole[1338] within the plant's trichomes and other parts to fight an infestation. This terpene, also known as eucalyptol, could result in an increased minty flavor in the harvested strain.

Other plants are known to increase production of terpenes under insect attack. Passion vine plants increase by many times the emissions and production of ocimene, farnesene, and β-caryophyllene after herbivory by the butterfly heliconius erato[1339]. Royal Gala apple seedlings increase production of ocimene, nerolidol, linalool, β-caryophyllene, and farnesene in response to herbivory by light brown apple moth larvae, a terpene blend that in turn attracts parasitic wasps to feed on the larvae[1340]. Cucumber plants increase emissions of ocimene and farnesene in response to herbivory by spider mites[1341].

1336 Huang, Mengsu, Adela M. Sanchez-Moreiras, Christian Abel, Reza Sohrabi, Sungbeom Lee, Jonathan Gershenzon, and Dorothea Tholl. The major volatile organic compound emitted from Arabidopsis thaliana flowers, the sesquiterpene (E)-Beta-caryophyllene, is a defense against a bacterial pathogen. New Phytologist 193.4 (2012): 997-1008.

1337 Achotegui-Castells, Ander, Roberto Danti, Joan Llusia, Gianni Rocca, Sara Barberini, and Josep Pefiuelas. Strong Induction of Minor Terpenes in Italian Cypress, Cupressus sempervirens, in Response to Infection by the Fungus Seiridium cardinale. Journal of Chemical Ecology 41.3 (2015): 224-243.

1338 Identification of Insecticidal Constituents of the Essential Oil of Curcuma wenyujin Rhizomes Active against Liposcelis bostrychophila Badonnel.

1339 de O. Borges, E., Martins, C.B.C., R. da Silva, R. et al. Terpenoids dominate the bouquet of volatile organic compounds produced by Passiflora edulis in response to herbivory by Heliconius erato phyllis (Lepidoptera: Nymphalidae). Arthropod-Plant Interactions 12, 123-131 (2018).

1340 Suckling DM, Twidle AM, Gibb AR, Manning LM, Mitchell VJ, Sullivan TE, Wee SL, El-Sayed AM. Volatiles from apple trees infested with light brown apple moth larvae attract the parasitoid Dolichogenidia tasmanica. J Agric Food Chem. 2012 Sep 26; 60(38):9562-6.

1341 Kappers, I.F., Verstappen, F.W.A., Luckerhoff, L.L.P. et al. Genetic Variation in Jasmonic Acid- and Spider Mite-Induced Plant Volatile Emission of Cucumber Accessions and Attraction of the Predator Phytoseiulus persimilis . J Chem Ecol 36, 500-512 (2010).

All of these chemical responses can be exploited through various experiments. For an example that will be familiar to most cannabis professionals, consider the pesky aphid. Assuming that - like other plants using the molecule - cannabis increases emissions of farnesene in response to aphid infestation, these insects could be released into a particular crop, allowed to proliferate for 24-48 hours (during which time the plants will theoretically increase farnesene and other emissions), then the aphids could be eradicated by a subsequent release of harmless yet aphid-hungry ladybugs. Plants harvested soon afterward will hypothetically feature higher amounts of farnesene, a therapeutically valuable compound.

CO-PLANTING FOR ALLELOCHEMICAL STRESS

Plants that compete for the same resources are generally engaged in constant chemical warfare called allelopathy. These chemicals often consist of terpenes, which are used to delay, retard, or stop the growth of other plants and organisms. It is likely that different plants cause different chemical reactions in cannabis when competing for the same resources. For instance, a reaction between cannabis and tomato plants, both of which require copious amounts of water, seems reasonable. What this reaction and interaction consists of and how it changes the phytochemical content in cannabis is unknown, but it's logical to theorize that this must happen. We need both trained and citizen scientists to test this theory and catalog the related phytochemical changes.

MECHANICAL WOUNDING STRESS

Although it is probable that significant mechanical wounding changes the terpene and flavonoid profile of cannabis, growing seasons are typically too short to allow plants to recover from such damage and remain viable through harvest. Nevertheless, researchers can test terpene and flavonoid content before and after mechanical wounding and compare results across different varieties.

However, experimentation in this regard should be undertaken with the caveat that terpene or other phytochemical changes caused by deliberate mechanical wounding are likely to be localized and not systemic. For example, a broken branch may not change the terpene content in developing buds on the same or nearby branches, whereas the chemical response to an insect infestation is more likely to be systemic. Research here is interesting but probably not as fruitful as other methods of terpene and flavonoid manipulation in cannabis.

CHEMICAL TREATMENT STRESS

Changing the chemical profile of cannabis may be possible through the application of phytonutrients like cis-Jasmone. A volatile molecule found in the essential oil of jasmine flowers, cis-Jasmone has been shown to activate plant defenses in maize plants[1342], stimulating a significant increase in bergamotene, a bicyclic sesquiterpene that often occurs in trace amounts in cannabis. Experimentation with this compound could produce a strain that is high in bergamotene, a variety profile that as of late 2021 is not known to exist.

Methyl jasmonate exhibits similar effects, and limonene - a top five terpene in cannabis - has been shown to activate plant defenses in thale cress against a common pathogenic fungus, with researchers suggesting that plants are capable of recognizing gaseous limonene[1343].

Cannabis plants can be primed to defend against a variety of diseases, infections, and infestations using these or other natural compounds. This priming will change the phytochemical content of a treated crop, including terpenes, possibly flavonoids, and other constituents.

[1342] Oluwafemi, Sunday et al. Priming of Production in Maize of Volatile Organic Defence Compounds by the Natural Plant Activator cis-Jasmone. PloS one vol. 8,6 e62299. 26 Jun. 2013.

[1343] Fujioka, K., Gotoh, H., Noumi, T. et al. Protection induced by volatile limonene against anthracnose disease in Arabidopsis thaliana . J Gen Plant Pathol 81, 415-419 (2015).

pH MANIPULATION

A plant used traditionally in China for centuries – Atractylodes lancea – has been shown to increase production of beta-eudesmol in roots after low-pH water treatment, while also increasing the density and dry weight of the roots, leading researchers to conclude that a more acidic pH stimulates production of beta-eudesmol in A. lancea[1344]. This work means that it may be possible to alter the phytochemical profile of cannabis plants via experimentation with different pH levels.

[1344]Yuan, Y., Liu, Y.J., Huang, L.Q. et al. Soil acidity elevates some phytohormone and β-eudesmol contents in roots of Atractylodes lancea . Russ J Plant Physiol 56, 133–137 (2009).

The Big Book of Terps Final Exam

Question #1: Describe Wallach's Law:

Question #2: The majority of early terpene scientists were mostly chemists from which country:

a. Romania
b. United Kingdom
c. Switzerland
d. Germany
e. Russia

Question #3: Draw a line to connect each terpene with the proper category:

a. Farnesene	1. Acyclic monoterpene
b. Beta-caryophyllene	2. Acyclic sesquiterpene
c. Pinene	3. Bicyclic monoterpene
d. Linalool	4. Bicyclic sesquiterpene

Question #4: Which scientist discovered the THC molecule?

a. Raphael Mechoulam
b. Susan Trapp
c. Wilhelm Triebs
d. Ethan Russo

Question #5: Which researcher is noted for their studies on the interactions and putative synergy between THC and CBD?

a. Raphael Mechoulam
b. Susan Trapp
c. Wilhelm Triebs
d. Ethan Russo

Question #6: What is the top monoterpene in cannabis?

Question #7: What is the top sesquiterpene in cannabis?

Question #8: Name at least 3 isomeric designations:

1 ______________________ 2 ___________________________ 3 _________________________

Question #9: True terpenes contain only what two components:

1 ______________________________________ 2 ______________________________________

Question #10: Other than oxygenation, what else can define a terpene as a terpenoid?

____

Question #11: Some terpenes can influence the growth cycles of other plants and some insects. Terpenes used by plants for this purpose are called:

a. Allelochemicals
b. Phytochemicals
c. Phytonutrients
d. Alleyoopicals

Question #12: Pinene and myrcene are found in high concentrations in which trees?

a. Deciduous trees
b. Coniferous trees
c. Ficus trees
d. Synthetic trees
e. Sigillaria trees

Question #13: How many rings does gurjunene contain?

a. Four
b. Three
c. Two
d. Five

Question #14: This cannabinoid is both an endocannabinoid and a phytocannabinoid:

a. Delta-11-tetrahydrocannabinol
b. Delta-8-tetrahydrocannabinol
c. Anandamide
d. Beta-caryophyllene
e. Fenchol alcohol

Question #15: The addition of an oxygen atom changes the ending of "terpene" to what designation?

Question #16: Explain the "A" designation of THCA, CBDA, CBCA, CBGA:

Question #17: Define this acronym: CB1

Question #18: Name the top ten terpenes in cannabis:

1 ______________________________
2 ______________________________
3 ______________________________
4 ______________________________
5 ______________________________
6 ______________________________
7 ______________________________
8 ______________________________
9 ______________________________
10 ______________________________

Question #19: Name 5 "Top 40" terpenes/enoids in cannabis:

1 ______________________________
2 ______________________________
3 ______________________________
4 ______________________________
5 ______________________________

Question #20: To retain the most monoterpenes possible, when is the best time to harvest cannabis?

Question #21: Which of the following is a top ten terpene in cannabis:

a. Terpineol
b. Terpinolene
c. Terpinene
d. Turpentine

Question #22: Which top ten terpene is named after the hops plant?

a. Valencene
b. Terpinolene
c. Myrcene
d. Humulene
e. Humulus

Question #23: Which terpene is the primary constituent responsible for the smell of roses?

a. Geranyl acetate
b. Guaiol
c. Bergamotene
d. Geraniol

Question #24: Which bicyclic sesquiterpene is responsible for the spicy characteristics of black pepper?

a. Beta-caryophyllene
b. Caryophyllene oxide
c. 4-ol-caryophyllene
d. 28-caryophyllene

Question #25: What is the primary factor determining the terpene and cannabinoid profile of a cannabis chemovar?

a. Number of rings
b. Genetics
c. Oxygenated state
d. Indica and sativa

Question #26: Phytol is a byproduct of what plant compound?

a. Plant cortisone
b. Chlorophyll
c. Phytosterol
d. Phytol

Question #27: Name three experimental methods of manipulating the terpene profile of a cannabis strain / chemovar:

Question #28: d-limonene and l-limonene, as well as D8-THC and D9-THC are examples of:

a. Stereometrics
b. Functional groups
c. Geranyl substitutions
d. Isomers

Question #29: Myrcene is a well known:

a. Sedative
b. Diuretic
c. Antispasmodic
d. Analgesic

Question #30: Commercially produced limonene is sourced primarily from:

a. Lemons
b. Oranges
c. Limes
d. Grapefruit

Question #31: Which scientists first proposed the entourage effect?

a. Russo and McPartland
b. Pickens and Fairbanks
c. Ben-Shabat and Mechoulam
d. Cogan and Duggan

Question #32: Most research into the entourage effect or synergy has focused on which two compounds?

a. THC and CBG
b. CBGA and THC
c. THC and CBD
d. CBC and THC

Question #33: Which compound is the precursor to most of the cannabinoids?

a. CBCA
b. CBGA
c. CBDA
d. THCV

Question #34: Name one of Russo's three papers that are heavily cited as evidence of synergy:

Question #35: The entourage effect takes place in:

a. Geranyl phosphate pathway
b. Limbic system
c. Mevalonate pathway
d. Endocannabinoid system

For the answer key to this final exam, please go here: http://thebigbookofterps.com/bbot-final-exam-answer-key/

What's Next?

Because much of this book relies on research conducted with plants other than cannabis, the author intends to open a research laboratory and testing facility. This facility will study terpenes, flavonoids, ketones, esters, aldehydes, thiols, sulfurs, and other compounds in cannabis to provide research for future editions of this and other texts. Studies carried out at this facility will be used to expand the scientific knowledge of cannabis and its chemical constituents, and to bolster existing research into the medical and therapeutic implications of these and other natural compounds. To get involved, contact the author at Russ@TheBigBookOfTerps.com

Suggested Reading

There are few texts that specifically focus on terpenes, however, the books listed here are excellent starting points for research on information and data not reported in this text. Works that discuss small molecules in general or those that are specific to phytochemicals may also be good sources of information about terpenes and flavonoids.

A Comprehensive Treatise on Terpenes - Sameksha Koul

A thorough view on naturally occurring organic compounds, including classes of metabolites and the drugs that produce them, origins, structural data, and therapeutic uses of major and minor terpenes and terpenoids.

The Chemistry of the Terpenes - Freidrich Heusler

A student of Otto Wallach, Heusler's 1902 work, *The Chemistry of the Terpenes*, is dedicated to the discoverer of the isoprene rule, and expands upon many early terpene breakthroughs, serving as a comprehensive experimental guide of the time. Heusler was the first to propose that terpene oxidation products should be classified distinctly as terpenes.

Biosynthesis of Terpenes, Steroids, and Acetogenins - John H Richards and James B. Hendrickson

A broad monograph that traces the biogenetic routes from acetate to natural products including terpenes and flavonoids. Rich with images and heavily cited, this work retains much of its relevance despite long being out of print.

Handbook of Naturally Occurring Compounds – Volume II, Terpenes - by T.K. Devon and A.I. Scott

A large handbook detailing monoterpenes, sesquiterpenes, diterpenes, sesterterpenes, triterpenes, and related compounds. This work covers most of the known or elucidated terpenes as of the early 1970s.

Terpenes: Flavors, Fragrances, Pharmaca, Pheromones – Breitmaier, Eberhard

A thorough introductory text covering the origins of terpenes, terpene structures, selected total syntheses, biological and pharmacological properties, and industrial processes, among other related topics.

The Secret of Scent – Luca Turin

An exploration of two competing theories of smell. Is scent determined by molecular shape or molecular vibrations? Turin describes the science, evidence, and the often contentious scent debate from the beginnings of organic chemistry to the present day.

Acknowledgements

PubChem

One of the best resources for information about terpenes is PubChem, which has a massive database of information on these and other molecules, including structural and other molecular/atomic data, biological activity, toxicity, patents, and references to thousands of studies carried out with each compound. PubChem is an open chemistry database at the National Institutes of Health. The author thanks PubChem for providing such a valuable resource to the public, and refers readers to contribute to and utilize the data found there at: https://pubchem.ncbi.nlm.nih.gov/.

The Good Scents Company

This book contains information from TGSC including boiling and flash points, aroma and flavor profiles, and plant sources. Whenever possible, this information is cross-referenced with other data or research. For more than 4 decades, The Good Scents Company has been the leading source of information about volatile molecules for use in the foods, flavors, and fragrances industries. The website consists of a huge database describing the appearance, smell, and taste profile of various small molecules, their plant sources, and commercial usage data. The Good Scents Company often features valuable information that is not available on PubChem or other scientific sources. The author expresses significant appreciation for the work carried out by TGSC and refers readers to the site for supplemental or other interesting data not reported here.

Please visit this extremely comprehensive website at: http://www.thegoodscentscompany.com/index.html.

Collaborators & Supporters

The author expresses appreciation for the candor, kindness, and participation of Mitch Earleywine, Raphael Mechoulam, Ed Rosenthal, Ethan Russo, Rob Mejia, Susan Trapp, Gloria Fuentes, Zandra Chisolm-el, and Colleen Quinn. We owe the continued advancement of terpene and cannabis science in part to these individuals and their colleagues and supporters.

Special Thank You

The author wishes to express a special thank you to King Bishop of Belfast, Maine, USA. King has been instrumental in the advancement of cannabis legislation and its application in Maine, and has operated a cannabis processing laboratory and academy there for several years. King is also the founder of Your Coastal Connection, a cannabis caregiver storefront perched on the Atlantic Ocean in Midcoast Maine. King provided education, counsel, materials, and funding for research and laboratory testing for the original edition of this text. King's interest in the science of cannabis and his support of the author's work on The Big Book of Terps is potently appreciated.

Finally, the author expresses deep gratitude to the editor, Jackie Graddon, for her endless support, keen eye, tenacity, and patient encouragement, which were immensely important in getting this book published and into the hands of readers. Thank you, Jackie.

More From Russ Hudson

Russ Hudson is a cannabis researcher and consultant specializing in small molecules including terpenes and flavonoids, and in cannabis legislation, licensing, and business operations, particularly in Spain and the United States. Hudson's websites include TheBigBookOfTerps at www.thebigbookofterps.com, Cannabis Consultants Group at www.cannabisconsultants.group, Molecules, Inc. at www.moleculesincorporated.com, and the CannaBastard at www.cannabastard.com.

Other books and work by Hudson include:

Weed Deeds: From Seed to Sage. Hudson's autobiography detailing three decades working with cannabis in the black, grey, and white markets. Available on Amazon and in bookstores worldwide.

What's That Weed. Hudson's children's book about cannabis features his daughter Stella as the main character, and discusses non-recreational drug uses of the plant. Available on Amazon and in bookstores worldwide.

TerpTalk TV. Hudson's television and Youtube series focused on terpenes, flavonoids, and cannabinoids, among other valuable compounds in cannabis. Available on YouTube, Facebook, and Instagram TV.

Upcoming work:

The Formidable Book of Flavonoids: Understanding Flavonoids and Synergy in Cannabis. This book will be a companion text to The Big Book of Terps, and will cover the 20 most common flavonoids in cannabis. Chapters will include Friends of Flavonoids (outlining scientists working with these compounds,) Organic Chemistry 101, Flavonoids 101, Synergy in Flavonoids, 20 individual compound chapters modeled precisely after this text, and more, to be published in 2024.

The Compendium of Cannabs Constituents: The Big Book of Terps, The Formidable Book of Flavonoids, and other upcoming works including The Eensy Book of Esters (with future edtions written under the title The Enormous Book of Esters as the cannabis industry develops more testing data regarding the occurrence of these compounds), The Small Book of Sulfurs (with future edtions written under the title The Substantial Book of Sulfurs as the cannabis industry develops more testing data regarding the occurrence of these compounds), The Tiny Book of Thiols (with future edtions written under the title The Tremendous Book of Thiols as the cannabis industry develops more testing data regarding the occurrence of these compounds), The Little Book of Keys (Ketones), The Ancillary Book of Aldehydes, and The Abridged Book of Alcohols will be released in stages over the coming years. The goal with these works is to develop a rich understanding of the constituents of cannabis.

INDEX CONTINED BELOW AFTER A SPECIAL ARTISTIC PAGE BY FATNUGS MAGAZINE! >>>>>>

The Terpene Terrarium

An illustration inspired by conversations with Russ Hudson and created by the Fat Nugs Magazine Team of Rebekah Jenks, Casey Renteria, Isabella DeChard, Daniel Crawford, and Dustin Hoxworth. The Terpene Terrarium represents a look into a world full of countless species of plants being studied for their terpene content. Creating the feeling of a living lab, one inviting enough to not only work in, but to live in forever. A place that could provide for your every need. A wonderous and romantic, almost dreamy existence where science, nature, and humanity exist in perfect harmony.

Born from a deep love and respect for the arts and nature, Fat Nugs Magazine is a global perspective of cannabis and stoner culture as a bridge to the rest of the world. It's where you'll find true cannabis-lovers showcasing culture and community through authentic journalism, compelling stories, comedy, parody, science, education, and phenomenal art. All with one goal in mind of uplifting the plant into its rightful place in our society.

Dustin Hoxworth

Founder/Creator & Editor-at-Large

FAT NUGS Magazine

FLOWCODE

PRIVACY.FLOWCODE.COM

www.ingramcontent.com/pod-product-compliance
Lightning Source LLC
LaVergne TN
LVHW071243210926
848486LV00007B/335
* 9 7 8 0 9 8 6 3 5 7 1 4 5 *